17th International Symposium on Magnetic Bearings (ISMB17)

Rio de Janeiro, Brazil
18-21 August 2021

ISBN: 978-1-7138-9515-2

Printed from e-media with permission by:

Curran Associates, Inc.
57 Morehouse Lane
Red Hook, NY 12571

Some format issues inherent in the e-media version may also appear in this print version.

Printed by Curran Associates, Inc. (2024)

For permission requests, please contact Linz Center of Mechatronics GmbH
at the address below.

Linz Center of Mechatronics GmbH
Altenberger Straße 69
4040 Linz
Osterreich

Phone: +43 732 2468-6002
Fax: +43 732 2468-6005

office@lcm.at

Additional copies of this publication are available from:

Curran Associates, Inc.
57 Morehouse Lane
Red Hook, NY 12571 USA
Phone: 845-758-0400
Fax: 845-758-2633
Email: curran@proceedings.com
Web: www.proceedings.com

FOREWORD

The International Symposium on Magnetic Bearings (ISMB) was held for the first time in 1988. Since then, the community of universities, research institutions and industrial companies working in the field of magnetic bearings has organized the meeting every two years. The conference is usually held alternately in the Americas, China, Europe, and Japan and brings together global experts and leaders in the progressing field of magnetic bearing technology.

All scientific papers undergo a peer-review process and are available digitally on the web-portal www.magneticbearings.org after the conferences. The Linz Center of Mechatronics launched this portal in 2014 and operates it in coordination with the international community.

Special thanks go to all the authors of the scientific contributions, the organizers of ISMB 17, Prof. R. Stephan and his team, the members of the Advisory Committee, the sponsors of the web portal and the involved funding agency: The publication work has been supported by the COMET-K2 Center of the Linz Center of Mechatronics (LCM) funded by the Austrian federal government and the federal state of Upper Austria.

TABLE OF CONTENTS

Switching Stiffness Control of Lateral Motion in Magnetic Suspension System Using Lateral Displacement Detection with Hall Elements

Taku Egawa[1], Takeshi Mizuno[1], Masaya Takasaki[1], Yuji Ishino[1], and Daisuke Yamaguchi[2]

[1] Saitama University, Saitama, Japan
[2] Okayama University, Okayama, Japan
mizar@mech.saimtama-u.ac.jp

Abstract

The suppression of lateral vibration is investigated based on switching stiffness control strategy for partially active magnetic suspension system. In such magnetic suspension system, the motion in the normal direction of the suspended object (floator) is controlled actively while the motion in the lateral direction is passively supported. Therefore, it is often subject to vibration in the lateral direction. The application of switching stiffness control has been proposed to suppress such vibration by electromagnets placed in the normal direction. However, the conventional system used a displacement sensor in the lateral direction for detecting the lateral motion of the floator. It restricts the design of suspension system and causes high cost. In this work, it is proposed to detect the lateral motion of the floator with Hall elements instead of installing a sensor in the lateral of the floator. The proposed method of detecting the lateral motion is studied experimentally. The switching stiffness control is achieved based on the output of the fabricated detection device.

1 Introduction

Most of magnetic suspension systems use controlled DC electromagnets (Jayawant, 1981). In this magnetic suspension system, stable levitation is achieved by feedback control. The current flowing in the coil is increased when the suspended object (floator) moves to increase the gap between the electromagnet and the floator. In contrast, it is usually stable in the lateral directions due to the edge effects in the magnetic circuits. Such suspension is referred to as passive suspension. To achieve complete noncontact suspension with less cost and smaller size, passive suspension is often used in combination with active suspension. Such a system is referred to as *partially active magnetic suspension system*. One of the problems of partially active system is small damping in the passively suspended direction. Vibration is easily induced and hardly attenuated in this direction.

In active suspension system, two electromagnets operating differentially are often used to control a single degree of freedom of motion mainly because an electromagnet produces only attractive force for a ferromagnetic floator. This work focuses on such differentially operated magnetic suspension system. It has been proposed to suppress the lateral vibration by applying switching stiffness control to the electromagnets installed in the normal direction (Mizuno, Takasaki, & Ishino, 2016). The efficacy of the proposed strategy has been confirmed experimentally (Javed, et al., 2018).

However, an optical displacement sensor was installed in the lateral direction in the previous works (Mizuno, Takasaki, & Ishino, 2016) (Javed, et al., 2018). It caused the system to be large and costly. To detect the lateral displacement without any sensor in the lateral of the floator, it is proposed to install Hall elements beside the electromagnets in the stator. This configuration leads to several benefits:

- design flexibility
- lower cost
- reduction of size.

This paper organized as follows. First, the principle of switching stiffness control is explained in a differentially operated magnetic suspension. Second, the principle of detecting the lateral displacement with Hall elements is presented. Third, the switching stiffness controls is carried out based on the signals detected by the Hall elements to demonstrate the efficacy of the proposed detection method.

2 Principle of Switching Stiffness Control

Figure 1 shows a magnetic suspension system with a pair of electromagnets (E_1 , E_2). This system is inherently unstable in the normal direction (z) . Stable action can be achieved by an active control. In contrast, it can be stable in the lateral direction (x) due to the edge effects in the magnetic circuits. An example of the edge effects is shown in Figure 2. This is the same effect producing torque is reluctance motors (Boldea & Naser, 1997). Because restoring force is generated in the lateral direction, the system is modeled as shown by Figure 3 (Mizuno, Takasaki, & Ishino, 2016). The equation of motion is given by

$$m\ddot{x}(t)+kx(t)=0\,, \tag{1}$$

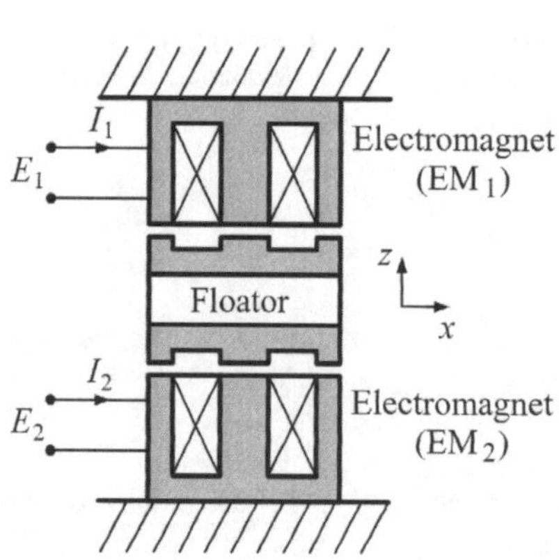

Figure 1 Differentially operated magnetic suspension system

where m : mass of the floator, k : effective stiffness of the system and x : displacement in the lateral direction. Figure 4 shows trajectories in the phase plane when (a) k is high, (b) k is low, and (c) k is switched according to the following control law (Winthrop, Baker, & Goob, 2005):

$$k = \begin{cases} k_0 + \Delta k & (x\dot{x} \geq 0) \\ k_0 & (x\dot{x} < 0) \end{cases} \tag{2}$$

In Equation (2), $x\dot{x} > 0$ means that the floator moves away from the equilibrium position where the flux lines align. In contrast, $x\dot{x} < 0$ means that the floator approaches to the equilibrium position where the flux lines align. By this switching stiffness control, the displacement (x) and velocity ($\dot{x}$) of the floator converge to zero as shown by Figure 4(c). In the magnetic suspension system, the value of k increases as the coil current I increases. Therefore, Equation (2) is modified to Equation (3) (Mizuno, Takasaki, & Ishino, 2016):

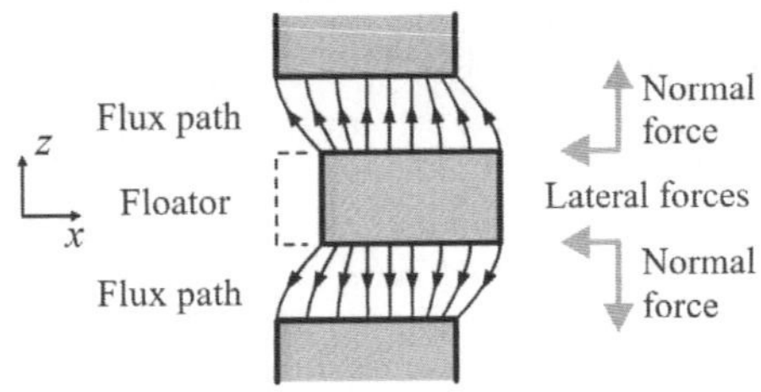

Figure 2 Edge effect

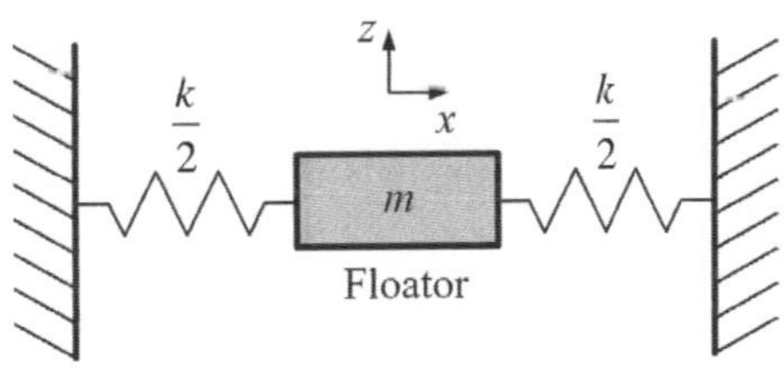

Figure 3 Model of magnetic suspension in the lateral direction

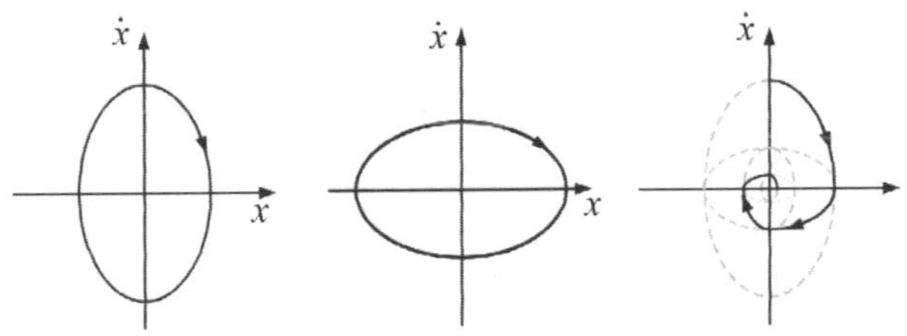

(a) High stiffness (b) Low stiffness (c) Switching stiffness

Figure 4 trajectories in the phase plane

$$I = \begin{cases} I_b + \Delta I & (x\dot{x} \ge 0) \\ I_b & (x\dot{x} < 0) \end{cases} \tag{3}$$

where I_b is a bias current and ΔI is a switching current. This is the principle of switching stiffness control in the differentially operated magnetic suspension system. It is to be noted that the forces in the normal (vertical) direction cancel each other out when the electromagnets are same in characteristics. Therefore, the vertical motion (z) will not be disturbed by the switching stiffness control.

3 Detection of Lateral Displacement

In the first work (Mizuno, Takasaki, & Ishino, 2016), a laser sensor was installed in the lateral direction of the floator to detect the displacement of the floator. However, it is preferable in applications to install nothing in the lateral direction of the floator. Then, another method of measuring the lateral displacement of the floator was introduced (Javed, et al., 2016). The measurement was based on the detection of force exerted by the floator to the electromagnets for suspension due to the magnetized force between them. In this method, however, the electromagnets in the stator were suspended by flexible joints for force detection. It led to some complexity in designing the stator.

In this work, Hall elements are introduced to detect the lateral motion. A Hall element is a sensor element that can detect the magnetic flux passing through the element. Figure 5 shows the flux lines between the electromagnet and the floator when the floator displaced from the equilibrium position. There is leakage flux on the edge of the electromagnet. This leakage flux is correlated with the lateral distance of the electromagnet and the floator. Therefore, Hall elements installed beside the electromagnet can detect the lateral displacement.

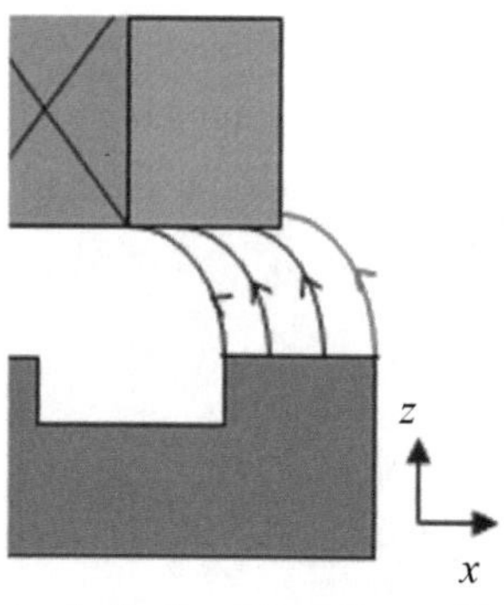

Figure 5 Flux lines when floator is displaced from the equilibrium position (x=0).

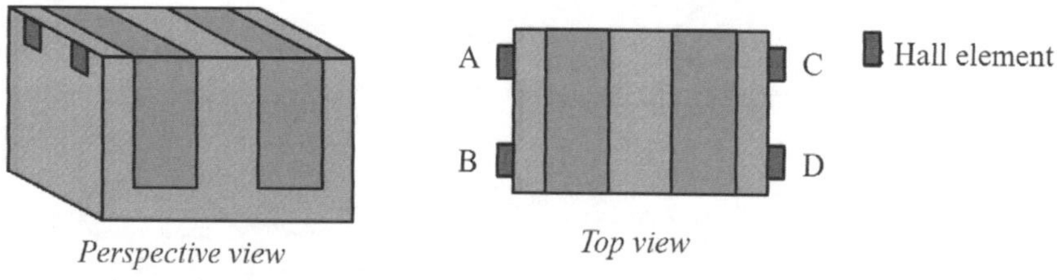

Figure 6 Arrangement of Hall elements on an electromagnet

In experiment, Hall elements are arranged as shown by Figure 6. To increase the sensibility of detection, two outputs in the same side are added. Then the difference between the two sums is calculated. The final output of the Hall-element detection circuit is

$$V_{\mathrm{H}} = (V_{\mathrm{C}} + V_{\mathrm{D}}) - (V_{\mathrm{A}} + V_{\mathrm{B}}) \tag{4}$$

where V_{α} is the output of the element α (α =A, B, C or D), and the lateral position of the floator is estimated from the value of V_{H}. This signal is referred to as the output of Hall-element sensor.

4 Experiment

4.1 Experimental apparatus

Figure 7 (a) shows a picture of the experimental apparatus and Figure 7(b) shows a schematic view of the experimental apparatus. Two electromagnets are arranged above and below the floator. The floator is suspended by four leaf springs to constrain the motion of the floator to a single-degree-of-freedom translation in the lateral direction. A laser displacement sensor is used to confirm the effect of

(a) Picture

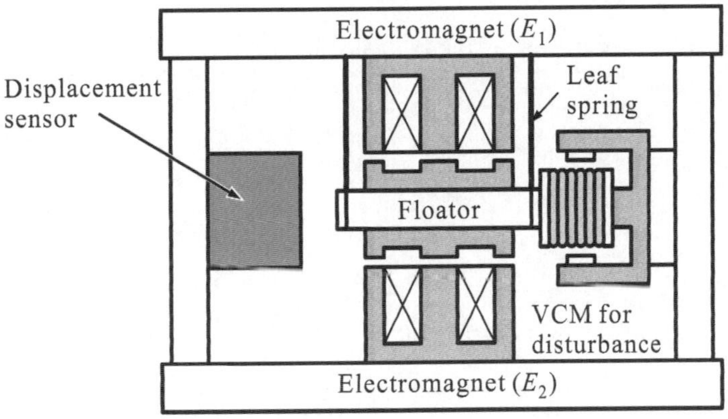

(b) Schematic drawing

Figure 7 Experimental apparatus

vibration control. A voice coil motor (VCM) is used to generate a disturbance or adjust the initial position of the floator in the lateral direction. It is to be mentioned that these device in the lateral direction can be removed in practical situations.

The outputs of the laser sensor and the Hall-element sensor are inputted to a DSP-based controller. The controller determines the commands to two power amplifiers according to Equation (3). Each power amplifier energizes the connected electromagnet.

4.2 Detecting lateral displacement by Hall elements

First, the performances of the fabricated sensor device are studied experimentally. Figure 8 compares the output of the laser sensor (reference) and that of the Hall-element sensor. They agree well with each other. It indicates that the lateral displacement of the floator can be detected with the device with Hall elements.

4.3 Vibration control with switching stiffness control

Figure 9 shows a block diagram of the switching stiffness control in the following experiments. Experimental procedures are as follows. First, the floator is set to an initial position (x=0.8mm) with the VCM. Second, the VCM is powered off and the lateral displacement is observed during switching stiffness control. Figure 10 shows the time histories of the output of the Hall-element sensor and the

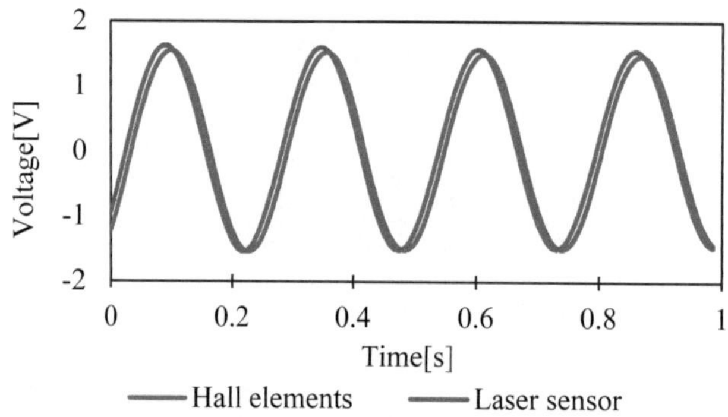

Figure 8 Comparison of the laser sensor and the Hall-element sensor

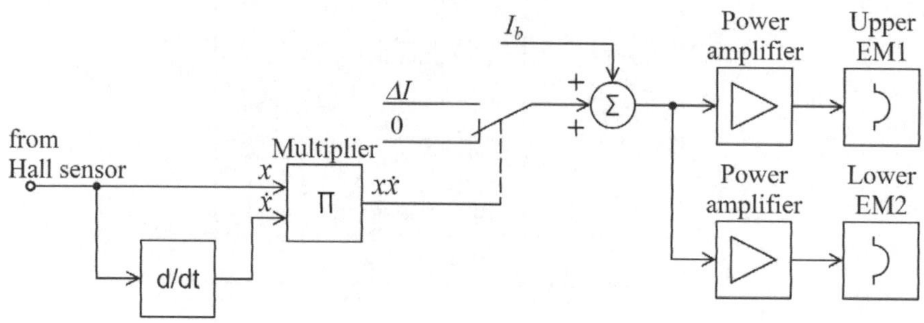

Figure 9 Block diagram

current command to the two electromagnets. As shown by Figure 10 (a), the output of the Hall-element sensor oscillates, and the switching's occur rapidly. The noises generated by a current switching cause

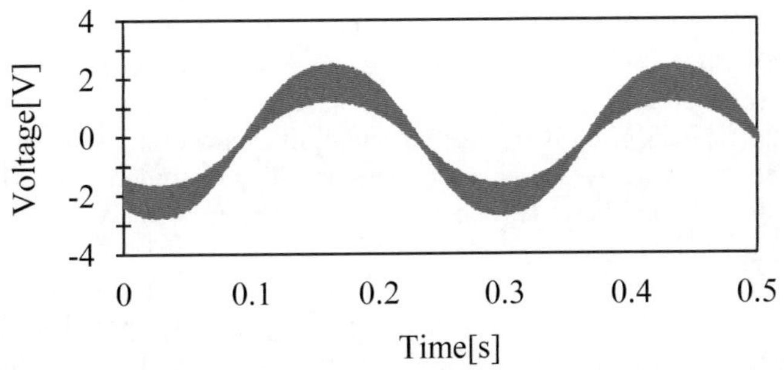

(a) Output of Hall sensor

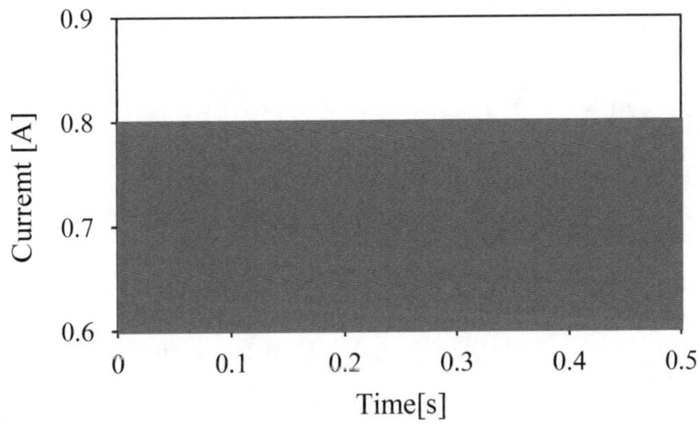

(b) Current command

Figure 10 Time histories when switching stiffness control works

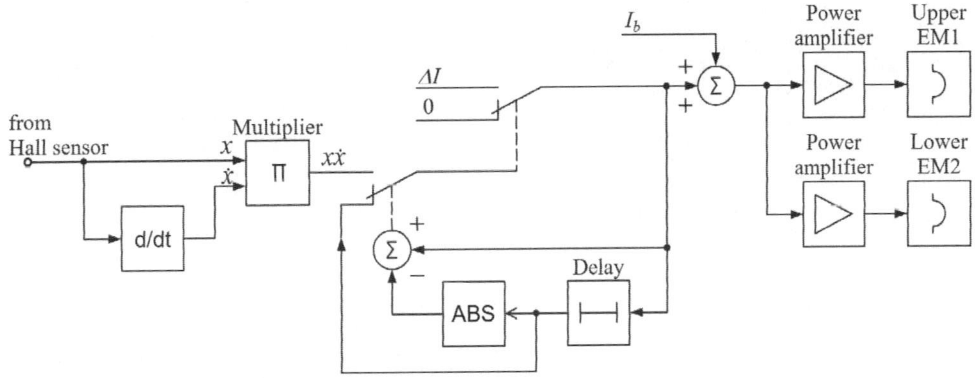

Figure 11 Block diagram

the next current switching and this process repeats. Therefore, the output of the Hall-element sensor oscillates, and the current is switched continuously. It makes the vibration control ineffective.

The control system is modified to avoid such mis-switching's. Figure 11 shows a block diagram of the modified control system. When the currents of the two electromagnets are switched, the previous value is maintained to determine switching in the next sampling. By this modified process, repeating current switching's caused by switching noises can be removed. Therefore, the noises in the output of the Hall-element sensor decrease.

Figure 12 show the time histories of the output of the Hall element sensor and the current command to the power amplifiers, respectively. As shown from Figure 12(a), the output of the Hall-element sensor includes less noises. In addition, Figure (b) shows that the current is switched properly according to the value of $x\dot{x}$.

Vibration control with the modified control system is conducted. Figure 13 shows the time histories of the lateral displacement of the floator (a) without switching stiffness control and (b) with switching stiffness control (b), respectively. The comparison of Figure 13 (b) with Figure 13 (a) demonstrates the effectiveness of the switching stiffness control for attenuating the vibration in the lateral direction.

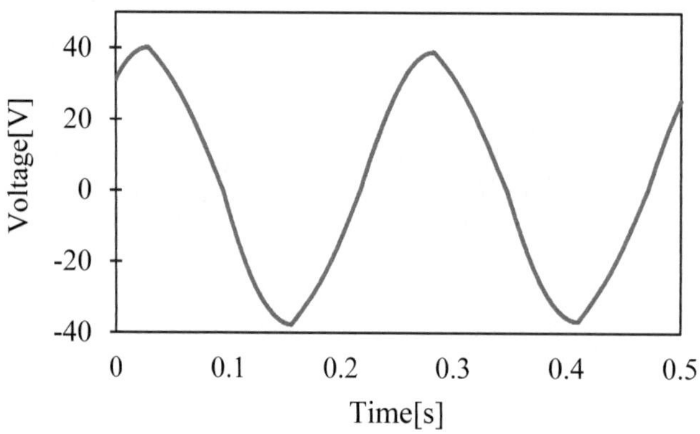

(a) Output of Hall sensor

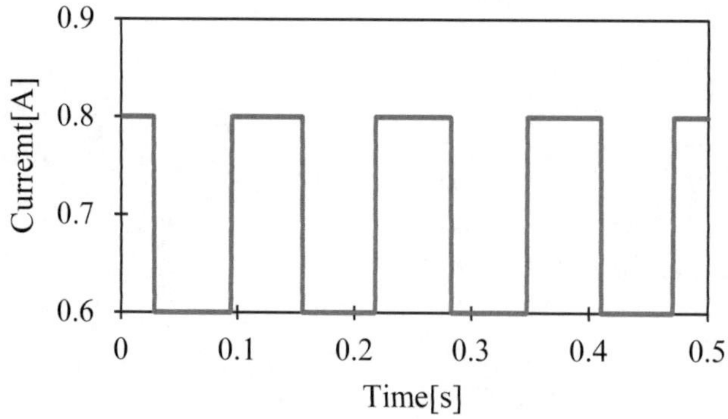

(b) Current command

Figure 12 Time histories when modified switching stiffness control works

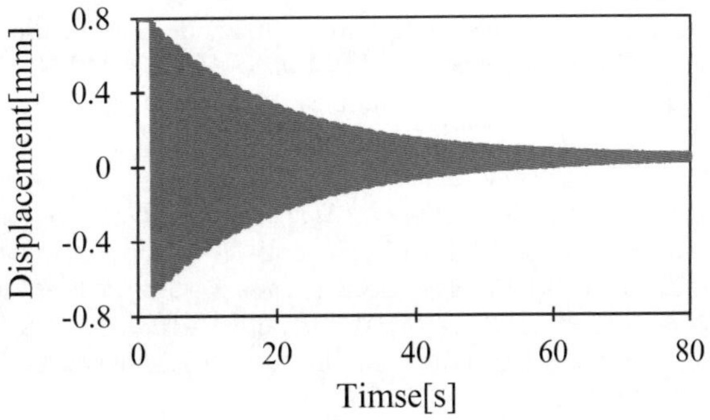

(a) Without switching stiffness control

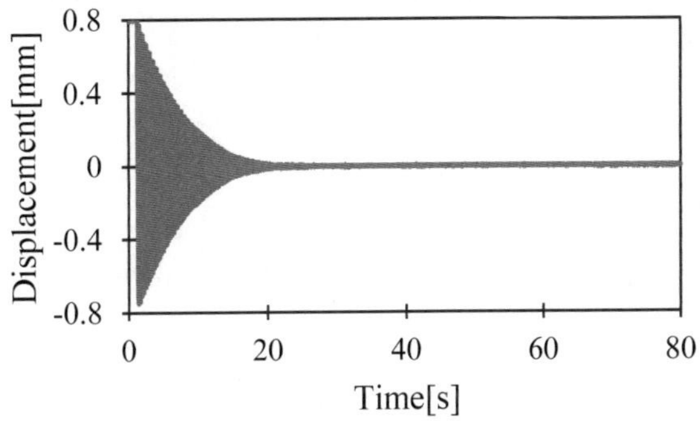

(b) With switching stiffness control

Figure 13 Effects of switching stiffness control

5 Conclusions

A method of detecting the lateral displacement with Hall elements was proposed. The method is characterized by locating Hall elements besides the stator electromagnets. The signal detected with the fabricated device with Hall elements included some noises due to current switching. It caused mis-switching's. Then the controller was modified to use the pre-sampled value to determine the control current. The modified control system succeeded to suppress the lateral vibration.

References

Boldea, I., & Naser, S. A. (1997). *Linear Electric Actuators and Generators.* Cambridge University Press.

Javed, A., Mizuno, T., Takasaki, M., Ishino, Y., Hara, M., & Yamaguchi, D. (2016). Proposal of Lateral Vibration Control Based on Force Detection in Magnetic Suspension System. *International Journal of Applied Electromagnetics and Mechanics, 52*, 223-229. doi:10.3233/JAE-162064

Javed, A., Mizuno, T., Takasaki, M., Ishino, Y., Hara, M., & Yamaguchi, D. (2018). Lateral Vibration Suppression by Varying Stiffness Control in a Vertically Active Magnetic Suspension System. *Actuators, 7*, 21. doi:10.3390/act7020021

Jayawant, B. (1981). *Electromagnetic Levitation and Suspension Technique.* London: Edward Arnold.

Mizuno, T., Takasaki, M., & Ishino, Y. (2016). Lateral Vibration Attenuation by the Dynamic Adjustment of Bias Currents in Magnetic Suspension System. *Journal of Physics: Conference Series 744: MOVIC2016 & RASID2016*, 012005. doi:10.1088/1742-6596/744/1/012005

Winthrop, M., Baker, W., & Goob, R. (2005). A variable stiffness device selection and design tool for lightly damped structures. *Journal of Sound and Vibration, 287*(4-5), 667-782.

Axial Synchronous Current Suppression of Maglev High-speed Motor Using Adaptive Notch Filter

Yue Zhang[1], Jin Zhou[1], Xudong Guan[1], Yuanping Xu[1], Chaowu Jin[1] and Yingzhe Lin[2]
[1] Nanjing University of Aeronautics and Astronautics, Nanjing, China
[2] Nanjing CIGU Technology Corp., Ltd., Nanjing, China
zhangyue08@nuaa.edu.cn

Abstract

Affected by the radial coupling disturbance, the axial displacement of the active magnetic bearings (AMBs) system will contain the rotation speed signal, which may lead the power amplifier saturation when the maglev high-speed motor rotates at high speed. To solve this issue, an axial synchronous current suppression method based on adaptive notch filter (ANF) combined with speed estimation algorithm is proposed in this paper and the influence of ANF parameters on the estimation result is analyzed. Simulation and experimental results show that the proposed method can estimate the rotational speed accurately and suppress the axial synchronous current effectively, ensuring the high speed and stable operation of the maglev high-speed motor.

1 Introduction

AMBs have been increasingly applied in high-speed motors and drawn attention in the field of turbine machinery because of its non-contact and non-lubricating [1]. The maglev high-speed motor is connected to the load in a direct drive manner. Affected by airflow disturbances, the vibration amplitude of the rotor near the impeller is larger than that of the rotor away from the impeller, which will cause the coupling of radial and axial AMBs. As a result, the rotor axial displacement signal contains a sinusoidal signal with the same frequency as the speed. Due to the difference in structure and design, the differential coefficient of the axial AMB system is generally several times that of the radial AMB system when the decentralized PID control is used. The differential link has an amplification effect on the high frequency signal. The amplitude of the axial control current will be very large at high speed, which makes the axial power amplifier saturated earlier than radial power amplifier. That is not conducive to the high speed and stable operation of the maglev high-speed motor. The suppression of axial synchronous current plays an important role in the high speed rotation of the maglev high-speed motor.

The notch filter is widely used in the unbalance control of AMBs and its function is to extract and filter out the synchronous signal. However, it requires the rotating speed sensor to provide the rotational frequency. Rotating speed sensors are not available generally in the maglev high-speed motors due to the structure limitation and high operating temperature. In addition, the sensorless vector control (SLVC) method is generally used in high-speed motors. Frequency estimation and signal extraction for sinusoidal signals with unknown parameters are feasible, such as phase-locked loop [3], frequency-locked loop [4], Kalman filter [5], LMS algorithm [6], short-time Fourier transform [7] and so on. In the application of magnetic bearings, reference [8] proposed two adaptive observers to estimate the unbalanced signal frequency and amplitude of the magnetic levitation system, but did not consider the influence of signal harmonics and noise on the estimation process. Reference [9] proposed an adaptive frequency estimator that uses the displacement signals of one radial AMB in two directions to estimate the speed. Reference [10] proposed an improved method on the basis of [9] that added phase shift compensation to keep the system stable in a large operating speed range. However, the adaptive frequency estimator requires the amplitudes of the displacement signals in the two directions to be the same, which is difficult to achieve in the AMB system. In response to this problem, reference [11] proposed an improved Hilbert transform method, which only uses one direction displacement signal for estimation. This method has high accuracy, but the algorithm is complex and the calculation amount is large.

In addition, reference [12] proposed an improved adaptive estimation method to estimate the surge frequency and amplitude of the maglev blower, eliminating the influence of signal amplitude on the estimated speed. However, this method is only used for open-loop detection, and is not used for closed-loop control. The above references all use the radial displacement signal for frequency estimation and unbalance compensation, and do not consider the influence of the axial Synchronous current.

This paper proposes a method of speed estimation and axial synchronous current suppression based on discrete ANF for maglev high-speed motor without a rotational speed sensor. The axial displacement signal is used to estimate the rotor speed, and the axial synchronous control current is filtered out to avoid the saturation of the power amplifier. The experimental results measured on the designed maglev high-speed motor prove the effectiveness of this method.

2 Speed Estimation and Synchronous Current Suppression

Affected by the radial coupling disturbance, the rotor axial displacement of the AMB system will produce sinusoidal fluctuations. Let the rotor axial displacement signal be

$$x(k) = A\sin(\omega_0 k + \theta) + v_0(k), \quad k = 1,2,\dots,N \tag{1}$$

where ω_0 is the speed frequency, A and θ is the amplitude and initial phase of the synchronous displacement signal respectively, $v_0(k)$ is an additive gaussian white noise.

The transfer function of the second-order IIR ANF is

$$H(z,a) = \frac{N(z,a)}{D(z,a)} = \frac{1+az^{-1}+z^{-2}}{1+\rho a z^{-1}+\rho^2 z^{-2}} \tag{2}$$

where ρ controls the notch bandwidth of the ANF, $0 < \rho < 1$. The larger the ρ, the smaller the notch bandwidth is. $a = -2\cos\tilde{\omega}$ is the adaptive adjustment parameter of the ANF. In the adaptive adjustment process, a keeps approaching $a_0 = -2\cos\omega_0$, making the center frequency of the ANF equal to the frequency of the input signal. Define $w(k)$ as the internal state of the ANF, and $y(k)$ as the output of the ANF, then

$$w(k) = \frac{1}{D(z,a)}x(k) = x(k) - \rho a w(k-1) - \rho^2 w(k-2) \tag{3}$$

$$y(k) = H(z,a)x(k) = w(k) + aw(k-1) + w(k-2) \tag{4}$$

The estimated recursive formula for the adaptive adjustment parameter a is [13]

$$a(k+1) = a(k) - \mu D(k) \tag{5}$$

$$D(k) = \arctan\left[\frac{y(k)}{x(k)+c}\right] \tag{6}$$

where μ is a positive number that controls the convergence speed, and c is a small constant to prevent a singularity of the ratio. In the steady state, $|y(k)|$ is always smaller than $|x(k)|$, then $-1 \leq y(k)/x(k) \leq 1$. At this time, Eq. (6) can be approximately expressed as

$$D(k) \approx \frac{\frac{y(k)}{x(k)}}{1+\alpha\left(\frac{y(k)}{x(k)}\right)^2} = \frac{x(k)y(k)}{x^2(k)+\alpha y^2(k)} \tag{7}$$

where $\alpha = 9/32$. Then Eq. (5) can be approximately expressed as

$$a(k+1) = a(k) - \mu\frac{x(k)y(k)}{x^2(k)+\alpha y^2(k)} \tag{8}$$

The rotational speed frequency can be obtained by

$$f = \frac{f_s}{2\pi}\arccos\left(-\frac{a}{2}\right) \tag{9}$$

where f_s is the sampling frequency.

From the transfer function of the ANF, the amplitude-frequency response and phase can be obtained as

$$|H(z,a)| = \frac{|2\cos\omega_0+a|}{\sqrt{[(\rho^2+1)\cos\omega_0+\rho a]^2+[(1-\rho^2)\sin\omega_0]^2}} \tag{10}$$

$$\phi = \begin{cases} -\arctan\frac{(1-\rho^2)\sin\omega_0}{(\rho^2+1)\cos\omega_0+\rho a} & \omega_0 \leq \frac{\pi}{2} \\ \pi + \arctan\frac{(1-\rho^2)\sin\omega_0}{(\rho^2+1)\cos\omega_0+\rho a} & \omega_0 > \frac{\pi}{2} \end{cases} \tag{11}$$

The real frequency characteristic is

$$\mathrm{Re}\{H(z,a)\} = \frac{(2\cos\omega_0+a)[(\rho^2+1)\cos\omega_0+\rho a]}{[(\rho^2+1)\cos\omega_0+\rho a]^2+[(1-\rho^2)\sin\omega_0]^2} \tag{12}$$

Let

$$D(k) = \arctan\left[\frac{y(k)}{x(k)}\right] - \arctan(\mathrm{Re}\{H(z,a)\}) = 0 \tag{13}$$

Then

$$(2\cos\omega_0 + a)[(\rho^2+1)\cos\omega_0 + \rho a] = 0 \tag{14}$$

It is easy to prove that $(\rho^2+1)\cos\omega_0 + \rho a \neq 0$. If and only if $a = -2\cos\omega_0$, Eq. (14) holds, that is, $\tilde{\omega} = \omega_0$. When $\tilde{\omega} > \omega_0$, $D(k) > 0$, a tends to decrease, and $\tilde{\omega}$ decreases; when $\tilde{\omega} < \omega_0$, $D(k) < 0$, a tends to increase, and $\tilde{\omega}$ increases. After constant adjustment, $\tilde{\omega}$ will eventually stabilize at the optimal solution ω_0, that is, the estimated frequency converges to the speed frequency.

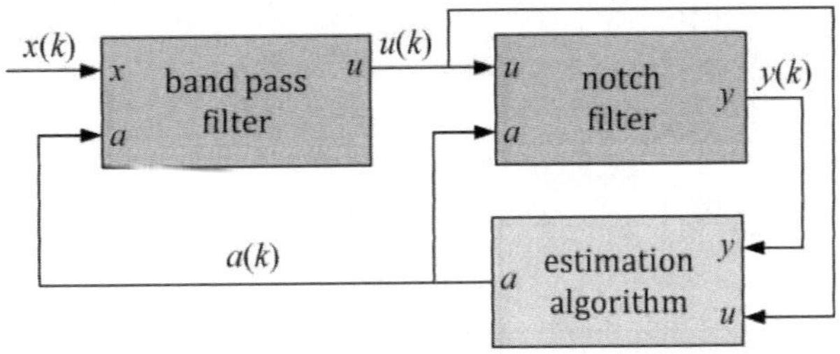

Fig.1 Structure of adaptive notch filter

There are harmonics and noise signals superimposed on the displacement signal. This will cause a certain deviation and a small oscillation of the estimated frequency. In order to reduce the interference of harmonics and noise, a band-pass filter can be added to the signal input for pre-filtering. As shown in Fig.1, the center frequencies of the band-pass filter and the notch filter are the same, which change with the iteration of the estimation algorithm. The output $u(k)$ of the bandpass filter is the synchronous component in the input signal at steady state, which can be expressed as:

$$u(k) = [1 - H(z,a)]x(k) \tag{15}$$

Fig.2 is the diagram of axial AMB control system with ANF. The difference between the axial sensor output and the reference position is used as the input of the ANF. After calculation, the rotation speed and the extracted synchronous component are output, and the synchronous component is subtracted from the controller input signal. Then after the controller and the power amplifier, the output control current does not include the current with the same frequency as the speed, and the synchronous current suppression is realized.

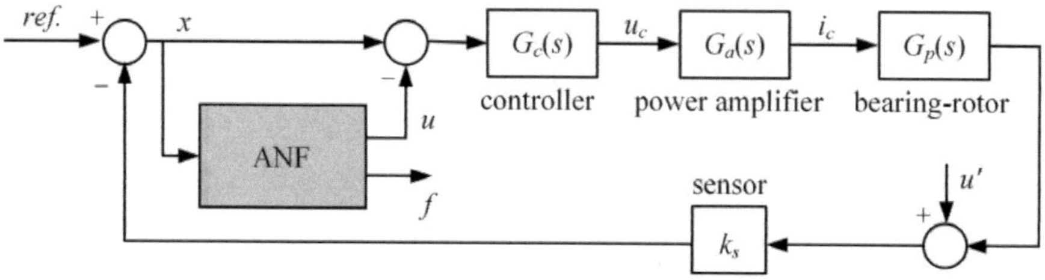

Fig.2 Diagram of axial AMB system with ANF

3 Simulation results

The axial displacement signal of the magnetic levitation rotor not only contains the sinusoidal signal with the same frequency as the speed, but also contains harmonics and noise signals. Since the harmonic signal has a small amplitude compared with the fundamental frequency signal, the input signal of the ANF used in the simulation is

$$x = \sin(2\pi ft) + 0.3\sin(6\pi ft) + 0.1\sin(10\pi ft) + v \tag{16}$$

Take $\rho = 0.97$, $\mu = 0.001$. The speed frequency f is initially 300Hz, after 1s it becomes 400Hz, and after 3s it becomes 500Hz. The frequency estimation is carried out using the proposed method, as shown in Figure 3. It can be seen that the estimated frequency can converge to the speed frequency within 0.4s. Figure 4 is a section of input and output signals when the rotation speed is 400Hz. After the input signal passes through the ANF, the fundamental frequency signal of the rotation speed can be extracted.

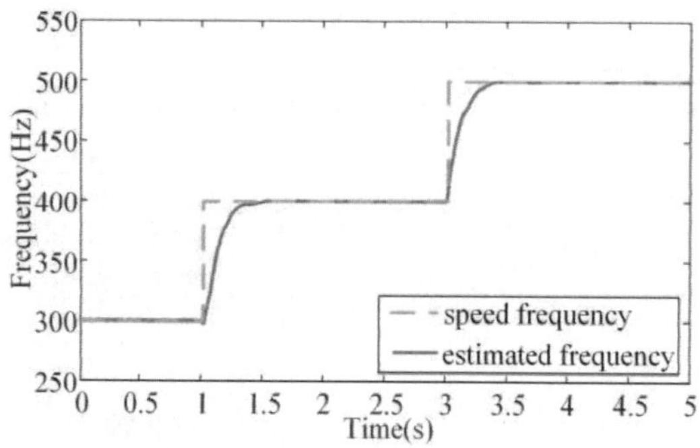

Fig.3 Simulation result of frequency estimation

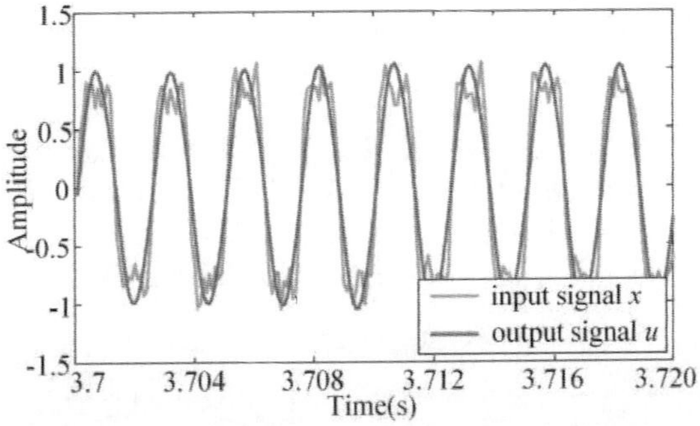

Fig.4 Input and output signal comparison

There are two adjustable parameters ρ and μ in the ANF, which will affect the estimation results. When $\mu = 0.001$ and the rotational speed frequency is 300 Hz, the estimation results of different ρ are shown in Fig.5. When ρ is too small, errors will occur in the estimation results. When ρ is too large, the closer to 1, the higher the estimation accuracy is. However, at this time, the notch width is too narrow, and the robustness of the algorithm will be reduced, or even no convergence. Therefore, the value of ρ is generally 0.92~0.98.

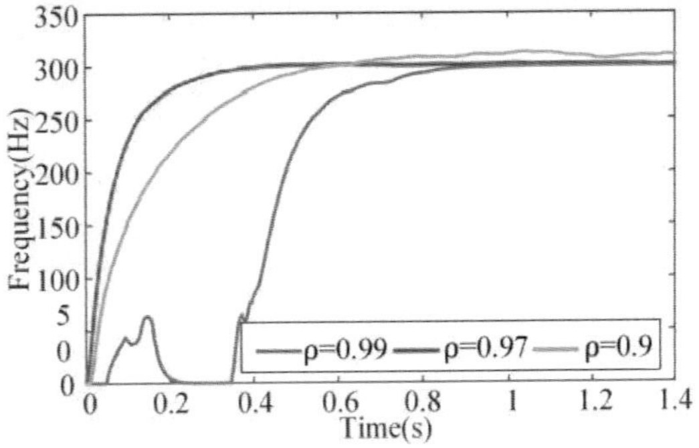

Fig.5 The influence of different ρ on the estimation results

When $\rho = 0.97$ and the rotating speed frequency is 300Hz, the estimation results of different μ are shown in Fig.6. When μ is too large, the response speed is faster, but the result will be oscillation and have poor anti-interference ability. When μ is too small, the steady-state performance is better and the accuracy is higher, but the response speed is slower. Therefore, it is necessary to choose the appropriate μ to meet the requirements of accuracy and speed.

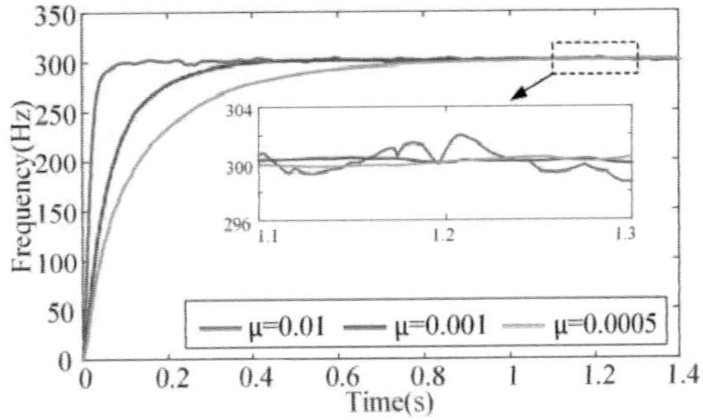

Fig.6 The influence of different μ on the estimation results

4 Experimental results

In order to verify the effectiveness of the proposed method, a maglev high-speed motor is used for experiments. The experimental rig is shown in Fig.7. The controller uses dSPACE 1202 with a sampling frequency of 20kHz, which outputs a control voltage signal. The power amplifier outputs control current signals to control the AMB coils so that the rotor is suspended at a set position. Decentralized PID control is adopted for the radial and axial AMBs to realize the rotor stable suspension.

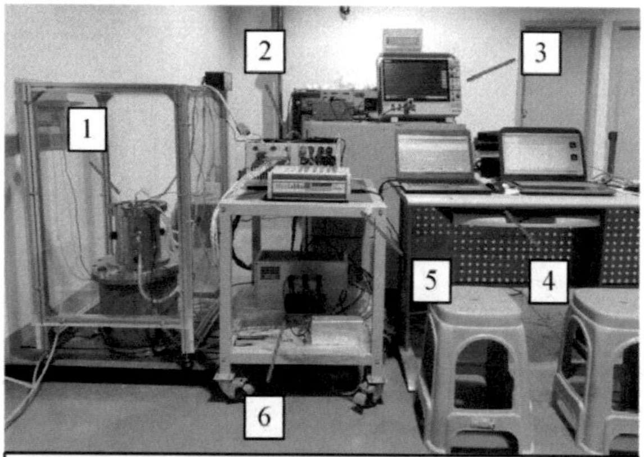

1.maglev high-speed motor 2.power amplifier
3.oscilloscope 4.PC 5.dSPACE 1202 6.motor controller

Fig.7 Maglev high-speed motor test rig

The parameters of the axial AMB system are shown in Table 1. The transfer functions of the axial controller, power amplifier, and bearing-rotor are respectively

$$G_c(s) = k_P + \frac{k_I}{s} + \frac{k_D s}{T_D s+1} \tag{17}$$

$$G_a(s) = \frac{k_a}{T_a s+1} \tag{18}$$

$$G_p(s) = \frac{k_i}{ms^2+k_x} \tag{19}$$

Table 1 The parameters of the axial AMB system

Parameters	Value	Unit
m	3.58	Kg
k_i	264.12	N/A
k_x	-0.73	N/μm
k_P	1.4	/
k_I	1	/
k_D	0.005	/
T_D	0.0001	/
k_a	0.4	A/V
T_a	0.0001	/
k_s	3450	V/m
i_0	2	A

The motor rotor and AMBs configuration used in the experiment are shown in Fig.8. Due to the limitation of the motor internal space, the speed sensor cannot be installed. The axial displacement sensor is installed at the end of the impeller, and the measured axial displacement is greatly affected by the radial direction.

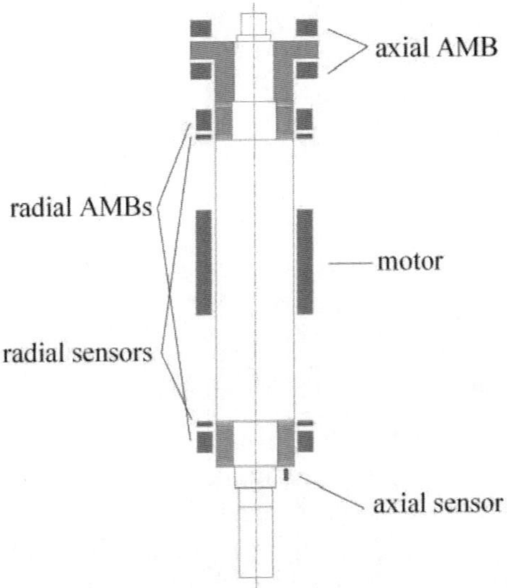

Fig.8 Structure diagram of maglev high-speed motor

Fig.9 and Fig.10 are the time-domain and frequency-domain diagrams of the axial displacement and radial displacement when the ANF is not used. It can be seen that the main components of the axial and radial displacement signals are sinusoidal signals with a frequency of 500 Hz. As we all know, the main component in the radial displacement signal is the unbalanced vibration caused by the rotor unbalanced mass, and its frequency is the same as the speed frequency. Therefore, the main component in the axial displacement signal is also the same as the speed frequency. It can be seen from the time domain diagram that, compared with the radial displacement signal, the axial displacement signal has less harmonics and noise. Using the axial displacement signal to estimate the rotor speed will be more accurate.

Firstly, the feasibility of the proposed method is investigated at a fix speed. Because the bias current of the axial AMB is 2A, theoretically the maximum amplitude of the coil control current is 2A, otherwise it will cause the saturation of the power amplifier. In fact, because of the electromagnetic noise, sensor measurement error and other reasons, the maximum allowable control current amplitude will be smaller. The maglev high-speed motor used in the experiments is a vertical structure. In order to counteract gravity, there will still have an offset of about 0.35A in the axial control current, which leads to further restrictions on the maximum allowable control current.

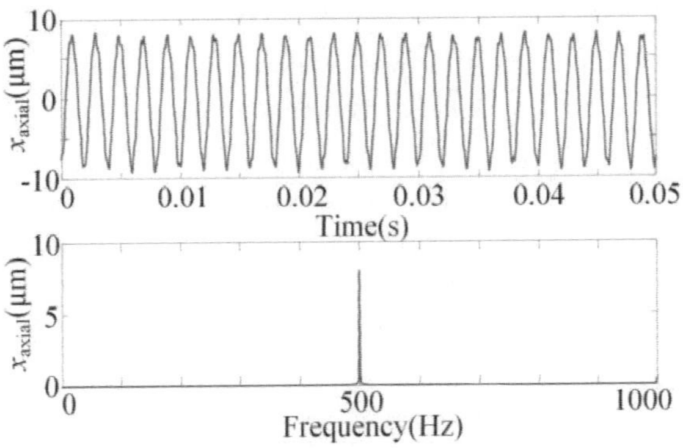

Fig.9 Time and frequency domain diagrams of the axial displacement

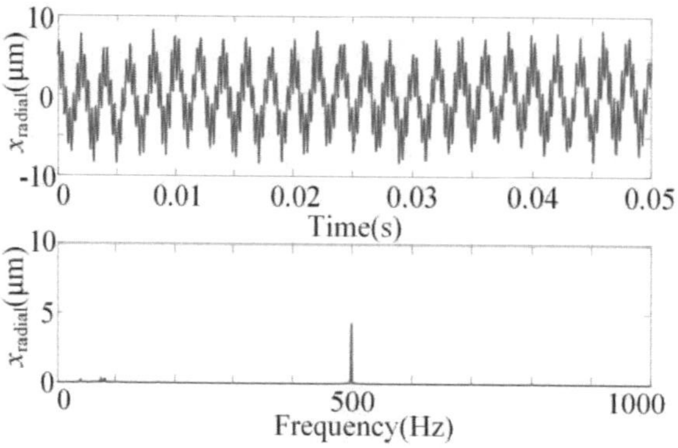

Fig.10 Time and frequency domain diagrams of the radial displacement

Fig.11 is a comparison diagram of the time domain and frequency domain of the axial AMB control current without and with ANF. It can be seen from the frequency domain diagram that the rotor speed at this time is 30000rpm (500Hz). Fig.12 shows the speed frequency estimated by the ANF. It fluctuates around 500Hz with a fluctuation range of ±0.1Hz, which proves the accuracy of the proposed method. When the ANF is not used at 30000rpm, the axial AMB control current is close to 2A. There is a risk of instability if the speed continues to increase. After using the ANF, the amplitude of the synchronous control current is reduced by 78%, from 1A to 0.22A.

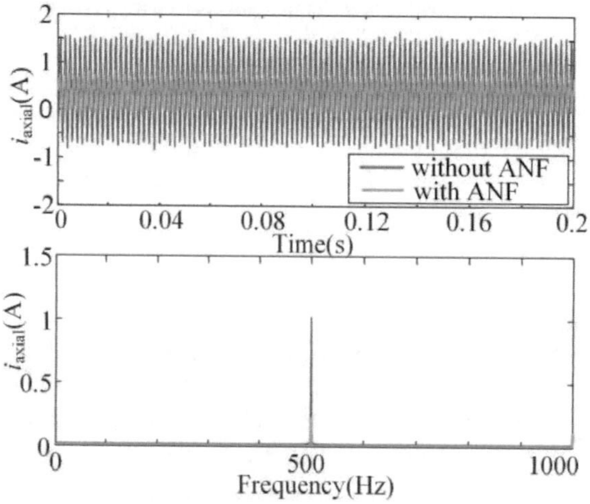

Fig.11 The coil current without and with ANF

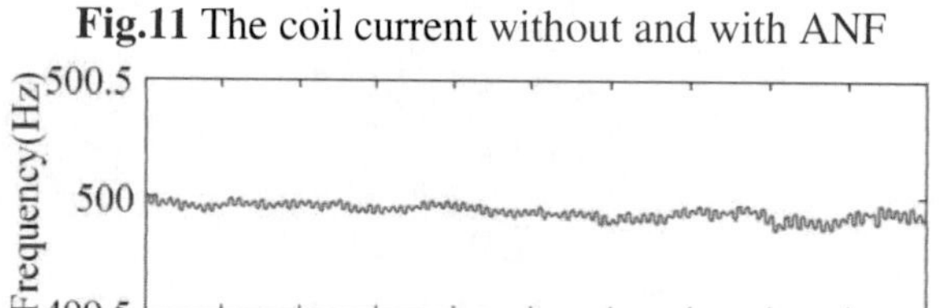

Fig.12 Experimental result of speed estimation

To make the rotor run at high speed, it is necessary to use the ANF during the speed-up process. Since the notch filter may cause the rotor unstable below the rigid body modal frequency, it is set to suppress the axial current when the estimated speed is above 20,000 rpm. Fig.13 shows the synchronous control current of the axial AMB without ANF. It can be seen that the amplitude of the axial synchronous current becomes larger and larger as the rotational speed increases, reaching 1.1A at 35,000 rpm.

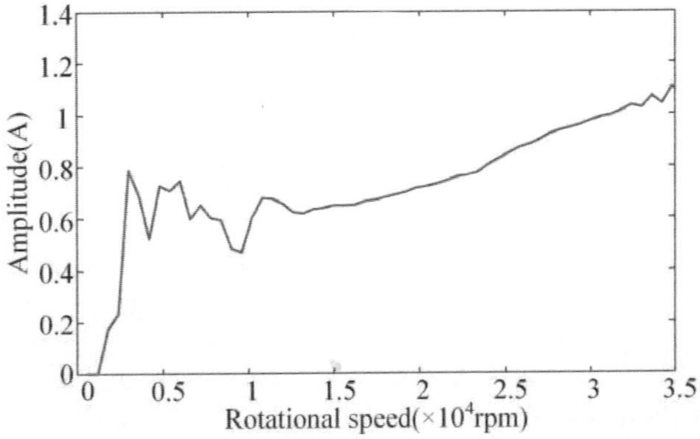

Fig.13 Axial synchronous current amplitude without ANF

As shown in Fig.14, the amplitude of the axial synchronous current still increases with the increase of the rotational speed after 20,000rpm. However, it is much smaller than when the ANF is not used. Experiments have shown that the use of ANF for axial synchronous current suppression avoids the premature saturation of the axial power amplifier. As a result, the maglev high-speed motor can run stably at 80,000rpm and the maximum speed reaches 85,000 rpm.

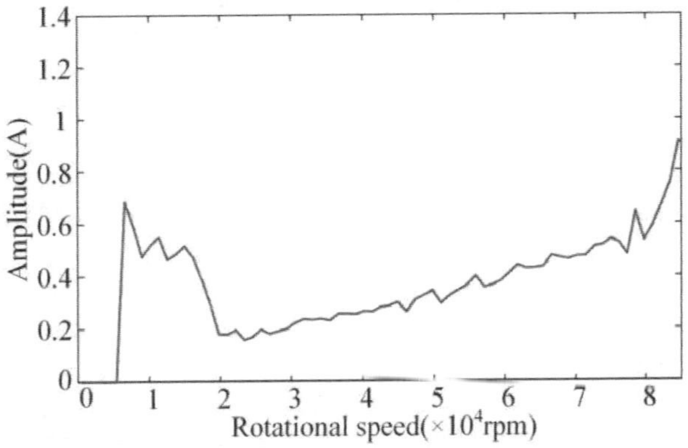

Fig.14 Axial synchronous current amplitude with ANF

5 Conclusion

This paper presents a method of speed estimation and axial synchronous current suppression for maglev high-speed motors using adaptive notch filter. This method uses the rotor axial displacement signal to estimate the rotor speed without a rotating speed sensor. In addition, the axial AMB synchronous current is suppressed to avoid the power amplifier saturation and ensure the high speed and stable operation of the maglev high-speed motor with a maximum speed of 85,000 rpm.

References

[1] Schweitzer G, Maslen E, Bleuler H, et al (2009). *Magnetic Bearing: Theory, Design, and Application to Rotating Machinery*. Berlin, Germany: Springer.

[2] Herzog R, Buhler P, Gahler C, et al (1996). *Unbalance compensation using generalized notch filters in the multivariable feedback of magnetic bearings*. IEEE Transactions on Control

Systems Technology, 4(5), pp. 580-586.

[3] Karimi-Ghartemani M, Karimi H, Iravani M R (2004). *A magnitude/phase-locked loop system based on estimation of frequency and in-phase/ quadrature-phase amplitudes.* IEEE Transactions on Industrial Electronics, 51(2), pp. 511-517.

[4] Rodriguez P, Luna A, Candela I, et al (2011). *Multiresonant Frequency-Locked Loop for Grid Synchronization of Power Converters Under Distorted Grid Conditions.* IEEE Transactions on Industrial Electronics, 58(1), pp. 127-138.

[5] Dhaouadi R, Mohan N, Norum L (1991). *Design and implementation of an extended Kalman filter for the state estimation of a permanent magnet synchronous motor.* IEEE Transactions on Power Electronics, 6(3), pp. 491-497.

[6] Wang J, Shen W, Zhao J (2007). *Adaptive adjustment of maneuvering frequency in target tracking*[J]. Transaction of Beijing Institute of Technology, 1, pp. 42-45.

[7] Borkowski J, Kania D, Mroczka J (2014). *Interpolated DFT-Based Fast and Accurate Frequency Estimation for the Control of Power.* IEEE Transactions on In-dustrial Electronics, 61(12), pp. 7026-7034.

[8] Vahedforough E, Shafai B, Beale S (2007). *Estimation and rejection of unknown sinusoidal disturbances using a generalized adaptive forced balancing Method.* IEEE American Control Conference.

[9] Chen Q, Liu G, Han B (2015). *Suppression of Imbalance Vibration in AMB-Rotor Systems Using Adaptive Frequency Estimator.* IEEE Transactions on Industrial Electronics, 62(12), pp. 7696-7705.

[10] Chen Q, Liu G, Han B (2017). *Unbalance vibration suppression for AMBs system using adaptive notch filter.* Mechanical Systems and Signal Processing, 93, pp. 136-150.

[11] Zhang R, Liu H, Fan Y (2017). *Estimation method of rotor speed using signal of displacement sensor in magnetic suspension flywheel.* Journal of Astronautics, 38(12), pp. 1314-1323.

[12] Han B, Wang K, Zheng S, Zhang Y (2017). *Surge detection of magnetically suspended high-speed centrifugal blower.* Optics and Precision Engineering, 4.

[13] Punchalard R (2014). *A modified inverse tangent based adaptive algorithm for a second-order constrained adaptive IIR notch filter.* Signal Processing, 94, pp. 350-358.

Robust Control of Rigid Rotor Active Magnetic Bearing System Based on Signal Compensation

Yichen Yao[1*], Yixin Su[2†], Tianye Yu[3‡], Honglei Sha[4§] and Suyuan Yu[5¶]

[1] Department of Energy and Power Engineering, Tsinghua University, Beijing, 100084, P. R. China.
yaoyc18@mails.tsinghua.edu.cn
[2] Department of Energy and Power Engineering, Tsinghua University, Beijing, 100084, P. R. China.
syx16@mails.tsinghua.edu.cn
[3] Tianjin Emaging Technology Co. Ltd,Tianjin, 300457, P. R. China.
yutianye@emaging.com.cn
[4] Tianjin Emaging Technology Co. Ltd,Tianjin, 300457, P. R. China.
amb@emaging.com.cn
[5] Department of Energy and Power Engineering, Tsinghua University, Beijing, 100084, P. R. China.
suyuan@mail.tsinghua.edu.cn

Abstract

Active magnetic bearings (AMBs) use the controlled magnetic force to achieve frictionless relative motion between the stator and the rotor. In the case of a large operating speed range, the gyroscopic effect, the unbalance force, the unmodeled dynamics and disturbances in the rotor-AMBs system has become a non-negligible factor that may influence the stability and the dynamic behavior of the system. To avoid this problem, robust control methods based on signal compensation are applied to design the controllers of the rotor-AMBs system in this paper. Firstly, the rigid rotor-AMBs system is described with lumped uncertainties under sensor general coordinates system. On the basis of the model, the control system is designed. The control system contains nominal controllers and compensators. The target of the nominal controller is to stabilize the nominal part of the system. The robust compensator aims to generate the compensation signal to suppress the lumped uncertainties equivalent disturbance robustly. According to the design, we have done laboratory experiments on an AMB-supported industrial permanent magnet synchronous motor (PMSM). The results show that the design method based on signal compensation has good performance.

1 Introduction

Active magnetic bearings (AMB) is widely used in rotating machinery because of its advantages of no friction, no lubrication or sealing requirements, long lifespan, low maintenance and active vibration control [11, 12, 14]. However, the design of magnetic bearing controller relies on accurate rotor-AMBs system modeling as well as precise external disturbance modeling, which restrict the industrial application and popularization of magnetic bearing. In fact, the AMB controller often requires a long time to adjust related parameters while in operation.

In recent years, many researches have focused on AMB controller design in order to achieve better vibration suppression performance and improve the robustness of the controller. As

*Proposed the idea and designed the controller.
†Did numerous tests and provided a lot of suggestions.
‡Helped establish the test rig.
§Helped adjust the test rig and the control parameters.
¶Guided the research and provided a lot of suggestions.

for vibration suppression, plentiful good controllers have been proposed. Shiqiang Zheng et al. [2] proposed a feed-forward control strategy combined with a novel adaptive notch filter to solve the rotor unbalance problem in magnetic suspended centrifugal compressors. To the same problem, Qi Chen, et al. [1] put forward a double-loop compensation design approach based on the AMBs. Fang et al. [4] designed a gain phase modifier to compensate the gain and phase errors caused by the power amplifier and completely achieved automatic balance of the rotor. Zhang Kai, et al. [8] designed a comprehensive controller for the AMBs on a turbo molecular pump, using different methods to suppress different kinds of vibration. Jiang et al. [6] realized multi-frequency periodic vibration suppressing in active magnetic bearing-rotor systems through response matching in frequency domain. However, the premise of good performance of these controllers is the exact model of the external forces and the system model.

To achieve enough robustness, adaptive control strategy, robust control strategy, active disturbance rejection strategy and many other methods have been applied on AMBs. Dhyani et al.[3] applied heuristic moth-flame optimization algorithm to optimize the scaling factors of the fuzzy-PID controller in a AMBs system. Under this algorithm, the AMBs showed great robustness to external periodic excitations and step input. Su et al. [13] proposed a PID-surface sliding mode control method, which possessed high tracking accuracy and anti-jamming capacity. In [10], an adaptive back-stepping control based on sliding mode approach is applied to a linearized model of an active magnetic bearing system in order to deal with the external disturbance. In [7] and [5], an PID controller with extended state observer (ESO) is applied on decentralized control of AMB-rotor system and corresponding experiments have been performed. In [9], an robust controller with feedback-linearization ESO is employed on a MSCMGs system and achieves good performance. However, these controllers have relatively more complex structure and the parameters adjustment procedure.

This paper proposes a robust controller based on signal compensation. It contains a PD controller as nominal controller and a simple-structure robust compensator. Corresponding experimental verification has been carried out on an industrial AMB-supported permanent magnet synchronous motor (PMSM) test rig. The proposed strategy is compared with a PID controller with the same nominal controller. The results show that the proposed control strategy is effective.

2 System modeling

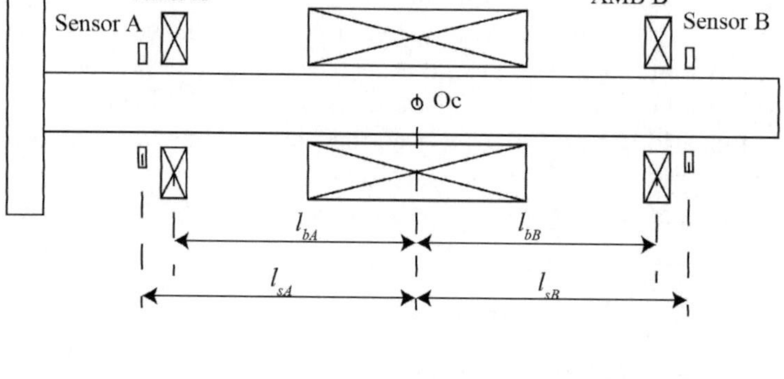

Figure 1: Structure of the AMB-rigid rotor system.

The structure of the AMB-rigid rotor system is shown in Fig.1. The terms x_c, y_c represent rotor's centroid's displacement; α, β are the rotor's angular displacement around x and y axes. The terms $l_{bA}, l_{bB}, l_{sA}, l_{sB}$ show the distance between the A/B bearing/sensor and the centroid. In AMB-rigid rotor system, the coupling between the axial DOF and radial DOF is negligible. Thus, axial DOF is not mentioned in this model and merely radial 4 DOF and axial angle γ are considered. There are three coordinates to illustrate the motion of rotor, which are centroid coordinate $\mathbf{y}^c = \begin{bmatrix} x_c & \alpha & y_c & \beta \end{bmatrix}^T$, sensor coordinate $\mathbf{y}^s = \begin{bmatrix} y^s_{Ax} & y^s_{Ay} & y^s_{Bx} & y^s_{By} \end{bmatrix}^T$, and bearing coordinate $\mathbf{y}^b = \begin{bmatrix} y^b_{Ax} & y^b_{Ay} & y^b_{Bx} & y^b_{By} \end{bmatrix}^T$. The relation between them can be expressed as $\mathbf{y}^s = \mathbf{T}^{sc}\mathbf{y}^c, \mathbf{y}^c = \mathbf{T}^{cb}\mathbf{y}^b$, in which

$$\mathbf{T}^{cb} = \begin{bmatrix} 1 & 0 & 1 & 0 \\ 0 & -l_{bA} & 0 & l_{bB} \\ 0 & 1 & 0 & 1 \\ l_{bA} & 0 & -l_{bB} & 0 \end{bmatrix}$$

$$\mathbf{T}^{sc} = \begin{bmatrix} 1 & 0 & 0 & l_{sA} \\ 0 & -l_{sA} & 1 & 0 \\ 1 & 0 & 0 & -l_{sB} \\ 0 & l_{sB} & 1 & 0 \end{bmatrix}$$

The bearing force is defined as $\mathbf{f}^b_b = \begin{bmatrix} f_{bAx} & f_{bAy} & f_{bBx} & f_{bBy} \end{bmatrix}^T$, with the load defined as $\mathbf{f}^c_g = \begin{bmatrix} f^c_{gx} & f^c_{g\alpha} & f^c_{gy} & f^c_{g\beta} \end{bmatrix}^T$. According to the second Lagrange equation, the equation of motion (EOF) of the rotor is obtained:

$$\begin{cases} m\ddot{x}_c = f_{bAx} + f_{bBx} + f^c_{ex} + f^c_{gx} \\ J_r\ddot{\alpha} + J_z\dot{\gamma}\dot{\beta} = -l_{bA}f_{bAy} + l_{bB}f_{bBy} + f^c_{e\alpha} + f^c_{g\alpha} \\ m\ddot{y}_c = f_{bAy} + f_{bBy} + f^c_{ey} + f^c_{gy} \\ J_r\ddot{\beta} - J_z\dot{\gamma}\dot{\beta} = l_{bA}f_{bAx} - l_{bB}f_{bBx} + f^c_{e\beta} + f^c_{g\beta} \end{cases} \tag{1}$$

where the unbalance force can be rewritten in vector form

$$\mathbf{f}^c_e = \begin{bmatrix} f^c_{ex} \\ f^c_{e\alpha} \\ f^c_{ey} \\ f^c_{e\beta} \end{bmatrix} = \dot{\gamma}^2 \begin{bmatrix} -me_y & me_x \\ -J_{xz} & -J_{yz} \\ me_x & me_y \\ -J_{yz} & J_{xz} \end{bmatrix} \begin{bmatrix} \sin\gamma \\ \cos\gamma \end{bmatrix}$$

Transform (1) to sensor coordinate yields (2)

$$\begin{cases} \ddot{y}^s_{Ax} = -\frac{J_z\Omega l_{sA}}{J_r(l_{sA}+l_{sB})}\left(\dot{y}^s_{Ay} - \dot{y}^s_{By}\right) + \left(\frac{1}{m} + \frac{l_{bA}l_{sA}}{J_r}\right)f_{bAx} + \left(\frac{1}{m} - \frac{l_{sA}l_{bB}}{J_r}\right)f_{bBx} + f^s_{eAx} + f^s_{gAx} \\ \ddot{y}^s_{Ay} = -\frac{J_z\Omega l_{sA}}{J_r(l_{sA}+l_{sB})}\left(\dot{y}^s_{Bx} - \dot{y}^s_{Ax}\right) + \left(\frac{1}{m} + \frac{l_{bA}l_{sA}}{J_r}\right)f_{bAy} + \left(\frac{1}{m} - \frac{l_{sA}l_{bB}}{J_r}\right)f_{bBy} + f^s_{eAy} + f^s_{gAy} \\ \ddot{y}^s_{Bx} = -\frac{J_z\Omega l_{sB}}{J_r(l_{sA}+l_{sB})}\left(\dot{y}^s_{By} - \dot{y}^s_{Ay}\right) + \left(\frac{1}{m} - \frac{l_{bA}l_{sB}}{J_r}\right)f_{bAx} + \left(\frac{1}{m} + \frac{l_{bB}l_{sB}}{J_r}\right)f_{bBx} + f^s_{eBx} + f^s_{gBx} \\ \ddot{y}^s_{By} = -\frac{J_z\Omega l_{sB}}{J_r(l_{sA}+l_{sB})}\left(\dot{y}^s_{Ax} - \dot{y}^s_{Bx}\right) + \left(\frac{1}{m} - \frac{l_{bA}l_{sB}}{J_r}\right)f_{bAy} + \left(\frac{1}{m} + \frac{l_{bB}l_{sB}}{J_r}\right)f_{bBy} + f^s_{eBy} + f^s_{gBy} \end{cases} \tag{2}$$

The rotor centroid's precise position is difficult to determine due to its complex shape and its heterogeneous material, which indicate (2) has uncertainty. This uncertainty can be regarded as contained in $\mathbf{f}_g$.

As for AMBs, no matter what control strategy is applied, the bearing force can always be considered as the expected control force generated by a specific controller:

$$f_{bn} \approx u_n, \quad n = Ax, Ay, Bx, By \tag{3}$$

where u_n is the expected control force.

So that (2) can be rewritten as (4). In (4), each sensor coordinate's differential equation is decoupled formally and the plant model of the decentralized controller can be obtained from this.

$$\begin{cases} \ddot{y}^s_{Ax} = \left(\frac{1}{m} + \frac{l_{bA} l_{sA}}{J_r}\right) u_{Ax} + d_{Ax} \\ \ddot{y}^s_{Ay} = \left(\frac{1}{m} + \frac{l_{bA} l_{sA}}{J_r}\right) u_{Ay} + d_{Ay} \\ \ddot{y}^s_{Bx} = \left(\frac{1}{m} + \frac{l_{bB} l_{sB}}{J_r}\right) u_{Bx} + d_{Bx} \\ \ddot{y}^s_{By} = \left(\frac{1}{m} + \frac{l_{bB} l_{sB}}{J_r}\right) u_{By} + d_{By} \end{cases} \tag{4}$$

In (4), d represents the static coupling (coupling of u) and the dynamic coupling (coupling of ,which means the gyroscopic effect), the specific expression is

$$\begin{cases} d_{Ax} = \left(\frac{1}{m} - \frac{l_{sA} l_{bB}}{J_r}\right) u_{Bx} - \frac{J_z \Omega l_{sA}}{J_r (l_{sA} + l_{sB})} \left(\dot{y}^s_{Ay} - \dot{y}^s_{By}\right) + f^s_{eAx} + f^s_{gAx} \\ d_{Ay} = \left(\frac{1}{m} - \frac{l_{sA} l_{bB}}{J_r}\right) u_{By} - \frac{J_z \Omega l_{sA}}{J_r (l_{sA} + l_{sB})} \left(\dot{y}^s_{Bx} - \dot{y}^s_{Ax}\right) + f^s_{eAy} + f^s_{gAy} \\ d_{Bx} = \left(\frac{1}{m} - \frac{l_{bA} l_{sB}}{J_r}\right) u_{Ax} - \frac{J_z \Omega l_{sB}}{J_r (l_{sA} + l_{sB})} \left(\dot{y}^s_{By} - \dot{y}^s_{Ay}\right) + f^s_{eBx} + f^s_{gBx} \\ d_{By} = \left(\frac{1}{m} - \frac{l_{bA} l_{sB}}{J_r}\right) u_{Ay} - \frac{J_z \Omega l_{sB}}{J_r (l_{sA} + l_{sB})} \left(\dot{y}^s_{Ax} - \dot{y}^s_{Bx}\right) + f^s_{eBy} + f^s_{gBy} \end{cases} \tag{5}$$

3 Controller design

The AMB controller can be divided into three parts: driven mode of AMB and centralized controller. The centralized controller then contains nominal controller and robust compensator.

3.1 Driven mode of AMB

The driven mode of AMB refers to the realization of the expected AMB force. The differential driven mode is a traditional application in industrial AMBs, shown in Fig.2. In differential driven mode, the current in the magnet pole contains bias current i^0 and control current i^c. A pair of opposite magnet poles together control one degree of freedom (DOF). The current in this pair of opposite magnet poles share the same bias current and and inverse control current, which can be written as $i_{n1}=i^0_n+i^c_n$, $i_{n2}=i^0_n-i^c_n$. In one magnet pole, the electromagnetic force generated by the single pole of the magnetic bearing can be expressed as

$$f_p = k\frac{i^2}{s^2}, k = \frac{\mu_0 A N^2}{4}\cos\theta \tag{6}$$

where i is the current, s is the gap, μ_0 is the magnetic field constant in vacuum, N is the number of coils turns, A is the cross-section area of the pole and θ is the angle of the pole. In differential driven mode, the AMB force can be written as

$$f_{bn} = \frac{k(i_{n1})^2}{\left(s_0 - y^b_k\right)^2} - \frac{k(i_{n2})^2}{\left(s_0 + y^b_k\right)^2}, \quad n = Ax, Ay, Bx, By \tag{7}$$

The expected AMB force can be designed by linearizing (7) at the balance position and the bias current.

$$\begin{aligned} &u_n = k_i i_n^c + k_x y_n^b \\ &k_i = 4k\frac{i^0}{s_0^2}, k_x = 4k\frac{(i^0)^2}{s_0^3} \quad n = Ax, Ay, Bx, By \end{aligned} \tag{8}$$

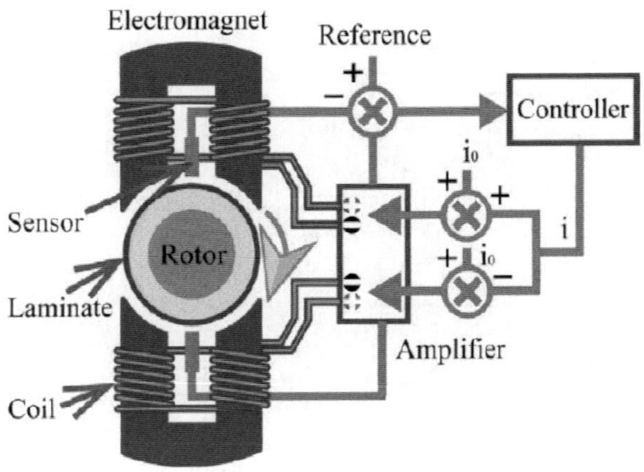

Figure 2: The differential driven mode of single DOF AMB-rotor system.

The framework of the differential driven mode in the controller can be described as Fig.3. In Fig.3, $G_c(s)$ refers to the transfer function of the centralized controller in each channel.

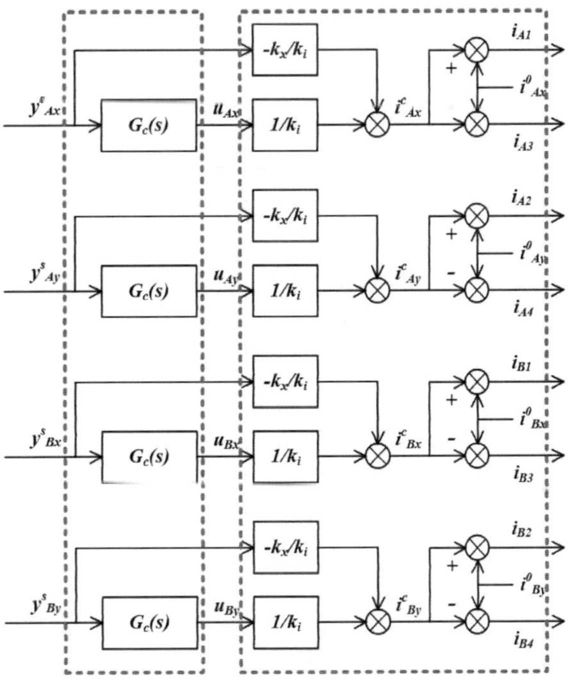

Figure 3: The framework of the differential driven mode in the controller.

3.2 Centralized controller

The centralized controller contains contains nominal controller and robust compensator. The framework of the centralized controller is shown in Fig.4. The nominal controller aims to stabilized the nominal system and the compensator aims to estimate the lumped uncertainties d according to the response signal y as well as the control signal u.

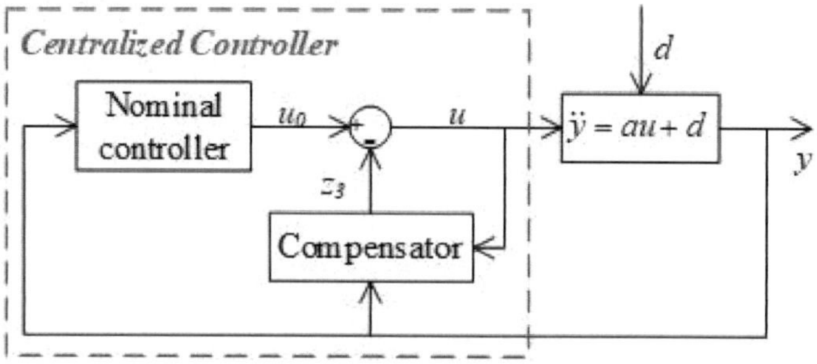

Figure 4: The framework of the centralized controller.

According to (4) and Fig.4, it is obvious that the nominal plant of the system is $ddoty = au$. In fact, proportional-differential (PD) form of nominal controller (shown in (9)) is enough to stabilize the nominal system.

$$G_n(s) = \frac{u_0(s)}{y(s)} = k_P + \frac{k_D s}{\tau_D s + 1} \tag{9}$$

As for compensator, it requires compensate the influence of lumped uncertainties. According to (4) and Fig.4, the plant with lumped uncertainties can be written as:

$$y(s) = \frac{a}{s^2} u(s) + \frac{1}{s^2} d(s) \tag{10}$$

The total control signal u is

$$u(s) = u_0(s) + v(s) \tag{11}$$

Taken (11) into (10), it obtains

$$y(s) = \frac{a}{s^2} G_n(s) y(s) + \frac{1}{s^2} [av(s) + d(s)] \tag{12}$$

Equation (12) indicates that the best choice to compensate the lumped uncertainties is

$$v^*(s) = -\frac{1}{a} d(s) = u(s) - \frac{s^2}{a} y(s) \tag{13}$$

However, (13) is a non-causal signal. Thus, a robust filter $F(s)$ is required and the compensation signal can be written as:

$$v(s) = F(s) v^*(s) \tag{14}$$

Taken $F(s)$ as

$$F(s) = \frac{\beta_3}{s^3 + \beta_1 s^2 + \beta_2 s + \beta_3} \tag{15}$$

Then, the compensator has a specific expression

$$v(s) = \frac{\beta_3}{s^3+\beta_1 s^2+\beta_2 s+\beta_3}u(s) - \frac{1}{a}\frac{\beta_3 s^2}{s^3+\beta_1 s^2+\beta_2 s+\beta_3}y(s) \tag{16}$$

The characteristics function of the compensator is

$$s^3+\beta_1 s^2+\beta_2 s+\beta_3 = 0 \tag{17}$$

To ensure the stability of the compensator, corresponding parameters can taken as

$$\beta_1 = 3\omega_0, \quad \beta_2 = 3\omega_0^2, \quad \beta_3 = \omega_0^3 \tag{18}$$

4 Experimental verification

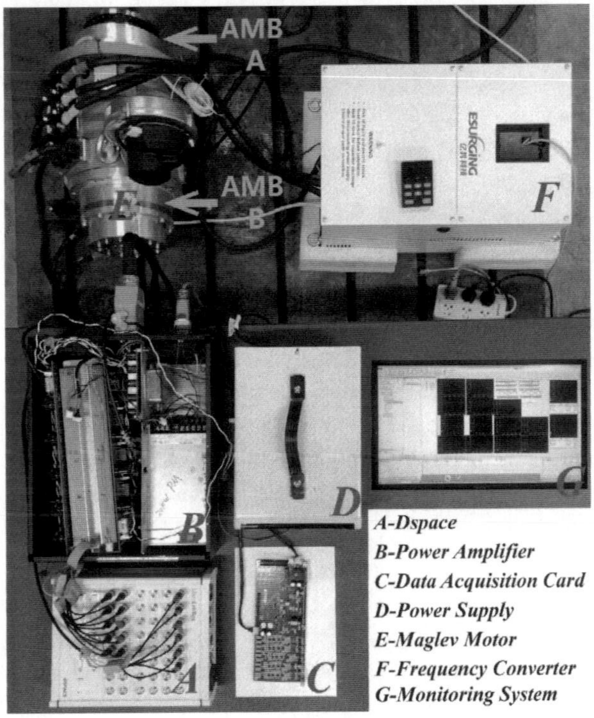

Figure 5: Description of the test rig.

In order to verify the effectiveness of the proposed control method, verification experiments are carried out on the magnetic bearing test rig. The experimental test rig for this study is a PMSM at Tsinghua University, pictured in Fig.5. The rotor of the test rig is 0.4866 m long with a total mass of 13.98 kg and is horizontally supported by two radial AMBs and one axial AMB. The radial clearance of the AMBs is 0.4 mm and the clearance of touchdown bearings is 0.2 mm. Radial and axial displacements of the rotor were measured using five inductive sensors. Ten current-controlled pulse width modulated amplifiers power the magnet coils to generate the expected bearing force.

The test rig works originally under a decentralized differential PID controller with a bias current of 2.5 A. This provides an open-loop bearing negative stiffness of 3.08×10^5 N/m and a current gain of 49.34 N/A. The equivalent stiffness brought by the original PID controller is 2×10^5 N/m. For comparison, the PD parameter in the designed controller takes the same value as the original PID.

The displacement response at sensor of original PID controller and the proposed PD-Compensator controller at the rotating speed from 0 to 100 Hz (6000 r/min) are shown in Fig.6. In Fig.6, the peak value under PID appears at 37 Hz while that of PD-Compensator controller appears at 47 Hz. Besides, from 50 Hz to 70 Hz, the radius response under PD-Compensator controller is larger than that of PID. In fact, the performance of the compensator depends on filter $F(s)$. The crossover frequency of $F(s)$ was designed at 45 Hz in this test rig, which means the compensator will not compensate the unbalance force when the rotating speed is over 45 Hz. From another perspective, the compensator can be considered as a revision of equivalent stiffness and damping on the basis of PD controller. This indicates that compensator, in a way, enlarge the bearing stiffness below the crossover frequency and narrow it above the crossover frequency. In other words, compensator will not compensate the unbalance force any more in high frequency range and will not influence the realization of self-balance at high speed.

5 Conclusion

In view of the gyroscopic effect, the unbalance force, the unmodeled dynamics and disturbances in the rotor-AMBs system, a robust controller based on signal compensation was proposed in this study to increase the robustness of the controller.Firstly, a decentralized controller plant model was developed with lump uncertainties. On the basis of the model, the control system is designed. The control system contains nominal controllers and compensators. The target of the nominal controller is to stabilize the nominal part of the system. The robust compensator aims to generate the compensation signal to suppress the lumped uncertainties equivalent disturbance robustly. Corresponding experimental verification has been carried out on an industrial AMB-supported PMSM test rig. The proposed strategy is compared with a PID controller with the same nominal controller. The results show that the proposed control strategy is effective.

References

[1] Qi Chen, Gang Liu, and Bangcheng Han. Unbalance vibration suppression for ambs system using adaptive notch filter. *Mechanical Systems and Signal Processing*, 93:136–150, 2017.

[2] Qi Chen, Gang Liu, and Shiqiang Zheng. Suppression of imbalance vibration for ambs controlled driveline system using double-loop structure. *Journal of Sound and Vibration*, 337:1–13, 2015.

[3] Abhishek Dhyani, Manoj Kumar Panda, and Bhola Jha. Moth-flame optimization-based fuzzy-pid controller for optimal control of active magnetic bearing system. *Iranian Journal of Science and Technology, Transactions of Electrical Engineering*, 42(4):451–463, 2018.

[4] Jiancheng Fang, Xiangbo Xu, and Jinjin Xie. Active vibration control of rotor imbalance in active magnetic bearing systems. *Journal of vibration and control*, 21(4):684–700, 2015.

[5] Xudong Guan, Jin Zhou, Chaowu Jin, and Yuanping Xu. Disturbance suppression in active magnetic bearings with adaptive control and extended state observer. *Proceedings of the Institution of Mechanical Engineers, Part I: Journal of Systems and Control Engineering*, 234(2):272–284, 2019.

[6] Kejian Jiang and Changsheng Zhu. Multi-frequency periodic vibration suppressing in active magnetic bearing-rotor systems via response matching in frequency domain. *Mechanical Systems and Signal Processing*, 25(4):1417–1429, 2011.

[7] Chaowu Jin, Kaixuan Guo, Yuanping Xu, Hengbin Cui, and Longxiang Xu. Design of magnetic bearing control system based on active disturbance rejection theory. *Journal of Vibration and Acoustics*, 141(1), 2019.

[8] Z. Kai, J. Dong, X. Dai, and X. Zhang. Vibration control of a turbo molecular pump suspended by active magnetic bearings. *AMER SOC Mechanical Engineers*, pages 795–799, 2011.

[9] Chao Liu, Gang Liu, and Jiancheng Fang. Feedback linearization and extended state observer-based control for rotor-ambs system with mismatched uncertainties. *IEEE Transactions on Industrial Electronics*, 64(2):1313–1322, 2017.

[10] Hai Rong and Kai Zhou. Nonlinear zero-bias current control for active magnetic bearing in power magnetically levitated spindle based on adaptive backstepping sliding mode approach. *Proceedings of the Institution of Mechanical Engineers, Part C: Journal of Mechanical Engineering Science*, 231(20):3753–3765, 2016.

[11] Gerhard Schweitzer and Eric H Maslen. Magnetic bearings. theory, design, and application to rotating machinery. 2009.

[12] R. Siva Srinivas, R. Tiwari, and Ch Kannababu. Application of active magnetic bearings in flexible rotordynamic systems – a state-of-the-art review. *Mechanical Systems and Signal Processing*, 106:537–572, 2018.

[13] Te-Jen Su, Wen-Pin Kuo, Van-Nam Giap, H Quan Vu, and Quang-Dich Nguyen. Active magnetic bearing system using pid-surface sliding mode control. In *2016 Third International Conference on Computing Measurement Control and Sensor Network (CMCSN)*, pages 5–8. IEEE.

[14] Weiyu Zhang and Huangqiu Zhu. Radial magnetic bearings: An overview. *Results in Physics*, 7:3756–3766, 2017.

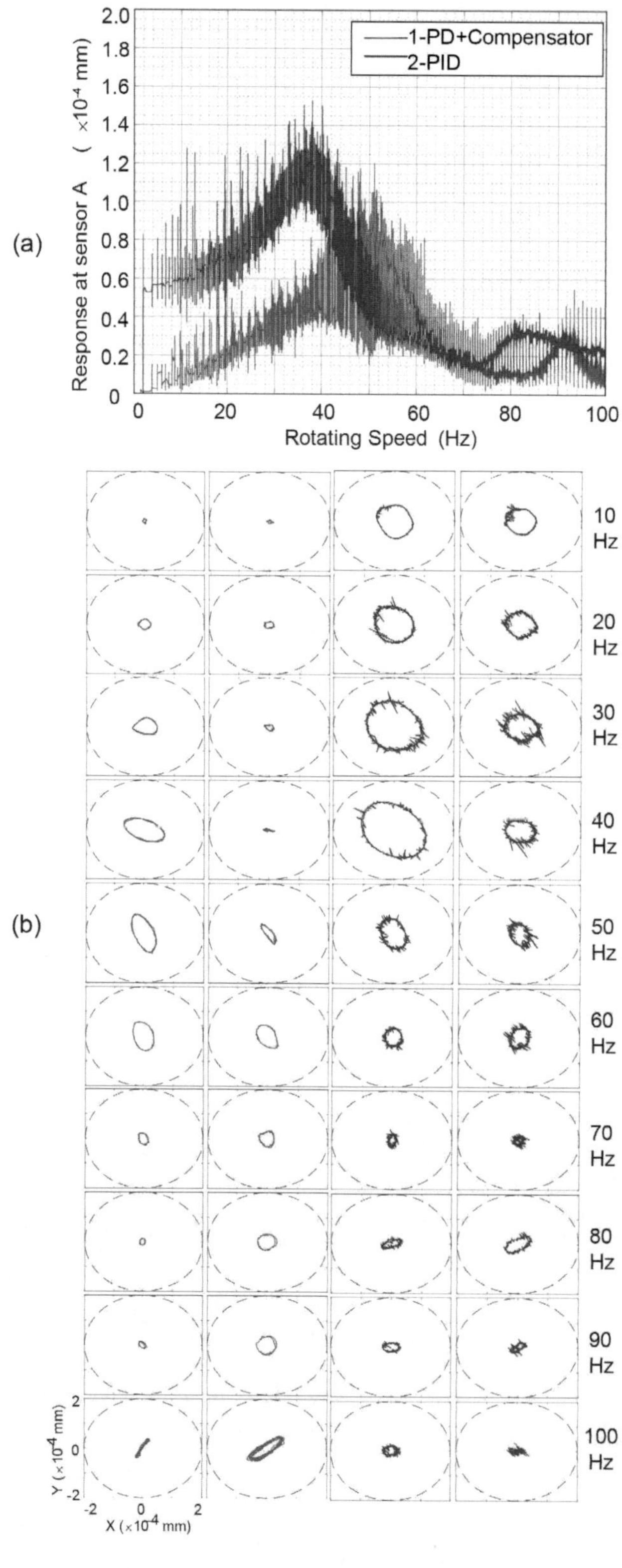

Figure 6: Response of the original PID controller from 0 to 100 Hz. (a) Displacement response at sensor A (b) Rotor trajectory at AMB A and AMB B during different rotating speed.

A New Generation High-Temperature Superconducting Maglev Vehicle Developed in Chengdu, China

Zigang Deng*, Jinkai Zhang, Wen Wang, Yuhang Yuan, Wuyang Lei, Jingzhong Zhao, Zhehao Liu, Minghui Wei, Penghui Zhang, Penghui Wang, Xu Zhang, and Jun Zheng

State Key Laboratory of Traction Power, Southwest Jiaotong University, Chengdu 610031, China

*E-mail: deng@swjtu.cn

Abstract: In order to further develop the manned potential and curving performance of the high-temperature superconducting (HTS) Maglev vehicle and verify the carrying capability of the system, a new HTS Maglev vehicle named "KENYON" has been successfully developed at Southwest Jiaotong University (SWJTU) in December 2019. The maglev vehicle (2.9 m in length, 1.1 m in width) with a levitation height of 10 mm was designed for 8 seated passengers which is run on a 45-m-long racetrack-type evacuated tube (Ring Line) with a 2-m-diameter circular cross section. Compared with the previous maglev vehicle “Super-Maglev”, the “KENYON” vehicle decreases dead weight by 30% and increases the propulsion and braking forces by 75%. Meanwhile, the amount of HTS bulks used in this vehicle is the same as that of the previous generation. By the efficient usage of the limited space, it has the design characteristics of 56% low floor area of the whole vehicle which is convenient for the whole 8 passengers seating, getting on and getting off. Moreover, the curve passing performance of the vehicle on the Ring Line has been specially improved by adjusting running mechanism. The experimental results of dynamic performance show that the vehicle has a good riding comfort and small radius curve negotiation. The system component and test data are reported in detail in this paper. The progress strongly demonstrates the HTS maglev has the potential of load capacity and curving performance for practical application.

Keywords: Bulk high-temperature superconductors, HTS Maglev, Prototype vehicle, Test line, Dynamic running.

Stability of Electrodynamic Maglev Vehicles Propelled by Permanent-Magnet Linear Electric Motors

Salvatore Circosta[1]*, Renato Galluzzi[2], Andrea Tonoli[1], Nicola Amati[1], Angelo Bonfitto[1], Torbjörn A. Lembke[3], and Milan Kertész[3]

[1] Department of Mechanical and Aerospace Engineering, Politecnico di Torino
Corso Duca degli Abruzzi 24, 10129 Turin, Italy
salvatore.circosta@polito.it
andrea.tonoli@polito.it
nicola.amati@polito.it
angelo.bonfitto@polito.it
[2] School of Engineering and Sciences, Tecnologico de Monterrey
Calle del Puente 222, 14380 Mexico City, Mexico
renato.galluzzi@tec.mx
[3] HyperloopTT
135 Avenue du Comminges, 31270 Cugnaux, France
torbjorn.lembke@hyperlooptt.com
milan.kertesz@hyperlooptt.com

Abstract

The search for fast and efficient transportation systems has raised the interest in magnetic levitation technologies over the last decades. In this context, the Hyperloop concept has been conceived as a solution for future mobility. However, the stability of the electrodynamic levitation system represents a key enabling technology for the Hyperloop implementation. In this context, the state of the art has addressed the full stabilization of downscaled vehicles, where levitation and guidance are provided by electrodynamic means. This is achieved passively by introducing a proper amount of damping into the system. Nevertheless, this system stability could be affected when using permanent-magnet propulsion motors. In this perspective, we propose the stabilization of a downscaled vehicle under the influence of its propulsion system. To this end, a permanent-magnet linear synchronous motor is designed. Its stiffness contribution is evaluated and introduced into the vehicle model. Then, its impact on the stability and performance of the system is discussed in detail.

1 Introduction

In the last decades, the mobility paradigm has received a strong innovative push. Globalization and environmental concerns have driven the transport industry towards the search for sustainable, efficient, fast, and safe systems. Worldwide governments strongly support this trend through programs aimed at funding research in this field [1, 2].

In this framework, the Hyperloop concept has been conceived as a breakthrough solution for future mobility [3]. Based on Robert Goddard's vactrain [4], it features magnetically levitated capsules traveling inside a low-pressure tube under the propulsion of linear electric motors [5]. This enables cruise speeds up to 1200 km/h and an ideal zero-emission feature. Recently,

*Corresponding author.

several companies have focused on the development and implementation of Hyperloop. The academic community has also been involved in this effort through direct collaborations with these companies or through student teams participating in the SpaceX Hyperloop Pod Competition.

The magnetic levitation of Hyperloop relies on the Inductrack technology [6]. It uses the electrodynamic principle, where lift and drag force components originate from the interaction between capsule-mounted permanent magnets and a conducting track yields. Electrodynamic levitation perfectly suits very high-speed applications since it features increasing lift and decreasing drag forces as the speed escalates. Other appealing features of this technology are its fully passive layout, the large air gaps attained through levitation, its self-reliability, simplicity and low component count.

The electrodynamic phenomenon has been widely studied in the past for electrodynamic bearings (EDBs). The literature proposes different EDB configurations and modeling approaches. Simulations and experimentation have identified the unstable nature of this principle [7, 8, 9, 10, 11]. In fact, its interaction with the mechanical domain is intrinsically unstable and requires the addition of damping for stabilization purposes.

Post *et al.* first proposed the application of electrodynamic levitation to translational systems in the Inductrack project frame [6, 12, 13]. As its rotating counterpart, linear motion levitation is unstable. This was observed during an experimental campaign with a full-scale prototype [14, 15], where increasing vertical vibrations were limited by safety auxiliary wheels.

Extensive work has dealt with the modeling and identification of the electrodynamic forces [16, 17, 18, 19, 20, 21, 22, 23, 24]. Although these efforts have proposed experimentation and diverse modeling approaches, stability issues for systems of translational nature have been often oversimplified and not treated properly. Furthermore, the stabilization of such systems was not properly addressed.

To close this gap, Galluzzi *et al.* proposed a multidomain lumped-parameter modeling approach that reproduces the strong interaction between electrical and mechanical dynamics [25]. The unstable nature was identified and tackled by introducing passive damping through a secondary suspension. Nevertheless, the work dealt only with the stabilization of the vertical dynamics of a quarter-car system.

Stable operation of every degree of freedom (DOF) is required for the successful implementation of systems featuring full electrodynamic levitation. In this perspective, a subsequent effort by Circosta *et al.* addressed the full stabilization of a downscaled vehicle, where levitation and guidance are provided by electrodynamic means [26].

Although the latest works pose a solid base for the future implementation of systems levitated through electrodynamic means, they investigate and address stability without accounting for the propulsion system. The latter consists of a linear electric motor that could introduce stiffness contributions along the degrees of freedom of the levitated system. Consequently, the stability of the system could be affected.

In this context, the present work addresses the stability of a downscaled vehicle under the influence of its propulsion system. The proposed vehicle is taken from reference [26] and enhanced with a permanent-magnet linear synchronous motor (PMLSM). Furthermore, system performance is investigated in terms of electrodynamic drag.

Although numerous Hyperloop proposals focus on the use of linear induction machines, a PMLSM solution has been proposed by Gurol *et al.* [15]. PMLSMs offer greater force capability, larger power factor and reduced overheating problems when compared to induction machines. However, they pose a challenge in terms of stability because they introduce significant stiffness contributions in the vehicle DOFs.

The remainder of this work is structured as follows. Section 2 describes the system layout.

Section 3 discusses the modeling of the system. At first, the electric machine is modeled and designed to fulfill required performance. Then, its stiffness contributions on the vehicle degrees of freedom are evaluated and included in the system model. Section 4 presents the system performance assessment in terms of stability and electrodynamic drag. Finally, Section 5 concludes the work.

2 System layout

The retained system (Fig. 1) is derived from Circosta *et al.* [26], where the sprung and unsprung bodies are connected by four secondary suspensions, and the unsprung body is equipped with four levitation (vertical) and four guidance (lateral) pads. The described configuration is enhanced with a permanent-magnet linear synchronous motor (PMLSM) added to the unsprung body to provide propulsion force.

The geometrical dimensions of the different bodies are set to roughly reproduce the track and wheelbase of a Hyperloop Pod Competition vehicle [27]. The height of the vehicle is considered relatively short when compared to the other dimensions.

The levitation and guidance pads are located at the four corners of the body: the former are mounted on the bottom of the unsprung body, whereas the latter appear on its sides. As in [26], the pad consists of a Halbach array of NdFeB permanent magnets (PM) interacting with a continuous aluminum slab that follows the track specifications of the Hyperloop Pod Competition [28].

The secondary suspension provides stiffness and damping in the vertical (z) and lateral (y) directions. Conversely, the suspension is infinitely rigid in the longitudinal direction (x). The secondary suspension is simplified as a prismatic joint acting independently along each direction. Its positioning and stiffness features were defined to set the natural frequencies of the sprung body in the range 1.2 to 1.44 Hz. Damping features were derived from an optimization procedure aimed at maximizing overall system stability in terms of a root locus [26].

The general layout of the electric machine is taken from [29], where a linear induction machine features the stator mounted on the pod. To worsen system stability, we replaced the linear induction machine with a PMLSM. It presents a magnetic rail which is magnetized by N45SH-grade NdFeB PMs, and a stator with two magnetic circuits (left and right) and three-phase windings. The former is fixed to the track, the latter is placed on the bottom of the unsprung body. The centroid of the stator's active part is located at a vertical offset (z direction) of 60 mm with respect to the unsprung body's center of gravity (COG).

The geometrical and inertial features along with the secondary suspension parameters of the retained system are reported in Table 1.

3 Modeling

The proposed system travels along the x direction with speed v. For each body, the translation DOFs y (sway) and z (heave) along with the rotation DOFs around x axis (roll, ϕ), y axis (pitch, θ) and z axis (yaw, ψ) are investigated. The dynamics of the translation DOF x is not considered.

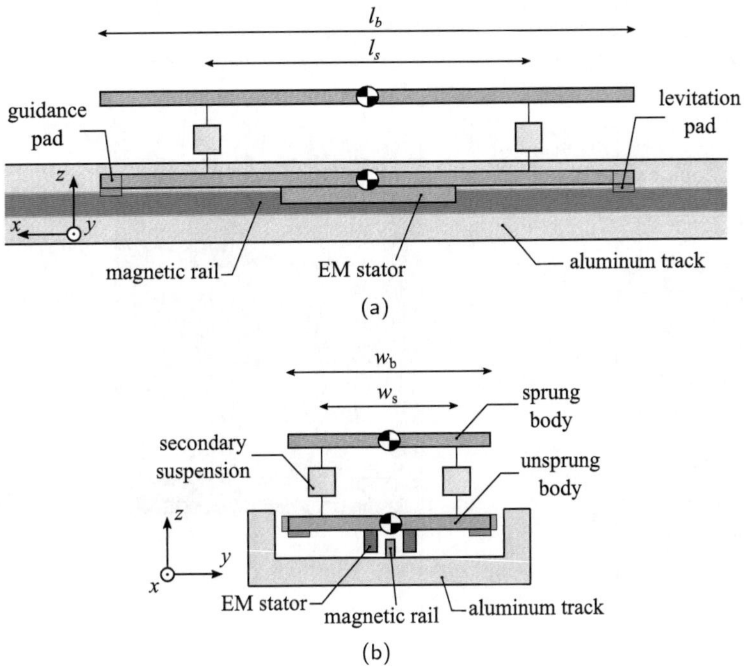

Figure 1: System layout: (a) side view and (b) front view

Table 1: Geometrical and inertial features of the system

	Feature	Symbol	Value	Unit
Secondary suspension	Positioning x	l_s	1732	mm
	Positioning y	w_s	416	mm
	Stiffness y	$k_{s,y}$	1.137	kN/m
	Stiffness z	$k_{s,z}$	1.137	kN/m
	Damping y	$c_{s,y}$	265.3	Ns/m
	Damping z	$c_{s,z}$	262.7	Ns/m
Sprung and unsprung bodies	Length	l_b	2500	mm
	Width	w_b	600	mm
Sprung body*	Mass	m_s	80	kg
	Moment of inertia x	$I_{s,x}$	2.40	$kg \cdot m^2$
	Moment of inertia y	$I_{s,y}$	41.67	$kg \cdot m^2$
	Moment of inertia z	$I_{s,z}$	44.07	$kg \cdot m^2$
Unsprung body*	Mass	m_u	8	kg
	Moment of inertia x	$I_{u,x}$	0.24	$kg \cdot m^2$
	Moment of inertia y	$I_{u,y}$	4.17	$kg \cdot m^2$
	Moment of inertia z	$I_{u,z}$	4.41	$kg \cdot m^2$

* Inertial properties are referred to the body center of gravity

3.1 Permanent-Magnet Linear Synchronous Motor

To yield a propulsion force to the proposed vehicle, a PMLSM was custom-sized. For versatility, a single elementary segment is proposed.

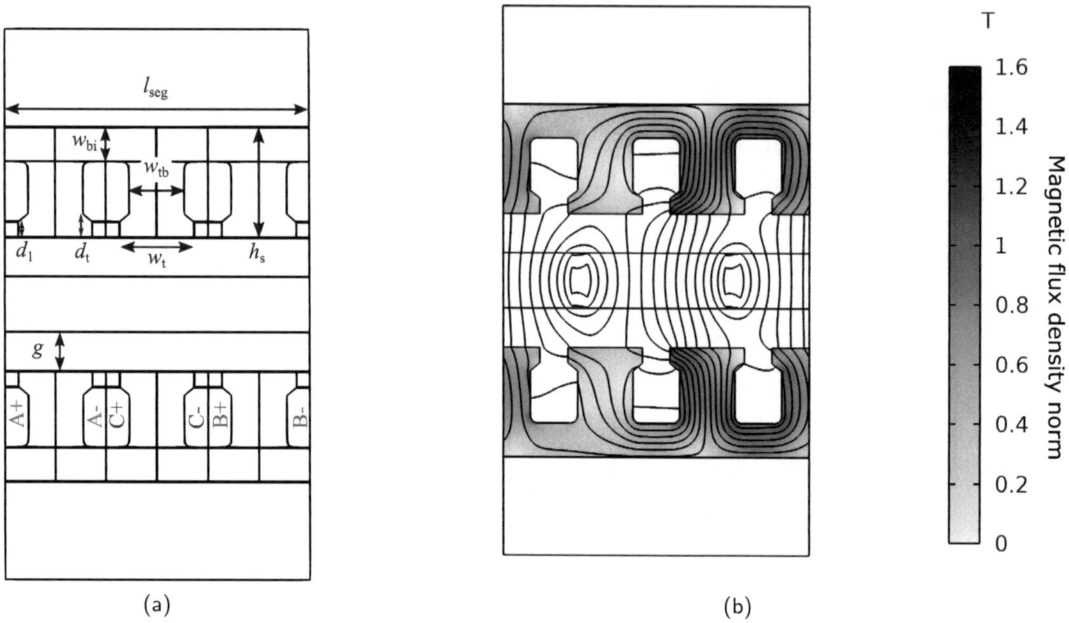

Figure 2: On-board stator of the permanent-magnet linear synchronous motor: 2D view of (a) geometric features and (b) finite-element model results in terms of stator magnetic flux density norm (color map) and out-of-plane magnetic vector potential contours. Windings are fed by a root-mean-square three-phase current density of $6\,\mathrm{A/mm^2}$.

Table 2: Permanent-magnet linear synchronous motor parameters

Feature	Symbol	Value	Unit
Segment length	l_{seg}	200	mm
Segment nominal force	F_{seg}	400	N
Back iron width	w_{bi}	21.67	mm
Tooth base width	w_{tb}	36	mm
Tooth tip width	w_t	48.67	mm
Shoe tip height	d_1	10	mm
Shoe total height	d_t	5	mm
Stator height	h_s	70	mm
Nominal propulsion air gap	g	25	mm
Active length	l_{act}	30	mm

Figure 2 illustrates the layout of a single PMLSM segment, which consists of two symmetrical on-board stator cores made of soft iron and a part of the magnetic rail which is magnetized by N45SH-grade PMs. Both stators have three-phase concentrated windings to reduce end-turn length. In nominal conditions, the magnetic rail is perfectly centered with respect to the stator structures with a nominal air gap g.

A design procedure was followed to establish the geometry of the proposed electric machine. It consists in selecting the proper back iron and tooth dimensions to maximize propulsion forces while avoiding magnetic saturation. Thereafter, parameters listed in Table 2 are obtained, thus leading to a segment able to yield 400 N of propulsion force.

Given the total mass of the vehicle (88 kg), a propulsion force of 2 kN is needed to attain a

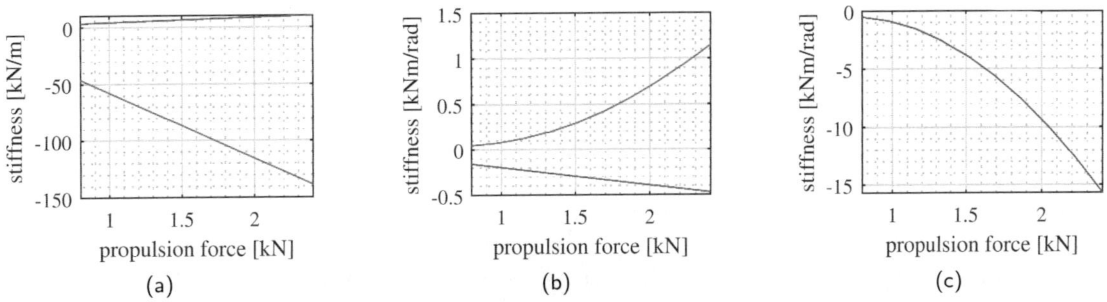

Figure 3: Linear permanent-magnet synchronous motor stiffness terms as a function of the propulsion force: (a) heave (blue) and sway (orange); (b) roll (blue) and pitch (orange); (c) yaw.

stall acceleration of 2.4 g. These specifications comply with those of a prototype competing at the Hyperloop Pod Competition [30]. To achieve this force output, the magnetic rail must be adapted with the necessary poles. The designed segment with two poles is able to provide a propulsion force of 400 N, thus, ten poles or five segments are needed to attain 2 kN.

The interaction among ferromagnetic and PM domains will inevitably yield additional force and torque components. Specifically, efforts will arise in relevant degrees of freedom: heave, sway, roll, pitch and yaw. From a modeling perspective, these contributions can be represented as stiffness parameters—positive or negative—depending on the degree of freedom of interest. They are evaluated through 3D finite-element models. Moreover, these simulations can be repeated for different propulsion forces, which can be obtained by varying the length of the magnetic rail. This yields the sensitivity of elastic effects to motor force capability. Results for all the relevant stiffness terms (k_{heave} k_{sway} k_{roll} k_{pitch} k_{yaw}) are presented in Fig. 3. Do note that these contributions can be changed also by changing the active length of the PMLSM. However, $l_{\text{act}} = 30\,\text{mm}$ is required for optimal operation within the levitation air gaps of interest, so this parametric variation was not explored herein.

The motor provides negative stiffness contributions to the sway, yaw and roll DOFs, thus affecting system stability in these dynamics. Conversely, it provides positive stiffening effects to the heave and pitch DOFs.

3.2 System model

The state-space representation reported in [26] is used to model the entire system. The electrodynamic levitation principle is represented through a multi-domain lumped parameter approach, first proposed by [25]. After proper tuning, this model is able to reproduce the desired electrodynamic characteristic at a small fraction of the computational cost needed in a more complex finite-element simulation.

In the following, the problem is formulated for a generic pad. The electrical parallel of N_{b} branches reproduces the dynamic behavior of the current path inside the track conductor. Each branch features specific resistance and inductance values. For the kth branch, the electrical circuit is governed by

$$\frac{\mathrm{d}i_{\mathrm{d},k}}{\mathrm{d}t} = -\omega_{\mathrm{p},k} i_{\mathrm{d},k} + \omega i_{\mathrm{q},k} - \frac{1}{L_k}\frac{\partial \Lambda}{\partial q_{\mathrm{p}}}\dot{q}_{\mathrm{p}} \tag{1}$$

$$\frac{\mathrm{d}i_{\mathrm{q},k}}{\mathrm{d}t} = -\omega_{\mathrm{p},k} i_{\mathrm{q},k} - \omega i_{\mathrm{d},k} - \frac{1}{L_k}\Lambda\omega \tag{2}$$

where i_{d} and i_{q} are the direct- and quadrature-axis current components, q_{p} is the generic-pad air gap, $\omega_{\mathrm{p},k} = R_k/L_k$ the electromagnetic pole frequency, and R_k and L_k the branch resistance and inductance, respectively. The pad longitudinal speed v and the mechanical frequency are linked through the pole pitch ratio of the PM array γ:

$$\omega = \frac{v}{\gamma} \tag{3}$$

The Λ term denotes the amplitude of the PM array flux linkage acting on the track lumped coils.It is an exponential function of the air gap [6]:

$$\Lambda = \Lambda_0 \mathrm{e}^{-\frac{q_{\mathrm{p}}}{\gamma}} \tag{4}$$

being Λ_0 the flux linkage constant.

When N_{b} circuit branches superimpose, lift and drag force components are written as

$$F_{\mathrm{lift}} = \frac{\partial \Lambda}{\partial q_{\mathrm{p}}}\sum_{k=1}^{N_{\mathrm{b}}} i_{\mathrm{d},k} \tag{5}$$

$$F_{\mathrm{drag}} = -\frac{\Lambda}{\gamma}\sum_{k=1}^{N_{\mathrm{b}}} i_{\mathrm{q},k} \tag{6}$$

At constant air gap q_{p} and longitudinal speed v, lift and drag static forces are obtained from the combination Eqs. 1,2, 5 and 6:

$$F_{\mathrm{lift},0} = \frac{\Lambda_0^2}{\gamma}\mathrm{e}^{-\frac{2q_{\mathrm{p}}}{\gamma}}\sum_{k=1}^{N_{\mathrm{b}}}\frac{\omega^2/\omega_{\mathrm{p},k}^2}{L_k\left(1+\omega^2/\omega_{\mathrm{p},k}^2\right)} \tag{7}$$

$$F_{\mathrm{drag},0} = \frac{\Lambda_0^2}{\gamma}\mathrm{e}^{-\frac{2q_{\mathrm{p}}}{\gamma}}\sum_{k=1}^{N_{\mathrm{b}}}\frac{\omega/\omega_{\mathrm{p},k}}{L_k\left(1+\omega^2/\omega_{\mathrm{p},k}^2\right)} \tag{8}$$

The levitation pads feature the layout proposed by Post [12]. For this configuration, the lumped-parameter modeling approach was proposed by Galluzzi *et al.* [25], who demonstrated a feasible set of parameters with $N_{\mathrm{b}} = 3$ branches (Table 3). The subscript l refers to the levitation pad.

As in [26], the guidance pads are a scaled version of the levitation ones, where the width is scaled by a factor $\beta = 0.75$. The flux linkage constant, the inductance and the resistance of the

Table 3: Levitation pad parameters

Feature	Symbol	Value	Unit
Branch pole frequency	$\omega_{p,1}$	355.45	rad/s
	$\omega_{p,2}$	$2.57 \cdot 10^3$	
	$\omega_{p,3}$	$1.73 \cdot 10^4$	
Branch inductance	$L_{1,l}$	$7.72 \cdot 10^{-10}$	H
	$L_{2,l}$	$2.06 \cdot 10^{-9}$	
	$L_{3,l}$	$4.12 \cdot 10^{-9}$	
Flux linkage constant	$\Lambda_{0,l}$	0.33	mWb
Pole pitch ratio	γ	15.9	mm

kth branch are derived from the levitation pad parameters as follows:

$$\Lambda_{0,g} = \beta\Lambda_{0,l} \tag{9}$$
$$R_{k,g} = \beta R_{k,l} \tag{10}$$
$$L_{k,g} = \beta L_{k,l} \tag{11}$$

where subscript g refers to the guidance pad.

Each pad of the system is represented through Eqs. 1, 2, 5 and 6 by replacing the generic-pad air gap (q_p) with the levitation-pad (z_l) or the guidance-pad (y_g) air gaps along with the respective pad parameters. The air gaps are expressed as a linear combination of the unsprung body DOFs by following geometrical considerations.

Lift and drag forces enter the equations of motion of the unsprung body, thus coupling the electrical and mechanical domains. The electrodynamic levitation description is nonlinear. By following the analytical approach developed in [26], the above reported equations are linearized around the air gap value $q_{0,p}$. It is expressed as $z_{0,l}$ and $y_{0,g}$ for the levitation and guidance pads, respectively.

The levitation pads counteract the weight of the system. Hence, they undergo an imposed constant load. The air gap adapts to guarantee static equilibrium between the weight and the lift forces, as discussed in [25]. Hence, such value is obtained by matching the total lift to the total weight of the system. Four pads work in parallel in this case, hence:

$$z_{0,l}(\omega) = -\frac{\gamma}{2}\ln\left(\frac{(m_s + m_u)g\gamma}{4\Lambda_{0,l}^2\Gamma_l(\omega)}\right) \tag{12}$$

On the other hand, the air gap of the guidance pads is controlled to have a pad equivalent stiffness $k_{eq,g}$, as discussed in [26]. According to the reference, the air gap that satisfies this constraint is given by

$$y_{0,g}(\omega) = -\frac{\gamma}{2}\ln\left(\frac{k_{eq,g}\gamma^2}{2\Lambda_{0,g}^2\Gamma_g(\omega)}\right) \tag{13}$$

The linearized electrodynamic description is introduced in the mechanical equations of motion. The latter relies on the assumption that the pitch, yaw and roll dynamics are decoupled. Hence, the three systems with reduced number of DOFs—heave-pitch, sway-yaw and heave-sway-roll—are defined. Finally, the obtained description is arranged into the state-space form.

For the sake of brevity, the full description along with the state-space dynamic matrices are skipped. The reader can find this information in reference [26]. However, when compared to the referred configuration, dynamic matrices of the present formulation must be updated to account for the electric motor stiffness contributions. Accordingly, the following elements are modified:

$$[A_{\mathrm{m},1}]_{8\times 8} \quad \begin{cases} a_{1,2} = -\dfrac{4k_{\mathrm{s,z}}}{m_{\mathrm{u}}} - \dfrac{k_{\mathrm{heave}}}{m_{\mathrm{u}}} \\ a_{3,4} = -\dfrac{k_{\mathrm{s,z}} l_{\mathrm{s}}^2}{I_{\mathrm{u,y}}} - \dfrac{k_{\mathrm{pitch}}}{I_{\mathrm{u,y}}} \end{cases} \tag{14}$$

$$[A_{\mathrm{m},2}]_{8\times 8} \quad \begin{cases} a_{1,2} = -\dfrac{4k_{\mathrm{s,y}}}{m_{\mathrm{u}}} - \dfrac{k_{\mathrm{sway}}}{m_{\mathrm{u}}} \\ a_{3,4} = -\dfrac{k_{\mathrm{s,y}} l_{\mathrm{s}}^2}{I_{\mathrm{u,z}}} - \dfrac{k_{\mathrm{yaw}}}{I_{\mathrm{u,z}}} \end{cases} \tag{15}$$

$$[A_{\mathrm{m},3}]_{12\times 12} \quad \begin{cases} a_{1,2} = -\dfrac{4k_{\mathrm{s,z}}}{m_{\mathrm{u}}} - \dfrac{k_{\mathrm{heave}}}{m_{\mathrm{u}}} \\ a_{3,4} = -\dfrac{4k_{\mathrm{s,y}}}{m_{\mathrm{u}}} - \dfrac{k_{\mathrm{sway}}}{m_{\mathrm{u}}} \\ a_{5,6} = -\dfrac{k_{\mathrm{s,z}} w_{\mathrm{s}}^2}{I_{\mathrm{u,x}}} - \dfrac{k_{\mathrm{roll}}}{I_{\mathrm{u,x}}} \end{cases} \tag{16}$$

where $A_{\mathrm{m},j}$ [26] are the mechanical sub-matrices whose sub-index j refers to the particular model: heave-pitch ($j = 1$), sway-yaw ($j = 2$) and heave-sway-roll ($j = 3$) models. The added stiffness terms of the PMLSM have been already defined in Sec. 3.1.

4 Results and discussion

4.1 System stability

The state-space description reported in Section 3.2 is used to assess system stability. To this end, the linearization air gaps—$z_{0,\mathrm{l}}$ and $y_{0,\mathrm{g}}$ for the levitation and guidance pads, respectively—are computed through Eqs. 12 and 13, respectively. The latter is defined to set the sway natural frequency of the unsprung body $\omega_{\mathrm{n,su}} \cong 15 \cdot \omega_{\mathrm{n,ss}}$, being $\omega_{\mathrm{n,ss}}$ the sway natural frequency of the sprung body [26]. The unsprung body is constrained by four secondary suspensions and four pads. Furthermore, it must compensate for the electric motor stiffness k_{sway}. Hence, the required equivalent stiffness of the guidance pad is given by

$$k_{\mathrm{eq,g}} = \frac{1}{4}\left(\omega_{\mathrm{n,su}}^2 m_{\mathrm{u}} - 4k_{\mathrm{s,y}} - k_{\mathrm{sway}}\right) \tag{17}$$

Figure 4 shows the linearization air gap for the levitation (dashed) and guidance (solid) pads. The latter depends on the sway stiffness contribution of the motor and, consequently, on its size. As the motor negative stiffness increases, the guidance pads must provide larger positive stiffness, thus leading to thinner guidance air gaps. At speed values below 0.89 m/s, the levitation pads are not sufficient to balance the vehicle weight. Therefore, a backup system is needed during take-off and landing operations. The same occurs for the guidance pads: they cannot provide the required stiffness below speed values ranging from 1.05 m/s to 1.44 m/s, according to the motor size.

Then, the poles of the three analytical models (Sec. 3.2) are calculated at different longitudinal speeds and motor sizes. To this end, the eigenvalues of the state matrices (Eqs. 15,16 and 16) are computed. Within the whole speed range, the pole with the largest real part is

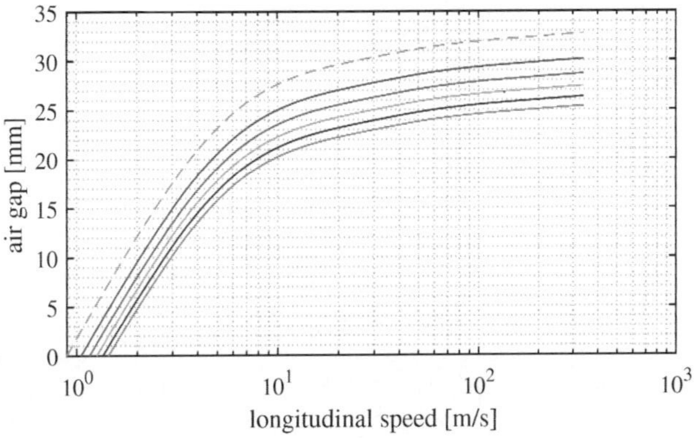

Figure 4: Air gap as function of the longitudinal speed: levitation (dashed) and guidance (solid) pads. The latter is a function of the propulsion force: 800 N (blue), 1200 N (orange), 1600 N (yellow), 2000 N (purple) and 2400 N (green).

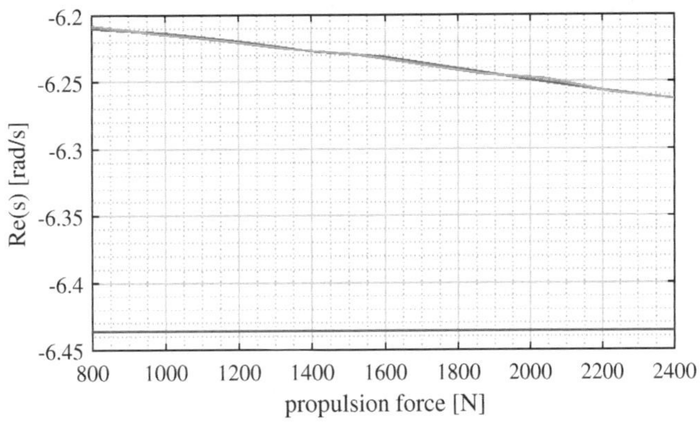

Figure 5: Real part of the most stable pole as function of the propulsion force: heave-pitch (blue), sway-yaw (orange) and heave-sway-roll (yellow) models.

identified and its real part is plotted against the motor force capability, which is proportional to its size. These results are depicted in Fig. 5 for the heave-pitch, sway-yaw, and heave-sway-roll models.

The sway-yaw and heave-sway-roll models have superimposed trends. Hence, they feature the same poles with largest real part, namely the sway ones. The motor stiffness contribution has little influence on the system stability therein. As shown in [26], the poles with largest real part are related to the sprung body dynamics. These are almost decoupled to the unsprung body ones, as the natural frequencies of the two bodies are at least one decade apart. This is achieved thanks to the secondary suspension design [26] and the guidance-pad air gap. The latter is fundamental for the system stability since it must compensate for the sway negative stiffness. Otherwise, sway poles of the unsprung body would move to lower frequency bands, couple with the sprung body dynamics and push the system to instability.

4.2 Drag force assessment

In this section, the drag force due to the electrodynamic levitation system is evaluated to assess the propulsion performance. Four levitation and four guidance pads contribute to the total drag

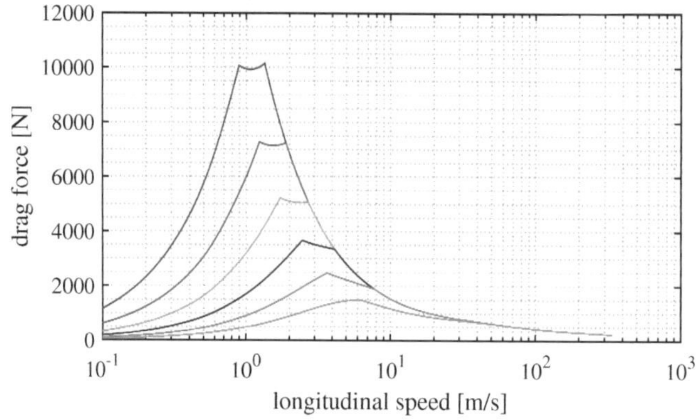

Figure 6: Total drag as a function of the longitudinal speed when the air gap is low-bounded at 0 mm (blue), 5 mm (orange), 10 mm (yellow), 15 mm (purple), 20 mm (green) and 25 mm (cyan).

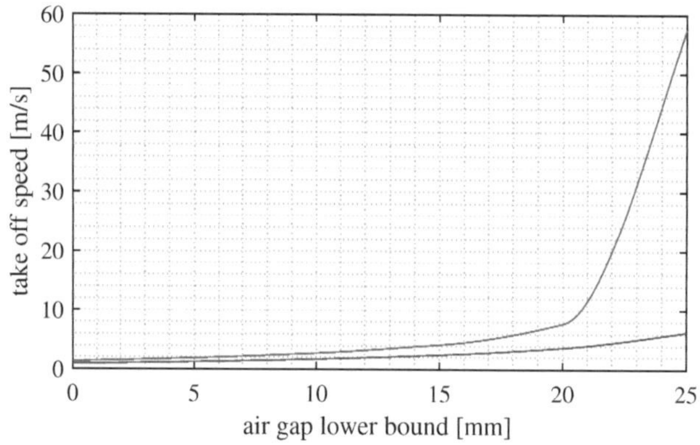

Figure 7: Take-off speed as a function of the air gap lower bound for the levitation (blue) and guidance (orange) pads.

according to Eq. 8. This expression is a function of the longitudinal speed and the linearization air gaps, i.e. equations 12 and 13. The latter is influenced by the motor stiffness contribution according to Eq. 17. For demonstration purposes, the following analysis considers the stiffness contribution given by a motor force capability of 2 kN.

Figure 6 shows the total drag when the air gap profile is bounded at values ranging from 0 to 25 mm. This is achieved by means of a backup-wheel system that provides levitation and guidance during take-off and landing operations. The motor force capability (2 kN) must be larger than the total drag to guarantee available force to accelerate the vehicle at 2.4g. Thereby, the air gap must be bounded to at least 25 mm.

The take-off speed is the value at which the electrodynamic lift is sufficient to counteract the vehicle weight (levitation pads) or provide sufficient guidance stiffness (guidance pads). Above this speed value, the backup-wheel system can be retracted. In this condition, the vehicle relies totally on electrodynamic lift.

Figure 7 shows the take-off speed for the levitation and guidance pads as function of the air gap lower bound. As the latter increases, the speed needed to develop sufficient electrodynamic lift increases.

A take-off speed of 57.5 m/s for the guidance pads occurs when the air gap is limited to a minimum of 25 mm. The latter is required to guarantee available force to accelerate the

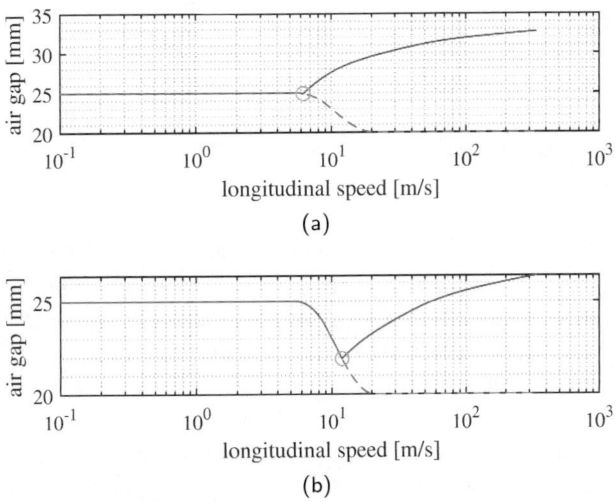

Figure 8: Actual (solid) and lower-bound (dashed) air gap profiles as a function of the longitudinal speed: (a) levitation and (b) guidance pads. The circle indicates the take-off point.

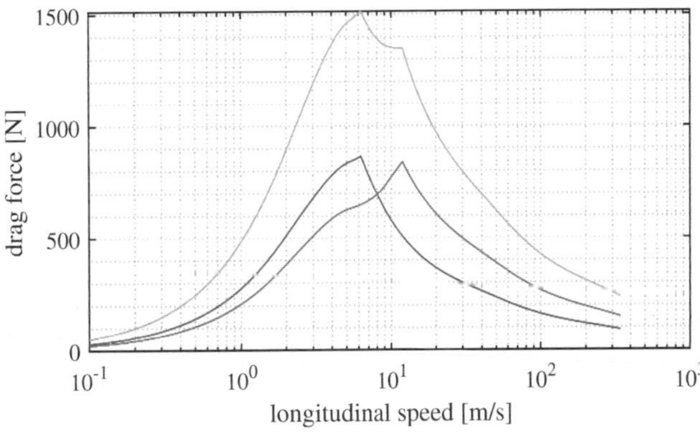

Figure 9: Drag force as function of the speed when the lower-bound air gap profile is imposed: total drag (yellow), levitation (blue) and guidance (orange) pads contribution.

vehicle. Large take-off speed values imply constraints on the design of the backup-wheel system. To this end, its value can be decreased by defining a suitable profile for the air gap lower bound. Specifically, the air gap is low-bounded to 25, mm at speed values below the speed that corresponds to the maximum drag force. Afterwards, the air gap lower bound is gradually decreased, as shown in Figs. 8a and 8b (dashed profile). By doing so, the take-off speed of the guidance pad is decreased to 11.9 m/s.

Figure 9 shows the drag force when the above-mentioned lower-bound profile of the air gap is imposed. Specifically, the levitation and guidance pads contribution, and the total drag are displayed. The latter is always smaller than the motor force (2 kN), thus guaranteeing sufficient propulsion.

5 Conclusions

The present paper dealt with the stability and performance assessment of downscaled vehicles levitated through electrodynamic means under the influence of permanent-magnet linear electric motors. The retained vehicle was taken from the literature and integrated with a permanent-magnet linear synchronous motor for propulsion.

To this end, the propulsion motor was custom-sized to fulfill a target stall acceleration. Then, its stiffness contributions to the vehicle DOFs were assessed through 3D finite-element simulations. Moreover, these simulations were repeated for different motor sizes to evaluate the sensitivity of such stiffness effects to motor force capability. As result, the motor provides stabilizing effect to the heave and pitch DOFs. Conversely, it contributes with negative stiffness to the sway, yaw and roll dynamics.

The stiffness contributions were included into the model of the system. The air gap of the guidance pads was set to compensate for the motor destabilizing effect. The system stability was investigated by extracting the poles of the linearized model. Specifically, the pole with largest real part was found at diverse motor sizes. Despite the stiffness contributions, the system was demonstrated to be stable thanks to the secondary suspension design and the selection of a suitable air gap for the guidance pads.

Finally, the propulsion performance of the system was assessed. As demonstrated, the peak electrodynamic drag must be limited to guarantee sufficient propulsion. To this end, the air gap was low-bounded and a suitable limit profile was proposed and validated.

References

[1] Michael E. Webber. Electric Highway. *Mechanical Engineering*, 141(06):32–37, 2019.

[2] Luisa Marelli, Anastasios Tsakalidis, Konstantinos Gkoumas, Anwar Haq, Monica Grosso, and Ferenc Pekar. Strategic transport research and innovation agenda (STRIA) roadmap factsheets, 2017.

[3] Konstantinos Gkoumas. Hyperloop Academic Research: A Systematic Review and a Taxonomy of Issues. *Applied Sciences*, 11(13):5951, June 2021.

[4] Natalie Burkhard. Why invent the Hyperloop?, 2014.

[5] Kenneth Decker, Jeffrey Chin, Andi Peng, Colin Summers, Golda Nguyen, Andrew Oberlander, Gazi Sakib, Nariman Sharifrazi, Christopher Heath, Justin Gray, and Robert Falck. Conceptual Feasibility Study of the Hyperloop Vehicle for Next-Generation Transport. page 22, Jan. 2017.

[6] R. F. Post and D. Ryutov. The Inductrack concept: A new approach to magnetic levitation. Technical Report UCRL-ID-124115, Lawrence Livermore National Lab., CA (United States), May 1996.

[7] Torbjörn A. Lembke. *Design and analysis of a novel low loss homopolar electrodynamic bearing.* PhD thesis, School of Electrical Engineering, KTH, Stockholm, Sweden, 2005.

[8] Andrea Tonoli, Nicola Amati, Fabrizio Impinna, and Joaquim Girardello Detoni. A Solution for the Stabilization of Electrodynamic Bearings: Modeling and Experimental Validation. *Journal of Vibration and Acoustics*, 133(2):021004, 2011.

[9] Joachim Van Verdeghem, Virginie Kluyskens, and Bruno Dehez. Stability and Performance Analysis of Electrodynamic Thrust Bearings. *Actuators*, 8(1):11, March 2019.

[10] J. Van Verdeghem, V. Kluyskens, and B. Dehez. Five degrees of freedom linear state-space representation of electrodynamic thrust bearings. *Journal of Sound and Vibration*, 405:48–67, September 2017.

[11] A.V. Filatov and E.H. Maslen. Passive magnetic bearing for flywheel energy storage systems. *IEEE Transactions on Magnetics*, 37(6):3913–3924, November 2001.

[12] R. F. Post. Inductrack demonstration model. Technical Report UCRL-ID-129664, Lawrence Livermore National Lab., CA (United States), February 1998.

[13] R.F. Post and D.D. Ryutov. The Inductrack: a simpler approach to magnetic levitation. *IEEE Transactions on Appiled Superconductivity*, 10(1):901–904, 2000.

[14] Husam Gurol, Robert Baldi, Phillip Jeter, In-Kun Kim, Daryl Bever, and General Atomics. General atomics low speed maglev technology development program (Supplemental# 3). Technical report, United States. Federal Transit Administration. Office of Technology, 2005.

[15] Sam Gurol and Bob Baldi. Overview of the General Atomics Urban Maglev Technology Development Program. In *Proceedings of the 2004 IEEE/ASME Joint Rail Conference*, pages 187–191, Baltimore, MD, USA, 04 2004.

[16] Hitoshi Tsunashima and Masato Abe. Static and dynamic performance of permanent magnet suspension for maglev transport vehicle. *Vehicle System Dynamics*, 29(2):83–111, 1998.

[17] O.F. Storset and B. Paden. Infinite dimensional models for perforated track electrodynamic magnetic levitation. In *Proceedings of the 41st IEEE Conference on Decision and Control, 2002.*, volume 1, pages 842–847, Las Vegas, NV, USA, 2002. IEEE.

[18] Roshan Pradhan and Aditya Katyayan. Vehicle dynamics of permanent-magnet levitation based Hyperloop capsules. In *ASME 2018 Dynamic Systems and Control Conference*, page V002T22A004, Atlanta, GA, USA, 2018.

[19] J Íñiguez and V Raposo. Laboratory scale prototype of a low-speed electrodynamic levitation system based on a Halbach magnet array. *European Journal of Physics*, 30(2):367–379, March 2009.

[20] TS Indraneel, Vignesh Jayakumar, Abhishek Soni, Dhaval R Shiyani, Kanishk Tyagi, and Shabaan Abdallah. Levitation array testing for Hyperloop pod design. In *AIAA Scitech 2019 Forum*, page 0787, San Diego, CA, USA, 2019.

[21] Guo, Li, and Zhou. Study of a Null-Flux Coil Electrodynamic Suspension Structure for Evacuated Tube Transportation. *Symmetry*, 11(10):1239, October 2019.

[22] Yin Chen, Wenlong Zhang, Jonathan Z. Bird, Subhra Paul, and Kunlun Zhang. A 3-D Analytic-Based Model of a Null-Flux Halbach Array Electrodynamic Suspension Device. *IEEE Transactions on Magnetics*, 51(11):1–5, November 2015.

[23] Jiaheng Duan, Song Xiao, Kunlun Zhang, Mihai Rotaru, and J. K. Sykulski. Analysis and Optimization of Asymmetrical Double-Sided Electrodynamic Suspension Devices. *IEEE Transactions on Magnetics*, 55(6):1–5, June 2019.

[24] Ruiyang Wang and Bingen Yang. A Transient Model of Inductrack Dynamic Systems. In *Volume 8: 31st Conference on Mechanical Vibration and Noise*, page V008T10A062, Anaheim, California, USA, August 2019. American Society of Mechanical Engineers.

[25] Renato Galluzzi, Salvatore Circosta, Nicola Amati, Andrea Tonoli, Angelo Bonfitto, Torbjörn A. Lembke, and Milan Kertész. A Multi-domain Approach to the Stabilization of Electrodynamic Levitation Systems. *Journal of Vibration and Acoustics*, pages 1–30, 2020.

[26] Salvatore Circosta, Renato Galluzzi, Nicola Amati, Andrea Tonoli, Angelo Bonfitto, Torbjörn A Lembke, and Milan Kertész. Passive Multi-Degree-of-Freedom Stabilization of Ultra-High-Speed Maglev Vehicles. *Journal of Vibration and Acoustics*, page 13.

[27] M. M. J. Opgenoord, C. Merian, J. Mayo, P. Kirschen, C. O'Rourke, G. Izatt, et al. MIT Hyperloop final report. Technical report, Massachusetts Institute of Technology, Cambridge, Massachusetts, 2017.

[28] SpaceX. SpaceX Hyperloop test-track specification, 2016.

[29] Available online: https://swissloop.ch/prototypes/.

[30] Available online: https://tumhyperloop.de/hyperloop-prototypes.

Radial Motion Control of Axial-Flux Self-Bearing Motor Using Plus and Minus Two-Pole Magnetic Fields

Satoshi Ueno[1], Yuki Yamamoto[2], and Changan Jiang[3]

[1] RItsumeika University, Kusatsu, Shiga, Japan
sueno@se.ritsumei.ac.jp
[2] Ritsumeikan University, Kusatsu, Shiga, Japan
[3] Osaka Institute of Technology, Osaka, Japan
changan.jiang@oit.ac.jp

Abstract

This paper introduces an axial-flux self-bearing permanent magnet motor that controls motor torque, axial force, radial forces, and tilting moments by a single stator. A rotating magnetic field with the same pole number as the rotor controls the motor torque and axial force. Rotating magnetic fields with the rotor' s pole number plus and minus two poles control the radial forces and tilting moments. These fields generate both radial forces and tilting moments, and the direction of one of them becomes the same, and the other becomes opposite. Hence, we can control the radial forces and tilting moments independently. This paper describes the theoretical analysis of the radial forces and tilting moments and derives the control currents of stator windings. Levitation and rotation tests of a prototype motor confirm the proposed method.

1 Introduction

An axial-flux self-bearing motor (AFSBM) is a combination of a disk type permanent magnet (PM) motor and active magnetic bearings and can simultaneously control both the motor torque and position of the rotor [5]. One-, three-, and five-degrees-of-freedom (DOF) active-controlled AFSBMs have been proposed.

The one-DOF active-controlled AFSBM controls motor torque and axial force by q- and d-components of a magnetic field with the same pole number of the rotor. The three-DOF active-controlled AFSBM utilizes the magnetic field with plus or minus two poles to control the tilting moments of the rotor [2]. However, these fields generate not only the tilting moment but also the radial force to the rotor. Therefore, it is necessary to use a sandwich structure to cancel the radial force or adjust the center of gravity (COG) of the rotor to ensure the stability of tilt control [3].

On the other hand, by utilizing this characteristic five-DOF active-control can be realized. Osa et al. have developed the five-DOF active-controlled AFSBM for small size artificial heart pump [1]. However, this motor requires a sandwich structure to control both the radial forces and tilting moments and restricts the design of the whole device. To solve this problem, we have proposed the different type five-DOF active-controlled AFSBM that uses unipolar and four-pole magnetic fields to control the radial force and tilting moment for a two-pole PM rotor [4]. Since the proposed motor does not require a sandwich structure, it can improve the flexibility of design. However, this AFSBM requires two pairs of the rotor and stator, then the mechanism and controller became complex.

This paper introduces a novel AFSBM that uses plus and minus two-pole magnetic fields to control the radial forces and tilting moments to realize a simple structure. Since the proposed method requires only a pair of the rotor and the stator, the structure becomes simple.

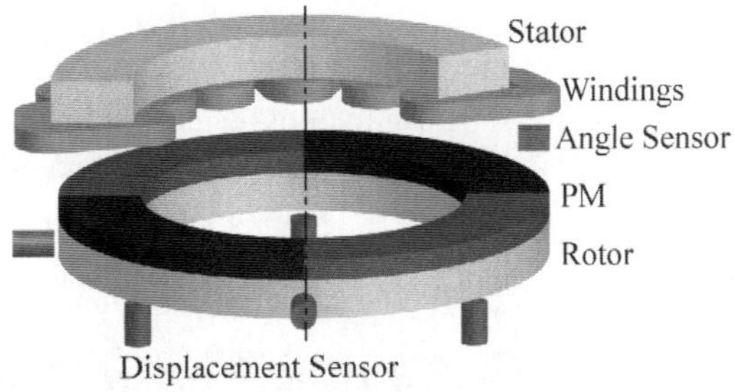

Figure 1: Structure.

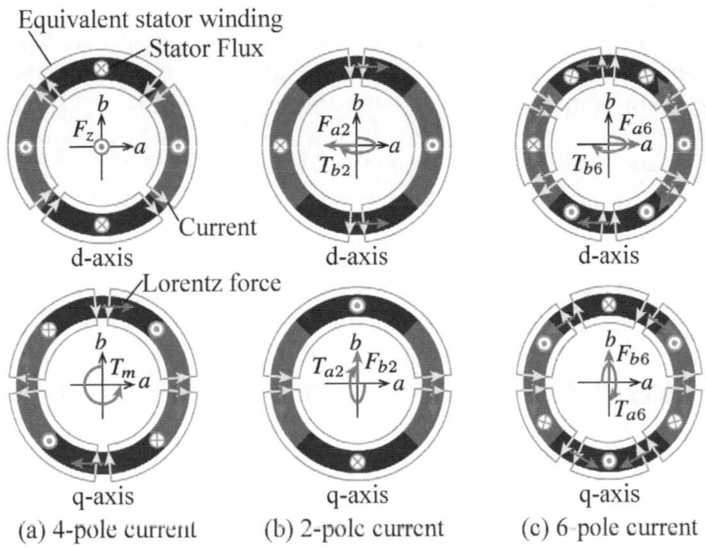

Figure 2: Principle.

In this paper, moments and forces generated by an AFSBM are analyzed theoretically by using a current sheet model. The stator currents are then calculated to control the motor torque and rotor position. Levitation and rotation tests of a prototype motor confirm the proposed method.

2 Axial-Flux Self-Bearing Motor

Fig. 1 shows the structure of the proposed AFSBM. A rotor is a flat disk, and PMs are attached to the upper side of the rotor. P denotes the number of pole pairs. A stator consists of an iron core and windings which generate rotating magnetic fields with $P-1$, P, and $P+1$ pole pairs. Sensors measure the displacements and rotation angles of the rotor, then the rotor position and posture are stabilized by feedback control.

Fig. 2 shows the principle of the generation of motor torque, axial force, radial force, and tilting moment. It illustrates a four-pole rotor as an example. The d-axis component of the four-pole current generates the magnetic field at the same as the magnetic field generated by the rotor PM. Then it controls the axial force by changing the amplitude of the magnetic flux density in the air gap. The q-axis component of the four-pole current generates the motor torque by the Lorentz force. Two- and six-pole currents generate both radial forces and tilting moments by magnetic force and Lorentz force. However, the direction of one of them is different, e.g., F_{a2} and F_{a6} is the opposite, while T_{a2} and T_{a6} are the same. Therefore, we can control the radial forces and tilting moments by combining two- and six-pole currents.

3 Analysis of Bearing Force and Motor Torque

3.1 Coordinate System

Motor torque, axial force, radial forces, and tilting moments are theoretically analyzed.

A coordinate system is defined as Figure 3. x-y is the horizontal plane, and θ_x and θ_y are rotation around them. z is vertical, and θ is rotation around z. Since the upper surface of the rotor is the acting point of magnetic force, that point is the origin of z.

ψ is the rotation angle of the rotor. The origin of θ is at ψ. a-b is the horizontal plane fixed on the rotor, and θ_a and θ_b are rotation around them.

d-axis is the same as a-axis, and q-axis depends on the pole-pair number of the magnetic field and is at $\theta = \pi/(2P)$.

We replace the stator windings and the rotor PMs with current sheets for simplification. The positive direction of the current is from outside to inside. The direction of the magnetic flux density is the same as z, then the direction of Lorentz force becomes counterclockwise. g_0 is the air gap between the stator and rotor cores. r_o, r_i, and r are outer, inner, and average radii, respectively. The surface area of the core is represented by $A = \pi(r_o^2 - r_i^2)$.

3.2 Current and Magnetic Flux Density

For simplicity, we assume current distribution to be sinusoidal as

$$i(\theta) = I \sin(P\theta + \phi) \tag{1}$$

where I and ϕ denote the amplitude and phase of current, respectively.

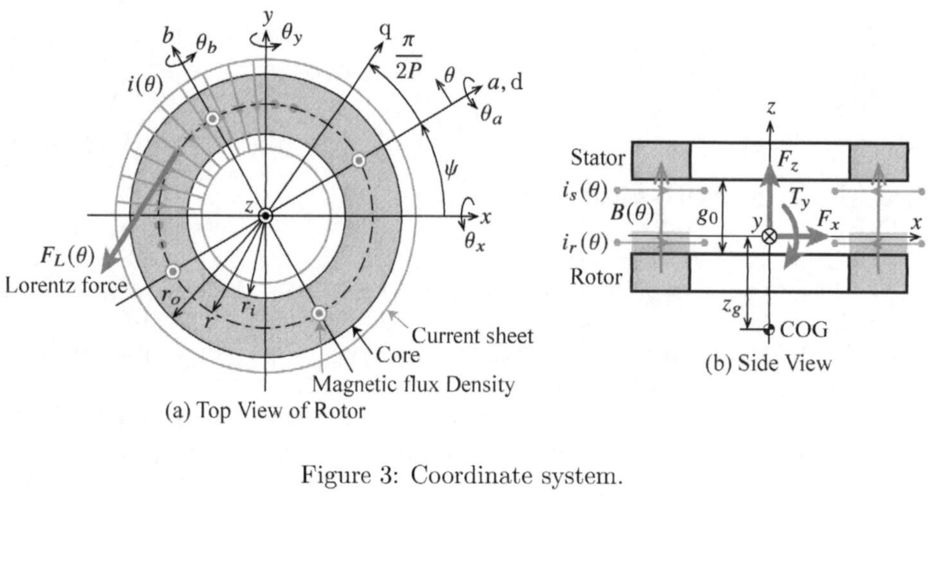

Figure 3: Coordinate system.

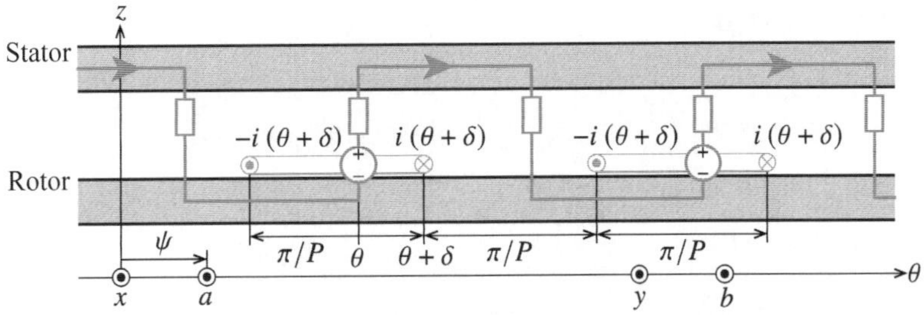

Figure 4: Magnetic circuit.

Figure 4 shows a magnetic circuit that considers only the current at $\theta+\delta+n\pi/P$, where $n = 0, 1, \cdots, P-1$. The currents at $\theta+\delta+2n\pi/P$ are $i(\theta+\delta)$, and the currents at $\theta+\delta+(2n+1)\pi/P$ are $-i(\theta+\delta)$. A blue line indicates a magnetic path generated by the currents. The cross section of the magnetic path is $A/(2P)$, and the number of air gaps in the magnetic path is $2P$, then the total magnetic resistance in the magnetic path becomes $4g_0P^2/(\mu_0 A)$. Since the number of current loops is P, the total magnetomotive force becomes $Pi(\theta+\delta)$. Consequently, the magnetic flux density at θ is calculated as

$$\Delta B(\theta) = \frac{\mu_0 A}{4g_0 P^2} Pi(\theta+\delta) \frac{2P}{A} = \frac{\mu_0}{2g_0} I \sin\{P(\theta+\delta)+\phi\} \tag{2}$$

We can calculate the magnetic flux density by the whole current by integrating $\Delta B(\theta)$ with respect to δ from 0 to π/P, then we have

$$B(\theta) = \int_0^{\pi/P} \Delta B(\theta) d\delta = \frac{\mu_0}{g_0}\frac{I}{P}\cos(P\theta+\phi) \tag{3}$$

The equivalent rotor current and magnetic flux density by the rotor PM are denoted by adding subscript r. ϕ of $i_r(\theta)$ is 0 because $\cos(P\theta+\phi)$ of $B_r(\theta)$ becomes 1 at d-axis ($\theta = 0$). Hence,

$$i_r(\theta) = I_r \sin(P\theta) \tag{4}$$

The stator current, denoted by $i_s(\theta)$, is assumed to be the sum of the currents with $P-1$, P, and $P+1$ pole pairs. Each current is decomposed into d- and q-axis components. The amplitude of each component is denoted by adding subscript d, q, $P-1$, P, and $P+1$. ϕ of the d-axis component is 0, on the other hand, ϕ of the q-axis component is $-\pi/2$ because q-axis is located at $\theta = \pi/(2P)$. Then, the stator current is expressed by

$$i_s(\theta) = i_{P-1}(\theta) + i_P(\theta) + i_{P+1}(\theta) \tag{5}$$

where

$$i_{P-1}(\theta) = I_{\mathrm{d}P-1}\sin\{(P-1)\theta\} - I_{\mathrm{q}P-1}\cos\{(P-1)\theta\} \tag{6}$$
$$i_P(\theta) = I_{\mathrm{d}P}\sin(P\theta) - I_{\mathrm{q}P}\cos(P\theta) \tag{7}$$
$$i_{P+1}(\theta) = I_{\mathrm{d}P+1}\sin\{(P+1)\theta\} - I_{\mathrm{q}P+1}\cos\{(P+1)\theta\} \tag{8}$$

3.3 Force and Torque

The attractive force between the stator and the rotor of a small angle is expressed by

$$\Delta F_M = \frac{A}{4\pi\mu_0}\left\{B_r(\theta) + B_s(\theta)\right\}^2 d\theta \tag{9}$$

The axial force is calculated as

$$F_z = \int_0^{2\pi} \Delta F_M = \frac{\mu_0 A}{4g_0^2 P^2}\left\{(I_r + I_{\mathrm{d}P})^2 + I_{\mathrm{q}P}^2 + I_{\mathrm{d}P-1}^2 + I_{\mathrm{q}P-1}^2 + I_{\mathrm{d}P+1}^2 + I_{\mathrm{q}P+1}^2\right\} \tag{10}$$

The tilting moments around a and b are calculated as

$$T_a = \int_0^{2\pi} r\sin\theta \Delta F_M = \frac{\mu_0 Ar}{4g_0^2 P^2}\left\{(-I_{\mathrm{q}P-1} + I_{\mathrm{q}P+1})(I_r + I_{\mathrm{d}P}) + (I_{\mathrm{d}P-1} - I_{\mathrm{d}P+1}) I_{\mathrm{q}P}\right\} \tag{11}$$

$$T_b = \int_0^{2\pi} -r\cos\theta \Delta F_M = \frac{\mu_0 Ar}{4g_0^2 P^2}\left\{(-I_{\mathrm{d}P-1} - I_{\mathrm{d}P+1})(I_r + I_{\mathrm{d}P}) + (-I_{\mathrm{q}P-1} - I_{\mathrm{q}P+1}) I_{\mathrm{q}P}\right\} \tag{12}$$

Assume that I_r is larger than the other currents, we can neglect the product of the stator currents. Therefore,

$$F_z \approx \frac{\mu_0 A}{4g_0^2 P^2} I_r^2 + \frac{\mu_0 A}{2g_0^2 P^2} I_r I_{\mathrm{d}P} = F_{z0} + K_z I_{\mathrm{d}P} \tag{13}$$

$$T_a \approx \frac{\mu_0 Ar}{4g_0^2 P^2} I_r (-I_{\mathrm{q}P-1} + I_{\mathrm{q}P+1}) = -K_t I_{\mathrm{q}P-1} + K_t I_{\mathrm{q}P+1} \tag{14}$$

$$T_b \approx \frac{\mu_0 Ar}{4g_0^2 P^2} I_r (-I_{\mathrm{d}P-1} - I_{\mathrm{d}P+1}) = -K_t I_{\mathrm{d}P-1} - K_t I_{\mathrm{d}P+1} \tag{15}$$

where F_{z0} is the attractive force with zero current, and K_z and K_t are the coefficients of the force and torque, respectively.

The Lorentz force at θ is expressed by

$$F_L = (r_o - r_i)\left\{B_s(\theta) i_r(\theta) - B_r(\theta) i_s(\theta)\right\} \tag{16}$$

The motor torque is calculated as

$$T_m = \int_0^{2\pi} rF_L d\theta = \frac{\mu_0 A}{g_0 P} I_r I_{\mathrm{q}P} = K_m I_{\mathrm{q}P} \tag{17}$$

where K_m is the coefficient of the motor torque. The radial forces are calculated as

$$\begin{aligned} F_a &= \int_0^{2\pi} -F_L \sin\theta d\theta = \frac{\mu_0 A}{4g_0 P^2 r} I_r \left\{-(2P-1) I_{\mathrm{d}P-1} + (2P+1) I_{\mathrm{d}P+1}\right\} \\ &= -K_{rP-1} I_{\mathrm{d}P-1} + K_{rP+1} I_{\mathrm{d}P+1} \end{aligned} \tag{18}$$

$$\begin{aligned} F_b &= \int_0^{2\pi} F_L \cos\theta d\theta = \frac{\mu_0 A}{4g_0 P^2 r} I_r \left\{(2P-1) I_{\mathrm{q}P-1} + (2P+1) I_{\mathrm{q}P+1}\right\} \\ &= K_{rP-1} I_{\mathrm{q}P-1} + K_{rP+1} I_{\mathrm{q}P+1} \end{aligned} \tag{19}$$

where K_{rP-1} and K_{rP+1} are the coefficients of the radial forces.

A rotational transformation calculates the radial forces toward x and y and tilting moments around x and y. By rearranging the transformed equations, we have

$$\begin{bmatrix} T_y \\ F_x \end{bmatrix} = \begin{bmatrix} -K_t & -K_t \\ -K_{rP-1} & K_{rP+1} \end{bmatrix} \begin{bmatrix} I_{T_yF_xP-1} \\ I_{T_yF_xP+1} \end{bmatrix} \tag{20}$$

$$\begin{bmatrix} T_x \\ F_y \end{bmatrix} = \begin{bmatrix} -K_t & K_t \\ K_{rP-1} & K_{rP+1} \end{bmatrix} \begin{bmatrix} I_{T_xF_yP-1} \\ I_{T_xF_yP+1} \end{bmatrix} \tag{21}$$

where

$$\begin{bmatrix} I_{T_yF_xP-1} \\ I_{T_xF_yP-1} \end{bmatrix} = \begin{bmatrix} \cos\psi & \sin\psi \\ -\sin\psi & \cos\psi \end{bmatrix} \begin{bmatrix} I_{\mathrm{d}P-1} \\ I_{\mathrm{q}P-1} \end{bmatrix} \tag{22}$$

$$\begin{bmatrix} I_{T_yF_xP+1} \\ I_{T_xF_yP+1} \end{bmatrix} = \begin{bmatrix} \cos\psi & -\sin\psi \\ \sin\psi & \cos\psi \end{bmatrix} \begin{bmatrix} I_{\mathrm{d}P+1} \\ I_{\mathrm{q}P+1} \end{bmatrix} \tag{23}$$

From Eqs. (6), (8), (22) and (23), $i_{P-1}(\theta)$ and $i_{P+1}(\theta)$ are rewritten as

$$i_{P-1}(\theta) = -I_{T_yF_xP-1}\sin\{\psi - (P-1)\theta\} - I_{T_xF_yP-1}\cos\{\psi - (P-1)\theta\} \tag{24}$$

$$i_{P+1}(\theta) = I_{T_yF_xP+1}\sin\{\psi + (P+1)\theta\} - I_{T_xF_yP+1}\cos\{\psi + (P+1)\theta\} \tag{25}$$

3.4 Control of Torque and Force

Eqs. (17) and (13) show that $I_{\mathrm{q}P}$ can control T_m and $I_{\mathrm{d}P}$ can control F_z. These are the same as one-DOF active-controlled AFSBMs.

Eqs. (20) and (21) show that $I_{TxFyP-1}$ and $I_{TxFyP+1}$ generate both T_x and F_y, while $I_{TxFyP-1}$ and $I_{TxFyP+1}$ generate both T_x and F_x. Moreover, F_x and F_y generate not only the radial force but also the tilting moment around the COG of the rotor; F_y generates $-z_gF_y$ around x-axis, and F_x generates z_gF_x around y-axis. Therefore, the following relationships are obtained.

$$\begin{bmatrix} T_{Gy} \\ F_{Gx} \end{bmatrix} = \begin{bmatrix} -(K_t + z_gK_{rP-1}) & -(K_t - z_gK_{rP+1}) \\ -K_{rP-1} & K_{rP+1} \end{bmatrix} \begin{bmatrix} I_{T_yF_xP-1} \\ I_{T_yF_xP+1} \end{bmatrix} \tag{26}$$

$$\begin{bmatrix} T_{Gx} \\ F_{Gy} \end{bmatrix} = \begin{bmatrix} -(K_t + z_gK_{rP-1}) & K_t - z_gK_{rP+1} \\ K_{rP-1} & K_{rP+1} \end{bmatrix} \begin{bmatrix} I_{T_xF_yP-1} \\ I_{T_xF_yP+1} \end{bmatrix} \tag{27}$$

Since both determinants of the coefficient matrices are non-zero, the solutions of the above equations are

$$\begin{bmatrix} I_{T_yF_xP-1} \\ I_{T_yF_xP+1} \end{bmatrix} = \frac{1}{-K_t(K_{rP-1} + K_{rP+1})} \begin{bmatrix} K_{rP+1} & K_t - z_gK_{rP+1} \\ K_{rP-1} & -(K_t + z_gK_{rP-1}) \end{bmatrix} \begin{bmatrix} T_{Gy} \\ F_{Gx} \end{bmatrix} \tag{28}$$

$$\begin{bmatrix} I_{F_yT_xP-1} \\ I_{F_yT_xP+1} \end{bmatrix} = \frac{1}{-K_t(K_{rP-1} + K_{rP+1})} \begin{bmatrix} K_{rP+1} & -(K_t - z_gK_{rP+1}) \\ -K_{rP-1} & -(K_t + z_gK_{rP-1}) \end{bmatrix} \begin{bmatrix} T_{Gx} \\ F_{Gy} \end{bmatrix} \tag{29}$$

These equation give the amplitudes of Eqs. (24) and (25).

4 Experimental Verification

We made and tested a prototype motor to verify the proposed control method.

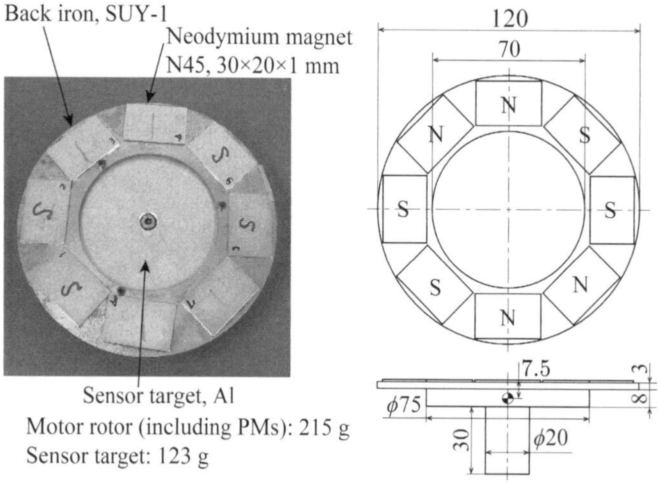

Figure 5: Rotor.

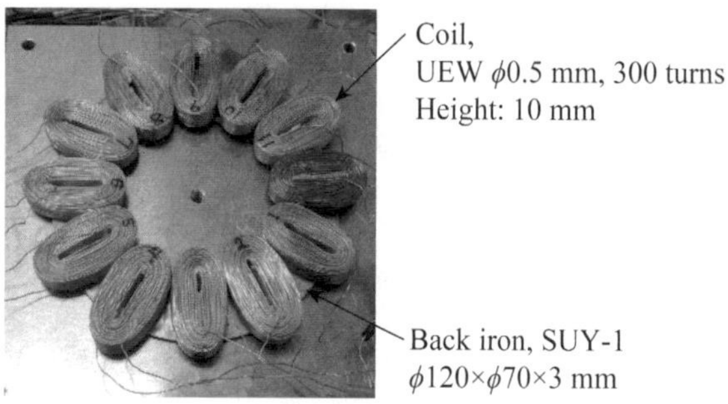

Figure 6: Stator.

4.1 Experimental Device and Control System

Fig. 5 shows the rotor. Rectangular PMs are adopted because it is easily accessible. Eight PMs are attached to a back iron to generate a four-pole magnetic field, then the pole pair number of the rotor is $P = 2$. We attached a sensor target for radial sensors under the disk.

Fig. 6 shows the stator. Twelve air core coils were made, and attached to a back iron. The location of coils are expressed as

$$\theta_k = \frac{2\pi}{12}(k-1) - \psi, \qquad k = 1, \cdots, 12 \tag{30}$$

Since the stator is the concentrated windings, the phase of the stator currents needs the phase

Figure 7: Whole device.

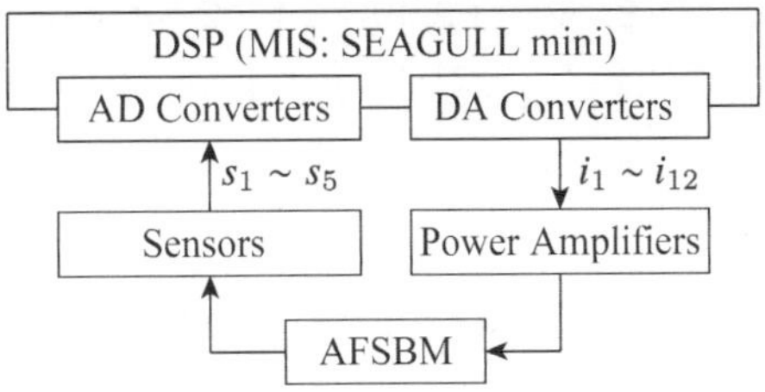

Figure 8: Control system.

lead of $\pi/2$. Therefore, each coil current can be calculated from Eqs. (7), (24), and (25) as

$$\begin{aligned} i_k = &- I_{T_yF_x1}\cos\left\{2\psi - \frac{2\pi}{12}(k-1)\right\} + I_{T_xF_y1}\sin\left\{2\psi - \frac{2\pi}{12}(k-1)\right\} \\ &+ I_{d2}\cos\left\{2\psi - \frac{4\pi}{12}(k-1)\right\} - I_{q2}\sin\left\{2\psi - \frac{4\pi}{12}(k-1)\right\} \\ &+ I_{T_yF_x3}\cos\left\{2\psi - \frac{6\pi}{12}(k-1)\right\} - I_{T_xF_y3}\sin\left\{2\psi - \frac{6\pi}{12}(k-1)\right\} \end{aligned} \tag{31}$$

Fig. 7 shows the whole device. Five eddy current sensors are installed under the rotor and measure the displacement of the rotor. The displacements and tilt angles about the COG of the rotor are calculated from the sensor values.

Fig. 8 shows the control system. A digital signal processor (DSP) controls the prototype motor. The DSP reads the sensor signals via analog to digital (AD) converters and calculates the current commands, then outputs the current commands to power amplifiers via digital to analog (DA) converters. The amplifiers that consist of a power operational amplifier (TI: OPA549S) with a current feedback circuit send the commanded currents through the coils. The sampling interval is 0.15 ms.

Fig. 9 shows the block diagram of the controller. The following proportional-derivative (PD) controller is adopted for position control.

$$G_{PD}(s) = K_P + K_D\frac{s}{s/\omega_b + 1} \tag{32}$$

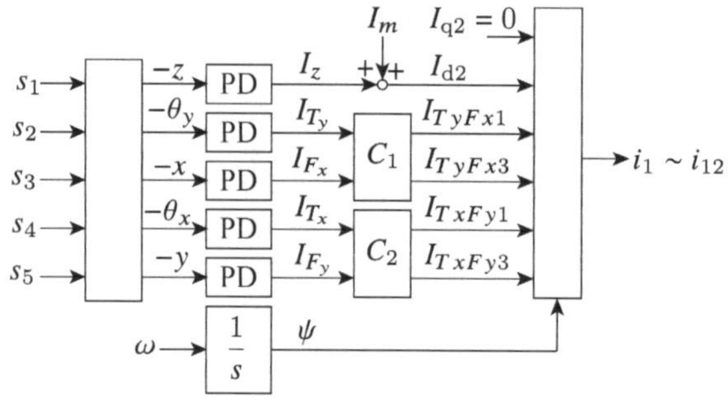

Figure 9: Controller.

	K_P	K_D	ω_b
z	2 A/mm	0.02 As/m	$2\pi \times 200$ rad/s
θ_x, θ_y	50 A/rad	0.5 As/rad	$2\pi \times 200$ rad/s
x, y	2 A/mm	0.05 As/m	$2\pi \times 200$ rad/s

Table 1: Gains of PD controllers.

where K_P is a proportional gain, K_D is a derivative gain, and ω_b is the brake frequency of the approximate differentiator. The PD controllers are converted to discrete-time transfer functions by the Tustin method and implemented to the DSP. To determine the gains experimentally, the outputs of the controllers are the control currents instead of the commands of the forces and tilting moments. Table 1 shows the gains of the controllers. C_1 and C_2 in Fig. 9 are transformation matrices, and described as

$$\begin{bmatrix} I_{T_yF_x1} \\ I_{T_yF_x3} \end{bmatrix} = C_1 \begin{bmatrix} I_{T_y} \\ I_{F_x} \end{bmatrix} = \begin{bmatrix} -c_T & -c_F \\ -1 & 1 \end{bmatrix} \begin{bmatrix} I_{T_y} \\ I_{F_x} \end{bmatrix} \tag{33}$$

$$\begin{bmatrix} I_{T_xF_y1} \\ I_{T_xF_y3} \end{bmatrix} = C_2 \begin{bmatrix} I_{T_x} \\ I_{F_y} \end{bmatrix} = \begin{bmatrix} -c_T & c_F \\ 1 & 1 \end{bmatrix} \begin{bmatrix} I_{T_x} \\ I_{F_y} \end{bmatrix} \tag{34}$$

where c_T and c_F were experimentally determined as $c_T = 1.7$ and $c_F = 0.9$. The rotation speed is controlled by the angular frequency of the stator currents.

4.2 Results of Levitation and Rotation Tests

We achieved non-contact levitation and rotation.

Fig. 10 shows step responses without rotation. These were measured by injecting step disturbance signals into the output of the PD controllers. The amplitude was 0.5 A and -0.5 A. The displacements and tilt angles settled without overshoot. These indicate that all axes are well actively controlled. There are interference between θ_x and y in case of $I_{T_x} = -0.5$ A, and θ_y and x in case of $I_{T_y} = 0.5$ A. We will investigate this problem in the future.

Fig. 11 shows the waveforms while rotation. The vibration in x and y directions increased by increasing the rotation speed. Since the radial vibration is caused by the unbalance of the rotor, the vibration became large at a high rotation speed. On the other hand, the vibration in θ_x and θ_y decreased by increasing the rotation speed due to a gyroscopic effect.

We stepped the rotation test at 900 rpm because the radial vibration became large. The rotation speed will increase by adjusting the rotor balance.

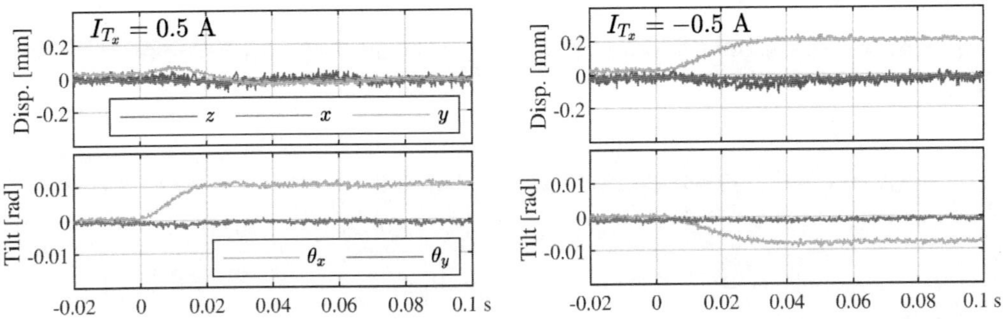

(a) I_{T_x}

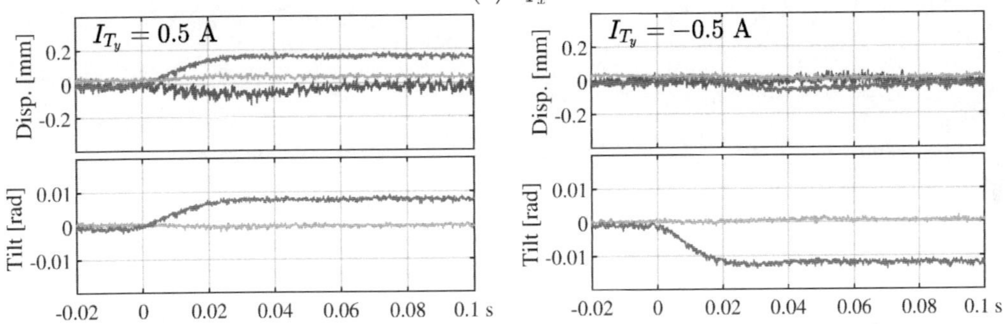

(b) I_{T_y}

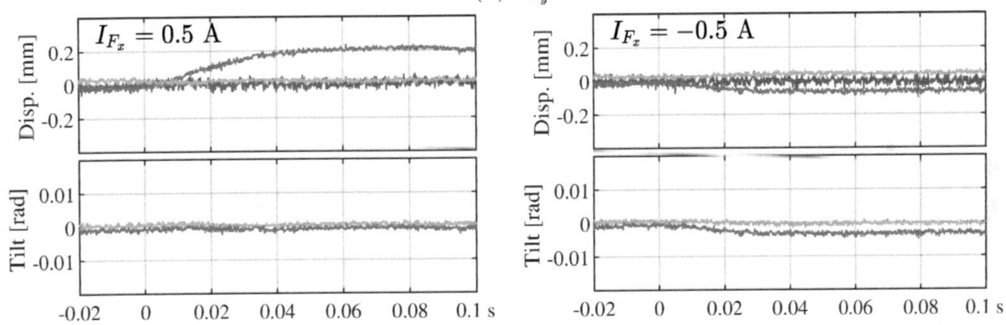

(c) I_{F_x}

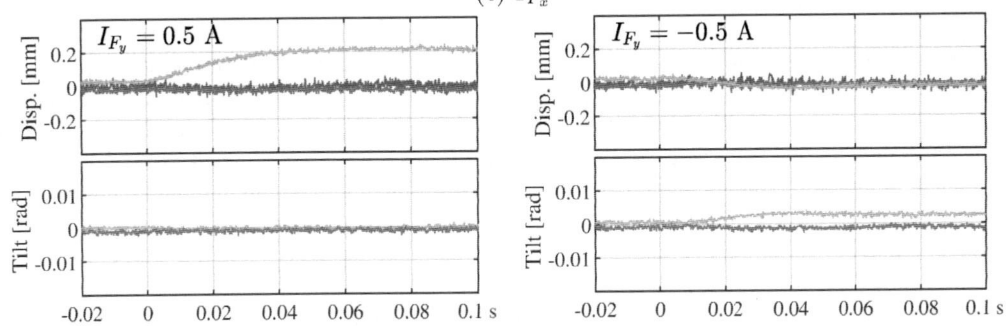

(d) I_{F_y}

Figure 10: Step responses.

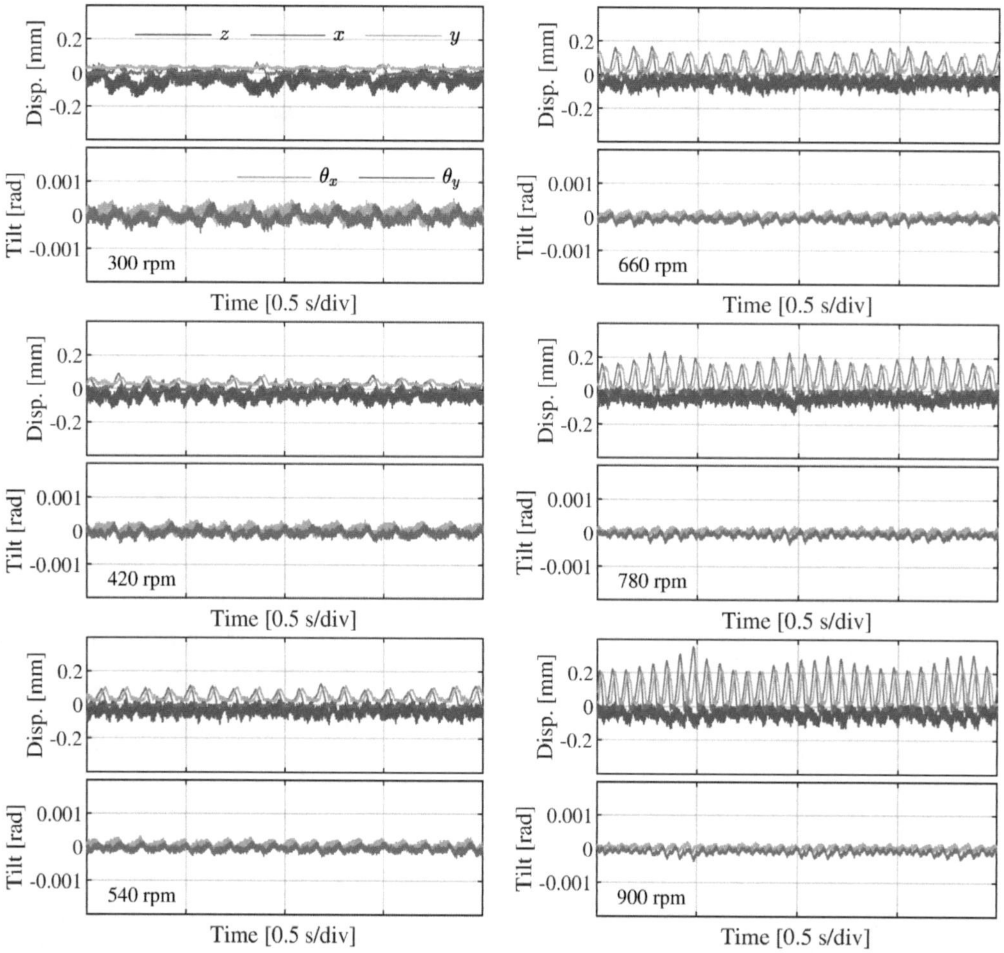

Figure 11: Waveforms while rotation.

5 Conclusions

In this paper, we introduced a five-DOF active-controlled AFSBM with a single stator. The theoretical analysis showed that the combination of the plus and minus two-pole magnetic fields controls the radial forces and tilting moments of the rotor. We succeeded in stable levitation and rotation until 900 rpm and confirmed that the proposed method is effective.

References

[1] Masahiro Osa, Toru Masuzawa, Takuya Saito, and Eisuke Tatsumi. Magnetic levitation performance of miniaturized magnetically levitated motor with 5-dof active control. *Mechanical Engineering Journal*, 2(4), 2017.

[2] Masayuki Sumino and Satoshi Ueno. Tilt control with axial self-bearing motor. *Transactions of the JSME (in Japanese)*, 2014.

[3] Satoshi Ueno, Ryo Iseki, and Changan Jiang. Stability of a tilt-controlling axial gap self-bearing motor with single-stator. *Mechanical Engineering Journal*, 4(4), 2017.

[4] Satoshi Ueno, Junichi Mameda, and Changan Jiang. Analysis and control of radial force and tilt moment for an axial-gap self-bearing motor. In *Proc. of IEEE International Electric Machines & Drives Conference, Miami, Florida, USA*, 2017.

[5] Satoshi Ueno and Yohji Okada. Characteristics of axial force and rotating torque and their control of pm type axial gap self-bearing motor. *Electrical Engineering in Japan*, 132(1):81–91, 2000.

Modified Excitation Signals for AMB Identification Procedures

Diego Diaz[1], Fernando Pinto[2], Thiago Ritto[3], David Maldonado[4], and Vinicius Côrtes[5]

[1] Acoustics and Vibrations Laboratory, Federal University of Rio de Janeiro, Brazil
dgodoy@ufrj.br
[2] fcpinto@ufrj.br
[3] tritto@mecanica.ufrj.br
[4] davidjuliang@hotmail.com
[5] viniciuscortes@poli.ufrj.br

Abstract

In this paper modified signals obtained by means of an optimization algorithm are used for identification procedures in a previous developed small-scale active magnetic bearing (AMB) test rig. Such an iterative algorithm reduces the crest factor of the original signal allowing higher energy injection into the system, yielding a higher signal-to-noise ratio if compared with general broadband excitation signals (swept, white noise, impulse, etc.). The results are compared with the well known stepped sine, which has the ability to concentrate the energy in a unique desired frequency, however at the cost of a significant increase of the total experiment time.

1 Introduction

The choice of excitation signals is an important step when an identification procedure, or a sensitivity analysis, is the scope of any system under test, even more when the allowed time to perform the experiment is a constrain. The bandwidth and its energy distribution are the most common aspects to have into account for an optimal excitation signal design. Due to the inherent presence of noise in experimental measurements, some strategies may be implemented aiming to increase the signal to noise ratio (SNR), and thus obtain more accurate estimations[9]. In the case of magnetic bearing systems, as well as various applications, the excitation signal amplitude is restricted by the maximum shaft displacement, namely, a part of the gap between the rotor and the touchdown bearing. A straightforward way, which consists in a longer experiment time to perform averaging techniques with a higher number of repetitions, as studied in detail by Verboben [21], may not be effective when the test stand is not able to maintain the test conditions for a long period, for instance, high pressure and flow ratio test rigs in oil and gas applications. In such a scenario, the study and assessment of different excitation signals become an interesting alternative to obtain a *balanced* solution between the time and amplitude restrictions.

In previous works, various excitation signals have been used in active magnetic bearing systems (AMBs) aiming a comparison of each signal performance. Perhaps the experiments developed by Hynynen [7] were the first to highlight the importance of well-designed signals for identification of an AMB test rig transfer function. In the same way, Khader et. al. [8] performed a similar analysis in order to assess different averaging methods, or also known in literature as *estimators* [5]. Multitone signals were tested in a high pressure test chamber devised by General Electric turbomachinery manufacturer [20] to evaluate the behavior of annular gas seals in presence of pressurized nitrogen. A reduction of total experiment time was one of the most important remarks in that paper. A qualitative comparison of the obtained frequency response functions (FRFs) by utilizing several excitation signals was performed by Vuojolainen et al. [22]. Likewise, Diaz et al. [2] obtained the FRFs of a small-scale AMB test rig by using different averaging methods in a non-rotating shaft condition and then at several rotation speeds [3]. With the addition of modified multitone excitation signals, the latter authors performed [4] an estimation of the equivalent stiffness and damping coefficients of the same AMB test rig, observing the benefits when the crest factor of an excitation signal is reduced, resulting in lower uncertainties.

A comparison between pure harmonic, broadband general purpose and modified excitation signals are performed in a small-scale AMB by applying the method utilized in[4], and presenting the results in a relative instead an absolute approach, which gives a better understanding of the uncertainties behavior of the different excitation signals applied to the test rig.

2 Identification Procedure

Due to the robust performance in presence of noise [19], identification method is made on the frequency domain approach, i.e. the use of frequency response functions (FRFs). The excitation signal $\mathbf{s}$ is summed to the control signal (see Figure 1) while the rotor is levitating. Both the excitation, current and position signals are measured, conditioned and stored in an acquisition system, represented by $\mathbf{u}$ for the currents and $\mathbf{y}$ for the shaft position vectors. Those vectors are composed by their respective *true* values ($\mathbf{u_0}$ and $\mathbf{y_0}$) and also by the noise contribution vectors ($\mathbf{n_u}$ and $\mathbf{n_y}$). The noise effect are considered to be independent for each sensor and has the proprieties of a normally distributed random variable. Then the Discrete Fourier Transform (DFT) technique is applied to those stored signals in order to obtain their spectrum. Finally, the ratio of the output to the input signal (excitation) can be defined as the FRF of the system.

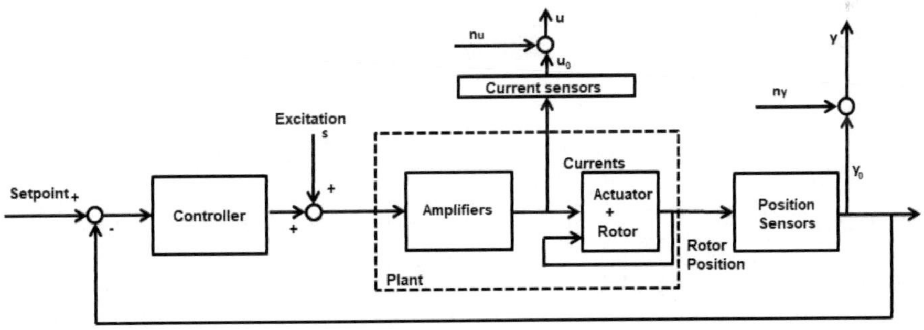

Figure 1: Block diagram for AMB identification (adapted from [17])

2.1 Identification Model

A two DOF mass-spring-damper model is employed to obtain the coefficients from the measured data. The tests are carried out with at rest condition (non-rotating shaft), and the governing equation can be represented as [23]:

$$\boldsymbol{M}\ddot{\mathbf{q}} + \boldsymbol{C}\dot{\mathbf{q}} + \boldsymbol{K}\mathbf{q} = \mathbf{f}, \tag{1}$$

where the vectors $\mathbf{q} = \begin{bmatrix} x & y \end{bmatrix}^T$ and $\mathbf{f} = \begin{bmatrix} f_x & f_y \end{bmatrix}^T \in \mathbb{R}^{2\times 1}$ represent the displacements and forces applied to the system, respectively. Therefore, stiffness $\boldsymbol{K}$, damping $\boldsymbol{C}$ and mass $\boldsymbol{M}$ matrices are expressed as:

$$\boldsymbol{M} = \begin{bmatrix} m & 0 \\ 0 & m \end{bmatrix}, \boldsymbol{C} = \begin{bmatrix} C_{xx} & C_{xy} \\ C_{yx} & C_{yy} \end{bmatrix}, \boldsymbol{K} = \begin{bmatrix} K_{xx} & K_{xy} \\ K_{yx} & K_{yy} \end{bmatrix}, \in \mathbb{R}^{2\times 2}. \tag{2}$$

The subscripts xx, yy and xy, yx represent the direct and cross-coupled effect in the system, respectively. m is related to the body (i.e. rotor) mass. By replacing the current and position vectors into their complex form (i.e. $\mathbf{q} = \bar{\mathbf{q}}_{\mathbf{u}} e^{j\omega t}$ and $\mathbf{f} = \bar{\mathbf{f}}_{\mathbf{u}} e^{j\omega t}$), equation 1 can be expressed as follows[11]:

$$\underbrace{\left[\boldsymbol{K} - \omega^2 \boldsymbol{M} + j\omega \boldsymbol{C}\right]}_{CDS} \bar{\mathbf{q}}_{\mathbf{u}} = \boldsymbol{H} \bar{\mathbf{q}}_{\mathbf{u}} = \bar{\mathbf{f}}_{\mathbf{u}}, \tag{3}$$

with $\boldsymbol{H} \in \mathbb{C}^{2\times 2}$ denoting the complex dynamic stiffness (CDS) of the system, $j = \sqrt{-1}$ and ω the excitation frequency. The force vector depends on the measured coil currents, shaft position and actuator topology (number of poles, air gap, pole area, etc.). For the system under test, as described in section 4, the force can be computed by the linear equation[10]:

$$\mathbf{f} = \begin{bmatrix} f_x \\ f_y \end{bmatrix} = \underbrace{\frac{4\cos(\pi/8)\mu_0 A_g N_v^2 i_b}{g_0^2}}_{K_{mag,i}} \begin{bmatrix} i_x \\ i_y \end{bmatrix} + \underbrace{\frac{-4\cos(\pi/8)\mu_0 A_g N_v^2 i_b^2}{g_0^3}}_{K_{mag,x}} \begin{bmatrix} x \\ y \end{bmatrix}, \tag{4}$$

where μ_0 represents the permeability of free space, A_g the pole cross-sectional area, N_v the number of turns in each coil, g_0 the nominal magnetic airgap, i_b the base current and $\begin{bmatrix} i_x & i_y \end{bmatrix}^T$ the vector of the control currents. Note that at least two *different* experiments ($\bar{\boldsymbol{Q}} = \begin{bmatrix} \bar{\mathbf{q}}_{\mathbf{u1}} & \bar{\mathbf{q}}_{\mathbf{u2}} \end{bmatrix}^T$ $\bar{\boldsymbol{F}} = \begin{bmatrix} \bar{\mathbf{f}}_{\mathbf{u1}} & \bar{\mathbf{f}}_{\mathbf{u2}} \end{bmatrix}^T$)are required to obtain the CDS matrix, which finally is written as:

$$\boldsymbol{H} = \begin{bmatrix} K_{xx} - m\omega^2 + j\omega C_{xx} & K_{xy} + j\omega C_{xy} \\ K_{yx} + j\omega C_{yx} & K_{yy} - m\omega^2 + j\omega C_{yy} \end{bmatrix} = \bar{\boldsymbol{F}} \bar{\boldsymbol{Q}}^{-1}. \tag{5}$$

A typical way to obtain the two independent experiments is firstly apply only the excitation force on the x axis and then perform the second one applying the force on the y axis.

2.2 Estimator

Aiming to reduce the noise effect, various repetitions or *blocks*[14] are carried out, and averaging techniques are utilized to obtain the force and displacement matrix. Pintelon and Shouckens[13] describe in detail the proprieties of the averaging methods used in the frequency domain. For the analysis made in this paper, the *error in variables* estimator was selected, which for each excitation frequency ω_k and a defined number of blocks N_b, can be expressed as:

$$\boldsymbol{H_{EIV}}(\omega_k) = \left(\frac{1}{N_b}\sum_{l=1}^{N_b} \bar{\boldsymbol{F}}(\omega_k)^{(l)}\right)\left(\frac{1}{N_b}\sum_{l=1}^{N_b} \bar{\boldsymbol{Q}}(\omega_k)^{(l)}\right)^{-1}. \tag{6}$$

3 Excitation Signals

The stepped sine, which has an amplitude A and can be expressed as $s_t = A\cos(\omega_k t)$, has a high SNR since the energy is stored in only the excited frequency [15], and is one of the common excitation signals used in system identification procedures. Its major limitation lies in the total time needed to perform the experiment. On the other hand, some broadband signals, as the widely used sweep, impulse or white noise are a practical solution to tackle the time limitation issue. Nevertheless, such signals distribute the energy in a great quantity of frequencies, yielding a lower SNR. The linear swept signal can be defined as follows[14]:

$$s(t) = A\sin((at+b)t) \qquad 0 \le t \le T_0, \tag{7}$$

with $a = \pi(k_2 - k_1)f_0^2$ and $b = 2\pi k_1 f_0$, $f_1 = k_1 f_0$ and $f_2 = k_2 f_0$ and $f_0 = 1/T_0$. The random signals are generated by a computer algorithm, and are able to satisfy the noise theory proprieties. In this case, the common used *pseudo-white* noise is chosen to be one of the excitation signal for the comparison.

An intermediate alternative is related to *multisine* or *multitone* signals, which are defined as the sum of N_f number of discrete harmonics, and are capable to distribute the energy into the chosen frequencies. Such a signal can be represented as:

$$s(t) = \sum_{k=1}^{N_f} A_k \cos(\omega_k t + \phi_k), \tag{8}$$

3.1 Modified multisine signals

The shape of a multisine signal relies on the set of its amplitudes (A_k), frequencies (ω_k) and phases (ϕ_k) . Impulsive signals trend to allow a lower amount of energy into the system under test[15]. Such characteristic can be quantified by the crest factor, denoted as[14]:

$$Cr(u) = \frac{s_{pk}}{s_{rmse}} = \frac{\max\limits_{t\in[0,T]} |s(t)|}{s_{rms}\sqrt{P_{int}/P_{tot}}}, \tag{9}$$

where $s_{rms} = \sqrt{\frac{1}{N}\sum_{i=1}^{N} s_i^2}$ is the signal RMS, s_{pk} its peak value, P_{tot} and P_{int} are the total and the contained power in the frequency band of interest, respectively. In general words, the lower the crest factor, the less impulsive the signal is. A second indicator, known as time factor[14], indicates how long takes for each frequency component to have the same SNR when compared to a stepped signal with the same frequency.

There exist several techniques to reduce the crest factor by *manipulating* the signal phases, for instance, the equation proposed by Schroeder [16]. Some iterative algorithms, like the Chebyshev method[6], also have an effective performance. Such a method consists in an optimization problem, where the objective is the minimization of the cost function l_p, which is dependent of the $p-th$ norm of the signal excitation vector, yielding a lower crest factor. The algorithm flow chart can be observed in Figure 2. The phases can be initialized randomly, and then a Levenberg-Marquardt algorithm[12] is applied in order to update the set of phase values. As the crest factor converges, the $p-th$ norm is increased twice until a minimum possible Cr is obtained or the maximum defined $p-th$ is reached.

As an example, a multitone signal with 17 frequencies with flat spectrum (i.e. $A_1 = A_2 = ... = A_{17}$), is used to assess the optimization algorithm. The frequencies are selected to be evenly spaced and distributed following the expression $\omega_k = 2\pi(4k+1)$ with $k = 1, 2, ..., 17$. The spectrum of the original signal, as well as its last iteration step is presented in Figure 3. Note that the spectrum amplitudes are conserved whilst the phases are modified to reduce the peaks of the signal. The crest factor evolution can be observed in figure 4. The numbers located

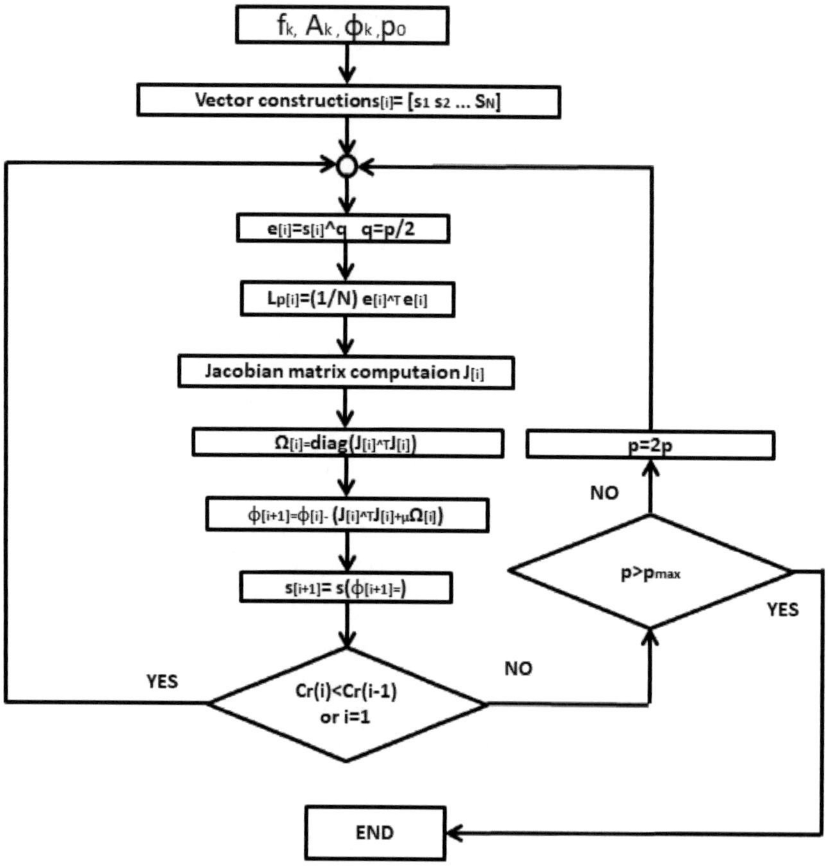

Figure 2: Chebyshev method flow chart [6]

close to some points of the curve indicate the $p-th$ norm value and in which iteration they were increased by two. In this case, the lowest possible crest factor was 1.61, and the algorithm took 73 iterations to converge on the final value. The lasted time was approximately 4 seconds, with a Matlab™ script and a computer with basic specifications (Intel®core i5™ and 8 GB of RAM).

3.2 Experiment time

The total demanded time to perform the data acquisition, in a defined frequency band, when a stepped signal is used [15], can be estimated by using the equation:

$$T_{step} = N_e \left(N_f \left[T_w + N_b \sum_{i=0}^{N_f - 1} \left(\frac{1}{f_i} \right) \right] \right), \tag{10}$$

where T_w refers to the needed time to leave the transient effect when the frequency is modified, and N_e are the number of independent experiments to be performed (for this case, as mentioned in subsection 2.1, $N_e = 2$). In the same way, a time needed to acquire the data, within the

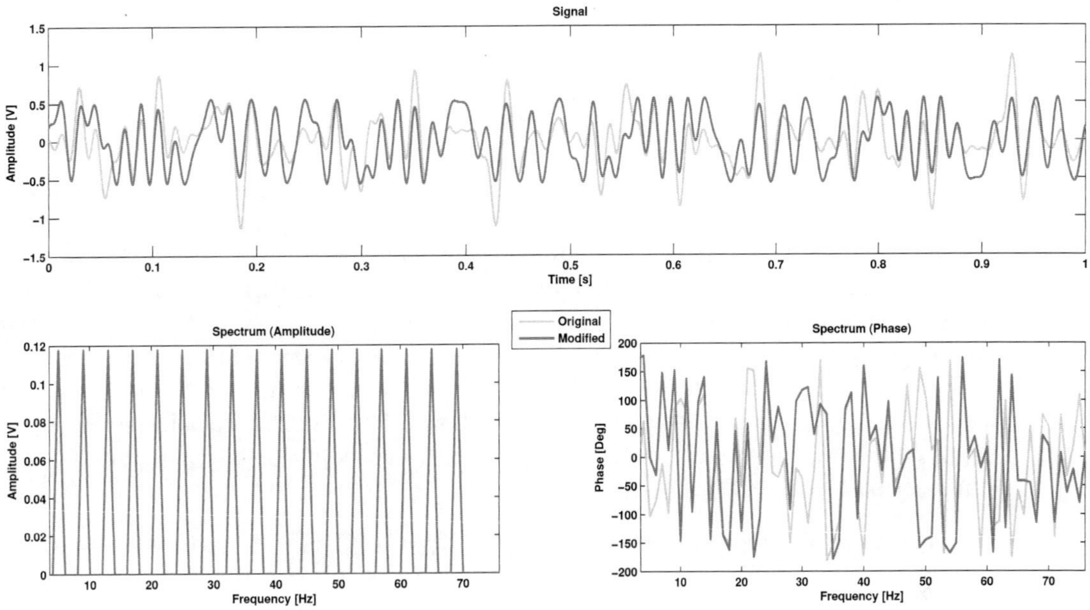

Figure 3: Original and modified multisine signal with Chebyshev method

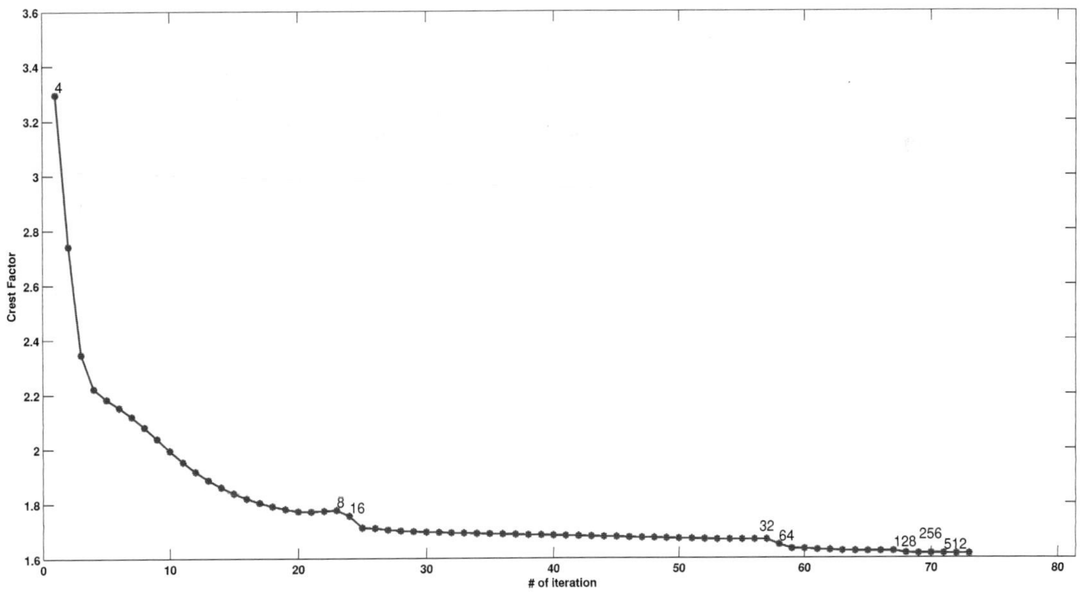

Figure 4: Crest factor evolution for signal presented in Figure 3

same frequency band, when a multisine signal is used, can be calculated by[15]:

$$T_{ms} = N_e \left(\frac{N_b}{f_0} + T_w \right), \tag{11}$$

with f_0 the lowest frequency excited value.

4 Small Scale AMB Test Rig

A rigid rotor supported by an active magnetic bearing system, previously developed in the Acoustics and Vibrations Laboratory [1][18] at Federal University of Rio de Janeiro, was used to assess the excitation signals for the identification procedure, which comprises the estimation of the equivalent AMB stiffness and damping coefficients. Such a test rig can be observed in Figure 5. Each magnetic actuator has an eight heteropolar *coupled pair* configuration. To levitate the rotor, it was employed a closed loop decentralized PID control algorithm, implemented in a development FPGA module by means of the Tustin approximation, with a discrete control period of $100\mu s$. Furthermore, a differential mode is adopted to reduce the complexity of the control strategy scheme. The algorithm also allows to add the excitation signals after the control block output as shown in Figure 1. The main dimensional and functional characteristics are summarized in Table 1.

Figure 5: Small scale AMB test rig used for experiments

Because the identification model structure is governed by only lateral displacements, both sides A and B are excited in the same way to mimic only translational motion. Then the measurements are obtained from the both side sensors, and averaged to obtain the resulting displacement data. Therefore, the magnetic forces from the actuator A and B are summed to compute the resultant system force.

5 Results

For the experiments, it was selected a total of six excitation signals (see Table 2), comprised of the widely used stepped sine, white noise, sweep and the last three are composed of multisine signals. The first one of that latter group has the phases obtained randomly within an interval $[0 \quad 2\pi]$ with a uniform distribution, and the spectrum shape adjusted in such a way that the

Parameter	Symbol	Value	Unit
Shaft mass	m	1	kg
Shaft length	L_s	400	mm
Shaft diameter	d_s	14.28	mm
Journal diameter	d_j	36.4	mm
Touchdown bearing diameter	d_{to}	1	mm
Distance between actuators	$2a$	330	mm
Distance between sensors	$2c$	315	mm
Number of turns per pole	N_v	130	
Bias current	i_b	1	A
Flux density saturation	B_{sat}	1.7	T
Nominal radial airgap	g_0	1	mm
Pole cross-sectional area	A_g	235	mm^2
Open-loop gain	$K_{mag,x}$	-2.11e4	Nm^{-1}
Actuator gain	$K_{mag,i}$	21.14	NA^{-1}
Nominal inductance	L_0	4.3	mH
Coil wire gauge		20	AWG
Number of poles per actuator	N_p	8	

Table 1: Small AMB test rig characteristics

frequencies that require more energy, have their respective input amplitudes increased aiming to maintain the output spectrum amplitude flat. For inductive loads as magnetic bearing applications, this occurs at higher frequencies. To obtain the spectrum shape, the FRF amplitude of the AMB system can be obtained by utilizing any broadband excitation signal, considering the ratio of the input to the displacement spectrum amplitude. Since the objective is to know only the amplitude, it is not required many blocks, indeed, one block should be sufficient. With the same spectrum amplitude shape, the second multisine signal has its phases optimized by the l_p algorithm (i.e. Chebyshev method), and finally, the last multitone signal also has its phases optimized, but instead adapted amplitudes, its shape consists of a flat spectrum. The band of

	Signal	Spectrum amplitudes	Spectrum phases
1	Stepped Sine	Adapted	-
2	Sweep	Flat	-
3	White noise	-	-
4	Modified multisine with l_p algorithm	Flat	Optimized
5	Multisine with random phases	Adapted	Random
6	Modified multisine with l_p algorithm	Adapted	Optimized

Table 2: Excitation signal characteristics

interest was defined well below the maximum force slew rate, starting in 5 Hz with 17 components. In order to separate the nonlinear distortions, at least, the fourth order contribution, the $k-th$ frequency follows the expression $\omega_k = 2\pi(4k+1)$.

A dedicated acquisition module, which stores synchronously the excitation signals, currents and positions measurements from the AMB is used at the rate of $5kHz$. As an example, a period of the excitation signal *6-Modified multisine with l_p algorithm*, when the x axis is excited, as well as its resulting force and shaft position can be observed in Figure 6. The maximum displacement is set to not be more than $\pm 100\mu m$ from the center for all cases.

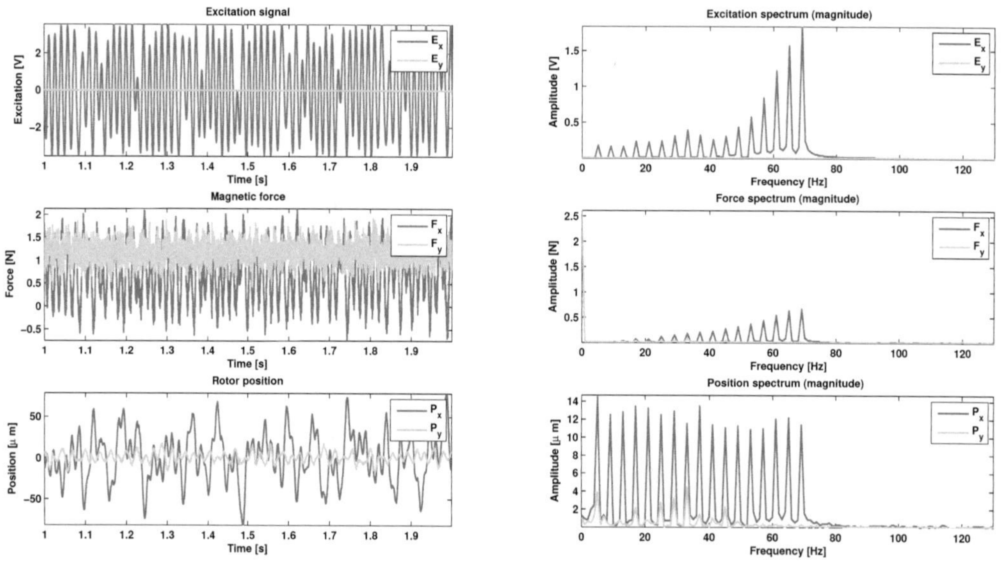

Figure 6: Excitation signal, force and shaft position for the signal *6-Modified multisine with* l_p *algorithm*, when axis x is excited

With an established number of blocks $N_b = 10$, the estimated stiffness coefficients are presented in Figure 7. Note that for the direct terms (i.e. K_{xx} and K_{yy}), all the values are close due to the contribution of the known mass and excitation frequency $m\omega^2$ (see Equation 5). The cross-coupled terms exist due the nonlinear magnetic actuator and its dependence of the shaft position rotor, and also the gravity contribution. The estimated damping coefficients can be observed in Figure 8. In general, the direct terms as well as cross-coupled damping terms seem to have no well defined pattern, and the estimated coefficients differences are more evident.

In order to analyze the uncertainties, the relative standard deviation concept is applied to the results presented previously in Figures 7 and 8, taking as reference the mean coefficient estimated in each frequency. For the stiffness and damping coefficients, as shown in Figure 9 and 10, respectively, the relative standard deviation curves confirm the outstanding performance of the stepped signal, which for almost all frequencies, had the lowest standard deviation. At low components (e.g. $5Hz$), it is possible to notice an increment of uncertainties, mainly in the direct terms of stiffness and damping, probably due to the low amplitude excitation signal required to move the shaft below the maximum allowed displacement, combined with the noise of the current sensors. There are several points obtained with the white noise and sweep signal that are not plotted in the figures, since they are located out of the graphic scale. Moreover, the signals with flat spectrum trend to have a higher standard deviation values at high frequencies, as expected, since the applied energy yields a lower system excitation in such frequencies. One can also observe the higher deviation in the x axis, associated likely to the gravity effect when the horizontal axis (i.e. x) is excited.

Aiming to have an overall sight of the assessed excitation signals, an average of the deviations over the band of interest is calculated for each term, and summarized in Table 3. It is possible to

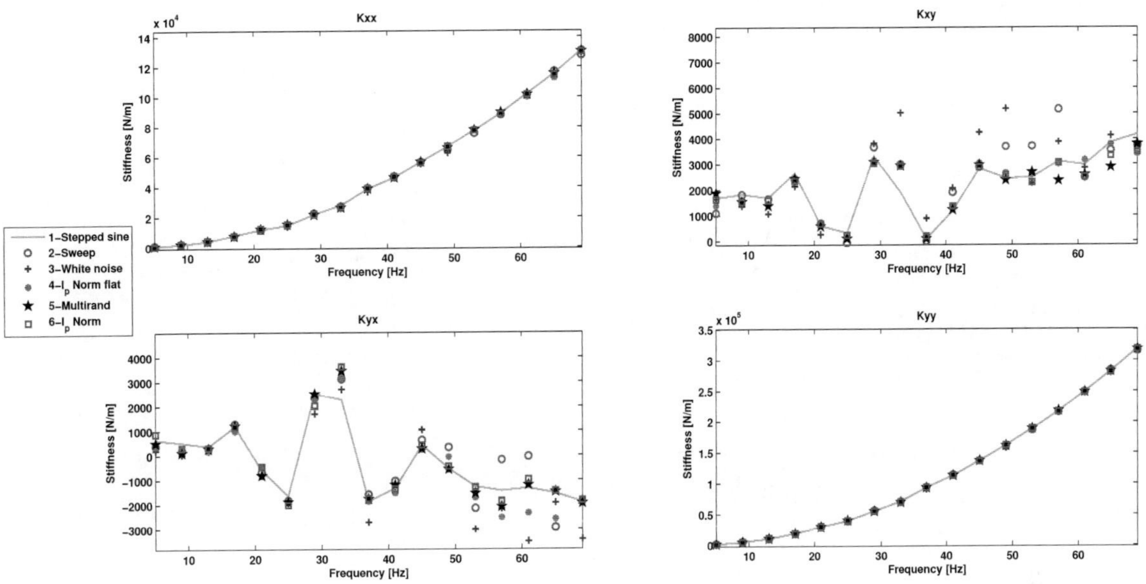

Figure 7: AMB estimated stiffness coefficients

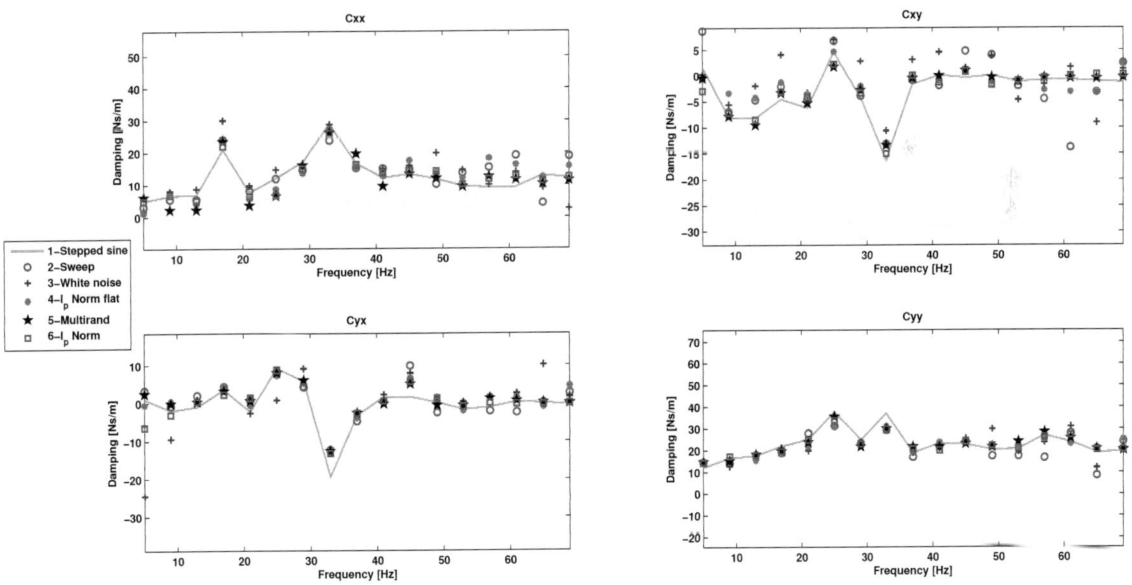

Figure 8: AMB estimated damping coefficients

observe the significantly lower deviation in multisine signals, if compared with sweep and white noise. The multitone signals with their phases modified by the Chebyshev method performed better than the multisine with random phases.

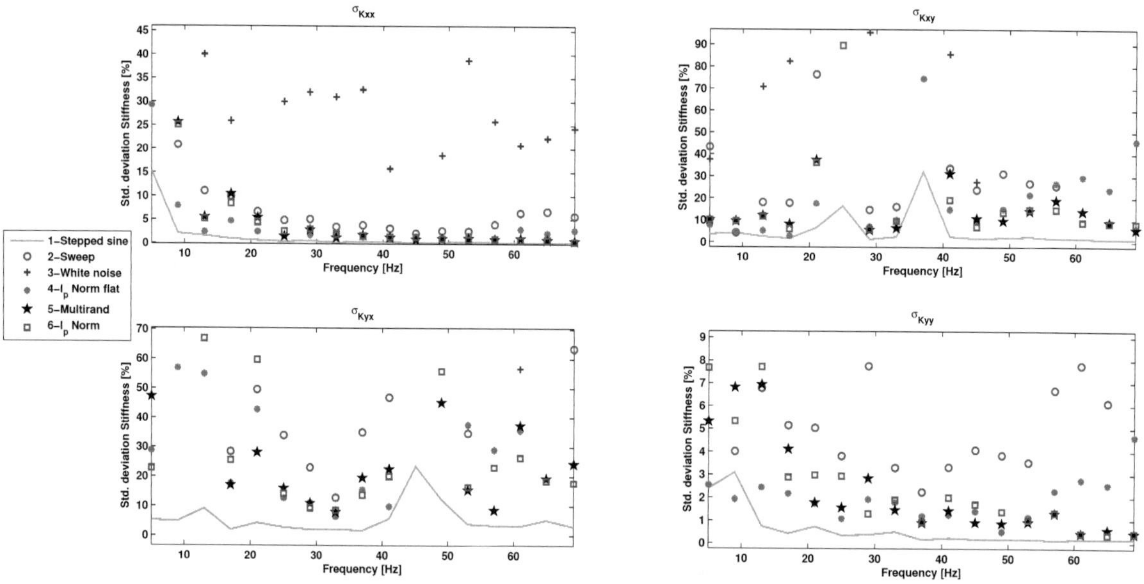

Figure 9: Relative standard deviation of the estimated AMB system stiffness coefficients

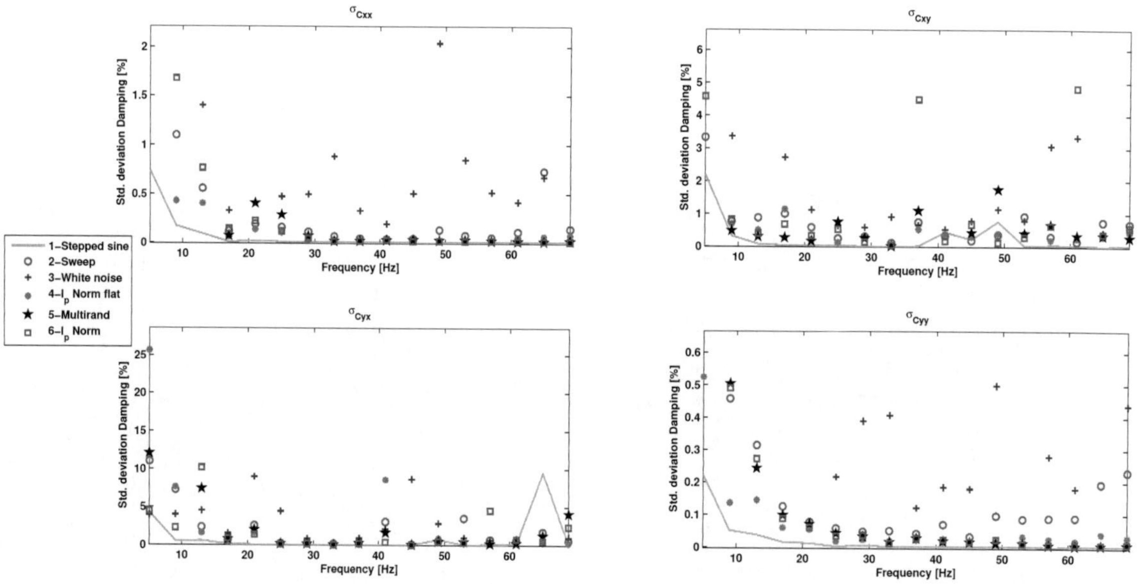

Figure 10: Relative standard deviation of the estimated AMB system damping coefficients

Because this paper intends not only to assess the uncertainties, but also the error with respect to the *true* values, a comparison of the estimated coefficients is carried out. The stepped sine, which had the lowest relative standard deviation, is chosen as the reference to

Excitation signal	σ_{Kxx}	σ_{Kxy}	σ_{Kyx}	σ_{Kyy}	σ_{Cxx}	σ_{Cxy}	σ_{Cyx}	σ_{Cyy}
	%	%	%	%	%	%	%	%
1- Stepped sine	1.36	5.04	5.51	0.59	0.06	0.25	0.95	0.02
2- Sweep	21.99	223.84	606.30	6.44	0.68	0.68	2.08	0.36
3- White noise	76.24	519.60	253.91	28.03	1.66	7.81	15.94	1.11
4- Multisine (l_p) flat shape	3.94	24.47	74.08	1.96	0.56	1.48	2.86	0.07
5- Multisine with random phases	8.07	40.37	47.74	2.26	0.82	35.88	12.34	0.19
6- Multisine (l_p) adapted shape	7.58	23.59	33.57	2.49	0.69	1.16	1.71	0.15

Table 3: Relative standard deviation average for stiffness and damping coefficients, $N_b = 10$

calculate the relative error of the estimated stiffness and damping coefficients with the other five signals, presented in Figures 11 for stiffness and 12 for damping. The multitone signals obtained a lower error trend, when compared with the white noise and swept signals. It is also possible to observe a lower error at high frequencies for multisines with adapted spectrum amplitude than the obtained with flat amplitude signals.

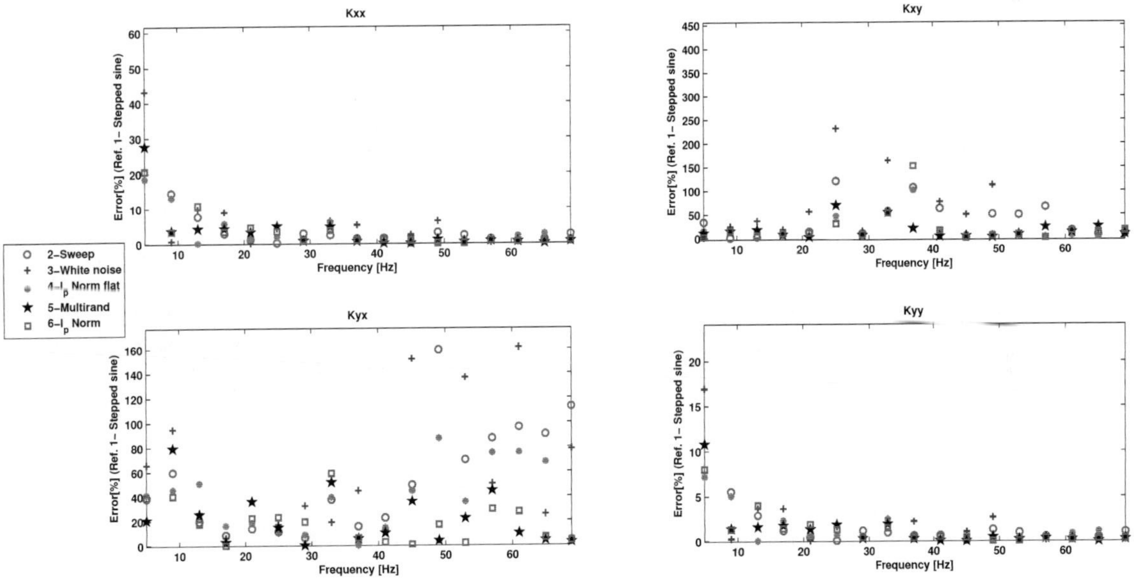

Figure 11: Relative error of the AMB estimated stiffness coefficients

As previously proposed for uncertainties, a general interpretation can be achieved by computing the mean relative error coefficients in the band of interest, which is summarized in Table 4. The Multisine signal with optimized phases and adapted spectrum shape presented the lowest errors (principally in the direct damping terms), with the exception of the cross-coupled damping C_{xy} and C_{yx}.

For the $N_f = 17$ frequencies, the total experiment times, which can be obtained by using Equation 10 for the stepped sine, and Equation 11 for the selected multisines (as well for white noise and sweep), where approximately 40 minutes and 40 seconds, respectively, meaning a ratio of 60 times. One can imply that the higher the number of frequencies, the greater the

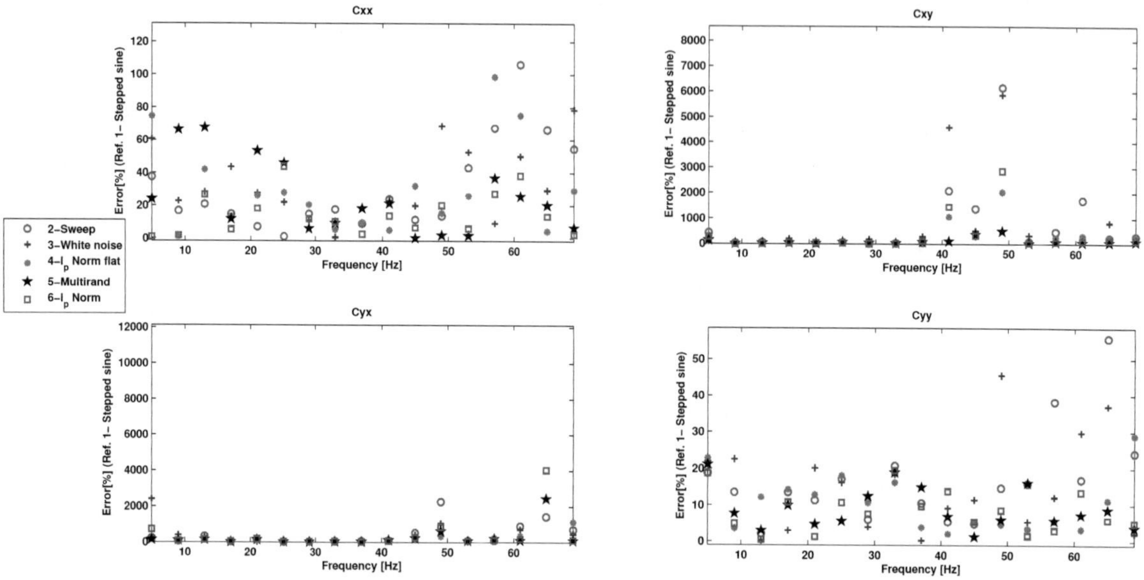

Figure 12: Relative error of the AMB estimated damping coefficients

Excitation signal	$\widehat{Kxx}$	$\widehat{Kxy}$	$\widehat{Kyx}$	$\widehat{Kyy}$	$\widehat{Cxx}$	$\widehat{Cxy}$	$\widehat{Cyx}$	$\widehat{Cyy}$
	%	%	%	%	%	%	%	%
2- Sweep	7.42	36.63	52.72	2.90	30.89	769.39	410.20	17.03
3- White noise	5.44	114.38	56.58	2.14	33.24	805.37	8446.3	15.27
4- Multisine (l_p) flat shape	3.29	19.80	37.46	1.29	29.69	292.95	1014.6	10.54
5- Multisine with random phases	3.37	17.70	21.63	1.31	24.46	92.24	259.81	9.06
6- Multisine (l_p) adapted shape	3.29	11.81	18.48	1.28	14.56	328.11	402.22	8.26

Table 4: Relative error deviation average for stiffness and damping coefficients, $N_b = 10$ Ref. 1- Stepped signal

total time experiment difference between the stepped and broadband excitation signal, however the lower the latter signal crest and time factor.

6 Conclusions and Future Works

Basically, the general purpose signals (sweep and white noise) performance were lower than multisines in the defined set of frequencies. In the case of white noise, it could be due to its energy distribution out of the frequencies of interest, thus yielding a reduction of applied power to the *desired* harmonics. In general, the best performed excitation signal for the tests carried out in this paper was the multisine with adapted spectrum amplitude and modified phases by using the Chebyshev method.

The direct terms of uncertainties and errors values were lower in the y axis (see Tables 3 and 4) than x axis. Such behavior may be caused by the gravity contribution, which can be represented as a vector parallel to the y axis.

The need of adapted amplitudes becomes more indispensable when the system under test spectrum has a low flatness (e.g. flexible shaft supported by an AMB system), namely, the signal should excite with lower amplitudes the frequencies located close to the system resonances.

As a suggestion, the excitation signal design starts with the band of frequencies definition, as well as its number of frequencies to cover such band. Then, the system spectrum amplitude is obtained in order to establish its shape. This procedure can be achieved by using a general purpose broadband signal, and computing the FRF magnitude of the excitation signal to the system output (it should not use more than two blocks). Finally, a reduction of the crest factor is made by means of algorithms as the presented in this paper.

Since the spectrum shape was obtained manually before applying the crest factor reduction, the authors are currently working on a new strategy aiming to include this step in an automatized way. Furthermore, future experiments are intended to compare the number of excited frequencies N_f and blocks N_b, which are strongly related to the total experiment time and signal excitation quality (crest and time factors).

References

[1] Jefferson Coelho. Aspectos mecânicos de sistemas de mancais magnéticos. M.sc. dissertation, Universidade Federal do Rio de Janeiro, Rio de Janeiro, RJ, Brasil, 2016.

[2] Diego Diaz, Fernando Pinto, Thiago Ritto, and David Maldonado. Stepped sine and multisine signal excitation for identification in a small amb test rig. In *Proceedings of the 24th International Congress of Mechanical Engineering*, page 9, Curitiba, Brazil, 2017.

[3] Diego Diaz, Fernando Pinto, Thiago Ritto, David Maldonado, and Vinicius Côrtes. Nonparametric identification of a small amb test rig at several rotating speeds. In *Proceedings of the 16th International Symposium on Magnetic Bearings (ISMB 16)*, page 7, Beijing, China, 2018.

[4] Diego Diaz, Fernando Pinto, Thiago Ritto, David Maldonado, and Vinicius Côrtes. Comparison of different excitation signals in a rotordynamic system. In *Proceedings of the 25th International Congress of Mechanical Engineering COBEM*, page 10, Uberlandia, Brazil, 2019.

[5] Patrick Guillaume. Frequency response measurements of multivariable systems using nonlinear averaging techniques. *IEEE Transactions on Instrumentation and Measurement*, 47:796–800, 1998.

[6] Patrick Guillaume, Johan Schoukens, Rik Pintelon, and Istvin Kollar. Crest-factor minimization using nonlinear Chebyshev approximation methods. *IEEE Transactions on Instrumentation and Measurement*, 40(6):982–989, 1991.

[7] Katja Hynynen. *Broadband Excitation in the System Identification of Active Magnetic Bearing Rotor Systems.* Ph.D. thesis, Lappeenranta University of Technology, Lappeenranta, Finland, 2011.

[8] Shahbaz Khader, Bin Liu, and Johan Sjöberg. System identification of active magnetic bearing for commissioning. In *Proceedings of International Conference on Modelling, Identification and Control*, pages 289–294, Melbourne, Australia, 2014.

[9] Lennare Ljung. *System identification: Theory for User*. Prentice Hall, 1987.

[10] Eric Maslen. *Magnetic Bearings*. University of Virginia, UVA. Lecture Notes, 2000.

[11] A Muszynska. *Rotordynamics*. Mechanical Engineering. CRC Press, 2005.

[12] J Norton. *An Introduction to Identification*. Dover publications, 2009.

[13] Rik Pintelon and Johan Schoukens. Measurement of frequency response functions using periodic excitations, corrupted by correlated input/output errors. *IEEE Transactions on Instrumentation and Measurement*, 50:1753–1760, 2001.

[14] Rik Pintelon and Johan Schoukens. *System Identification. A Frequency Domain Approach*. Wiley, 2012.

[15] Johan Schoukens, Rik Pintelon, and Yves Rolain. Broadband versus stepped sine frf measurements. *IEEE Transactions on Instrumentation and Measurement*, 49:275–278, 2000.

[16] Manfred Schroeder. Synthesis of low-peak-factor signals and binary sequences with low autocorrelation (Corresp.). *IEEE Transactions on Information Theory*, 16(1):85–89, 1970.

[17] Gerhard Schweitzer and Eric Maslen. *Magnetic Bearings. Theory, Design and Application to Rotating Machinery*. Springer, 2009.

[18] Renan Siqueira. Projeto e implementação de um mancal magnético ativo com controle por modos deslizantes. M.sc. dissertation, Universidade Federal do Rio de Janeiro, Rio de Janeiro, RJ, Brasil, 2013.

[19] O. Turkay and A.G. Ulsoy. Frequency versus time domain parameter estimation: Application to a slot milling operation. *Mechanical Systems and Signal Processing*, 2(3):265–277, 1988.

[20] Giuseppe Vannini, Stefano Cioncolini, Vincenzo Calicchio, and Francesco Tedone. Development of a high pressure rotordynamic test rig for centrifugal compressors internal seals characterization. In *Proceedings of the Fortieth Turbomachinery Symposium*, pages 46–59, Houston, USA, 2011.

[21] Peter Verboben. *Frequency-Domain System Identification for Modal Analysis*. D.sc. thesis, Vrije Universiteit Brussel, Buxelas, Belgium, 2002.

[22] Jouni Vuojolainen, Niko Nevaranta, Rafal Jastrzebski, and Olli Pyrhönen. Comparison of excitation signals in active magnetic bearing system identification. *Modeling, Identification and Control (MIC)*, 38:11, 2017.

[23] Toshio Yamamoto and Yukio Ishida. *Linear and Nonlinear Rotordynamics: A Modern Treatment with Applications*. Wiley, 2001.

Design Considerations for 2-Level Bearingless Homopolar Motors

Tomislav Strinić[1,2]*, Wolfgang Gruber[1,3]†, and Fadil Omeragic[1,4]‡

[1] Institute of Electrical Drives and Power Electronics,
Johannes Kepler University Linz
4040 Linz, Austria
[2] tomislav.strinic@jku.at
[3] wolfgang.gruber@jku.at
[4] fadil.omeragic@jku.at

Abstract

The simple rotor structure of 2-level bearingless homopolar motors provides a high mechanical and thermal robustness due to its saliency-based design without permanent magnets. The lack of permanent magnet materials in the rotor significantly reduces costs, but results in lower torque densities. To provide a qualitive overview of viable motor topologies for different stator and rotor pole combinations, this paper provides force and torque performance factors for 2-level bearingless homopolar motors with combined windings. The effects of different tangential displacement angles between the top and the bottom rotor level on the performance factors are investigated and numerically evaluated. Additionally, measurement results of a selected prototype are presented in the bearingless operation.

1 Introduction

Recent developments in the field of electric drives have led to the development of so-called bearingless motors, which are essentially electric motors with magnetic bearings combined into one unit [1]. The magnetic forces for the rotor suspension in such a motor are generated within the motor itself and not in separate magnetic bearings. Nowadays, there is a bearingless counterpart for almost every type of an electric motor [2]. Permanent magnet synchronous machines (PMSM) are widely used for this purpose due to their high power density. Such machines can be designed with either a heteropolar or homopolar permanent magnet air gap field.

The bearingless homopolar motor presented in [3] has an intriguingly simple rotor design, since it is designed as a reluctance rotor without any permanent magnet. This results in a high mechanical and termal robustness. Although the overall cost can be dramatically reduced by eliminating the magnetic material from the rotor, such a design results in lower torque densities.

The motor topology of the bearingless 2-level homopolar reluctance motor introduced in [4] and shown in Figure 1 consists of a stator with a cylindrical permanent magnet in the middle and iron stator teeth, that is, salient poles with concentric coils at each end (top and bottom). The top and bottom teeth arrangement is electrically displaced by 180° for an improved torque generation. This type of a motor can be built either with separate windings in which some phases are dedicated only for force generation, while the remaining phases are used for torque

*Performed measurements, processed the data, reviewed the literature, and wrote the manuscript.

†Administrated, supervised, and guided the project, reviewed and edited the manuscript.

‡Designed, optimized, and built the motor, implemented the control, performed measurements, and wrote the digest.

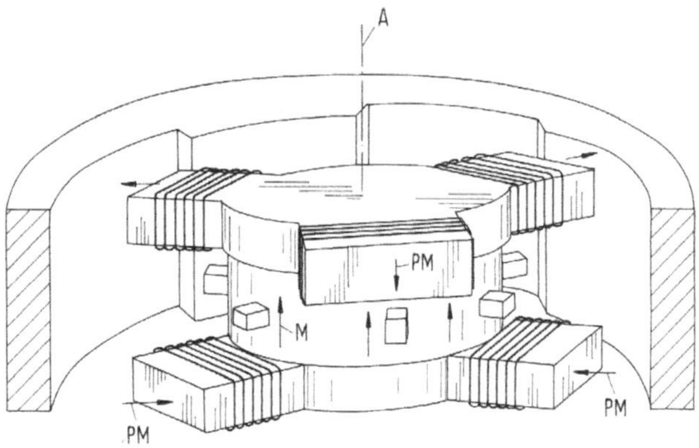

Figure 1: The bearingless 2-level homopolar reluctance motor presented in [4].

generation, or with combined windings in which each phase is used for both force as well as torque generation simultaneously. Combined windings result in reduced losses, but their design and control is more complex [5]. Hence, this paper investigates bearingless 2-level homopolar reluctance motor topologies equipped with combined windings.

Normalized performance factors are used for a qualitative evaluation of torque and force generation based on the force and torque single-phase characteristics introduced in [3]. High performance factor values indicate favorable designs, whereas small values indicate unfavorable designs. This paper therefore presents analytically obtained performance factors calculated for different electrical displacement angles between the top and the bottom rotor teeth of bearingless 2-level homopolar motors that have the top and the bottom stator teeth electrically displaced by 180° relative to each other. The obtained performance factors provide an insight into viable motor topologies for different stator and rotor pole combinations. Additionally, measured force and torque single-phase characteristics of one selected motor design are presented along with measurement results of reference position step responses and disturbance force position responses in the bearingless operation.

2 Mathematical Modeling and Control Scheme

The basis for the analysis, optimization, and control of a bearingless motor is given by its mathematical model. Thus, a sufficiently accurate mathematical description of electro-mechanical relationships is needed for constructing such a model.

2.1 Force and Torque Generation

Among different methods of modeling force and torque generation, methods based on the virtual energy and Maxwell's stress tensor are most frequently used [6]. According to [7], radial forces F_{x} and F_{y} that act on the rotor of a magnetically non-saturated electrical machine in the x- and y-direction, respectively, together with the developed torque T_{z} can generally be obtained from the rotor position $\mathbf{x}_\mathrm{r} = \begin{bmatrix} x_\mathrm{r} & y_\mathrm{r} & \varphi_\mathrm{r} \end{bmatrix}^\mathsf{T}$ and the stator phase currents $\mathbf{i}_\mathrm{s} = \begin{bmatrix} i_{\mathrm{s}1} & i_{\mathrm{s}2} & \dots & i_{\mathrm{s}m} \end{bmatrix}^\mathsf{T}$ as

$$\begin{bmatrix} F_\mathrm{x} \\ F_\mathrm{y} \\ T_\mathrm{z} \end{bmatrix} = \begin{bmatrix} \mathbf{i}_\mathrm{s}^\mathsf{T} & \mathbf{0}_{1\times m} & \mathbf{0}_{1\times m} \\ \mathbf{0}_{1\times m} & \mathbf{i}_\mathrm{s}^\mathsf{T} & \mathbf{0}_{1\times m} \\ \mathbf{0}_{1\times m} & \mathbf{0}_{1\times m} & \mathbf{i}_\mathrm{s}^\mathsf{T} \end{bmatrix} \mathbf{M}_\mathrm{Q}(\mathbf{x}_\mathrm{r})\, \mathbf{i}_\mathrm{s} + \mathbf{M}_\mathrm{L}(\mathbf{x}_\mathrm{r})\, \mathbf{i}_\mathrm{s} + \mathbf{M}_\mathrm{C}(\mathbf{x}_\mathrm{r}). \tag{1}$$

$\mathbf{M}_\mathrm{Q}$ in (1) describes the armature reaction and the phase current induced reluctance forces and torques, $\mathbf{M}_\mathrm{L}$ describes the interaction between the magnetic field and the stator currents, while $\mathbf{M}_\mathrm{C}$ describes the cogging forces and the cogging torque. Moreover, $\mathbf{M}_\mathrm{Q}$ is independent of the presence of permanent magnet materials and is typically exploited in reluctance motors and active magnetic bearings, while the term with $\mathbf{M}_\mathrm{L}$ is linearly dependent on the stator phase currents and hence refers to Lorentz forces often predominantly responsible for the torque generation in machines with permanent magnets. Additionally, $\mathbf{M}_\mathrm{C}$ is only significant for machines with permanent magnets [8].

For small rotor displacements x_r and y_r in the x- and y-direction, respectively, the system can be linearized using Taylor series as

$$\begin{bmatrix} F_\mathrm{x} \\ F_\mathrm{y} \\ T_\mathrm{z} \end{bmatrix} \approx \begin{bmatrix} \mathbf{i}_\mathrm{s}^\mathsf{T} & \mathbf{0}_{1\times m} & \mathbf{0}_{1\times m} \\ \mathbf{0}_{1\times m} & \mathbf{i}_\mathrm{s}^\mathsf{T} & \mathbf{0}_{1\times m} \\ \mathbf{0}_{1\times m} & \mathbf{0}_{1\times m} & \mathbf{i}_\mathrm{s}^\mathsf{T} \end{bmatrix} \mathbf{M}_\mathrm{Q}(\varphi_\mathrm{r})\, \mathbf{i}_\mathrm{s} + \mathbf{M}_\mathrm{L}(\varphi_\mathrm{r})\, \mathbf{i}_\mathrm{s} + \mathbf{M}_\mathrm{C}(\varphi_\mathrm{r}) + \mathbf{K}(\mathbf{i}_\mathrm{s}, \varphi_\mathrm{r})\, \mathbf{x}_\mathrm{r}, \tag{2}$$

where $\mathbf{x}_\mathrm{r} = \begin{bmatrix} x_\mathrm{r} & y_\mathrm{r} \end{bmatrix}^\mathsf{T}$, while the newly introduced matrix $\mathbf{K}$ is referred to as the stiffness matrix because it describes the negative radial stiffness and consequently the unstable radial behavior of the system. For that reason, a closed control loop is necessary to stabilize the system.

However, in the case of bearingless motors with an air gap field pre-magnetized by permanent magnets, the forces and the torque resulting from $\mathbf{M}_\mathrm{Q}$ are often negligible. In such a case, forces and torque can be generated by the stator coils by means of $\mathbf{M}_\mathrm{L}$, which is referred to as the current-force matrix and represented by $\mathbf{T}_\mathrm{m}(\varphi_\mathrm{r})$ in the unstable rest position $x_\mathrm{r} = y_\mathrm{r} = 0\,\mathrm{mm}$. Due to the linear dependency of force and torque generation on $\mathbf{i}_\mathrm{s}$, superposition of force and torque generating currents is feasible [9].

2.2 Structure of the Current-Force Matrix

For a stator with m phases, $\dim(\mathbf{T}_\mathrm{m}(\varphi_\mathrm{r})) = 3 \times m$, whereby $\mathbf{T}_\mathrm{m}(\varphi_\mathrm{r})$ can generally be written as

$$\mathbf{T}_\mathrm{m}(\varphi_\mathrm{r}) = \begin{bmatrix} T_{\mathrm{m},1,1}(\varphi_\mathrm{r}) & T_{\mathrm{m},2,1}(\varphi_\mathrm{r}) & \cdots & T_{\mathrm{m},m,1}(\varphi_\mathrm{r}) \\ T_{\mathrm{m},1,2}(\varphi_\mathrm{r}) & T_{\mathrm{m},2,2}(\varphi_\mathrm{r}) & \cdots & T_{\mathrm{m},m,2}(\varphi_\mathrm{r}) \\ T_{\mathrm{m},1,3}(\varphi_\mathrm{r}) & T_{\mathrm{m},2,3}(\varphi_\mathrm{r}) & \cdots & T_{\mathrm{m},m,3}(\varphi_\mathrm{r}) \end{bmatrix}. \tag{3}$$

In case of symmetrical stator and rotor geometries, the overall $\mathbf{T}_\mathrm{m}$ matrix can be calculated from the first column as

$$\begin{bmatrix} T_{\mathrm{m},n,1}(\varphi_\mathrm{r}) \\ T_{\mathrm{m},n,2}(\varphi_\mathrm{r}) \\ T_{\mathrm{m},n,3}(\varphi_\mathrm{r}) \end{bmatrix} = \begin{bmatrix} \cos(\nu) & -\sin(\nu) & 0 \\ \sin(\nu) & \cos(\nu) & 0 \\ 0 & 0 & 1 \end{bmatrix} \begin{bmatrix} T_{\mathrm{m},1,1}(\varphi_\mathrm{r} - p_\mathrm{z}\nu) \\ T_{\mathrm{m},1,2}(\varphi_\mathrm{r} - p_\mathrm{z}\nu) \\ T_{\mathrm{m},1,3}(\varphi_\mathrm{r} - p_\mathrm{z}\nu) \end{bmatrix}, \tag{4}$$

where $\nu = 2\pi(n-1)/m$ for $n \in \mathbb{N} \,|\, 1 < n \leq m$ and p_z represents the rotor pole pair number. The first column of $\mathbf{T}_\mathrm{m}$ constitutes the force and torque single-phase characteristics. If the x_r direction coincides with the magnetic axis of the first stator coil, then $T_{\mathrm{m},1,1}$ corresponds to F_x, $T_{\mathrm{m},1,2}$ to F_y, and $T_{\mathrm{m},1,3}$ represents T_z generated by that coil. In (4), the nth column of $\mathbf{T}_\mathrm{m}$ is calculated using a rotation matrix to account for the stator fixed force direction and an adapted rotor angle of each phase by assuming the mathematically positive direction of the coil sequence.

The force and torque single-phase characteristics can be determined by means of analytical models, finite element simulations, or measurements. The analytical calculation of $\mathbf{T}_{\mathrm{m}}$ drastically reduces simulation effort, while the only requirement is absence of magnetic saturation, so that the components of $\mathbf{i}_{\mathrm{s}}$ that generate the forces can be superimposed on those that generate the torque [9].

2.3 Control Scheme

The control of a bearingless motor requires properly decoupled components of $\mathbf{i}_{\mathrm{s}}$ responsible for generating F_{x}, F_{y}, and T_{z}. Those components can be obtained from

$$\begin{bmatrix} F_{\mathrm{x}} \\ F_{\mathrm{y}} \\ T_{\mathrm{z}} \end{bmatrix} = \mathbf{T}_{\mathrm{m}}(\varphi_{\mathrm{r}})\,\mathbf{i}_{\mathrm{s}}, \tag{5}$$

but $\mathbf{T}_{\mathrm{m}}$ needs to be inverted. The inverse of $\mathbf{T}_{\mathrm{m}}$ is denoted by $\mathbf{K}_{\mathrm{m}}$ in terms of which $\mathbf{i}_{\mathrm{s}}$ can be expressed as

$$\mathbf{i}_{\mathrm{s}} = \mathbf{K}_{\mathrm{m}}(\varphi_{\mathrm{r}}) \begin{bmatrix} F_{\mathrm{x}} \\ F_{\mathrm{y}} \\ T_{\mathrm{z}} \end{bmatrix}. \tag{6}$$

According to [7, 10], $\mathbf{K}_{\mathrm{m}}$ for the set of phase currents with minimal resistive power losses can be obtained as

$$\mathbf{K}_{\mathrm{m}}(\varphi_{\mathrm{r}}) = \mathbf{T}_{\mathrm{m}}^{\mathsf{T}}(\varphi_{\mathrm{r}}) \left(\mathbf{T}_{\mathrm{m}}(\varphi_{\mathrm{r}})\,\mathbf{T}_{\mathrm{m}}^{\mathsf{T}}(\varphi_{\mathrm{r}})\right)^{-1}, \tag{7}$$

from where it can be seen that $\mathbf{K}_{\mathrm{m}}$ is a Moore-Penrose inverse of $\mathbf{T}_{\mathrm{m}}$ because their product results in an identity matrix. In this way, the reference values of $\mathbf{i}_{\mathrm{s}}$ can be determined based on φ_{r} for certain reference values of F_{x} ($F_{\mathrm{x,ref}}$), F_{y} ($F_{\mathrm{y,ref}}$), and T_{z} ($T_{\mathrm{z,ref}}$), so that the rotor radial position (x_{r} and y_{r}) and the rotor angular speed ω_{r} can be separately controlled as shown in Figure 2. Hence, $\mathbf{K}_{\mathrm{m}}$ provides the reference values of $\mathbf{i}_{\mathrm{s}}$ used for the calculation of the error terms that are fed into the current controller in the inner control loop. The measured rotor position (x_{r} and y_{r}) as well as the measured rotor angle φ_{r} are used as feedback values to close the outer control loop. Furthermore, x_{r} and y_{r} are separately controlled by two independent PID controllers, while ω_{r} and $\mathbf{i}_{\mathrm{s}}$ are controlled by PI controllers [9].

3 Performance Comparison

From the forces and the torque generated by a single energized stator phase, the overall motor characteristics can be computed for cases when all stator phases are appropriately energized [3]. Based on that notion, performance factors can be defined with respect to the number of rotor poles and stator teeth to provide a way of evaluating different motor topologies.

3.1 Definition of Performance Factors

Performance factors are defined by comparing the forces and the torque generated by a single energized stator phase, which constitute the normalized force and torque single-phase characteristics shown in Figure 3 [9], with those generated by all stator phases appropriately energized

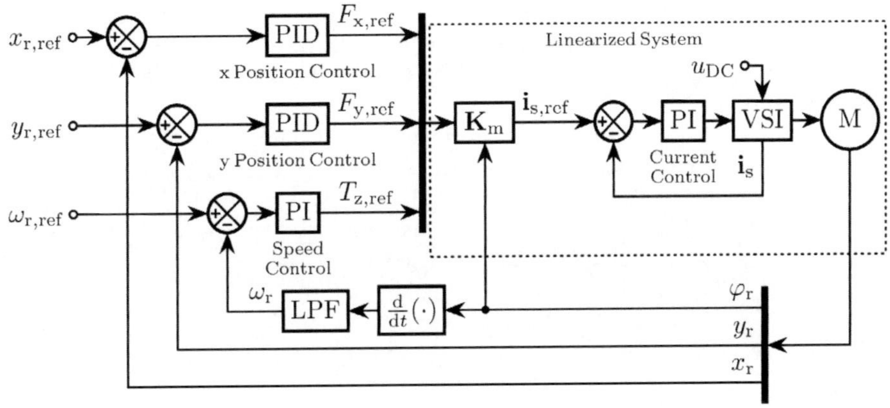

Figure 2: A simplified control block diagram.

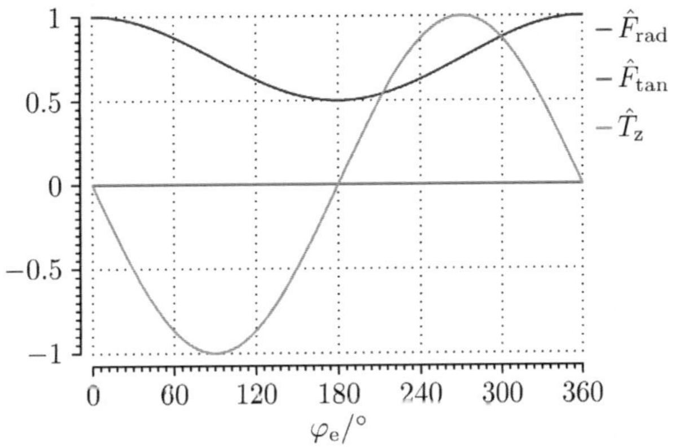

Figure 3: Normalized force and torque single-phase characteristics.

for the optimal force or torque generation. Those factors are of a great significance, especially for combined coil systems in which the load-bearing forces and the torque are jointly generated by the coils. The performance factors are based on the inverse of $\mathbf{T}_m$ (that is $\mathbf{K}_m$) obtained from the force and torque single-phase characteristics. They allow favorable motor topologies with different numbers of rotor and stator teeth to be found without a significant computational effort [7].

The force performance factor c_F describes the ability of a bearingless drive to generate forces and thus provides a comparison value between different machine topologies. It relates the smallest overall force $F_{overall,min}$ that can be generated by the motor over the rotor angle to the peak force of a single phase $F_{phase,max}$ as

$$c_F = \frac{2}{m_F}\frac{F_{overall,min}}{F_{phase,max}} = \frac{2}{m_F}\min_{j=1,2}\left(\frac{1}{\max\limits_{i,\varphi_r}|K_{m,i,j}(\varphi_r)|}\right), \tag{8}$$

where $2/m_F$ represents the normalization factor in which m_F stands for the force-generating stator phases, while $K_{m,i,j}$ represents the entry in the i-th row and j-th column of $\mathbf{K}_m$ [7, 9, 11].

In contrast to the torque generation, in this case an additional degree of freedom must be taken into account. In addition to the rotor rotation angle, the direction of the desired force effect also influences the generation of the load-bearing force. However, the expression for calculation of c_F given by (8) considers only F_x and F_y.

Similarly, the torque performance factor c_T correlates the smallest overall torque $T_{\text{overall,min}}$ that the motor can generate over the rotor angle with the peak torque of a single phase $T_{\text{phase,max}}$ as

$$c_T = \frac{2}{m_T}\frac{T_{\text{overall,min}}}{T_{\text{phase,max}}} = \frac{2}{m_T}\frac{1}{\max\limits_{i,\varphi_r}|K_{\text{m},i,3}(\varphi_r)|}, \tag{9}$$

where $2/m_T$ represents the normalization factor in which m_T stands for the torque-generating stator phases [7, 9].

3.2 Force and Torque Performance Factors for 2-Level Bearingless Homopolar Motors

Force and torque performance factors of a 2-level homopolar motor can be determined from the normalized force and torque single-phase characteristics shown in Figure 3, which represent first harmonic approximations. It can be seen that the normalized radial force $\hat{F}_{\text{rad}}$ is a cosine function of the electrical angle φ_e ($\varphi_e = p_z\varphi_r$) with an offset, whereas the normalized tangential force $\hat{F}_{\text{tan}}$ is zero, while the normalized motor torque $\hat{T}_z$ is a negative sine function of φ_e.

In contrast to the performance factors presented in [9], a stator with two levels is examined here. Additionally, the top and the bottom stator teeth are electrically displaced by 180°. The calculated force and torque performance factors listed in Table 1, Table 2, and Table 3 are calculated for the top and the bottom rotor teeth electrically displaced by 0°, 90°, and 180°, respectively, and provide an overview of viable motor topologies. The combinations of high performance factors reveals favorable topologies, whereas low performance factors indicate unfavorable designs. All combinations with $p_z = m$ can be excluded due to high cogging torques. In the observed case the choice of the motor topology is limited by the requirement of the motor to have six phases ($m = 6$) per each level. Thus, based on the required number of phases and according to the performance factors given in the presented tables, a combination with either eight ($p_z = 8$) or ten rotor teeth ($p_z = 10$) for six phases per each level is sensible. Since ten rotor teeth result in a higher electrical frequency than eight rotor teeth and a more powerful processor is therefore necessary for control, the rotor tooth number was chosen to be eight due to the good average motor characteristics in all three presented cases shown in Tables 1–3.

4 Prototype and Measurements

Although the following measurements were performed with tangentially aligned top and bottom rotor teeth, the rotor was built in a way to enable displacement of the top and the bottom rotor teeth as it can be seen in Figure 4.

4.1 Static Measurements

To verify the theoretical proposition of the force and torque single-phase characteristics, a static measurement was performed first by supplying a constant current linkage of 616 A-turns to a single stator phase of the bottom level of the motor. The resulting characteristics of the

	$m=4$	$m=5$	$m=6$	$m=7$	$m=8$
$p_z=4$	$c_F=0.6$ $c_T=0$	$c_F=0.6$ $c_T=1$	$c_F=0.6$ $c_T=0.9$	$c_F=0.6$ $c_T=1$	$c_F=0.6$ $c_T=1$
$p_z=5$	$c_F=0.6$ $c_T=1$	$c_F=0.6$ $c_T=0$	$c_F=0.6$ $c_T=1$	$c_F=0.6$ $c_T=1$	$c_F=0.7$ $c_T=1$
$p_z=6$	$c_F=0.7$ $c_T=1$	$c_F=0.6$ $c_T=1$	$c_F=0.6$ $c_T=0$	$c_F=0.6$ $c_T=1$	$c_F=0.6$ $c_T=1$
$p_z=7$	$c_F=0$ $c_T=0.1$	$c_F=0.6$ $c_T=1$	$c_F=0.6$ $c_T=1$	$c_F=0.6$ $c_T=0$	$c_F=0.6$ $c_T=1$
$p_z=8$	$c_F=0.5$ $c_T=0$	$c_F=0.7$ $c_T=1$	$c_F=0.6$ $c_T=0.9$	$c_F=0.6$ $c_T=1$	$c_F=0.6$ $c_T=0$
$p_z=9$	$c_F=0$ $c_T=0.1$	$c_F=0.1$ $c_T=0.1$	$c_F=0.6$ $c_T=1$	$c_F=0.6$ $c_T=1$	$c_F=0.6$ $c_T=1$
$p_z=10$	$c_F=0.7$ $c_T=1$	$c_F=0.5$ $c_T=0$	$c_F=0.7$ $c_T=1$	$c_F=0.6$ $c_T=1$	$c_F=0.5$ $c_T=0$
$p_z=11$	$c_F=0.6$ $c_T=1$	$c_F=0.1$ $c_T=0.1$	$c_F=0.1$ $c_T=0.1$	$c_F=0.6$ $c_T=0.9$	$c_F=0.6$ $c_T=1$
$p_z=12$	$c_F=0.6$ $c_T=0$	$c_F=0.7$ $c_T=1$	$c_F=0.5$ $c_T=0$	$c_F=0.7$ $c_T=1$	$c_F=0.6$ $c_T=1$

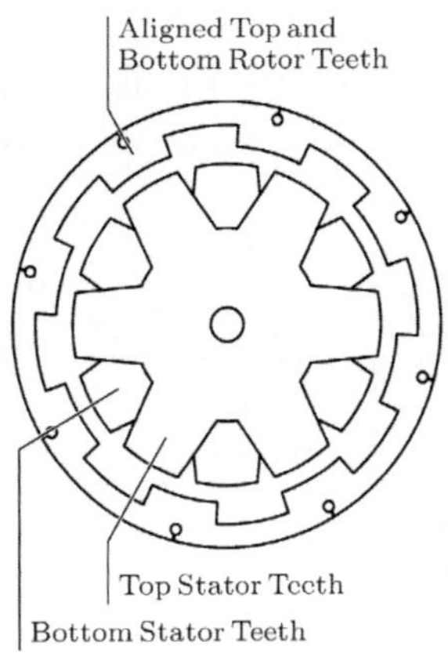

Table 1: Performance factors for the top and bottom rotor teeth electrically displaced by 0°.

	$m=4$	$m=5$	$m=6$	$m=7$	$m=8$
$p_z=4$	$c_F=0.5$ $c_T=1$	$c_F=0.4$ $c_T=0.6$	$c_F=0.6$ $c_T=1$	$c_F=0.6$ $c_T=0.9$	$c_F=0.6$ $c_T=0.1$
$p_z=5$	$c_F=0.4$ $c_T=0.7$	$c_F=0.6$ $c_T=1$	$c_F=0.4$ $c_T=0.7$	$c_F=0.6$ $c_T=1$	$c_F=0.6$ $c_T=1$
$p_z=6$	$c_F=0.7$ $c_T=0.1$	$c_F=0.4$ $c_T=0.6$	$c_F=0.5$ $c_T=1$	$c_F=0.4$ $c_T=0.7$	$c_F=0.6$ $c_T=1$
$p_z=7$	$c_F=0.4$ $c_T=0.6$	$c_F=0.6$ $c_T=0.9$	$c_F=0.4$ $c_T=0.7$	$c_F=0.5$ $c_T=1$	$c_F=0.4$ $c_T=0.7$
$p_z=8$	$c_F=0.5$ $c_T=1$	$c_F=0.6$ $c_T=1$	$c_F=0.6$ $c_T=1$	$c_F=0.4$ $c_T=0.7$	$c_F=0.5$ $c_T=1$
$p_z=9$	$c_F=0.4$ $c_T=0.7$	$c_F=0.4$ $c_T=0.6$	$c_F=0.6$ $c_T=0.1$	$c_F=0.6$ $c_T=1$	$c_F=0.4$ $c_T=0.7$
$p_z=10$	$c_F=0.7$ $c_T=0.1$	$c_F=0.5$ $c_T=1$	$c_F=0.6$ $c_T=0.9$	$c_F=0.6$ $c_T=1$	$c_F=0.6$ $c_T=1$
$p_z=11$	$c_F=0.4$ $c_T=0.7$	$c_F=0.4$ $c_T=0.7$	$c_F=0.4$ $c_T=0.7$	$c_F=0.6$ $c_T=0.9$	$c_F=0.6$ $c_T=1$
$p_z=12$	$c_F=0.5$ $c_T=1$	$c_F=0.6$ $c_T=1$	$c_F=0.5$ $c_T=1$	$c_F=0.6$ $c_T=1$	$c_F=0.6$ $c_T=0.1$

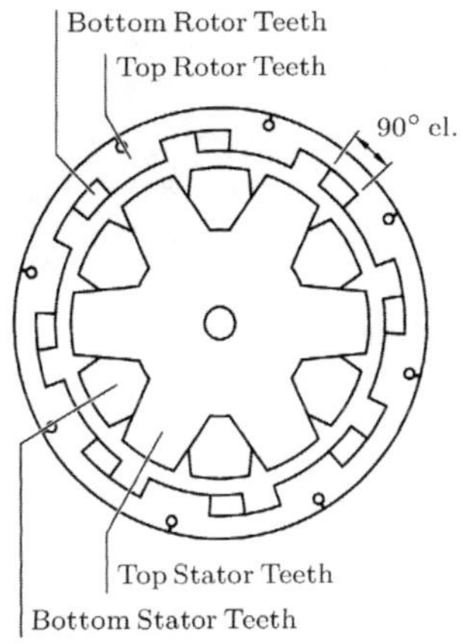

Table 2: Performance factors for the top and bottom rotor teeth electrically displaced by 90°.

	$m=4$	$m=5$	$m=6$	$m=7$	$m=8$
$p_z=4$	$c_F=0.6$ $c_T=1$	$c_F=0.6$ $c_T=1$	$c_F=0.6$ $c_T=0.9$	$c_F=0.6$ $c_T=1$	$c_F=0.6$ $c_T=1$
$p_z=5$	$c_F=0.6$ $c_T=1$	$c_F=0.6$ $c_T=0$	$c_F=0.6$ $c_T=1$	$c_F=0.6$ $c_T=1$	$c_F=0.6$ $c_T=1$
$p_z=6$	$c_F=0.7$ $c_T=1$	$c_F=0.6$ $c_T=1$	$c_F=0.6$ $c_T=0.1$	$c_F=0.6$ $c_T=1$	$c_F=0.6$ $c_T=1$
$p_z=7$	$c_F=0$ $c_T=0.1$	$c_F=0.6$ $c_T=1$	$c_F=0.6$ $c_T=1$	$c_F=0.6$ $c_T=0.1$	$c_F=0.6$ $c_T=1$
$p_z=8$	$c_F=0.5$ $c_T=0$	$c_F=0.7$ $c_T=1$	$c_F=0.6$ $c_T=0.9$	$c_F=0.6$ $c_T=1$	$c_F=0.6$ $c_T=0$
$p_z=9$	$c_F=0$ $c_T=0.1$	$c_F=0.1$ $c_T=0.1$	$c_F=0.6$ $c_T=1$	$c_F=0.6$ $c_T=1$	$c_F=0.6$ $c_T=1$
$p_z=10$	$c_F=0.7$ $c_T=1$	$c_F=0.5$ $c_T=0$	$c_F=0.7$ $c_T=1$	$c_F=0.6$ $c_T=1$	$c_F=0.6$ $c_T=1$
$p_z=11$	$c_F=0.6$ $c_T=1$	$c_F=0.1$ $c_T=0.1$	$c_F=0.1$ $c_T=0.1$	$c_F=0.6$ $c_T=0.9$	$c_F=0.6$ $c_T=1$
$p_z=12$	$c_F=0.6$ $c_T=0$	$c_F=0.7$ $c_T=1$	$c_F=0.5$ $c_T=0$	$c_F=0.7$ $c_T=1$	$c_F=0.6$ $c_T=1$

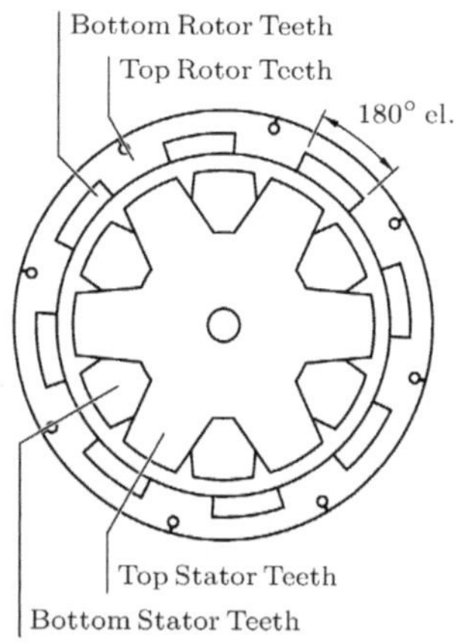

Table 3: Performance factors for the top and bottom rotor teeth electrically displaced by 180°.

Figure 4: The top stator windings (top left), the stator with the top windings (bottom left), the laminated rotor teeth with the pressed-in aluminum ring (top center), the rotor center piece (bottom center), and the final motor setup (right).

radial force F_{rad} and the tangential force F_{tan} shown with the characteristic of the torque T_{z} in Figure 5, aside from not being normalized, closely match the corresponding waveforms presented in Figure 3. Since oscillations in F_{tan}, which is perpendicular to the magnetic axis of the energized stator coil, is expected to be less than 10% of F_{rad}, F_{tan} in Figure 3 is taken as zero [3]. Based on the maximum value of F_{rad} in Figure 5, which corresponds to $F_{\mathrm{phase,max}}$ in

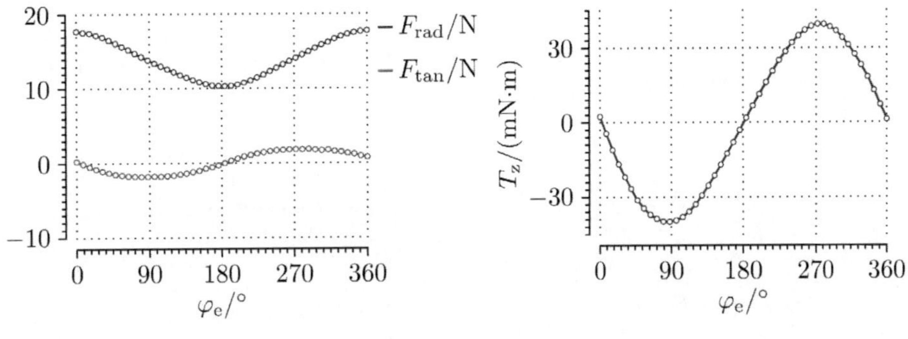

Figure 5: Measured force (left) and torque (right) single-phase characteristics.

(8), and the force performance factor highlighted in Table 1, $F_{overall,min}$ can be obtained as

$$F_{overall,min} = \frac{c_F m_F}{2} F_{phase,max}. \tag{10}$$

Similarly, $T_{overall,min}$ can be obtained based on the torque performance factor highlighted in Table 1 and the peak value of T_z, which corresponds to $T_{phase,max}$ in (9), as

$$T_{overall,min} = \frac{c_T m_T}{2} T_{phase,max}. \tag{11}$$

In both (10) and (11), $m_F = m_T = 12$ because the presented motor has twelve combined windings, that is, six windings on each level.

4.2 Measurements in the Bearingless Operation

Due to the unstable behavior of the rotor in the radial direction, active position and speed control are necessary for the operation as a bearingless drive. The control was implemented with a six-phase voltage source inverter (VSI) that has a DC-Link voltage of 350 V and delivers a maximum effective output phase current of 10 A. Since it has six half-bridges and the motor has twelve coils, two such inverters were required, one for the top and one for the bottom level. The control was designed under MATLAB® Simulink® [12] with X2C [13] rapid prototyping tool for code generation of real-time control algorithms for microprocessors. Since the motor is composed of two levels, the measurements for each level were carried out at a standstill but in the levitating state.

4.2.1 Reference Position Step Responses

To test the position control in both the x- and the y-direction, a step in the reference position of the rotor in the x-direction $x_{r,ref}$ from 0 mm to 0.05 mm was performed first, while keeping the reference position of the rotor in the y-direction $y_{r,ref}$ at 0 mm. Next, a step in $y_{r,ref}$ from 0 mm to 0.05 mm was performed, while keeping $x_{r,ref}$ at 0 mm. This procedure was done for each level of the motor and the results are shown in Figure 6, where the longer settling time of the top level is a result of lower dynamics of the PID controllers. In the observed case, 0 mm position in either direction represents the radially centered position of the rotor with respect to the stator, where the maximum radial displacement in any direction corresponds to the air gap and amounts to 0.5 mm. It can also be seen that a practically full decoupling of the x- and the y-direction can be achieved with the proposed control system shown in Figure 2.

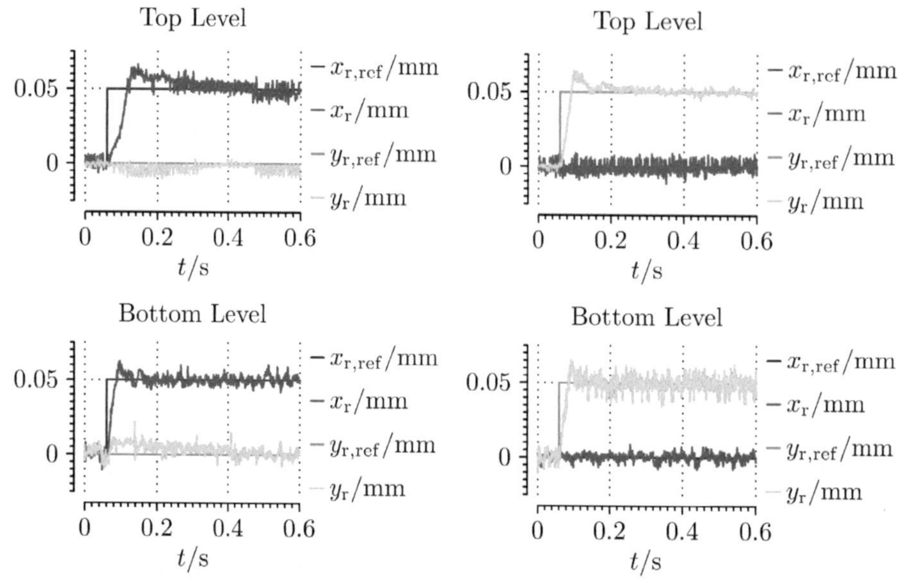

Figure 6: Reference position step responses.

4.2.2 Disturbance Force Position Responses

In order to examine the interference suppression capabilities of the position control, both $x_{r,ref}$ and $y_{r,ref}$ were kept at 0 mm and a step disturbance force with the magnitude of 3 N was first applied in the x- and then in the y-direction. The results are presented in Figure 7 for both the top and the bottom level. From those results it can be seen that although the rotor initially moves out of the center position in the direction of the disturbance force, the controllers quickly manage to compensate the disturbance so that the rotor returns back to the center.

5 Conclusion

This paper has presented normalized performance factors that provide a useful way of gaining an insight into the torque and radial force generation of 2-level bearingless homopolar motors [3]. The analytically obtained performance factors listed in Tables 1–3 are calculated for three different electrical displacement angles between the top and the bottom rotor teeth, with the top and the bottom stator teeth electrically displaced by 180°.

An optimal motor design was selected based on a combination of performance factors while taking into account the requirement of the motor to have six phases per each level and avoiding combinations in which the number of stator phases is equal to the number of the rotor pole pairs due to high cogging torques. Although the rotor of the presented motor was mechanically designed to enable electrical displacement of the top and the bottom rotor teeth by 0°, 90°, and 180°, the measurement results are presented for aligned top and bottom rotor teeth.

The characteristics of the measured force and torque single-phase characteristics presented in Figure 5 closely resembles the theoretical characteristics shown in Figure 3 and thus can be used for calculating the minimum overall force and torque as well as for generating the current-force matrix ($\mathbf{T}_m$) and consequently its inverse ($\mathbf{K}_m$) used in the control scheme of the motor. The

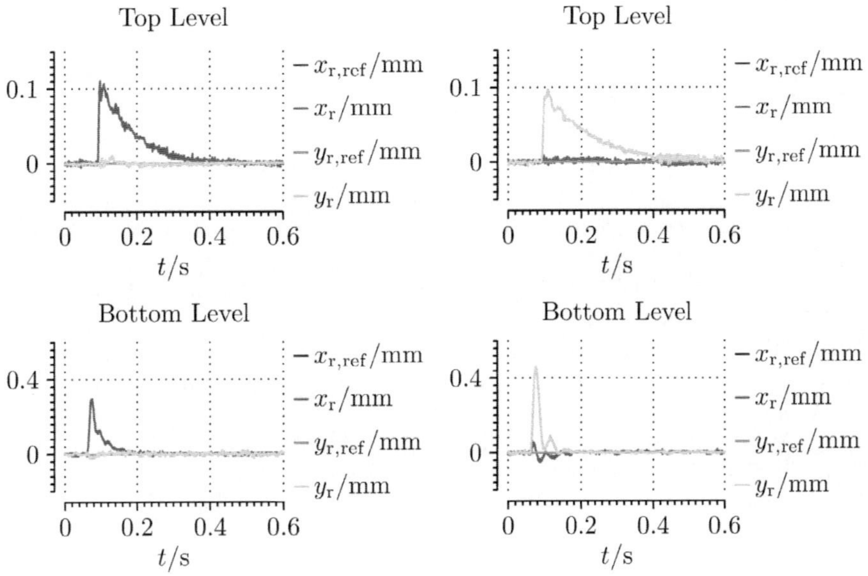

Figure 7: Disturbance force position responses.

measurement results of reference position step responses, disturbance force position responses in the bearingless operation show that a practically fully decoupled control of the x- and the y-direction is achieved with the linearization of the system shown in Figure 2.

6 Acknowledgments

This work was supported by the COMET-K2 "Center for Symbiotic Mechatronics" of the Linz Center of Mechatronics (LCM) funded by the Austrian Federal Government and the Federal State of Upper Austria.

References

[1] J. Bichsel, "The bearingless electrical machine," in *Proc. of the International Symposium on Magnetic Suspension Technology (ISMST).* NASA, Langley Research Center: NASA Conference Publication 3152, 1992, pp. 561–574.

[2] A. Chiba, T. Fukao, O. Ichikawa, M. Oshima, M. Takemoto, and D. Dorrell, *Magnetic Bearings and Bearingless Drives.* Elsevier Ltd, 2005.

[3] W. Gruber, M. Rothböck, and R. T. Schöb, "Design of a novel homopolar bearingless slice motor with reluctance rotor," *IEEE Transactions on Industry Applications*, vol. 51, no. 2, pp. 1456–1464, 2015, DOI: 10.1109/TIA.2014.2341739.

[4] R. Schöb, J. Hugel, and T. Holenstein, "Electromagnetic rotary drive," February 2017. [Online]. Available: https://patents.google.com/patent/EP3115616A1/en

[5] R. Klaus, J. W. Kolar, and T. Nussbaumer, "Comparison of winding concepts for bearingless pumps," in *Proc. of the 7th International Conference on Power Electronics (ICPE)*, 2007, pp. 1013–1020, DOI: 10.1109/ICPE.2007.4692535.

[6] D. J. Griffiths, *Introduction to Electrodynamics (3rd Edition).* Pearson Benjamin Cummings, 2008.

[7] H. Grabner, W. Amrhein, S. Silber, and W. Gruber, "Nonlinear feedback control of a bearingless brushless dc motor," *IEEE/ASME Transactions on Mechatronics*, vol. 15, no. 1, pp. 40–47, 2010, DOI: 10.1109/TMECH.2009.2014058.

[8] W. Gruber, W. Amrhein, S. Silber, H. Grabner, and M. Reisinger, "Theoretical analysis of force and torque calculation in magnetic bearing systems with circular airgap," in *Proc. of the 8th International Symposium on Magnetic Suspension Technology (ISMST8)*, Dresden, Germany, 2005, pp. 167–171.

[9] W. Gruber, *Bearingless Slice Motor Systems without Permanent Magnetic Rotors.* TRAUNER Verlag + Buchservice GmbH, 2017.

[10] H. Grabner, S. Silber, and W. Amrhein, "Bearingless torque motor – modeling and control," in *Proc. of 13th International Symposium on Magnetic Bearings 2012 (ISMB13)*, Arlington, Virginia, USA, 2012, pp. 1–13.

[11] W. Gruber, W. Briewasser, M. Rothböck, and R. T. Schöb, "Bearingless slice motor concepts without permanent magnets in the rotor," in *Proc. of the 2013 IEEE International Conference on Industrial Technology (ICIT)*, Cape Town, South Africa, 2013, pp. 259–265.

[12] The MathWorks Inc., "MATLAB® Simulink®." [Online]. Available: https://www.mathworks.com/products/simulink.html

[13] Linz Center of Mechatronics GmbH, "X2C." [Online]. Available: https://x2c.lcm.at/

Comparison Between NdFeB and SmCo Permanent Magnets Regarding Eddy Currents in Bearingless PM Synchronous Machines

Daniel Dietz and Andreas Binder

Institute for Electrical Energy Conversion, Technical University of Darmstadt, Germany
ddietz@ew.tu-darmstadt.de

Abstract

PM synchronous machines, used as high-speed drives, typically employ SmCo magnets instead of NdFeB due to the possibly high rotor temperature and the low air gap flux density. SmCo magnets exhibit a lower remanence and a higher electric conductivity than NdFeB material, leading to higher eddy current losses. Additionally, in bearingless PM machines, the asynchronously rotating suspension winding air gap field wave induces eddy currents in the rotor. These lead to a damping and deflection of the air gap field and, thus, to a disturbing rotor force, which may be crucial for the rotor position control.
We discuss the occurrence of the eddy current-originated disturbing force, present as an force error angle, for three PM materials: $SmCo_5$, Sm_2Co_{17} and NdFeB. By measurements on two geometrically identical rotors in a prototype machine, employing a Sm_2Co_{17}- and a NdFeB-PM in the rotor, we show that the force error angle is considerably bigger for the Sm_2Co_{17}-rotor. Still, for the considered small machine dimensions the error angle is small enough to not endanger the rotor position control stability.

1 Introduction

For high-speed machines, the permanent magnet (PM) material is not restricted to NdFeB material such as for highly utilized drives, since for the iron loss reduction low air gap flux

Table 1: Electromagnetic PM material properties (if not designated: data from [1, 2])

		$SmCo_5$	Sm_2Co_{17}	NdFeB
Remanence flux density B_{rem} / T	20 °C	0.85 ... 1.01	0.97 ... 1.12	1.09 ... 1.41
	100 °C	0.823 ... 0.98	0.947 ... 1.09	0.99 ... 1.28
	used[2]	0.947	0.947	1.096
Temperature coefficient[3] $\alpha_{th,rem}$ / $\% \cdot K^{-1}$		−0.04	−0.03	−0.114
Relative permeability $\mu_{r,PM}$		1.025 ... 1.07	1 ... 1.22	0.98 ... 1.145
	used[2]	1.056	1.056	1.068
Electric conductivity κ_{PM} / $MS \cdot m^{-1}$	20 °C	1.67 ... 2[1]	1.18 ... 1.33[1]	0.63 ... 0.71[1]
	used[2]	1.8	1.25	0.8

[1]) According to IEC Standard 60404-8-1[3]; [2]) used value for the analysis; [3]) regarding the remanence flux density at $B_{rem} = 20$ C°

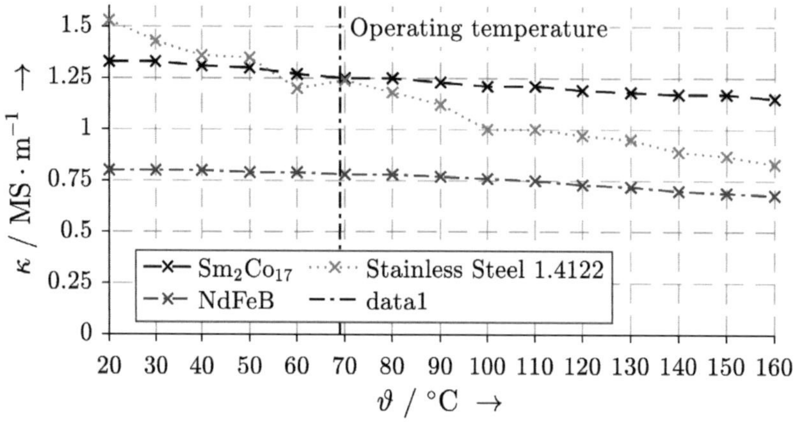

Figure 1: Measured electric conductivity of different PM materials and the shaft material for varying temperature

density values (typically $B < 0.6$ T) are employed. Thus, $SmCo_5$ and Sm_2Co_{17} are feasible alternatives. Also for bearingless motors (BM), which are currently in the focus of research [4], SmCo PMs are used. Prototypes of this topology are realized with high speed and high power values up to 100000 min^{-1} [5] and 60 kW [6]. Subsequently the term B is defined to be the air gap flux density in the radial middle of the mechanical air gap.
The electromagnetic property, such as the remanence flux density B_{rcm} and the electric conductivity κ_{PM}, vary much, depending on the manufacturer and the chemical composition. Table 1

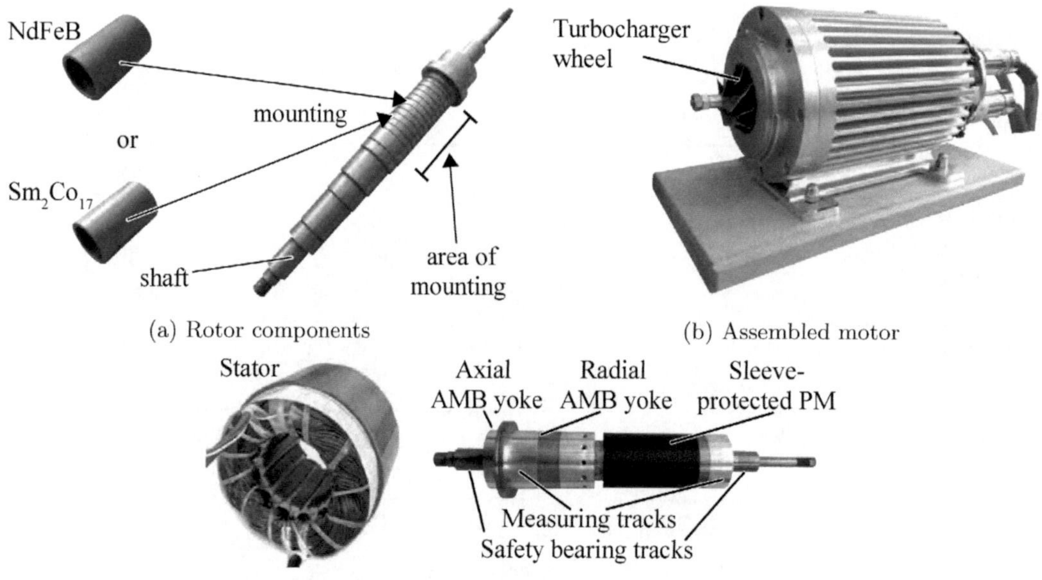

(a) Rotor components

(b) Assembled motor

(c) Wound stator and assembled rotor

Figure 2: Built 1 kW / 60000 min^{-1} bearingless PM synchronous machine

Table 2: Prototype machine properties ($P_N = 1$ kW, $n_N = 60000$ min^{-1})

Pole count $2p$ / Suspension winding pole count $2p_L$	2 / 4
Rated suspension force F_L (equal to gravitational force F_G) / N	5.3
Axial active length l_{Fe} / mm	40
Stator outer radius $r_{s,o}$ / stator inner radius $r_{s,i}$ / mm	37.5 / 17.5
Shaft radius r_{sh} / outer PM radius r_{PM} / mm	12.25 / 15
PM height h_{PM} / mm	2.75
Bandage height + mechanical air gap width: δ / mm	2.5
Shaft material: Stainless steel 1.4122, $\mu_{r,sh} = 100$	
PM material: two rotors, one with NdFeB and one with Sm_2Co_{17} (Table 1)	

shows this variety together with the used values for the analysis. The used remanence flux density B_{rem} is taken from the comparison between back-EMF measurements and 2D finite element simulations. The used conductivity values are taken from the measurement with demagnetized probes in Fig. 1, based on [7, 8].
In [2] a detailed analysis of a PM conductivity measurement is presented, showing the temperature dependence and anisotropy of the conductivity in and perpendicular to the magnetization direction (1). It reveals that the anisotropy has bigger impact on the conductivity (±20%) than the temperature variation (±5%) up to 150 °C. Unfortunately, it is not possible to model the anisotropy by 2-dimensional analytical or finite element simulations, since the decoupling of the vector potential components is not possible. However, eddy currents are mainly present in axial direction, so the electric conductivity perpendicular to the magnetization direction was measured and is used for the analysis.
The electric conductivity κ_{PM} determines the penetration depth $d_{E,PM}$ and thereby the eddy current distribution in the PM. High-speed machines and also the 1 kW / 60000 min^{-1} prototype machine (Fig. 2, Table 2) employ a solid PM ring for mechanical reasons, so that the eddy current loops are not intersected. Apart from the influence on the eddy current losses, the air gap field distortion by rotor eddy currents in BM leads to a suspension force error angle $\epsilon_{err,Ft}$ and to a reduction of the suspension force amplitude $G_{Ft,a,0}$. We discuss the occurrence of the disturbing force for different PM materials by calculation in Section 3 and by measurement in Section 4.

2 Air Gap Field Deformation by Eddy Currents

For the generation of a standstill, radial suspension force vector, acting on the rotor in a bearingless PM machine, e.g. to counteract the gravitational force, four air gap field waves (1)–(4) are necessary. The field waves are given in the stator reference frame with the circumference angle γ_m. The suspension winding is required to have a pole count of $2p_L = 2p \pm 2$ poles, where $2p$ is the rotor pole count. (1) and (3) are the radial and tangential components of the suspension winding air gap field fundamental B_L. The suspension winding is fed by a three-phase current system with synchronous frequency $f_s = n \cdot p$, where n is the rotational speed. Thus, the suspension winding fundamental synchronous velocity is $v_{s,L} = 2\tau_{p,L} \cdot f_s$, where $\tau_{p,L}$ is the suspension winding pole pitch.

(2) and (4) are the radial and tangential components of the rotor air gap field fundamental B_R with $2p$ poles. Its synchronous velocity is $v_\mathrm{s,R} = 2\tau_\mathrm{p} \cdot f_\mathrm{s} \neq v_\mathrm{s,L}$, where τ_p is the rotor pole pitch. v_s is different from $v_\mathrm{s,L}$ due to $2p_\mathrm{L} = 2p \pm 2$.

$$\begin{aligned} B_\mathrm{L,r}(r,\gamma_\mathrm{m},t) &= \mathrm{Re}\{\underline{B}_\mathrm{L,r}(r)\cdot e^{j\cdot p_\mathrm{L}\gamma_\mathrm{m}}\cdot e^{-j\cdot(p\cdot 2\pi\cdot n\cdot t+\varphi_\mathrm{L})}\} \\ &= |\underline{B}_\mathrm{L,r}(r)|\cdot\cos(p_\mathrm{L}\gamma_\mathrm{m} - p\cdot 2\pi\cdot n\cdot t - \varphi_\mathrm{L} - \Delta\gamma_\mathrm{Ft,r}) \end{aligned} \quad (1)$$

$$\begin{aligned} B_\mathrm{R,r}(r,\gamma_\mathrm{m},t) &= \mathrm{Re}\{\underline{B}_\mathrm{R,r}(r)\cdot e^{j\cdot p\gamma_\mathrm{m}}\cdot e^{-j\cdot(p\cdot 2\pi\cdot n\cdot t+\varphi_\mathrm{R})}\} \\ &= \hat{B}_\mathrm{R,r}(r)\cdot\cos(p\gamma_\mathrm{m} - p\cdot 2\pi\cdot n\cdot t - \varphi_\mathrm{R}) \end{aligned} \quad (2)$$

$$\begin{aligned} B_\mathrm{L,\gamma}(r,\gamma_\mathrm{m},t) &= \mathrm{Re}\{\underline{B}_\mathrm{L,\gamma}(r)\cdot e^{j\cdot p_\mathrm{L}\gamma_\mathrm{m}}\cdot e^{-j\cdot(p\cdot 2\pi\cdot n\cdot t+\varphi_\mathrm{L}+\frac{\pi}{2})}\} \\ &= |\underline{B}_\mathrm{L,\gamma}(r)|\cdot\sin(p_\mathrm{L}\gamma_\mathrm{m} - p\cdot 2\pi\cdot n\cdot t - \varphi_\mathrm{L} - \Delta\gamma_\mathrm{Ft,\gamma}) \end{aligned} \quad (3)$$

$$\begin{aligned} B_\mathrm{R,\gamma}(r,\gamma_\mathrm{m},t) &= \mathrm{Re}\{\underline{B}_\mathrm{R,\gamma}(r)\cdot e^{j\cdot p\gamma_\mathrm{m}}\cdot e^{-j\cdot(p\cdot 2\pi\cdot n\cdot t+\varphi_\mathrm{R}+\frac{\pi}{2})}\} \\ &= \hat{B}_\mathrm{R,p\gamma}(r)\cdot\sin(p\gamma_\mathrm{m} - p\cdot 2\pi\cdot n\cdot t - \varphi_\mathrm{R}) \end{aligned} \quad (4)$$

Due to the different synchronous velocity $v_\mathrm{s} \neq v_\mathrm{s,L}$, the asynchronously rotating suspension winding fundamental induces eddy currents in the rotor parts. The resulting suspension winding fundamental amplitude is damped and phase-shifted due to the eddy current reaction field. The damping of the field wave amplitudes is included in the absolute value $|\underline{B}|$ of the complex phasor $\underline{B} = |\underline{B}|\cdot e^{j\cdot\Delta\gamma_\mathrm{Ft}}$. The phase-shifting of the asynchronous field waves is considered by the angle $\Delta\gamma_\mathrm{Ft}$ of the complex phasor.
For the analytical magnetic field calculation, including rotor eddy currents, the formulation in rotor-fixed coordinates is needed. The circumference angle in rotor-fixed coordinates is $\gamma_\mathrm{R} = \gamma_\mathrm{m} - 2\pi\cdot n\cdot t$. The suspension air gap fundamental components in rotor-fixed coordinates are given in (5), (6).

$$B_\mathrm{L,r}(r,\gamma_\mathrm{R},t) = \mathrm{Re}\left\{\underline{B}_\mathrm{L,r}(r)\cdot e^{j\cdot p_\mathrm{L}\gamma_\mathrm{R}}\cdot e^{-j\cdot\left[\underbrace{(p-p_\mathrm{L})\cdot n}_{f_\mathrm{R}}\cdot 2\pi\cdot t+\varphi_\mathrm{L}\right]}\right\} \quad (5)$$

$$B_\mathrm{L,\gamma}(r,\gamma_\mathrm{R},t) = \mathrm{Re}\left\{\underline{B}_\mathrm{L,\gamma}(r)\cdot e^{j\cdot p_\mathrm{L}\gamma_\mathrm{R}}\cdot e^{-j\cdot\left[\underbrace{(p-p_\mathrm{L})\cdot n}_{f_\mathrm{R}}\cdot 2\pi\cdot t+\varphi_\mathrm{L}-\frac{\pi}{2}\right]}\right\} \quad (6)$$

(5), (6) contain the rotor frequency f_R, which is equal to the slip frequency. For the fundamental suspension winding air gap field wave, the rotor frequency is given in (7). It shows that f_R only depends on the rotor speed n for a standstill suspension force vector. Eddy currents are induced due to $f_\mathrm{R} = |-n|$ for topologies with $p_\mathrm{L} = p+1$.

$$f_\mathrm{R} = (p-p_\mathrm{L})\cdot n = f_\mathrm{s} - n\cdot(p\pm 1) = \begin{cases} -n & \text{for } p_\mathrm{L} = p+1 \\ n & \text{for } p_\mathrm{L} = p-1 \end{cases} \quad (7)$$

The rotor and suspension winding fundamental of a bearingless machine can be calculated 2-dimensionally (2D), either analytical [9, 10, 11] or by simulations based on the finite element method (FEM). The analytical calculation is based on a slotless model of concentric regions, whereas the FEM model includes the real, slotted, geometry. The calculation of the field waves is not in the scope of this paper. It was addressed in [12].

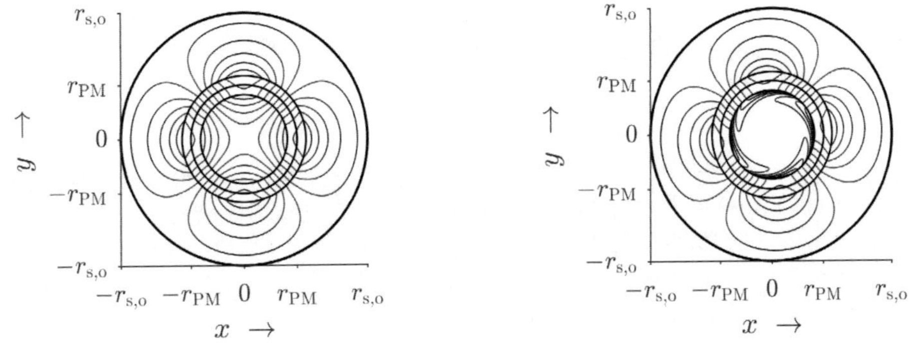

(a) Suspension field without eddy currents

(b) Suspension field with eddy currents (PM material: Sm_2Co_{17})

Figure 3: Analytically calculated magnetic field lines of the suspension winding field for the prototype (Table 2) at $n = 60000\ \mathrm{min}^{-1}$, $\left|\vec{F}_L\right| = 5.3\ \mathrm{N} = \left|\vec{F}_G\right|$

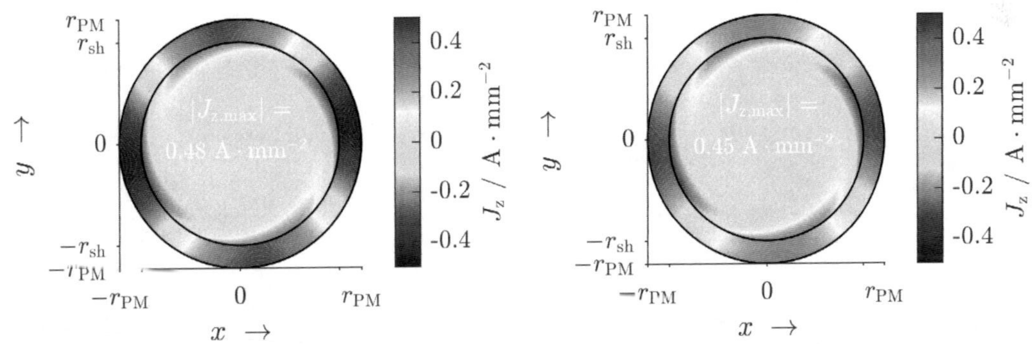

(a) Sm_2Co_{17}, $2p_L = 4$

(b) NdFeB, $2p_L = 4$

Figure 4: Analytically calculated current density J_z in the rotor parts of the prototype machine (Table 2) due to suspension field fundamental at $n = 60000\ \mathrm{min}^{-1}$, $\left|\vec{F}_L\right| = 5.3\ \mathrm{N} = \left|\vec{F}_G\right|$

Results are given here for the prototype machine (Fig. 2, Table 2). Fig. 3a (3b) shows the analytically calculated field lines of the suspension winding fundamental without (with) rotor eddy currents at rated speed $n = 60000\ \mathrm{min}^{-1}$ in case of Sm_2Co_{17} PM material. The eddy currents prevent the suspension winding field from entering the rotor shaft due to the small penetration depth of the shaft material ($d_{E,sh} < 0.3$ mm). Also they lead to a deflection of the air gap field (shift angle $\Delta\gamma_{Ft}$). The eddy current density in z-direction in the rotor parts is given in Fig. 4. The higher conductive Sm_2Co_{17}-material (Fig. 4a) leads to larger eddy current amplitudes in the permanent magnet than the NdFeB-material (Fig. 4b).

Table 3 gives a comparison of the fundamental air gap flux density components for the prototype machine, which are determined by 2D FEM simulation (*JMAG Designer 19.1*) and by the analytical calculation [12]. Since the iron parts of the prototype machine are not magnetically saturated, only the slot opening effect leads to deviations between analytical calculation and simulation, because it is not considered in the analytical calculation. The radial field quantities by analytical calculation overestimate the simulated solution.

Table 3: Comparison between analytically calculated and simulated air gap fundamental field wave amplitudes in the middle of the mechanical air gap (2D) for the prototype machine (Table 2) at $n = 60000\ \text{min}^{-1}$, $F_L = 5.3\ \text{N} = F_G$

	Without eddy currents		With eddy currents	
	Simulation	Analytical	Simulation	Analytical
$\hat{B}_{R,r}$ / mT	440.28	443.86 (+0.81%)	440.28	443.86 (+0.81%)
$\hat{B}_{R,\gamma}$ / mT	12.26	11.73 (−4.31 %)	12.26	11.73 (−4.31 %)
$\hat{B}_{L,r}$ / mT	8.48	8.80 (+4.53 %)	7.92	8.10 (+2.22 %)
$\hat{B}_{L,\gamma}$ / mT	4.98	5.03 (+0.95%)	5.02	5.08 (+1.12%)
$\Delta\gamma_{Ft,r}$	0	0	12.21°	12.88° (+5.54 %)
$\Delta\gamma_{Ft,\gamma}$	0	0	−1.08°	−1.06° (+1.85 %)

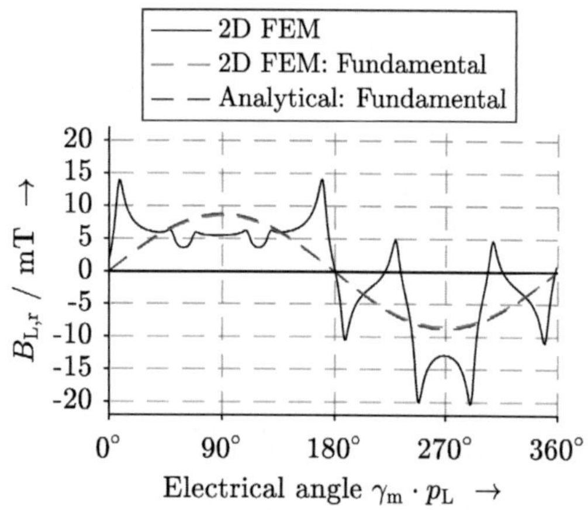

(a) Suspension winding radial air gap field without eddy currents

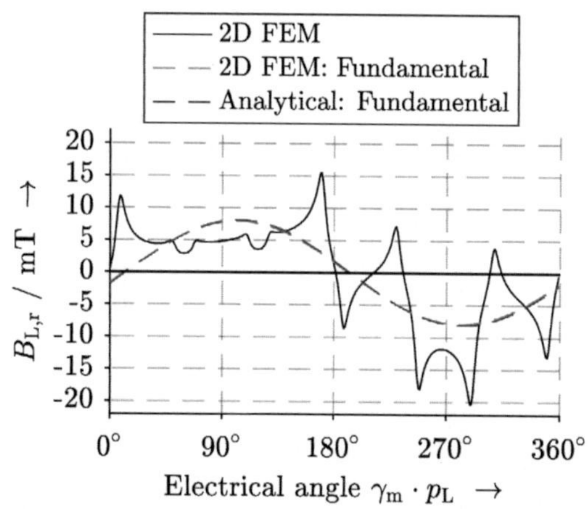

(b) Suspension winding radial air gap field with eddy currents ($\Delta\gamma_{Ft,r} \approx 12.5°$, Sm_2Co_{17})

Figure 5: Comparison between analytically calculated and simulated air gap flux density radial component for the prototype machine (Table 2) at $n = 60000\ \text{min}^{-1}$ in the middle of the mechanical air gap

There are many elaborate options to include the slot opening effect via permeance functions, such as [13, 14, 15]. The deviation between the analytical calculation and the simulation is $< 6\%$. Also the air gap field deformation by rotor eddy currents is well represented by the analytical calculation, so that the slot opening effect is generally neglected for the analytical calculations. Fig. 5 compares the spatial distribution of the radial air gap field waves to the determined field fundamentals from Table 3. By comparing Fig. 5a and 5b the effect of the rotor eddy currents becomes clearly visible as damping and shifting of the radial suspension field fundamental.

3 Eddy Current Influence on the Suspension Force

Due to the assumed material linearity for the analytical calculation, the overall radial and tangential field components B_r and B_γ are given by the superposition of the stator and rotor field waves. For the suspension force calculation, the 2-dimensional *Maxwell* stress tensor, including B_r and B_γ, in cylindrical coordinates (9) is integrated over a closed surface in the air gap at radius $r_{cal} = r_{s,i} - \delta/2$ (8). The forces are expressed in stator-fixed coordinates. Since we only consider a standstill force vector, the time-dependency is omitted here for clarity.

$$\begin{pmatrix} F_x \\ F_y \end{pmatrix} = \int_0^{l_{Fe}} \int_0^{2\pi} \begin{pmatrix} f_r(\gamma_m) \cdot \cos(\gamma_m) - f_\gamma(\gamma_m) \cdot \sin(\gamma_m) \\ f_r(\gamma_m) \cdot \sin(\gamma_m) + f_\gamma(\gamma_m) \cdot \cos(\gamma_m) \end{pmatrix} \cdot \mathrm{d}\gamma_m \mathrm{d}z \tag{8}$$

$$\left. \begin{pmatrix} f_r(\gamma_m) \\ f_\gamma(\gamma_m) \end{pmatrix} \right|_{r=r_{cal}} = \frac{1}{2 \cdot \mu_0} \cdot \begin{pmatrix} B_r^2(r_{cal}, \gamma_m) - B_\gamma^2(r_{cal}, \gamma_m) \\ 2 \cdot B_r(r_{cal}, \gamma_m) \cdot B_\gamma(r_{cal}, \gamma_m) \end{pmatrix} \tag{9}$$

Inserting the radial and tangential components of the rotor and the suspension winding air gap field waves, satisfying $p_L = p + 1$, from (3),(4) and (1),(2) into (8),(9) yields the expression for the standstill radial suspension force vector $\vec{F}_L = F_x \cdot \vec{e}_x + F_y \cdot \vec{e}_y$ (10), (11). The tangential component of the rotor air gap field is here neglected for clarity, since its contribution to the suspension force is $< 1\%$.

$$F_x = \frac{\pi \cdot r_{cal} \cdot l_{Fe}}{2 \cdot \mu_0} \cdot \hat{B}_{R,r}(r_{cal}) \cdot \left[\left| \underline{B}_{L,r}(r_{cal}) \right| \cdot \sin(\varphi_R - \varphi_L - \Delta\varphi_{Ft,r}) + \left| \underline{B}_{L,\gamma}(r_{cal}) \right| \cdot \sin(\varphi_R - \varphi_L - \Delta\varphi_{Ft,\gamma}) \right] \tag{10}$$

$$F_y = \frac{\pi \cdot r_{cal} \cdot l_{Fe}}{2 \cdot \mu_0} \cdot \hat{B}_{R,r}(r_{cal}) \cdot \left[\left| \underline{B}_{L,r}(r_{cal}) \right| \cdot \cos(\varphi_R - \varphi_L - \Delta\varphi_{Ft,r}) + \left| \underline{B}_{L,\gamma}(r_{cal}) \right| \cdot \cos(\varphi_R - \varphi_L - \Delta\varphi_{Ft,\gamma}) \right] \tag{11}$$

If no rotor eddy currents are present, (10), (11) can be simplified to (12), here only shown for the x-direction. In (12), $\varphi_{F,ref}$ is the angle, which determines the direction of the suspension force, e.g. $\varphi_{F,ref} = 0$ for $\vec{F}_L = F_y \cdot \vec{e}_y$.

$$F_x = \frac{\pi \cdot r_{cal} \cdot l_{Fe}}{2 \cdot \mu_0} \cdot \hat{B}_{R,r}(r_{cal}) \cdot \left[\hat{B}_{L,r}(r_{cal}) + \hat{B}_{L,\gamma}(r_{cal}) \right] \cdot \sin\Big(\underbrace{\varphi_R - \varphi_L}_{\varphi_{F,ref}} \Big) \tag{12}$$

According to (10), (11), the bearing force amplitude $\left|\vec{F}_L\right|$ is decreased ($\left|\vec{F}_L\right| < \left|\vec{F}_{L,ref}\right|$) by eddy currents, due to $\left|\underline{B}_{L,r}\right| < \hat{B}_{L,r}$ and the spatial alignment φ_F of the bearing force vector is different from the set-point bearing force vector ($\varphi_{F,ref} \neq \varphi_F$), due to $\Delta\gamma_{Ft,r} \neq 0$, $\Delta\gamma_{Ft,\gamma} \neq 0$. The occurring spatial error angle $\epsilon_{err,Ft}$ (Fig. 6) of a standstill suspension force vector $\vec{F}_{L,ref} = F_{y,ref} \cdot \vec{e}_y$ is given in (13), whereas (14) shows the decrease of the bearing force vector by the ratio $G_{Ft,a,0} = F_y/F_{y,ref}$. With a 2D analytical calculation ($2D_{an}$), with a 2D FEM simulation ($2D_{FEM}$) or with a 3D FEM simulation ($3D_{FEM}$), the resulting forces F_x and F_y are inserted into (13), (14) for the calculation of the rotor forces, accounting for the eddy

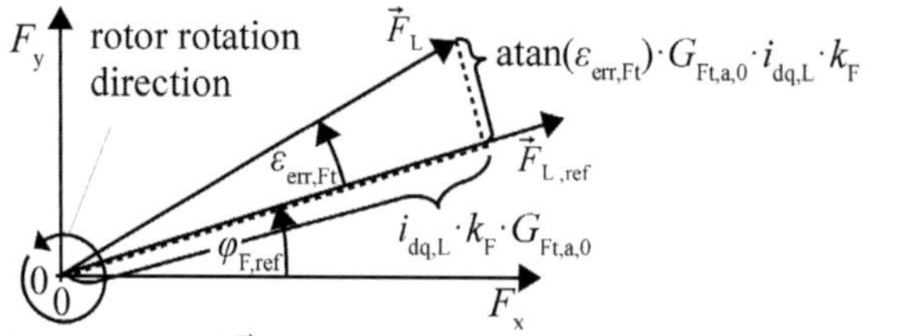

Figure 6: Suspension force vector ($\vec{F}_L$) deflection by $G_{Ft,a,0}$ and $\epsilon_{err,Ft}$ due to the eddy current effect ($i_{dq,L}$: suspension current, k_F: force-current coefficient without eddy currents)

current effect.

$$\epsilon_{err,Ft} = \operatorname{atan}\left(\frac{F_x}{F_y}\right) = \operatorname{atan}\left(\frac{|\underline{B}_{L,p_L,p,r}| \cdot \sin(\Delta\gamma_{Ft,r}) + |\underline{B}_{L,\gamma}| \cdot \sin(\Delta\gamma_{Ft,\gamma})}{|\underline{B}_{L,r}| \cdot \cos(\Delta\gamma_{Ft,r}) + |\underline{B}_{L,\gamma}| \cdot \cos(\Delta\gamma_{Ft,\gamma})}\right) \tag{13}$$

$$G_{Ft,a,0} = \frac{F_y}{F_{y,ref}} = \frac{\operatorname{Re}\left\{\underline{B}_{L,r}(r_{cal}) + \underline{B}_{L,\gamma}(r_{cal})\right\}}{\hat{B}_{L,r}(r_{cal}) + \hat{B}_{L,\gamma}(r_{cal})} \tag{14}$$

Fig. 7a shows that the suspension force amplitude decrease $G_{Ft,a,0}$ (14) decreases for rising PM electric conductivity κ_{PM}. Also the force error angle $\epsilon_{err,Ft}$ in Fig. 7b increases almost linear with rising PM electric conductivity κ_{PM} due to the resistance limitation of the eddy currents ($d_{E,PM} \approx 12\text{mm} \gg h_{PM} = 2.75$ mm). Thus, for non-segmented PM, NdFeB leads to the smallest suspension force degradation. The results are summarized in Table 4. Consequently, also the rotor eddy current losses $P_{d,Ft,R}$ are the smallest for the NdFeB material (Fig. 8). The big deviation between analytical calculation and FEM simulation is due to the neglect of the slot opening effect in the analytical model. It shows that the asynchronously rotating field waves due to the modulation of the rotor field by the slot openings account for approximately half of the total eddy current losses.

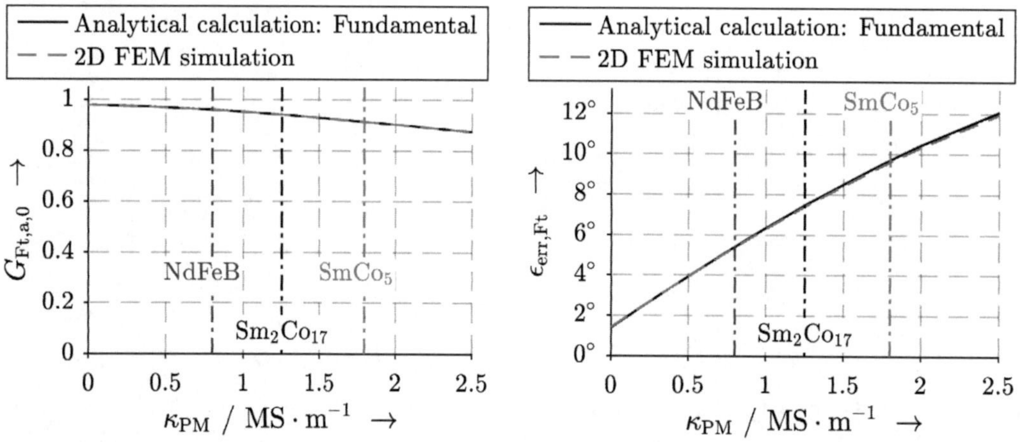

(a) Suspension force attenuation $G_{Ft,a,0}$

(b) Suspension force error angle $\epsilon_{err,Ft}$

Figure 7: Calculated and simulated suspension force amplitude attenuation $G_{Ft,a,0}$ and error angle $\epsilon_{err,Ft}$ for a standstill suspension force vector $\vec{F} = F_g \cdot \vec{e}_y$, depending on the electric PM conductivity κ_{PM}, used remanence flux density: $B_{rem} = 1$ T

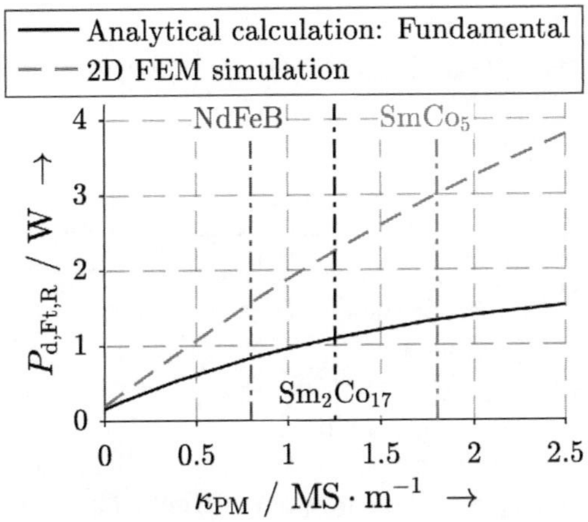

Figure 8: Calculated and simulated rotor eddy current losses $P_{d,Ft,R} = P_{d,Ft,PM} + P_{d,Ft,sh}$ depending on the electric PM conductivity κ_{PM}, used remanence flux density: $B_{rem} = 1$ T

Table 4: Suspension force attenuation $G_{Ft,a,0}$, error angle $\epsilon_{err,Ft}$ and rotor eddy current losses $P_{d,Ft,R}$ for a $SmCo_5$, a Sm_2Co_{17} and a NdFeB PM (prototype data from Table 2)

	$SmCo_5$			Sm_2Co_{17}			NdFeB		
	$G_{Ft,a,0}$	$\epsilon_{err,Ft}$	$P_{d,Ft,R}$	$G_{Ft,a,0}$	$\epsilon_{err,Ft}$	$P_{d,Ft,R}$	$G_{Ft,a,0}$	$\epsilon_{err,Ft}$	$P_{d,Ft,R}$
$2D_{an}$	0.91	9.6°	1.3 W	0.94	7.4°	1.1 W	0.96	5.4°	0.8 W
$2D_{FEM}$	0.92	9.5°	3.0 W	0.94	7.4°	2.3 W	0.96	5.4°	1.6 W

4 Identification of Eddy Current Disturbing Forces by Measurements

In order to identify the eddy current effect by measurement, an accurate determination of the rotor angle $\gamma_m = p \cdot 2\pi \cdot n$ is of high importance. Thus, the time span t_{cal} between current and rotor angle ($\gamma_{m,act}$) evaluation and PWM output must be compensated in the inverter by a feed-forward of the rotor angle ($\gamma_{m,cor}$) (15). The compensated rotor angle is $\tilde{\gamma}_m$.

$$\tilde{\gamma}_m = (\gamma_{m,act} + \gamma_{m,cor}), \bmod(2\pi) = (\gamma_{m,act} + t_{cal} \cdot \omega_s), \bmod(2\pi) \quad (15)$$

For the used inverter with a 150 MHz processor the calculation takes about 2500 cycles leading to $t_{cal} = 2500/150$ Hz $= 16.7$ ns. At $n = 60000$ min^{-1} this equals 1.7% of the electrical period, leading to an angle deviation of 6° at $n = 60000$ min^{-1}. The rotor is accelerated to half rated speed by speed control. Then the speed control is switched off so that the rotor slows down by no-load losses (zero drive current $i_{q,D} = 0$). In this situation the voltage drop in d_D-axis is zero and not speed dependent if the rotor angle is determined correctly. If this is not the case, t_{cal} has to be adapted in (15).
Additionally, the rotor angle determination for high-speed machines is prone to a non-linear sensor delay over the wide range of speed, here: $n = 0 \dots 60000$ min^{-1}. Despite the angle correction (15), the rotor angle $\tilde{\gamma}_m$ can differ from the real value γ_m. To analyze this error angle $\epsilon_{err,inv}$, a speed-down test by no-load losses (zero drive current $i_{q,D} = 0$) is conducted.

The error angle $\epsilon_{\mathrm{err,inv}}$ is calculated by (16). Fig. 9 shows the result for the prototype machine. For the accurate determination of $\epsilon_{\mathrm{err,Ft}}$, the angle $\epsilon_{\mathrm{err,inv}}$ must be considered as an offset (17). The big noise of $\epsilon_{\mathrm{err,inv}}$ in Fig. 9 at very low speed is due to the small voltage drop $u_{\mathrm{q,D}}$ in the q-axis of the drive voltage system.

$$\epsilon_{\mathrm{err,inv}} = \operatorname{atan}\left(\frac{u_{\mathrm{d,D}}\big|_{i_{\mathrm{q,D}}=i_{\mathrm{d,D}}=0}}{u_{\mathrm{q,D}}\big|_{i_{\mathrm{q,D}}=i_{\mathrm{d,D}}=0}}\right) \tag{16}$$

$$\gamma_{\mathrm{m}} = \tilde{\gamma}_{\mathrm{m}} - \epsilon_{\mathrm{err,inv}} \tag{17}$$

The suspension currents $i_{\mathrm{d,L}}$ and $i_{\mathrm{q,L}}$ give information about the radial bearing force vector $\vec{F}_{\mathrm{L}}$ which points in opposite direction to the gravitational force $\vec{F}_{\mathrm{G}}$ at stand still ($\vec{F}_{\mathrm{L}} = F_{\mathrm{y}} \cdot \vec{e}_{\mathrm{y}}$). As speed increases, the disturbing orthogonal force $F_{\mathrm{x}} \propto \epsilon_{\mathrm{err,Ft}}$ increases due to the eddy current effect ($f_{\mathrm{R}} = |n|$) and so the orthogonal suspension current $i_{\mathrm{d,L}}$ increases. In order to exclude the disturbing force which is caused by the interference of the drive winding air gap field with the rotor field [16], the rotor is accelerated to rated speed by speed control and slows down by no-load losses with $i_{\mathrm{q,D}} = i_{\mathrm{d,D}} = 0$.
The suspension currents ($i_{\mathrm{d,L}}$, associated to a force in x-direction, $i_{\mathrm{q,L}}$, associated to a force in y-direction) give access to the radial forces on the rotor at rotation. Rotational frequent oscillations are filtered ("force-free rotation" [17]), so that most of the geometrical inaccuracies in the machine do not influence the position controller output. Only DC effects, such as eddy current effect on a standstill bearing force vector, are visible. The tuning of the rotor initial angle is carefully done in a way that for an oscillation of the reference signal in y-position no variation in x-position occurs. Together with the aforementioned speed-dependent correction of the rotor angle γ_{m}, the error angle $\epsilon_{\mathrm{err,Ft}}$ is determined via (18) and the force amplitude reduction $G_{\mathrm{Ft,a,0}}$ via (19).

$$\epsilon_{\mathrm{err,Ft}} \approx \tan\left(\frac{i_{\mathrm{d,L,act}}}{i_{\mathrm{q,L,act}}}\right) - \epsilon_{\mathrm{err,inv}} \tag{18}$$

$$G_{\mathrm{Ft,a,0}} \approx \frac{i_{\mathrm{q,L,act}}\,(n=0)}{i'_{\mathrm{q,L}}} = \frac{\frac{i_{\mathrm{q,L,act}}(n=0)}{\cos(\epsilon_{\mathrm{err,inv}})}}{\frac{i_{\mathrm{q,L,act}}}{\cos(\epsilon_{\mathrm{err,inv}})} + \tan(\epsilon_{\mathrm{err,Ft}}) \cdot i_{\mathrm{d,L,act}}} \tag{19}$$

For the calculation of $\epsilon_{\mathrm{err,Ft}}$ (18), the measured currents $i_{\mathrm{d,L,act}}$ and $i_{\mathrm{q,L,act}}$ are directly used. The calculation of $G_{\mathrm{Ft,a,0}}$, requires the measured current $i_{\mathrm{q,L,act}}\,(n = 0)$ at zero speed and the

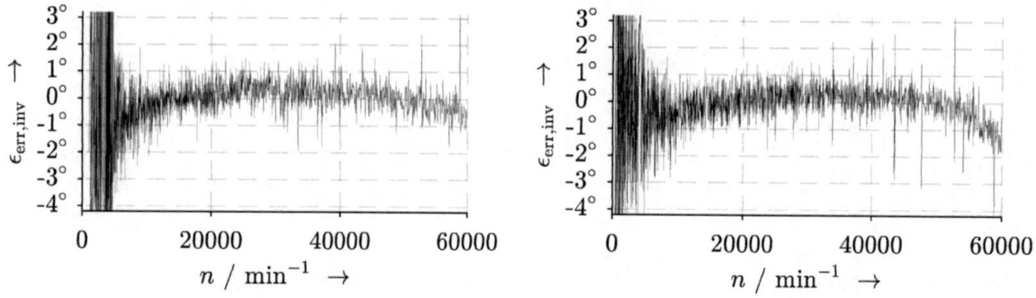

(a) Sm_2Co_{17}-rotor (b) NdFeB-rotor

Figure 9: Measured error angle $\epsilon_{\mathrm{err,inv}}$ (16) due to rotor angle sensor delay at varying speed n

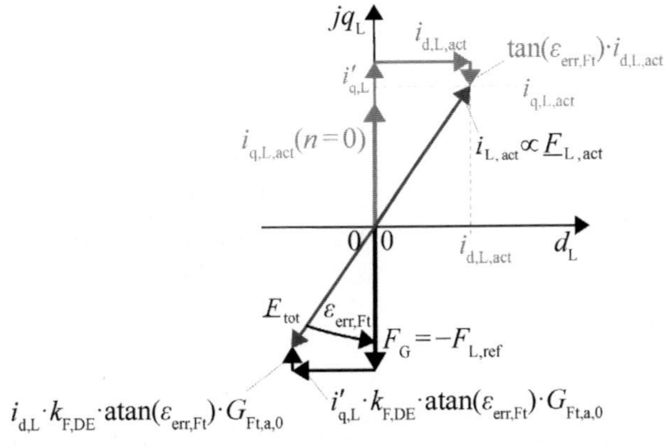

Figure 10: Schematic suspension current vector and rotor force vector equilibrium (correct rotor angle determination provided)

current $i'_{q,L}$ (Fig. 10). Fig. 10 shows schematically how the suspension current vectors and the rotor force vectors are composed in the stator-fixed coordinate frame.
The results for the prototype machine are shown in Fig. 11. Table 5 summarizes the eddy current related disturbing force angle $\epsilon_{err,Ft}$ and bearing force attenuation $G_{Ft,a,0}$ in the prototype machine at rated speed $n_N = 60000\ \mathrm{min}^{-1}$ for both analyzed PM materials: Sm_2Co_{17} and NdFeB.
Generally, the NdFeB-rotor exhibits a 34% smaller force error angle (measured value) due to the electric conductivity κ_{PM}, which is 35% smaller for the NdFeB than for Sm_2Co_{17} (Table 1). The proportionality $\kappa_{PM} \propto \epsilon_{err,Ft}$ holds for resistance-limited eddy currents. The 2D calculation and simulation results agree well but overestimate the error angle by 72% due to the neglected axial end effect. Even for slender rotors, the end effects are considerable, since the suspension winding pole pitch $\tau_{p,L}$ (here: $\tau_{p,L} = 27.5$ mm) may be in the range of the axial length l_{Fe} (here: $l_{Fe} = 40$ mm). Even the 3-dimensional FEM simulation overestimates the error angle by 22%. This deviation may be due to material degradation at the PM surface during the mounting of the bandage or by slightly orthotropic conductivity properties [2], which could not be identified in the simple conductivity measurement (Fig. 1). The amplitude attenuation of the 3-dimensional simulations agree well with the measured results. Following the results from Fig. 11, the speed dependency of $G_{Ft,a,0}$ and $\epsilon_{err,Ft}$ can be considered approximately linear

Table 5: Comparison of different determination techniques for the eddy current disturbing forces in the prototype machine (Table 2) at $n_N = 60000\ \mathrm{min}^{-1}$

	Sm_2Co_{17}		NdFeB	
	$G_{Ft,a,0}$	$\epsilon_{err,Ft}$	$G_{Ft,a,0}$	$\epsilon_{err,Ft}$
$2D_{an}$	0.951	7.09°	0.970	5.06°
$2D_{FEM}$	0.958	6.63°	0.975	4.63°
$3D_{FEM}$	0.968	5.06°	0.987	3.47°
Measurement	0.969	4.13°	0.982	2.73°

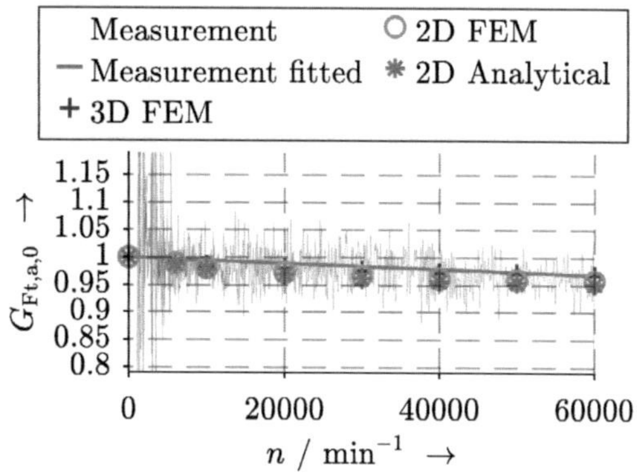

(a) Force amplitude attenuation $G_{Ft,a,0}$: Sm_2Co_{17}-rotor

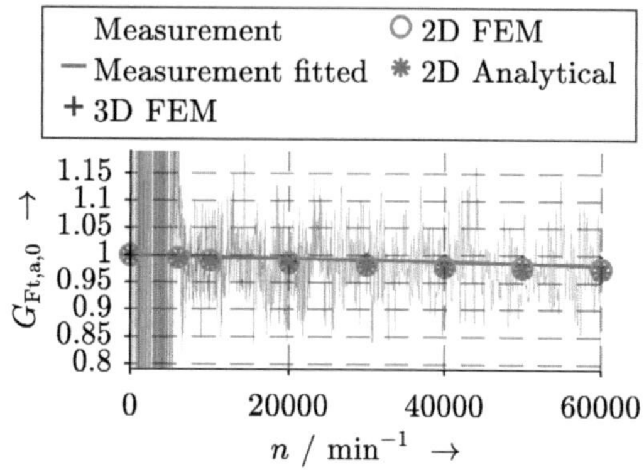

(b) Force amplitude attenuation $G_{Ft,a,0}$: NdFeB-rotor

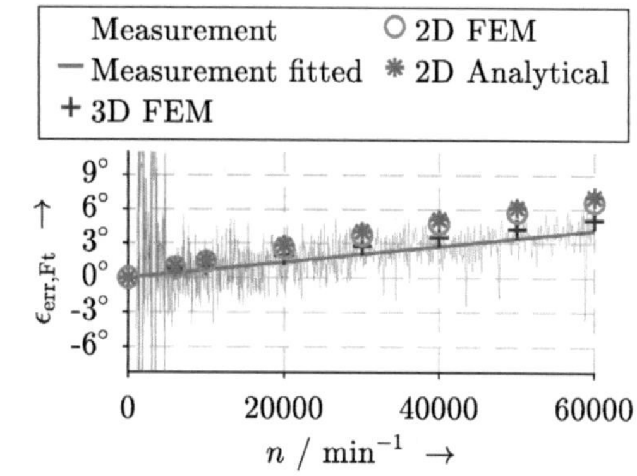

(c) Force error angle $\epsilon_{err,Ft}$: Sm_2Co_{17}-rotor

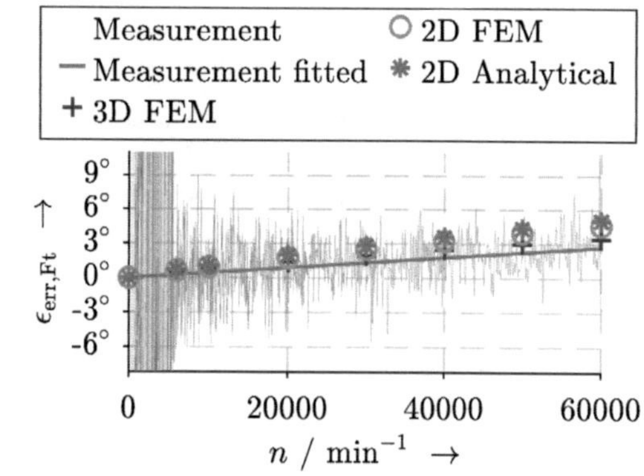

(d) Force error angle $\epsilon_{err,Ft}$: NdFeB-rotor

Figure 11: Comparison of different determination techniques for the force amplitude attenuation $G_{Ft,a,0}$ and the force error angle $\epsilon_{err,Ft}$ at the prototype machine (Table 2) for varying speed n

(20), (21). Since the speed is equal to the slip frequency $f_R = |n|$ (7), this holds as long as the eddy currents are resistance-limited.

$$G_{Ft,a,0}(n) \approx 1 - \frac{n}{n_N} \cdot [1 - G_{Ft,a,0}(n = n_N)] \tag{20}$$

$$\epsilon_{err,Ft}(n) \approx \frac{n}{n_N} \cdot \epsilon_{err,Ft}(n = n_N) \tag{21}$$

The measured suspension force attenuation $G_{Ft,a,0} > 0.95$ and error angle $\epsilon_{err,Ft} < 5°$ are not crucial for the rotor position control in the considered prototype machine. However, eddy current originated disturbance forces may be much bigger for larger machine dimensions, since the suspension winding fundamental intrudes deeper into the conductive rotor parts. Also for higher speed values n and, thus, higher slip frequency values f_R, the eddy current effect increases. Thus, especially for larger and faster rotating machines either low conductive PM material, such as NdFeB, or a segmentation is required to keep the eddy current effect small.

5 Conclusion

The influence of the PM material in bearingless machines on the suspension force degradation is discussed. The attenuation and phase-shifting of the suspension force due to rotor eddy currents is shown by calculation, simulation and measurements on a 1 kW / 60000 min^{-1} prototype machine. Low conductive PM material, such as NdFeB, leads to a smaller disturbing force by eddy currents. In the considered prototype machine even higher conductive PM materials, such as Sm_2Co_{17}, lead to a very small eddy current effect. However, for larger machines or at higher speed values the eddy current originated disturbance forces increase and may not be neglected anymore.

6 Acknowledgments

Funded by the Deutsche Forschungsgemeinschaft (DFG, German Research Foundation) – 437667923, BI 701/22-1 (Gefördert durch die Deutsche Forschungsgemeinschaft (DFG) – 437667923, BI 701/22-1). Supported by KEBA Industrial Automation Germany GmbH.

References

[1] {Vacuumschmelze GmbH & Co. KG}, "Rare Earth Permanent Magnets Vacodym Vacomax," 2014. [Online]. Available: https://vacuumschmelze.de/shared/quickNav/Downloads

[2] S. Ruoho, M. Haavisto, E. Takala, T. Santa-Nokki, and M. Paju, "Temperature Dependence of Resistivity of Sintered Rare-Earth Permanent-Magnet Materials," *IEEE Transactions on Magnetics*, vol. 46, no. 1, pp. 15–20, Jan. 2010, http://ieeexplore.ieee.org/document/5357499/.

[3] "Specifications of Individual Materials - Magnetically Hard Materials," in *IEC Standard 60404-8-1*, 2004, p. 65.

[4] H.-D. Stölting, E. Kallenbach, and W. Amrhein, Eds., *Handbook of fractional-horsepower drives.* Berlin, Heidelberg, New York: Springer-Verlag, 2008.

[5] Y. Fu, M. Takemoto, S. Ogasawara, and K. Orikawa, "Investigation of Operational Characteristics and Efficiency Enhancement of an Ultra-High-Speed Bearingless Motor at 100,000 r/min," *IEEE Transactions on Industry Applications*, vol. 56, no. 4, pp. 3571–3583, Apr. 2020.

[6] Z. Liu, A. Chiba, Y. Irino, and Y. Nakazawa, "Optimum Pole Number Combination of a Buried Permanent Magnet Bearingless Motor and Test Results at an Output of 60 kW With a Speed of 37000 r/min," *IEEE Open Journal of Industry Applications*, vol. 1, pp. 33–41, Feb. 2020.

[7] F. M. Smits, "Measurement of Sheet Resistivities with the Four-Point Probe," *The Bell System Technical Journal*, vol. 57, no. 3, pp. 711–718, 1958.

[8] O. Gutfleisch, M. Verdier, I. Harris, and A. Ray, "Characterisation of rare earth-transition metal alloys with resistivity measurements," *IEEE Transactions on Magnetics*, vol. 29, no. 6, pp. 2872–2874, Nov. 1993, http://ieeexplore.ieee.org/document/281081/.

[9] Z. Q. Zhu, D. Howe, E. Bolte, and B. Ackermann, "Instantaneous magnetic field distribution in brushless permanent magnet DC motors. I. Open-circuit field," *IEEE Transactions on Magnetics*, vol. 29, no. 1, pp. 124–135, Jan. 1993.

[10] A. Rahideh and T. Korakianitis, "Analytical magnetic field distribution of slotless brushless permanent magnet motors – Part I. Armature reaction field, inductance and rotor eddy current loss calculations," *IET Electric Power Applications*, vol. 6, no. 9, p. 628, 2012.

[11] M. Markovic and Y. Perriard, "Analytical Solution for Rotor Eddy-Current Losses in a Slotless Permanent-Magnet Motor: The Case of Current Sheet Excitation," *IEEE Transactions on Magnetics*, vol. 44, no. 3, pp. 386–393, Mar. 2008.

[12] D. Dietz and A. Binder, "Eddy Current Influence on the Control Behavior of Bearingless PM Synchronous Machines," *IEEE Transactions on Industry Applications*, vol. 57, no. 6, 2021, to be published.

[13] F. W. Carter, "The Magnetic Field of the Dynamo-electric Machine," *Journal of the Institution of Electrical Engineers*, vol. 64, no. 359, pp. 1115 – 1138, Nov. 1926.

[14] J. R. Anglada, S. M. Suleiman, and A. Qazalbash, "Influence of curvature on air-gap magnetic field distribution and rotor losses in PM electric machines," *COMPEL - The international journal for computation and mathematics in electrical and electronic engineering*, vol. 36, no. 4, pp. 871–891, Mar. 2017, http://www.emeraldinsight.com/doi/10.1108/COMPEL-05-2016-0200.

[15] D. Zarko, D. Ban, and T. A. Lipo, "Analytical calculation of magnetic field distribution in the slotted air gap of a surface permanent-magnet motor using complex relative air-gap permeance," *IEEE Transactions on Magnetics*, vol. 42, no. 7, pp. 1828–1837, 2006.

[16] G. Messager and A. Binder, "Derivation of forces and force interferences in a double conical high-speed bearingless permanent magnet synchronous motor," in *International Electric Machines and Drives Conference (IEMDC)*, Miami, USA, May 2017, pp. 1–8.

[17] G. Schweitzer and E. H. Maslen, Eds., *Magnetic Bearings: Theory, Design, and Application to Rotating Machinery.* Berlin, Dordrecht, Heidelberg, New York: Springer-Verlag, 2009.

Introduction of a Novel Highly Dynamic Thrust Bearing Control Based on a Fractional–Order Flux Estimator

Robert Seifert and Wilfried Hofmann

Technische Universität Dresden, Chair of Electrical Machines and Drives, Dresden, Germany
robert.seifert@tu-dresden.de, wilfried.hofmann@tu-dresden.de

Abstract

The dynamic forces of nonlaminated magnetic thrust bearings are highly impaired by the magnetic skin effect inside the solid core. As the modification of the core design only offers limited room for improvement at a high financial cost, control–based measures are desired. Currently, advanced control topologies try to overcome the problem with complex controllers and observers but hardly maintain the direct physical reference to essential bearing parameters like stiffness and damping, required for a simple bearing operation. This is one major reason why most of them are not established sustainably and the common decentralized and cascaded position control with subordinated current control remains the industry standard. The inability to cope with the skin effect and the accompanying loss of stability, dynamic and bandwidth is merely tolerated. Here we propose a new model–based control scheme based on a pre–calculated fractional–order flux estimator. It allows us to compensate the skin effect within the inner control loop by means of a simple IIR–filter–cascade in the feedback path and maintains all physical references and the simplicity of the standard topology. We show that it is possible to stably implement the known high–fidelity eddy current models, based on the diffusion equation, as real–time application on a microprocessor. Unlike other voltage–based flux estimators, our approach is based on the measured coil current as a more reliable true feedback signal. We conclude with a proof of concept of our new digital flux control, which significantly improved stiffness and damping and at least quadrupled the bandwidth of the axial position control over the industry standard.

Disclaimer

This digest is a brief summary of the experimental results shown in our publication titled *"Highly Dynamic Thrust Bearing Control Based on a Fractional–Order Flux Estimator"* in the IEEE IAS Special Issue on Magnetically Levitated Motor Systems [1, doi: 10.1109/TIA.2021.3076421].

1 Introduction and Control Concept

Over the last decades, research in magnetic bearing control has led to a multitude of manifold control algorithms. However, in most cases bearing manufacturers still rely on a rarely questioned industry standard: the common decentralized cascaded position control with subordinated current control [2] (X–I–control, Fig. 1b+d), even though it is impaired by an undesirable positive feedback and prone to perturbations like leakage and fringing fluxes, core saturation, hysteresis and especially eddy currents. But unlike most alternative topologies [2], it preserves the physical analogy to the mass–spring–damper–system, which generally describes all forms of bearings and simple tuning parameters like the stiffness k and damping d are available. Regarding radial bearings, there are no apparent issues with this standard approach, as the employed cores with dedicated electrical sheets are less susceptible for the mentioned perturbations. Thrust bearings, on the other hand, are commonly built from solid general–purpose steel compounds, as any lamination of the core is practically ineffective due to a 3D flux propagation.

Therefore, they suffer from high eddy current flows inside the iron core, resulting in a magnetic skin effect. As a consequence, one observes a substantial damping of the thrust force caused by a significant lag between the force–generating magnetic flux Φ_x and the measurable coil current i_{meas}, which hampers the dynamic, stability and bandwidth of the control. Flux–based topologies, which

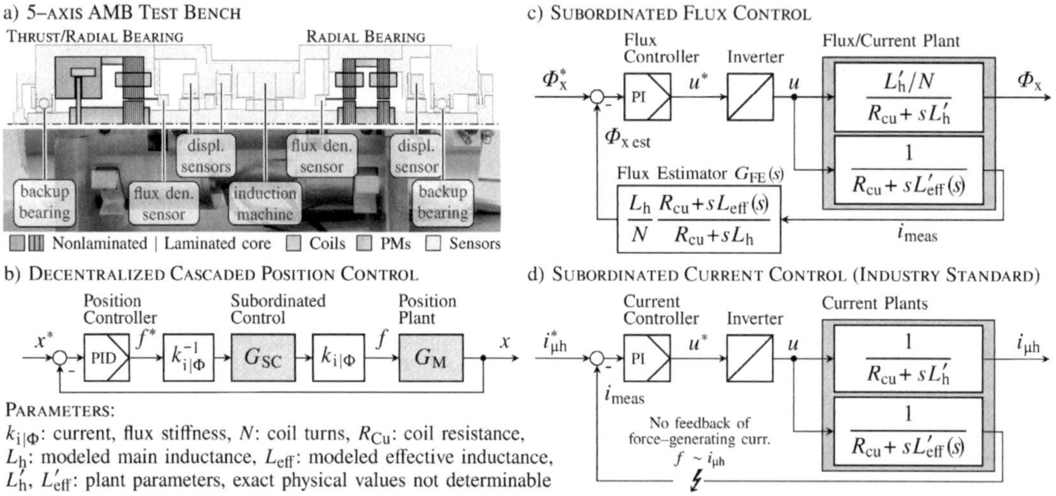

Figure 1: Drive and control concept: a) Test bench with radial bearings and solid thrust bearing controlled by b) Decentralized cascaded position control with subordinated c) flux control and d) current control, adapted from [1]

are mainly similar to the X–I–control, are known to effectively overcome these problems in theory, by directly using the force–generating flux as feedback signal. However, commercially available Hall sensors neither fit into the thin air gaps of modern magnetic bearings nor can they measure the actual total flux in the inhomogeneous air gap field of thrust bearings. To resolve this problem, we estimate the flux by employing the known analytical eddy current models in the feedback path. These models were established for thrust bearings by [3], but have not been used in magnetic bearing control yet.

In this digest and the main article [1], we present the proof of concept of a novel control approach: a decentralized cascaded position control with a subordinated flux control, based on a fractional–order flux estimator (X–Φ_{EST}–control, Fig. 1b+c). It allows us to eliminate the force–damping caused by the eddy currents and maintains the advantageous cascaded structure of the X–I–control. There is no need for additional sensors apart from the common displacement and current sensors (Fig. 1a). Additionally, our method preserves the physical reference of all control variables and compensates the eddy currents effects in the inner control loop in real–time, where they physically occur. The approach also significantly differs from voltage–based flux observers and hybrid forms known in the literature as discussed in [1].

The concept of the proposed flux estimator $G_{FE}(s)$ in Fig. 1c is simple and derives naturally from a proper modeling of the subordinated control loop. The term *"fractional"* illustrates the characteristic properties of the system, described by the effective inductance $L_{eff}(s)$, containing the Laplace–variable s with a fractional exponent $\gamma \in \mathbb{Q}$. The whole procedure from the analytical modeling of $L_{eff}(s)$, its rational integer–order approximation and the digital implementation of the estimator is thoroughly explained in [1]. We only briefly present the experimental results for the outer position control here.

2 Experimental Results and Conclusion

By means of frequency and step responses in Fig. 2, we compare the classic X–I–control with the novel X–Φ_{EST}–control for selected stiffness k/k_x and damping ratios D [1]. Furthermore, we use the analytical model to derive a *trend* for a range of parameters to illustrate the whole potential of our new approach, based on the following characteristic control properties:

The phase margin φ_{PR}, an important stability criterion, shows a significant limitation of the parameter range for the X–I–control, although the maximum achievable 20° are hardly sufficient for a robust controller design. The X–Φ_{EST}–control increases the phase margin by an average of 30 – 40° and therefore allows a securer bearing operation over the entire parameter range.

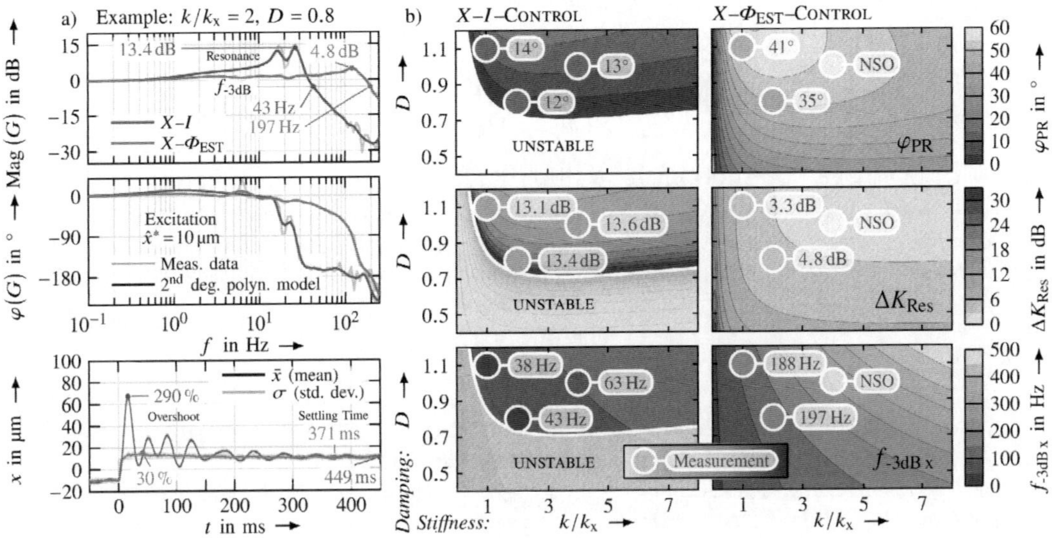

Figure 2: Control attributes derived from analytical model for various stiffness/damping parameters, validation by measurement for example parameters — NSO: No stable operation at resonance for $\hat{x}^* = 10\,\mu\text{m}$, adapted from [1]

The resonance amplitude ΔK_{Res} at the resonance frequency f_0 is a measure of the degree of oscillation occurring with an excitation of the reference position $\hat{x}^*$. The X–Φ_{EST}–control allows a much higher damping compared to the X–I–control by reducing the resonance amplitude by at least 5 – 25 dB.

The overshoot ϵ_{OS} is closely related to ΔK_{Res} in the time domain. It is partly caused by the unavoidable derivative gain and limited integral gain of the controller. While the X–Φ_{EST}–control with 30 – 40 % almost reaches the attainable minimum, the X–I–control yields an unacceptable overshoot of 150 – 400 %.

The cutoff frequency $f_{\text{-3dB}}$ of the position control is the attribute with the greatest potential of improvement. For the X–Φ_{EST}–control we observe at least a duplication if not even more than a quadruplication compared to the X–I–control. With a more potent inverter (here we are limited to $U_{\text{dc}} = 48\,\text{V}$), allowing higher voltages and stiffness ratios $k/k_x > 7$, a bandwidth up to 500 Hz is feasible.

From this proof of concept, we can conclude that the novel X–Φ_{EST}–control achieves significantly improved control properties compared to the industry standard, by compensating the eddy current effects in the inner control loop. A drawback of our approach is the increased voltage demand to compensate the magnetic skin effect, which may require an upgrade of the bearing supply to fully utilize the potential of this control. However, additional measures to reduce the actual causative eddy currents or the regulated control variant proposed in [1] mitigate the problem. Hence. the X–Φ_{EST}–control is a cost-efficient, possibly pure software–based measure to improve the dynamic behavior of magnetic thrust bearings. It neither requires additional sensors, nor state observers or an online parameter identification and is readily applicable even to existing magnetic bearing control implementations.

Acknowledgment The authors would like to thank the German Research Foundation (DFG) for funding this project under the grant HO 1483/78-1.

References

[1] Robert Seifert and Wilfried Hofmann. Highly dynamic thrust bearing control based on a fractional–order flux estimator. *IEEE Transactions on Industry Applications*, November 2021. doi: 10.1109/TIA.2021.3076421.

[2] G. Schweitzer, H. Bleuler, E. H. Maslen, P. Keogh, R. Larsonneur, R Nordmann, Y. Okada, et al. *Magnetic Bearings: Theory, Design, and Application to Rotating Machinery*. Springer, 2009.

[3] Lei Zhu and C. R. Knospe. Modeling of nonlaminated electromagnetic suspension systems. *IEEE Transactions on Mechatronics*, 15(1):59–69, February 2010.

Application of Active Disturbance Rejection Control (ADRC) in the Radial Position Control of a Bearingless Induction Machine With Split Winding

Werbet L. A. Silva[1]*, Elmer R. L. Villarreal[2]†, Jossana M. S. Ferreira[1]‡, José A. Paiva[3]§, Andrés O. Salazar[1]¶ and Andre L. Maitelli[1]||

[1] Universidade Federal do Rio Grande do Norte, Natal-RN, Brazil
werbethluizz@hotmail.com
jossana@ect.ufrn.br
maitelli@dca.ufrn.br
andres@dca.ufrn.br
[2] Universidade Federal Rural do Semi-Árido, Mossoró-RN, Brazil.
elmerllanos@ufersa.edu.br
[3] Instituto Federal de Educação, Ciência e Tecnologia do Rio Grande do Norte, Natal-RN, Brazil.
jalvaro.paiva@gmail.com

Abstract

In this work, the Active Disturbance Rejection Control (ADRC) technique was applied to control a simulated plant that represents the radial displacement of a bearingless induction machine with split winding. Active Disturbance Rejection Controllers have a robust operation when applied to systems with uncertainties, nonlinearities, and in the presence of disturbances. The work employed the linear and nonlinear versions of the ADRC to control the radial displacement. The simulations were performed in the Scilab software. The system response under parameter variation and external load are presented to compare the performance and implementation of both techniques.

1 Introduction

The emergence of electrical machines with magnetic suspension of the rotor meets industry demand for motors that could generate less noise, reach higher speeds and avoid process contamination. In the past two decades, magnetic bearing technology has received significant attention due to its potential application to high-speed and harsh environments applications, where maintenance routines are more complicated. In this way, different machine types have been tested under the magnetic suspension scheme, and new structures appeared to overcome the natural problems of this technology.

Electrical machines that combine motor functions for torque generation and magnetic bearings in the same system, known as bearingless machines, are considered the greatest advance in this area of knowledge [10][13]. The bearingless technology has been applied in different electrical machine types; the induction motor, e.g., is often chosen due to its robustness, simplicity, and low cost [17][1][20]. [12] proposed a bearingless induction motor in which the group of

*Designed and implemented the control system
†Run simulations and tests and studied the model
‡Did numerous tests and provided a lot of suggestions
§Did the literature review and o wrote the main sections
¶Literature, text review and orientation
||Literature, text review and orientation

coils that generate rotational torque also generates radial forces. This structure, known as split winding, aims to simplify the operation and improve the internal use of space of the motor. The same approach was used by [2] to propose the use of split windings in a traditional three-phase induction motor, obtaining a system capable of generating greater radial forces.

Since the magnetic shaft suspension is a naturally unstable system, the bearingless machine strongly depends on well-designed control routines. These systems are multivariable, non-linear, and coupled [20]. So linear position controllers, as the Proportional-Integral-Derivative (PID) [11], may not achieve a response fast and stable enough for proper shaft suspension. The use of microprocessors and power electronics devices enabled the application of advanced control theories, such as predictive control, sliding mode control, neural network control, Fuzzy control, and others [19]. In recent years, the Active Disturbance Rejection Control (ADRC) "is emerging as a promising alternative in many tough control scenarios, due to its structure simplicity and capability in dealing with the model uncertainties, nonlinearity and external disturbances within a unified framework" [16].

The ADRC has been widely studied in virtue of its anti-disturbance ability and its independence of a precise system model [6]. Its main idea is to group all the uncertainties of the plant – such as non-modeled dynamics, external disturbances, and variations in parameters – as an additional state variable to be estimated in real-time through the input and output variables. Thus, the additional state can be estimated and compensated in the control loop. This task is achieved thanks to the core element of the ADRC: the Extended State Observer (ESO). Employing the ESO, the output and its derivatives, as well as the lumped disturbance, the desired control performance can be achieved by a simple linear state feedback control algorithm with online disturbance rejection [18]. This works aims to apply the ADRC technique to control the rotor radial position of a four-pole bearingless induction machine with split winding.

The original concept of the ADRC controller proposed a completely non-linear structure. Despite being very efficient for non-linear systems, this structure is relatively complicated to be tuned because it has a large number of parameters. Gao's work sought to obtain a linear model of the ADRC controller that would maintain the good performance of the original controller.

This work is the first stage of a project that aims to apply the ADRC controller in the radial position control of an induction bearingless machine with split coil. Initially, we simulated the application of a linear and a non-linear ADRC structure to control a radial position system in order to understand the implementation, evaluate advantages and disadvantages, and assess the performance of the closed-loop control system. These results are the grounding foundation for the following stages of the project.

The results of these experiments are described in this text, which is organized as follows: in addition to this introductory section, section 2 provides a brief presentation of the radial displacement model used; section 3 describes the linear and non-linear structures of ADRC; section 4 shows the results of the simulations; finally, the conclusion section shows the most important points that were evaluated from this work.

2 Bearingless machine with split winding

2.1 Operation principle

In the stator of a three-phase bearing motor, there are three groups of windings divided to generate radial force and torque. The coils are split in the stator as shown in Figure 1. Considering a four-pole three-phase motor, Figure 2 can be used to explain the principle of generating radial magnetic forces in the rotor.

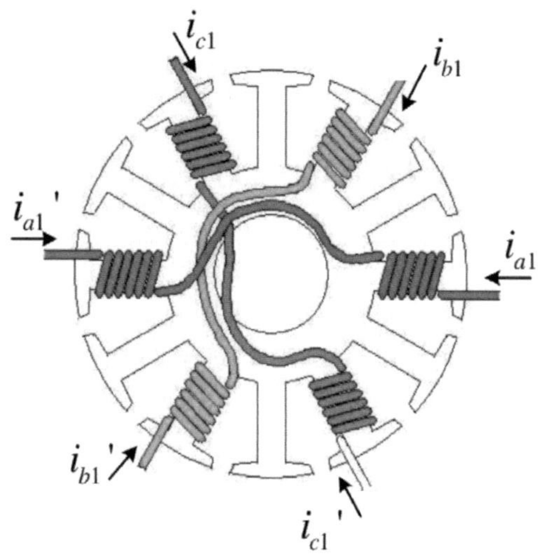

Figure 1: Winding's distribution.

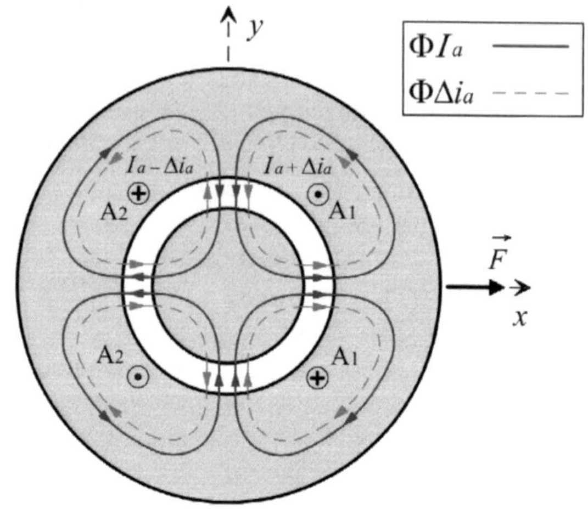

Figure 2: Radial forces generation.

Phase A is divided into two groups of coils: A1 and A2. When the same current (I_a) flows through coils A1 and A2, a magnetic flux is created, as represented in Figure 2 by solid blue lines. When the current increases by Δi_a in coil A1 and simultaneously decreases by Δi_a in coil A2, the flux density increases on one side of the rotor while it decreases on the opposite direction with the same magnitude. The unbalance in the original flux generated by Δi_a generates radial force in the positive direction of the axis X. Likewise, it occurs in the opposite direction when the additional current is added to coil A2 and subtracted from coil A1.

Extending this mode of operation to the three phases and decomposing the coordinate system of the position of the coils as shown in Figure 1 in the coordinate system xy, it is possible to control the radial position in the entire bounded plane for xy.

2.2 Radial displacement model

This prototype works vertically and is supported on the lower end of the shaft by a pivoted bearing that prevents radial and axial movements, but allows angular movements. This type of system has two degrees of freedom for position control and the rotor displacement occurs in the directions of x and y axes.

From the application of the second motion law to Figure 3, [15] developed a linear dynamical representation for the rotor radial displacement system:

$$\ddot{y} = 8374y + 3680000u + w \tag{1}$$

where y is the radial position, u is the input current signal that can vary from $-1A$ to $1A$ and w is an output disturbance. In the form of a transfer function that relates the control signal $U(s)$ to the output position $Y(s)$ of the rotor, the dynamics of the x-axis is equal to the y-axis and both are described by:

$$G(s) = \frac{3.68 \times 10^6}{(s + 91.51)(s - 91.51)} \tag{2}$$

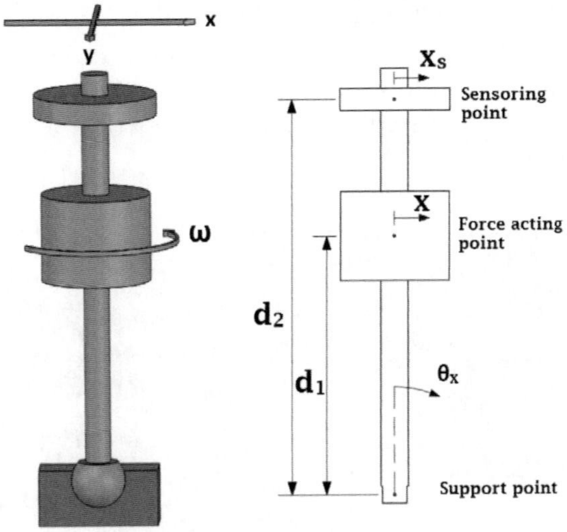

Figure 3: Rotor free body diagram.

Equation (2) shows that the radial position model has an unstable pole, thus a controller is fundamental to the system operation.

3 Active Disturbance Rejection Controller

The first version of the ADRC was presented in Chinese by its creator, Jingqing Han, in the work entitled "Auto disturbances rejection controller and its applications" in 1998 [7]. To promote advances in the theory and practice of control engineering, Jingqing Han researched the characteristics of nonlinear elements into closed-loop control problems and proved the efficiency of the controllers designed from them. According to [9], when compared to linear functions, nonlinear ones are potentially more effective in tolerating uncertainties and disturbances as well in improving systems dynamics.

In a more recent publication, Han described the ADRC as a digital closed-loop control technique, rooted in computer simulations, which inherits from the proportional-integral-derivative (PID) controller its characteristic of greatest success: a control law mainly based on error, and not just on the plant model; in addition, it uses state observers, a technique from the theory of modern controllers; and also makes use of nonlinear control tools for the realization of its structure [5].

Aiming to simplify the ADRC controller, [4] used only linear gains to propose a kind of linear ADRC that uses concepts of controller scaling and parameterization to make the controller parameters functions of the closed-loop bandwidth of the system. The version is called Linear ADRC (LADRC) and brought a new perspective to the industrial application of the technique once it maintained the controller abilities in rejecting disturbances and simplified the tuning process.

Both versions of ADRC (NADRC and LADRC) have exactly the same operation idea: using two control loops to emulate a simplified system, similar to a system composed only by cascaded integrators, in which the dynamic behavior can be controlled by a simple control law (feedback loop). To reach this task, an extended state observer estimates any unknown information (dynamics, disturbances, nonlinearities, parameter variations, time-delays, etc) by the input-output data of the controlled system and removes this information in the control loop (feedforward loop). This structure makes the ADRC independent of the plant model, where the same structure can handle a large variety of dynamic systems. Furthermore, a Reference Generator can be used to provide a smooth version of the reference and its derivatives. This structure is shown in Figure 4.

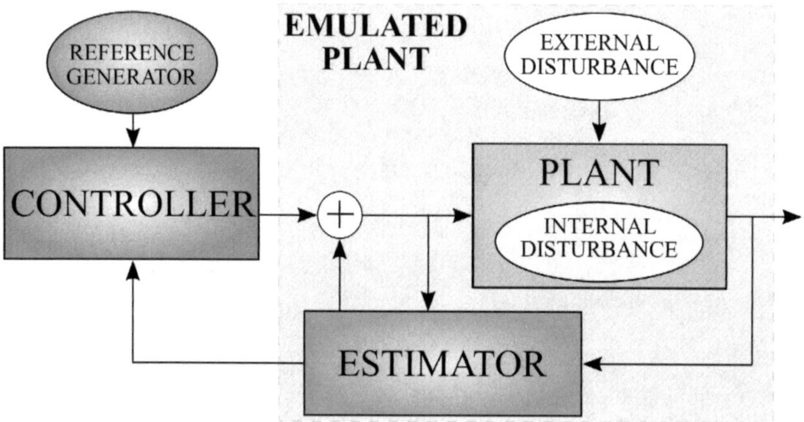

Figure 4: ADRC general idea.

The relative order of the plant is the only information necessary to build a controller based on ADRC. This information is needed because all the blocks described in Figure 4 are designed according to a specific dynamical order behavior. The following sections describe the design of the NADRC and the LADRC to a second-order system, once this is the relative order of our plant.

3.1 ADRC Structure

The mathematical description of the ADRC is based on [4]. A generic second-order system, to be controlled with ADRC, is represented by the following differential equation:

$$\ddot{y} = -a\dot{y} - by + w + bu \tag{3}$$

y is the output; u is the input; w represents an external disturbance; a and b are system parameters.

To apply the ADRC technique, there must be some information about the b parameter. Assuming an approximate value given by $b \approx b_0$, equation 3 can be rewritten as follows:

$$\ddot{y} = -a\dot{y} - by + w + (b - b_0)u + b_0 u = f + b_0 u \tag{4}$$

The term f in the above equation is denoted by generalized disturbance and is equal to $f = -a\dot{y} - by + w + (b - b_0)u$. The term $b - b_0$ represents the difference between the actual value of b and the approximation b_0, which must be specified to implement the algorithm. In state space form, assuming that $x_1 = y$ and $x_2 = \dot{x}_1$ the system given by 4 can be represented by:

$$\begin{cases} \dot{x}_1 = x_2 \\ \dot{x}_2 = b_0 u + f \\ y = x_1 \end{cases} \tag{5}$$

Assuming that f can be estimated, the ADRC technique cancels its effect on the system output by applying the following relationship to the control loop:

$$u = \frac{u_0 - f(t)}{b_0} \tag{6}$$

Hence the total (emulated) dynamics of the system 5 becomes:

$$\begin{cases} \dot{x}_1 = x_2 \\ \dot{x}_2 = u_0 \\ y = x_1 \end{cases} \tag{7}$$

This is a simple double integrator system, whose control has been widely performed in the literature.

The concepts described above indicate that an estimation method that provides an accurate approximation of the value of f is fundamental for an adequate application of the ADRC technique. Furthermore, the control block must be able to meet performance expectations dynamic. The following sections describe two possible frameworks for fully implementing an ADRC-based controller.

3.2 NADRC Structure

The nonlinear structure of an ADRC controller is usually composed by three blocks: a Tracking Differentiator (TD), an Extended State Observer (ESO), and the nonlinear controller (Figure 5).

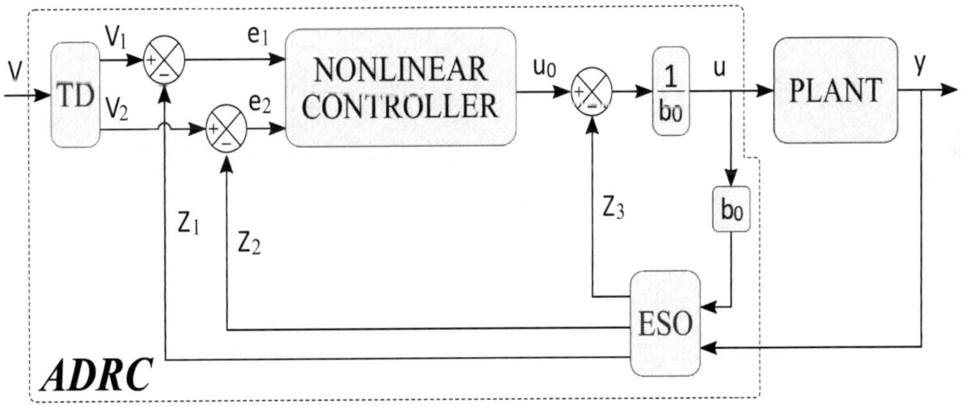

Figure 5: ADRC control loop structure.

3.2.1 Tracking Differentiator

The TD can be seen as a reference generator. It provides smooth versions of the reference signals that usually are put on the system input as a step function. Furthermore, it provides derivatives of the input without the noisy amplification of the traditional derivatives operations. For a second-order system given by:

$$\begin{cases} \dot{v}_1 = v_2 \\ \dot{v}_2 = u \end{cases} \tag{8}$$

an optimal solution is given by:

$$u = -rsign(v_1 - v + \frac{v_2|v_2|}{2r}) \tag{9}$$

This solution is named optimal because it can make the output of 8 reach the desired input fast and with low overshoot. The system can be rewritten as:

$$\begin{cases} \dot{v}_1 = v_2 \\ \dot{v}_2 = -rsign(v_1 - v(t) + \frac{v_2|v_2|}{2r}) \end{cases} \tag{10}$$

thus, $v_1(t)$ follows the input signal $v(t)$ with a speed adjusted by r. The variable $v_2 = \dot{v}_1$ is an approximation of the derivative of $v(t)$. The $sign$ is given by:

$$sgn(e) = \begin{cases} 1, & e \geq 0 \\ -1, & e < 0 \end{cases} \tag{11}$$

3.2.2 Nonlinear Extended State Observer

The ESO is the most important block of the ADRC. In its structure the generalized disturbance f is considered as an additional state of the system 5:

$$\begin{cases} \dot{x}_1 = x_2 \\ \dot{x}_2 = x_3 + b_0 u \\ \dot{x}_3 = h \\ y = x_1 \end{cases} \tag{12}$$

where $x_3 = f$ is the augmented state and $h = \hat{f}$, its derivative. h is seen as an unknown disturbance. The above system can be rewritten by:

$$\begin{aligned} \dot{x} &= Ax + Bu + Eh \\ y &= Cz \end{aligned} \tag{13}$$

where:

$$\mathrm{A} = \begin{bmatrix} 0 & 1 & 0 \\ 0 & 0 & 1 \\ 0 & 0 & 0 \end{bmatrix}, \mathrm{B} = \begin{bmatrix} 0 \\ b_0 \\ 0 \end{bmatrix}, \mathrm{C} = \begin{bmatrix} 1 & 0 & 0 \end{bmatrix}, \mathrm{E} = \begin{bmatrix} 0 \\ 0 \\ 1 \end{bmatrix}$$

From the augmented system 12, it is possible to construct an observer to estimate numerical values to z_1, z_2 e z_3, which are the estimated values of the states x_1, x_2 e x_3. The work [3] proposed a nonlinear structure in which the ESO is such as:

$$\begin{aligned} \dot{z}_1 &= z_2 - \beta_1 fal(e, \alpha_1, d_1), \\ \dot{z}_2 &= z_3 - \beta_2 fal(e, \alpha_2, d_2) + b_0 u \\ \dot{z}_3 &= -\beta_3 fal(e, \alpha_3, d_3), \end{aligned} \tag{14}$$

where β_1, β_2 and β_3 are the observer gains of the nonlinear fal functions, whose definition is given by:

$$fal(e, \alpha, d) = \begin{cases} |e|^\alpha sgn(e), & |e| > d \\ \frac{e}{d^{1-\alpha}}, & |e| \leq d \end{cases} \tag{15}$$

the value of e is the error between the system output and its estimate $(\hat{y} - y)$ and α and d are tuning parameters.

3.2.3 Nonlinear Control Law

Among the control law alternatives proposed by [5], we can highlight the following:

$$u_0 = -K_P fal(e_1, \alpha_4, d_4) - K_D fal(\dot{e}_1 \alpha_5, d_5) \tag{16}$$

where e_1 is the error signal from the difference between the reference and the system output; K_P and K_D are gains; α_4, α_5, d_4, and d_5 are tuning parameters.

3.3 LADRC

Since there is no linear implementation of the Tracking Differentiator on the LADRC version proposed by [4], only the linear ESO and control law of the LADRC are described below.

3.3.1 LESO

The linear ESO proposed by [4] is built by removing the nonlinear terms of 14, such as:

$$\begin{cases} \dot{z}_1 = z_2 - \beta_1 e \\ \dot{z}_2 = z_3 - \beta_2 e + b_0 u \\ \dot{z}_3 = -\beta_3 e \end{cases} \tag{17}$$

where $e = (z_1 - y)$. This is equivalent to a tradicional Luenberger observer and can be rewritten as:

$$\begin{aligned} \dot{z} &= Az + Bu - L(\hat{y} - y) \\ \hat{y} &= Cz \end{aligned} \tag{18}$$

where the vector L if composed by the observer gains:

$$\mathrm{L} = \begin{bmatrix} \beta_1 & \beta_2 & \beta_3 \end{bmatrix}^T$$

These gains can be found by any method of pole allocation. In order to simplify the determination of the gains of the vector L, the tuning method proposed by [4] allocates all the poles in $-\omega_0$. ω_0 is assumed as the bandwidth of the observer. Solving $|s\mathbf{I} - (\mathbf{A} - \mathbf{LC})| = (s + \omega_0)^3$ gives:

$$\mathrm{L} = \begin{bmatrix} 3\omega_0 & 3\omega_0^2 & \omega_0^3 \end{bmatrix}^T$$

This procedure offers a systematic way to obtain the tuning parameters of the LESO. The ESO has advantages over nonlinear ESO considering the control complexities [8].

3.3.2 Linear Control Law

The state feedback method can be applied to control the emulated system of the LADRC. Considering the feedback law $u = -Kx$ gives:

$$u_0 = -K_1(r - z_1) - K_2(r - z_2) \tag{19}$$

K_1 and K_2 are the controller gains. The closed loop poles can be found by the allocation of all poles in $-\omega_c$. Solving $|s\mathbf{I} - (\mathbf{A} - \mathbf{BK})| = (s + \omega_c)^2$ and putting the gains grouped in a vector K:

$$\mathrm{K} = \begin{bmatrix} \omega_c^2 & 2\omega_c \end{bmatrix}$$

As for the ESO, the control law of the LADRC has a systematic way to tune the controller parameters.

4 Results

We simulated the control of the plant given by (2) using LADRC and NADRC to compare the performance of both techniques. For the results described in this section, the following considerations must be made:

- Since the dynamic response for the x and the y axis are equal, the results for only one axis are shown;
- The control goal is to maintain the radial position of the machine's rotor in ($x = 0, y = 0$) position, so the reference signal v is kept in 0 and the Tracking Differentiator is not needed. Therefore the control loop implemented is showed on the Figure 6.

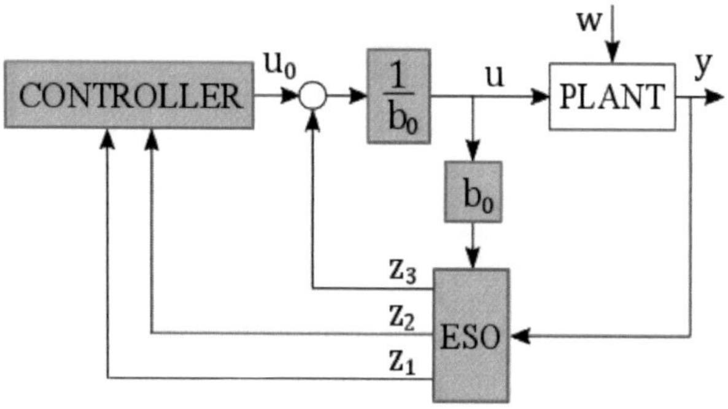

Figure 6: Control loop.

- The nonlinear ADRC was implemented using the equations (14) and the control law is given by (16);
- The linear ADRC was implemented using the equations (17) and the control law is given by (19);
- The controllers tuning was carried out by trial and error;
- The initial conditions are $x_1(0) = 1$ and $x_2(0) = 0$. This situation indicates that the machine rotor was stopped in an unwanted position before the controller starts acting;
- Two step-shaped disturbances were added at the output of the system at the instants $t = 0.1$ and $t = 0.3$ seconds. In a practical situation, disturbances of this type are caused by loads added to the axis of the machine in operation.
- Parameters variations in practice usually come from the natural heat by Joule effect of the circuits and power supplies. This condition is simulated in this work changing the plant dynamic according to:

$$\ddot{y} = (8374 + 8374 * 10 * t)y + 3680000u + w \tag{20}$$

Tables 1 and 2 show the controllers parameters used in the simulations for the LADRC and NADRC. The gains of the nonlinear ADRC blocks used the same of the LADRC.

Parameter	**Value**
step-size	0.0001 s
$x_1(0)$	1
$x_2(0)$	0
ω_c	147
ω_0	$5 * \omega_c$
b_0	2000000

Table 1: LADRC parameters.

Parameter	**Value**
step-size	0.0001 s
$x_1(0)$	1
$x_2(0)$	0
K_p	ω_c^2
Kd	$2\omega_c$
β_1	$3\omega_0$
β_2	$3\omega_0^2$
β_3	ω_0^3
b_0	2000000
$fal - ESO$	$d = 0,5;\ \alpha = 0,5$
$fal - Controller$	$d = 2;\ \alpha = 2$

Table 2: NADRC parameters.

Figure 7 shows the plant output for the two versions of the ADRC controller and the detailed views of relevant regions of the dynamical response.

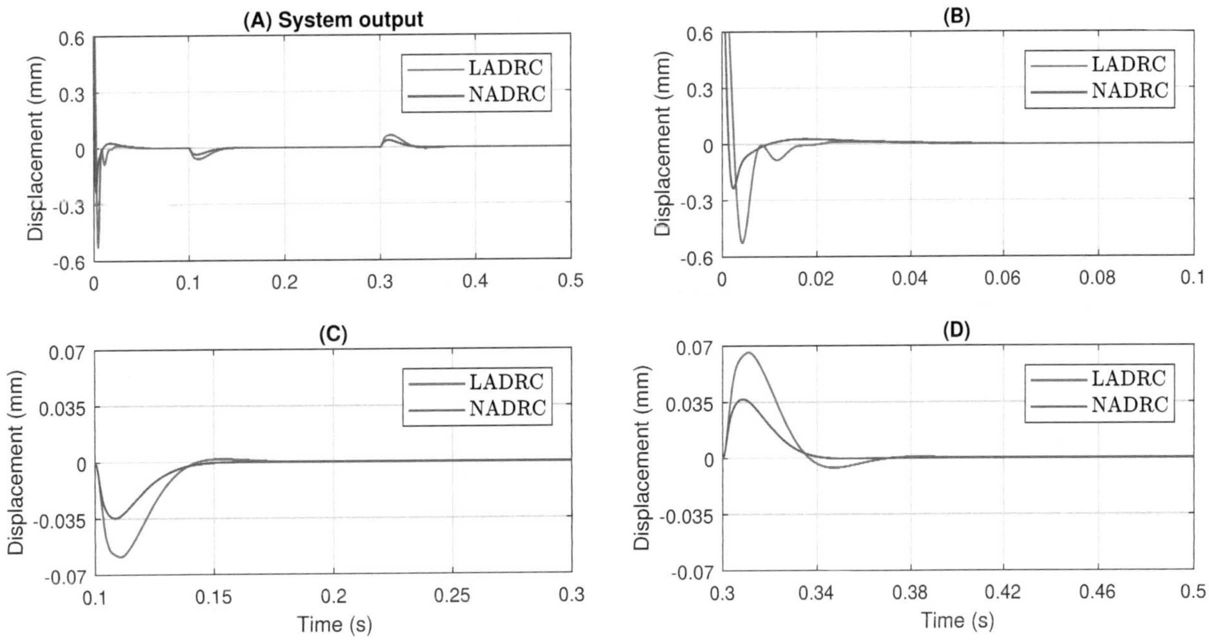

Figure 7: Plant response for LADRC and NADRC.

Figure 7(A) shows that both controllers corrected the plant output to the setpoint and were able to recover this condition even under a load change condition. Figures 7(B), 7(C) and 7(D) show a detailed view of the transitory regions. Figure 7 (B) indicates that the nonlinear model of ADRC can reach the reference faster and with less overshoot. Figures 7(C) and (D) are the response to the step external loads and indicates that the NADRC is more robust to disturbance.

The efficiency of the ADRC in controlling the output of the system depends directly on the ability of ESO to estimate the states of the plant, that is: $x_1 \rightarrow z_1$, $x_2 \rightarrow z_2$ e $F(t) \rightarrow z_3$. Figure 8 compares the dynamic response of the real states with their respective values estimated by the observer. It is clear the ability of the linear and nonlinear ESOs in providing proper estimations of the plant states.

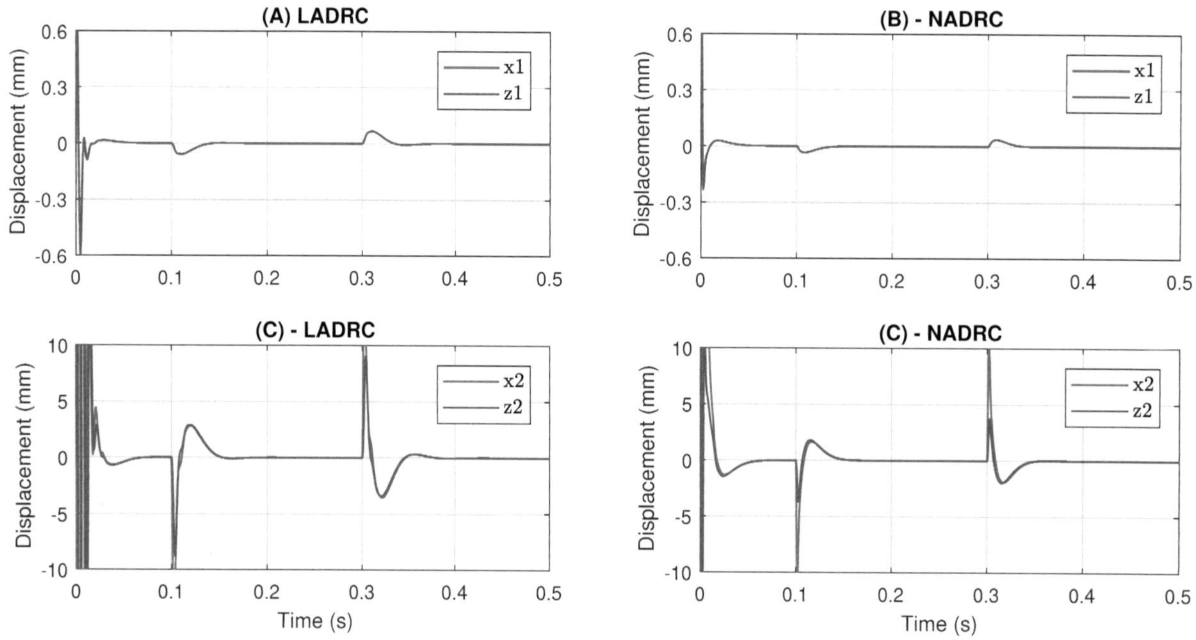

Figure 8: States estimation.

Figure 9 shows the generalized disturbance f and its estimation z_3. Despite some mismatch in the transitory period, the ESOs were successful in estimate the general behavior of the generalized disturbances.

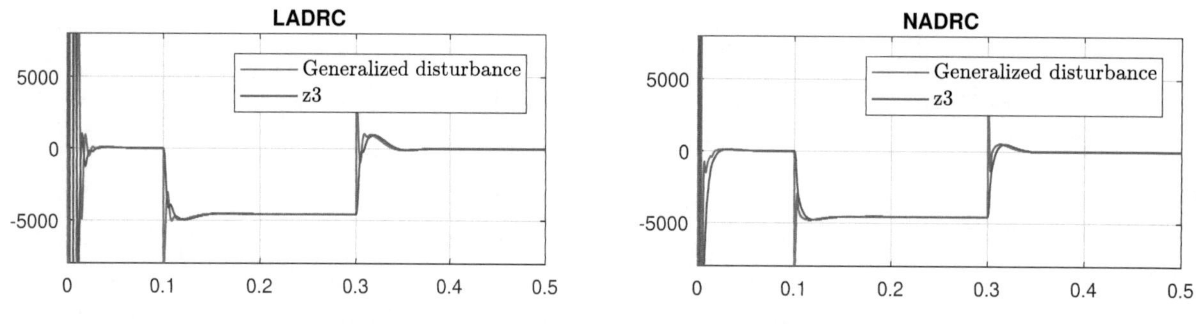

Figure 9: Generalized disturbance estimation.

5 Conclusions

This work studied the application of the ADRC technique to control a simulated plant that represent the rotor radial displacement of a bearingless machine with split winding.

From the simulated results, it was possible to notice that for both the linear and nonlinear controllers the ADRC was successful in providing a stable response to the system even in the presence of perturbations and parameters variations. Despite this, in a more precise point of view, the nonlinear ADRC structure presents a faster and more damped response when compared with the linear ADRC.

Since the performances for both controllers were similar, it is important to evaluate if it is necessary to use the NADRC, which is a more complex structure with more parameters to be tuned, instead of LADRC, which is easier to be built and tuned. Due the nonlinear nature of NADRC and the number of parameters of its implementation, in some cases researchers adopt optimization algorithms [14] instead of trial and error tuning. Again, it means an additional effort to use the NADRC technique that must be taken in account to make project decisions.

As the initial step of a project that aims to evaluate the ADRC to control a real prototype, this works provided the initial experience with this control paradigm and helped to test two different implementations of the ADRC. These results will be useful to future experimental implementations.

6 Acknowledgments

This work was supported by Comissão de Aperfeiçoamento de Pessoal do Nível Superior (CAPES), Programa de Pós-graduação em Engenharia Elétrica e de Computacão (PPgEEC/UFRN) and Conselho Nacional de Desenvolvimento Científico e Tecnológico (CNPq).

References

[1] Jiahao Chen and Eric L. Severson. Optimal design of the bearingless induction motor for industrial applications. In *2019 IEEE Energy Conversion Congress and Exposition (ECCE)*, pages 5265–5272, 2019.

[2] Jossana Maria de Souza Ferreira. Modelagem de máquina de indução trifásica sem mancais com bobinado dividido. 2006.

[3] Zhiqiang Gao. From linear to nonlinear control means: A practical progression. *ISA transactions*, 41(2):177–189, 2002.

[4] Zhiqiang Gao. Scaling and bandwidth-parameterization based controller tuning. In *Proceedings of the 2003 American Control Conference, 2003.*, volume 6, pages 4989–4996, 2003.

[5] Jingqing Han. From pid to active disturbance rejection control. *IEEE transactions on Industrial Electronics*, 56(3):900–906, 2009.

[6] Zhengjie Hao, Yang Yang, Yimin Gong, Zhengqiang Hao, Chenchen Zhang, Hongda Song, and Jiannan Zhang. Linear/nonlinear active disturbance rejection switching control for permanent magnet synchronous motors. *IEEE Transactions on Power Electronics*, 36(8):9334–9347, 2021.

[7] Deqing Huang, Da Min, Yupei Jian, and Yanan Li. Current-cycle iterative learning control for high-precision position tracking of piezoelectric actuator system via active disturbance rejection control for hysteresis compensation. *IEEE Transactions on Industrial Electronics*, 2019.

[8] Prashant Kumar, Abdul Rahiman Beig, Devara Vijaya Bhaskar, Khaled Ali Al Jaafari, Utkal Ranjan Muduli, and Ranjan kumar Behera. An enhanced linear active disturbance rejection controller for high performance pmbldcm drive considering iron loss. *IEEE Transactions on Power Electronics*, pages 1–1, 2021.

[9] Jie Li, Yuanqing Xia, Xiaohui Qi, and Zhiqiang Gao. On the necessity, scheme, and basis of the linear–nonlinear switching in active disturbance rejection control. *IEEE Transactions on Industrial Electronics*, 64(2):1425–1435, 2016.

[10] Ke Li, Guoyu Cheng, Xiaodong Sun, Dean Zhao, and Zebin Yang. Direct torque and suspension force control for bearingless induction motors based on active disturbance rejection control scheme. *IEEE Access*, 7:86989–87001, 2019.

[11] Yujun Liu, Hejin Xiong, and Heng Zhang. Research on active disturbance rejection control of active magnetic bearing-rotor system. In *2021 6th Asia Conference on Power and Electrical Engineering (ACPEE)*, pages 873–877, 2021.

[12] A.O. Salazar and R.M. Stephan. A bearingless method for induction machines. *IEEE Transactions on Magnetics*, 29(6):2965–2967, 1993.

[13] Eric Severson, Robert Nilssen, Tore Undeland, and Ned Mohan. Design of dual purpose no voltage combined windings for bearingless motors. In *2016 IEEE Energy Conversion Congress and Exposition (ECCE)*, pages 1–10, 2016.

[14] Suiyuan Shen and Jinfa Xu. Trajectory tracking active disturbance rejection control of the unmanned helicopter and its parameters tuning. *IEEE Access*, 9:56773–56785, 2021.

[15] Francisco Elvis Carvalho Souza. Otimização da estrutura de acionamento para o controle de posição radial de um motor de indução trifásico sem mancais com enrolamento dividido. 2018.

[16] Li Sun, Wenchao Xue, Donghai Li, Hongxia Zhu, and Zhi-gang Su. Quantitative tuning of active disturbance rejection controller for fopdt model with application to power plant control. *IEEE Transactions on Industrial Electronics*, pages 1–1, 2021.

[17] Xiaodong Sun, Long Chen, and Zebin Yang. Overview of bearingless induction motors. *Mathematical Problems in Engineering*, 2014, 2014.

[18] Thuy Vi Tran, Kyeong-Hwa Kim, and Jih-Sheng jason Lai. Optimized active disturbance rejection control with resonant extended state observer for grid voltage sensorless lcl-filtered inverter. *IEEE Transactions on Power Electronics*, pages 1–1, 2021.

[19] Zebin Yang, Jialei Ji, Xiaodong Sun, Huimin Zhu, and Qian Zhao. Active disturbance rejection control for bearingless induction motor based on hyperbolic tangent tracking differentiator. *IEEE Journal of Emerging and Selected Topics in Power Electronics*, 2019.

[20] Zebin Yang, Chengling Lu, Xiaodong Sun, Jialei Ji, and Qifeng Ding. Study on active disturbance rejection control of a bearingless induction motor based on an improved particle swarm optimization–genetic algorithm. *IEEE Transactions on Transportation Electrification*, 7(2):694–705, 2021.

Influence of Magnetic Eccentricity in Bearingless PM Synchronous Machines

Daniel Dietz and Andreas Binder

Institute for Electrical Energy Conversion, Technical University of Darmstadt, Germany
ddietz@ew.tu-darmstadt.de

Abstract

Bearingless permanent magnet (PM) synchronous machines, used as high-speed drives, typically employ a simple rotor structure. Machines of smaller size are equipped with a solid shaft on which a permanent magnet ring is mounted, protected by a carbon-fiber sleeve. Larger machines often require an additional rotor sheet package and use permanent magnet shells instead of a ring.
Since the permanent magnet exhibits a small yield strength, no shrink-fitting between the magnet ring and the shaft can be applied. Thus, the PM ring is required to be slightly bigger in inner diameter than the shaft in outer diameter, leading to a half-moon shaped air gap between magnet and shaft. Although this air gap is usually in the μm-range it can lead to a considerable rotational-frequent radial disturbing force. We discuss the influences, such as the rotor diameter and the pole count, which contribute to this magnetic eccentricity problem. We also show the disturbing influence on the rotor position control by measurements on a built prototype.

1 Introduction

Bearingless Motors (BM) have gained rising attractiveness in the past decade [3]. The most promising topology for industrial high-speed applications such as pumps or compressors is the bearingless permanent magnet (PM) synchronous machine. Several prototypes of this topology have reached high speed values up to 100 000 min^{-1} [4] and 60 kW [5].
One of the key properties that makes PM machines suitable for high-speed operation is the simple composition of the rotor. Typically, it is composed of a massive magnetic steel shaft on which a solid PM ring of low pole count is glued. For bigger diameters, often PM shells are used. The permanent magnet is retained by a protecting carbon- or glas-fiber bandage (Fig. 1c). The bandage is mounted onto the PM by shrink-fitting. Such a shrink-fitting is possible, since the PM exhibits a very high compressive ultimate strength (Table 1). The chosen undersize

Table 1: PM material properties, used for the prototype machine [1, 2]

	Sm_2Co_{17}	NdFeB
Remanence flux density (100 °C) B_{rem} / T	0.96	1.13
Relative permeability $\mu_{\text{r,PM}}$	1.056	1.068
Compressive ultimate strength $R_{\text{m,comp}}$ / $\text{N} \cdot \text{mm}^{-2}$	650	925
Tensile ultimate strength R_{m} / $\text{N} \cdot \text{mm}^{-2}$	95	90
Density ρ / $\text{kg} \cdot \text{m}^{-3}$	8300	7650
Undesired eccentricity air gap δ_{ecc} / μm	50	180

Table 2: Dimensions of the prototype machine

Axial active length l_{Fe} / mm	40
Stator outer / inner radius $r_{s,o}$/ $r_{s,i}$ / mm	37.5 / 17.5
Shaft radius r_{sh} / mm	12.25
PM height h_{PM} / mm	2.75
Bandage height + mechanical air gap width: δ / mm	2.5

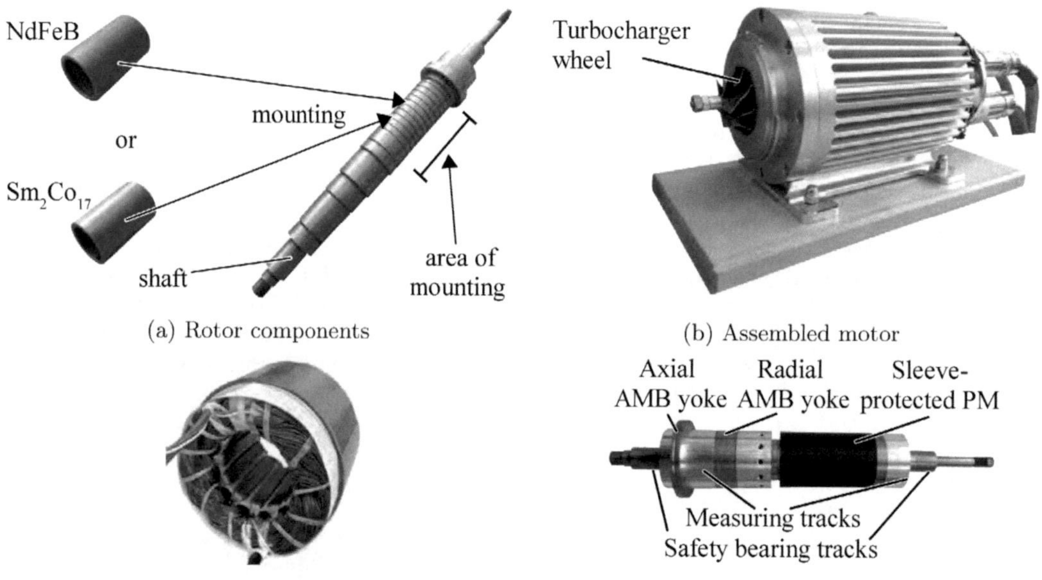

(a) Rotor components

(b) Assembled motor

(c) Wound stator

(d) Assembled rotor

Figure 1: Built 1 kW / 60000 min^{-1} bearingless PM synchronous machine and simplified 2D finite element model

between bandage and PM must be chosen to avoid a loss of contact between PM and shaft at high-speed and to stay below the bandage material's yield strength [6]. The choice of this shrink-fitting is by far the most crucial part of the rotor assembly. Thus, little attention is paid to the mounting of the permanent magnet onto the steel shaft. However, this mounting (Fig. 1a) is also problematic, since the PM only exhibits a very low tensile ultimate strength R_m (Table 1). Therefore, the PM must be mounted onto the shaft without any undersize, so that a small single-sided air gap δ_{ecc} (Fig. 2) between PM and shaft can occur, which is rotor-fixed. This single-sided air gap leads to a flux concentration at the opposite side of the maximum eccentricity air gap width δ_{ecc} (Fig.2b), resulting in a modulation of the fundamental air gap field (Fig. 2c).

This paper shows that even very small values $\delta_{ecc} < 200$ μm (here: < 1.6 % of the shaft radius r_{sh}, Table 2) lead to a considerable, disturbing radial force F_{ecc} on the rotor. We discuss, how this magnetic eccentricity influences the operation of bearingless machines. The investigations are carried out for two rotors, one with a Sm_2Co_{17} PM and one with a NdFeB PM, at a built prototype with 1 kW at a speed of $n = 60\ 000$ min^{-1}. The two PM rings differ in remanence

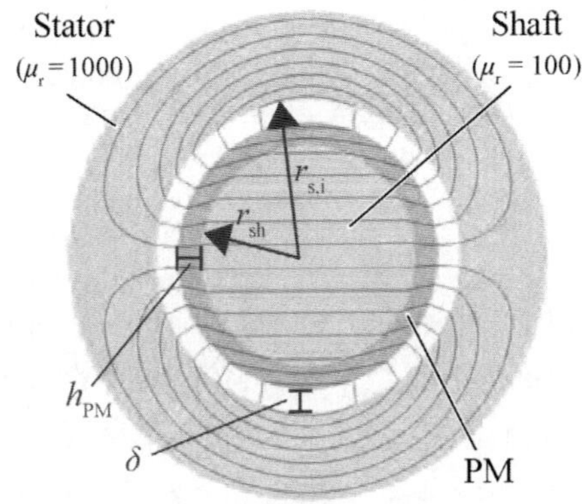

(a) "Healthy" machine: field plot and model geometry

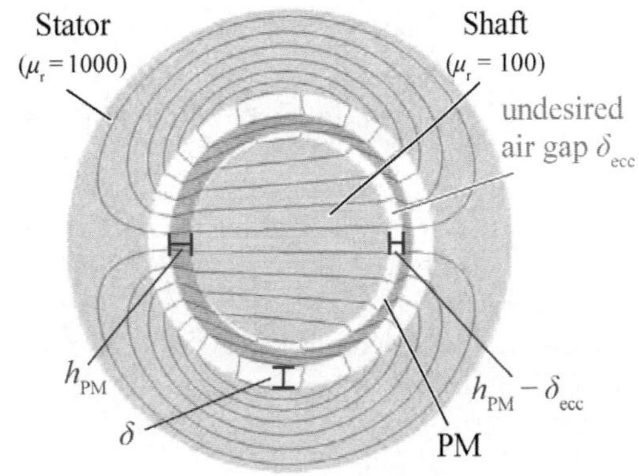

(b) "Unhealthy" machine: field plot and model geometry with undesired air gap δ_{ecc}

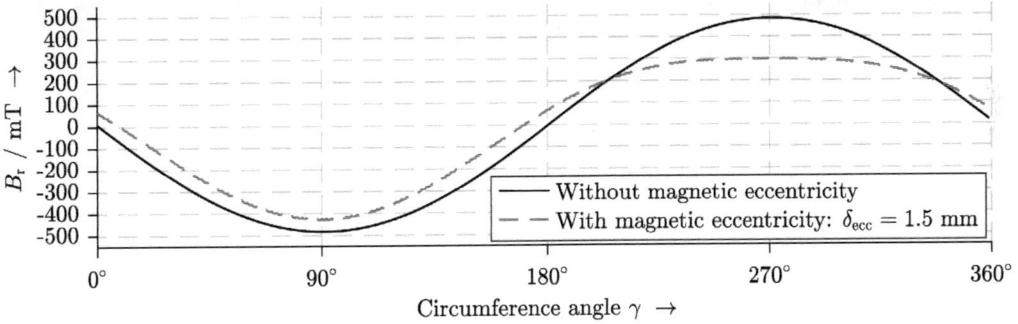

(c) Simulated radial component of the air gap flux density B_r with and without eccentricity

Figure 2: Visualization of the magnetic eccentricity, exemplary with an non-realistically big eccentricity air gap width $\delta_{ecc} = 1.5$ mm

flux density (Table 1). They also exhibit a different eccentricity air gaps δ_{ecc} which occurred unintentionally. We additionally show, how the eccentricity force is influenced by the machine's size, pole count and PM height.

2 Forces due to Magnetic Eccentricity

To calculate the forces on the rotor, the *Maxwell* stress tensor (2) is evaluated over a cylindrical surface of length l_{Fe} and radius $r_\delta = r_{s,i} - \delta/2$ (1). l_{Fe} is the machine's axial iron length, $r_{s,i}$ the bore radius and δ the sum of the air gap width and the bandage height (Table 2). The radial and tangential air gap field components B_r and B_γ are given as the sum of the rotor and stator field waves.

$$\begin{pmatrix} F_x(t) \\ F_y(t) \end{pmatrix} = \int_0^{l_{Fe}} \int_0^{2\pi} \begin{pmatrix} f_r(\gamma,t) \cdot \cos(\gamma) - f_\gamma(\gamma,t) \cdot \sin(\gamma) \\ f_r(\gamma,t) \cdot \sin(\gamma) + f_\gamma(\gamma,t) \cdot \cos(\gamma) \end{pmatrix} \cdot \mathrm{d}\gamma \mathrm{d}z \tag{1}$$

$$\left. \begin{pmatrix} f_r(\gamma,t) \\ f_\gamma(\gamma,t) \end{pmatrix} \right|_{r=r_\delta} = \frac{1}{2 \cdot \mu_0} \cdot \begin{pmatrix} B_r^2(r_\delta,\gamma,t) - B_\gamma^2(r_\delta,\gamma,t) \\ 2 \cdot B_r(r_\delta,\gamma,t) \cdot B_\gamma(r_\delta,\gamma,t) \end{pmatrix} \tag{2}$$

Usually in high-speed bearingless machines the radial component of the rotor air gap field amplitude $\hat{B}_{\mathrm{R},p,p,\mathrm{r}}$ with pole count $2p$ is much bigger than the stator field. In the considered bearingless machine prototype the rotor air gap field is about 15 times higher than the stator field. Therefore, it is neglected in the analysis. Eccentricity problems can be analytically calculated by using the so-called bi-linear conformal mapping method [7]. Alternatively, the perturbation method according to [8, 9] can be used. Here, a magnetostatic 2-dimensional finite element method simulation with magnetically linear and isotropic material assumption is used instead, based on the *JMAG Designer 19.1.* In this section, only the occurrence of the eccentricity-originated additional air gap field waves $B_{\mathrm{ecc},\mu,k,\mathrm{r}}$ is explained qualitatively, which lead to the eccentricity force F_{ecc}.

The radial (subscript: r) rotor (subscript: R) flux density wave $B_{\mathrm{R},\mu,k,\mathrm{r}}(r_\delta,\gamma,t)$ in the air gap of a PM machine at $r = r_\delta$ is approximately given by (3) if a 1-dimensional air gap field is assumed. γ is the mechanical circumference angle in stator-fixed-coordinates, k is the time harmonic order, which is always $k = p$ for the fundamental field wave in stator-fixed coordinates. μ is the absolute space harmonic order, which is always $\mu = p$ for the fundamental field wave. c is the ratio of the PM height h_{PM} to the air gap width δ.

$$B_{\mathrm{R},\mu,k,\mathrm{r}}(r_\delta,\gamma,t) = -\frac{h_{\mathrm{PM}}}{\delta}\cdot B_{\mathrm{PM},\mu,k,\mathrm{r}}(r_\delta,\gamma,t) = c(\gamma,t)\cdot B_{\mathrm{PM},\mu,k,\mathrm{r}}(r_\delta,\gamma,t) \tag{3}$$

The single-sided air gap δ_{ecc} (Fig. 2) leads to a PM height $h_{\mathrm{PM}}(\gamma,t)$ and air gap width $\delta(\gamma,t)$, which vary with the circumference angle γ and time t in stator-fixed coordinates (4), (5). Thus, the radio $c(\gamma,t)$ also varies (6), which can be approximated by (7).

$$h_{\mathrm{PM}}(\gamma,t) \approx h_{\mathrm{PM}} - \frac{\delta_{\mathrm{ecc}}}{2} + \frac{\delta_{\mathrm{ecc}}}{2}\cdot\cos(\gamma - 2\pi\cdot n\cdot t + \varphi_{\mathrm{ecc}}) \tag{4}$$

$$\delta(\gamma,t) \approx \delta + \frac{\delta_{\mathrm{ecc}}}{2} - \frac{\delta_{\mathrm{ecc}}}{2}\cdot\cos(\gamma - 2\pi\cdot n\cdot t + \varphi_{\mathrm{ecc}}) \tag{5}$$

$$c(\gamma,t) = -\frac{h_{\mathrm{PM}}(\gamma,t)}{\delta(\gamma,t)} = c_0 + \sum_{i=1}^{\infty}\hat{c}_i\cdot\cos(i\cdot\gamma - i\cdot 2\pi\cdot n\cdot t + \varphi_{\mathrm{ecc}}) \tag{6}$$

$$\approx \underbrace{-\frac{h_{\mathrm{PM}} - \frac{\delta_{\mathrm{ecc}}}{2}}{\delta + \frac{\delta_{\mathrm{ecc}}}{2}}}_{c_0} + \hat{c}_1\cdot\cos(\gamma - 2\pi\cdot n\cdot t + \varphi_{\mathrm{ecc}}) \tag{7}$$

By inserting (7) into (3), the PM magnetization wave is modulated by its shape via the ratio $c(\gamma,t)$. With the used simplification, the modulation leads to the rotor air gap radial field fundamental (8) in stator-fixed coordinates. Additionally, the air gap fields waves $B_{\mathrm{ecc},\mu,k,\mathrm{r}}$ occur (9). Owing to the model deficiency, (9) for 2-pole rotors also results in a zeroth order space harmonic, which does not exist in reality.

$$B_{\mathrm{R},\mu=p,k=p,\mathrm{r}}(r_\delta,\gamma,t) = \underbrace{c_0\cdot\hat{B}_{\mathrm{PM},p,p,\mathrm{r}}}_{\hat{B}_{\mathrm{R},p,p,\mathrm{r}}(r_\delta)}\cdot\cos(p\cdot\gamma - 2\pi\cdot n\cdot p\cdot t + \varphi_{\mathrm{R}}) \tag{8}$$

$$B_{\mathrm{ecc},\mu,k,\mathrm{r}}(r_\delta,\gamma,t) = \frac{\hat{B}_{\mathrm{PM},p,p,\mathrm{r}}\cdot\hat{c}_1}{2}\cdot\cos\left[\underbrace{(p\pm 1)}_{\substack{\text{spatial}\\ \text{order }\mu}}\cdot\gamma - \underbrace{(p\pm 1)}_{\substack{\text{time}\\ \text{order }k}}\cdot 2\pi\cdot nt + \varphi_{\mathrm{R}} \pm \varphi_{\mathrm{ecc}}\right] \tag{9}$$

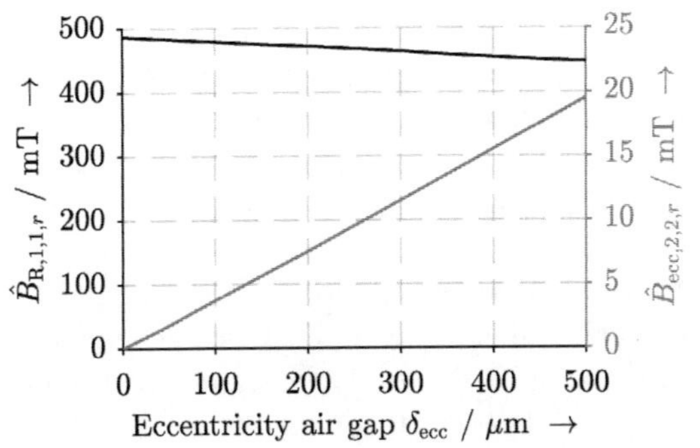

(a) Eccentricity air gap under pole: $\varphi_{ecc} = 0$

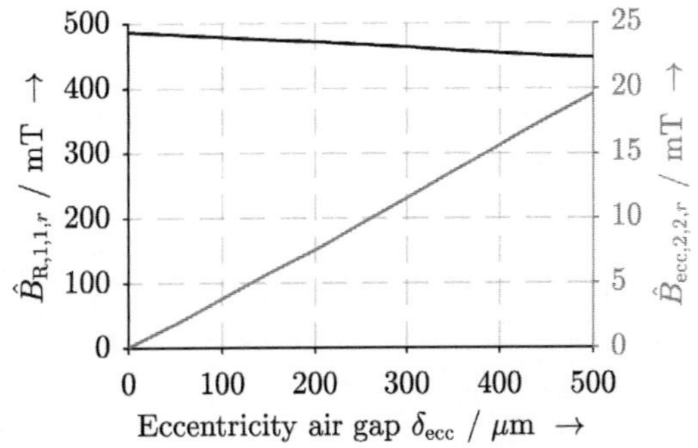

(b) Eccentricity air gap under pole gap: $\varphi_{ecc} = \frac{\pi}{2}$

Figure 3: Simulated radial magnetic field waves due to the eccentricity air gap δ_{ecc} for a 2-pole rotor with dimensions according to Figure 1c (magnetically linear material assumption according to Table 2 and $B_{rem} = 1$ T)

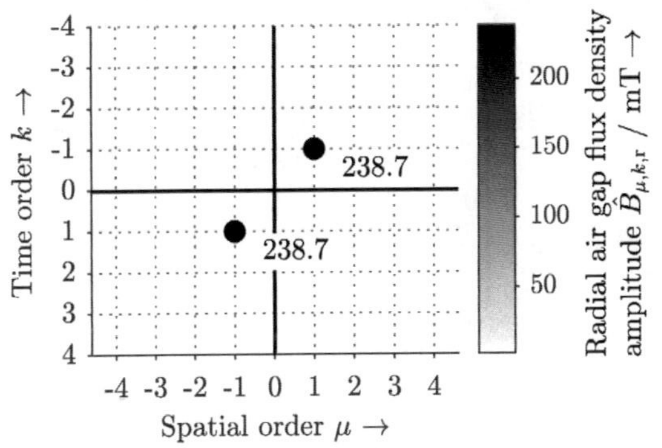

(a) Pole count $2p = 2$, $\delta_{ecc} = 0$

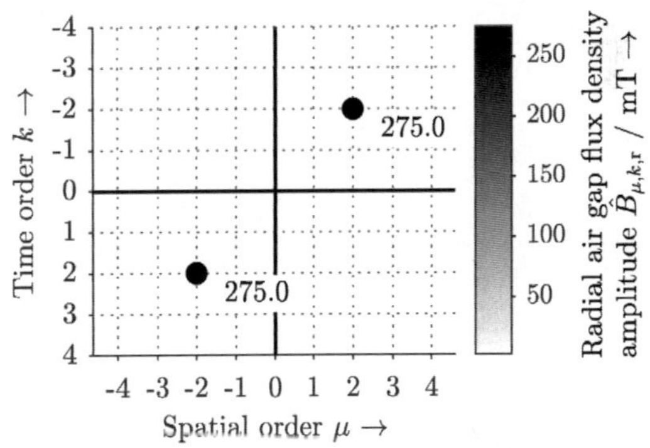

(b) Pole count $2p = 4$, $\delta_{ecc} = 0$

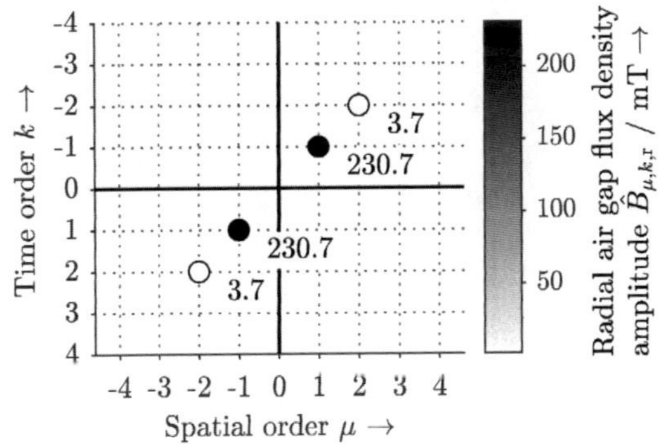

(c) Pole count $2p = 2$, $\delta_{ecc} = 100$ μm

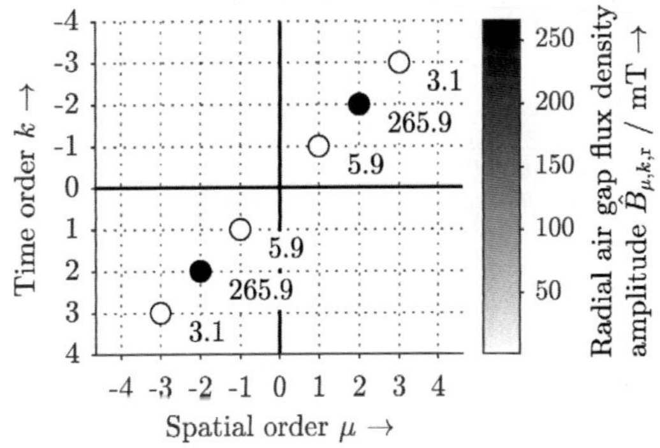

(d) Pole count $2p = 4$, $\delta_{ecc} = 100$ μm

Figure 4: 2D *Fourier* spectrum of the air gap flux density radial component $\hat{B}_r$ in stator reference frame from 2D finite element simulation (k: time harmonic order μ: space harmonic order, evaluated at radius $r_{s,i} - \delta/2$)

In the considered prototype machine with $2p = 2$ poles, the eccentricity air gap δ_{ecc} leads to a radial field wave $B_{\mathrm{ecc},2,2,\mathrm{r}}$ of time order $k = 2$ and space order $\mu = 2$. Under magnetically linear material assumptions, the amplitude $\hat{B}_{\mathrm{ecc},2,2,\mathrm{r}}$ of this field wave rises linearly with the width of the eccentricity air gap δ_{ecc} (Fig. 3). Note that $\hat{B}_{\mathrm{ecc},2,2,\mathrm{r}}$ does not depend on the location of the eccentricity air gap (angle φ_{ecc}), so that the values in Fig. 3a are equal to Fig. 3b. Thus, in the following analysis we set $\varphi_{\mathrm{ecc}} = 0$, so that the eccentricity air gap is located under the pole.
The amplitude of the fundamental rotor field wave $\hat{B}_{\mathrm{R},1,1,\mathrm{r}}$ decreases linearly in Fig. 3 according to c_0 (7). The rated suspension winding air gap field amplitude is $\hat{B}_{\mathrm{L},2,1,\mathrm{r}} = 8$ mT for the prototype machine. Since the modulated air gap field wave takes values $B_{\mathrm{ecc},2,2,\mathrm{r}} > 5$ mT for $\delta_{\mathrm{ecc}} > 130$ μm (Fig. 3), even small eccentricity values can lead to radial force amplitudes in the range of the rotor gravitational force. Compared to the rotor field fundamental, the modulated field waves due to eccentricity are approximately smaller by factor 50. Thus, they may cause crucial rotor forces but they certainly will not cause considerable additional iron losses. So, a loss analysis of the eccentricity effect is omitted.
A more general representation of the air gap field wave harmonics is given in Fig. 4, which shows the 2-dimensional *Fourier* spectrum of the air gap field radial component B_{r}, used in (2), with respect to time order k and space order μ [10]. Fig. 4b and 4d give the results for the 4-pole rotor with the same motor dimensions (Table 2). Fig. 4a and Fig. 4c show the 2-pole rotor results. Fig. 4c and Fig. 4d show that – due to the eccentricity – additional field waves $B_{\mathrm{ecc},\mu,k,\mathrm{r}}$ occur with the time and space harmonic orders according to (9). The integration of the *Maxwell* stress tensor (1) is zero for all combinations of two field waves with space orders μ_1, μ_2 for which $\mu_1 \neq \mu_2 \pm 1$ holds. For $\mu_1 = \mu_2 \pm 1$ the radial eccentricity force F_{ecc} on the rotor into an arbitrary direction, e.g. in x-direction ($F_{\mathrm{x,ecc}} = F_{\mathrm{ecc}}$), is given in (10) for k_1 and k_2 being the time harmonic orders of the involved field waves. The tangential field components are neglected here for simplicity. For 2-pole machines, F_{ecc} is given in (11), for 4-pole machines the eccentricity force is composed of (12), (13).

$$F_{\mathrm{ecc}}(t) = \frac{\pi \cdot r_\delta \cdot l_{\mathrm{Fe}}}{2 \cdot \mu_0} \cdot B_{\mu_1,k_1,\mathrm{r}}(r_\delta) \cdot B_{\mu_2,k_2,\mathrm{r}}(r_\delta) \cdot \sin((k_2 - k_1)\,2\pi n t + \varphi_1 \pm \varphi_2) \tag{10}$$

$$F_{\mathrm{ecc}}(t)\,|_{2p=2} = \frac{\pi \cdot r_\delta \cdot l_{\mathrm{Fe}}}{2 \cdot \mu_0} \cdot B_{\mathrm{R},1,1,\mathrm{r}}(r_\delta) \cdot B_{\mathrm{ecc},2,2,\mathrm{r}}(r_\delta) \cdot \sin(2\pi n t + \varphi_{\mathrm{R}} - \varphi_{\mathrm{ecc}}) \tag{11}$$

$$F_{\mathrm{ecc},-}(t)\,|_{2p=4} = \frac{\pi \cdot r_\delta \cdot l_{\mathrm{Fe}}}{2 \cdot \mu_0} \cdot B_{\mathrm{R},2,2,\mathrm{r}}(r_\delta) \cdot B_{\mathrm{ecc},1,1,\mathrm{r}}(r_\delta) \cdot \sin(2\pi n t + \varphi_{\mathrm{R}} + \varphi_{\mathrm{ecc}}) \tag{12}$$

$$F_{\mathrm{ecc},+}(t)\,|_{2p=4} = \frac{\pi \cdot r_\delta \cdot l_{\mathrm{Fe}}}{2 \cdot \mu_0} \cdot B_{\mathrm{R},2,2,\mathrm{r}}(r_\delta) \cdot B_{\mathrm{ecc},3,3,\mathrm{r}}(r_\delta) \cdot \sin(2\pi n t + \varphi_{\mathrm{R}} - \varphi_{\mathrm{ecc}}) \tag{13}$$

This shows that, that the eccentricity air gap δ_{ecc} always leads to a rotational frequent radial force F_{ecc} which depends on the fundamental rotor field wave amplitude $\hat{B}_{\mathrm{R},p,p,r}$ and on the bore radius $r_{\mathrm{s,i}} = r_\delta + \delta/2$. Also F_{ecc} is much higher for 4-pole machines, since two field wave combinations ($\mu_{\mathrm{R}} = 2/\mu_{\mathrm{ecc},1} = 1$ ($\rightarrow F_{\mathrm{ecc},-}$), $\mu_{\mathrm{R}} = 2/\mu_{\mathrm{ecc},2} = 3$ ($\rightarrow F_{\mathrm{ecc},+}$)) contribute to the force generation (Fig. 4d).
The additionally occurring field waves $B_{\mathrm{ecc},\mu,k}$ do no generate eddy current losses in the rotor parts, since these field waves move synchronously with the rotor: Transferring the field wave (5) into rotor-fixed coordinates with the circumference angle $\gamma_{\mathrm{R}} = \gamma - 2\pi \cdot n \cdot t$ leads to the time harmonic order k_{R} in rotor coordinates. Independent of the pole count, for both spectra in Fig. 4, it results $k_{\mathrm{R}} = k - \mu = 0$ due to $k = \mu$. Also in the stator the additional eddy current and hysteresis losses of $B_{\mathrm{ecc},\mu,k}$ may be neglected due to the very small field amplitude compared to the rotor field fundamental (Fig. 3).

3 Influencing Parameters on Eccentricity Forces

The prior section has shown that the eccentricity force oscillates with mechanical rotation frequency $f_{\mathrm{m}} = n$. In this section, we discuss which machine properties influence the amplitude of this eccentricity force $\hat{F}_{\mathrm{ecc}}$. Note that $\hat{F}_{\mathrm{ecc}}$ is not influenced by the angular position φ_{ecc} (7), (8), (9) of the maximum eccentricity air gap. That means, $\hat{F}_{\mathrm{ecc}}$ is identical for δ_{ecc} being located under a pole or under the pole gap (Fig. 3).

Fig. 5a (5b) indicates the eccentricity force amplitude $\hat{F}_{\mathrm{ecc}}$ for a 2-pole (4-pole) rotor topology in dependence of the eccentricity air gap width $\delta_{\mathrm{ecc}} = 0,\ 50\ \mu\mathrm{m},\ 100\ \mu\mathrm{m},\ ...,\ 500\ \mu\mathrm{m}$ and the PM remanence flux density $B_{\mathrm{rem}} = 0.8$ T, 0.9 T, ..., 1.3 T. The remanence flux density has low influence on $\hat{F}_{\mathrm{ecc}}$ for small values of δ_{ecc} and for large values of B_{rem}. The eccentricity force in a 4-pole topology is approximately bigger by factor 3 compared to 2-pole rotors which was explained in Section 2. The values of the two 2-pole rotors, used for the prototype machine, are marked in Fig. 5a. The NdFeB-rotor with the higher B_{rem}-value is more prone to eccentricity forces than the Sm_2Co_{17}-rotor. The eccentricity air gap is $\delta_{\mathrm{ecc}} = 50\ \mu\mathrm{m}$ for the Sm_2Co_{17}-rotor and $\delta_{\mathrm{ecc}} = 100\ \mu\mathrm{m}$ for the NdFeB-rotor. Thus, the simulated eccentricity force is $\hat{F}_{\mathrm{ecc}} = 0.9$ N for the Sm_2Co_{17}-rotor and $\hat{F}_{\mathrm{ecc}} = 3.8$ N for the NdFeB-rotor.

Fig. 6a (6b) indicates the eccentricity force amplitude $\hat{F}_{\mathrm{ecc}}$ for a 2-pole (4-pole) rotor topology in dependence of the eccentricity air gap width $\delta_{\mathrm{ecc}} = 0,\ 50\ \mu\mathrm{m},\ 100\ \mu\mathrm{m},\ ...,\ 500\ \mu\mathrm{m}$ and thickness $h_{\mathrm{PM}} = 1$ mm, 1.5 mm, ..., 5 mm of the PM ring. For small values of δ_{ecc} the PM height does hardly influence the eccentricity force. For high values of δ_{ecc} the eccentricity force is low if the PM height is rather small or rather big. This is because a big PM height increases the air gap flux density but also increases the magnetically effective air gap $\hat{B}_{\mathrm{R},p,p,r} \propto h_{\mathrm{PM}}/\left(h_{\mathrm{PM}} + \delta\right)$. This finding is independent of the pole count, but the eccentricity force is much bigger by approximately factor 3 for the 4-pole compared to the 2-pole topology.

Fig. 7a (7b) gives the eccentricity force amplitude $\hat{F}_{\mathrm{ecc}}$ for a 2-pole (4-pole) rotor topology in dependence of the eccentricity air gap width $\delta_{\mathrm{ecc}} = 0,\ 50\ \mu\mathrm{m},\ 100\ \mu\mathrm{m},\ ...,\ 500\ \mu\mathrm{m}$ and the stator inner diameter $2 \cdot r_{\mathrm{s,i}} = 20$ mm, 25 mm, ..., 100 mm at $B_{\mathrm{rem}} = 1$ T. The PM height h_{PM}

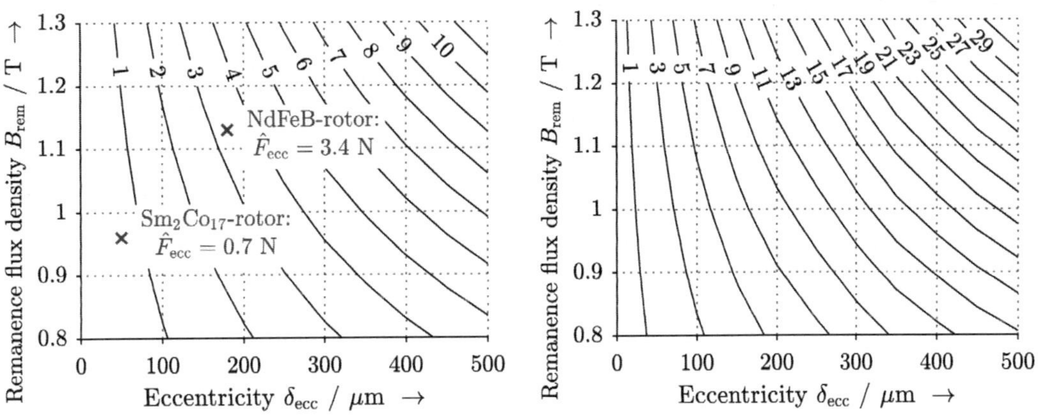

(a) Pole count $2p = 2$ (b) Pole count $2p = 4$

Figure 5: Simulated radial magnetic eccentricity force $\hat{F}_{\mathrm{ecc}}$ in N for varying magnetic eccentricity air gap width δ_{ecc} and varying remanence flux density B_{rem} (dimensions according to Table 2, assumption of magnetically linear materials)

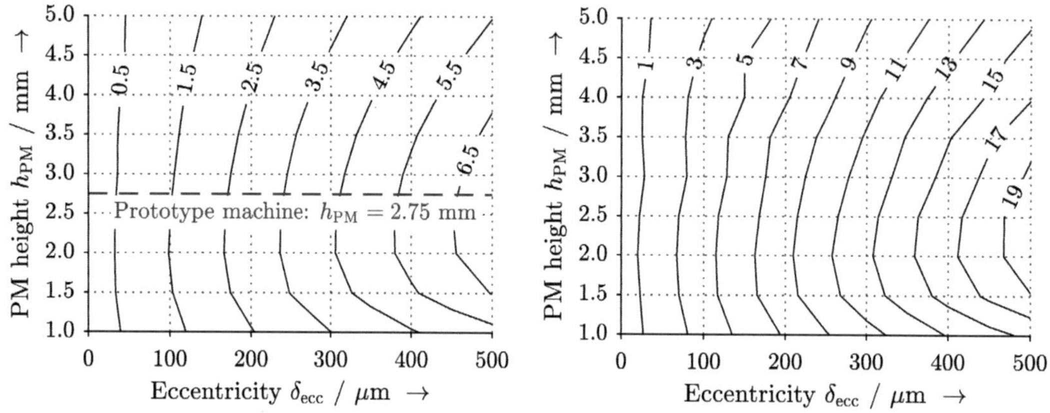

(a) Pole count $2p = 2$ (b) Pole count $2p = 4$

Figure 6: Simulated radial magnetic eccentricity force $\hat{F}_{\mathrm{ecc}}$ in N for varying magnetic eccentricity air gap width δ_{ecc} and varying PM height h_{PM} ($B_{\mathrm{rem}} = 1$ T,dimensions according to Table 2, assumption of linear materials)

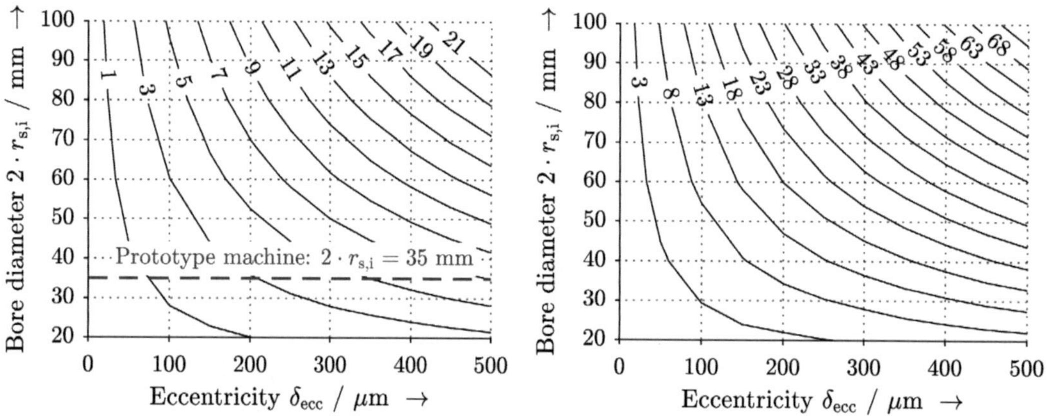

(a) Pole count $2p = 2$ (b) Pole count $2p = 4$

Figure 7: Simulated radial magnetic eccentricity force $\hat{F}_{\mathrm{ecc}}$ in N for varying magnetic eccentricity air gap width δ_{ecc} and varying stator bore diameter $2 \cdot r_{\mathrm{s,i}}$ ($B_{\mathrm{rem}} = 1$ T, $h_{\mathrm{PM}} = 2.75$ mm, $\delta + h_{\mathrm{b}} = 2.5$ mm, assumption of linear materials)

and the shaft radius is kept constant ($h_{\mathrm{PM}} = 2.75$ mm, $r_{\mathrm{sh}} = 12.25$ mm). The bore diameter $2 \cdot r_{\mathrm{s,i}}$ and the eccentricity air gap width δ_{ecc} both increase the $\hat{F}_{\mathrm{ecc}}$-values to the same extent. In (5) $r_{\mathrm{s,i}}$ is included linearly. Also in Fig. 3 the linear relationship between δ_{ecc} and the modulated air gap field wave $\hat{B}_{\mathrm{ecc},\mu,k,\mathrm{r}}$ is proved. Therefore, especially machines with large bore diameters are prone to magnetic eccentricity-originated forces. The findings are independent of the pole count, but the eccentricity force is much bigger by approximately factor 3 for the 4-pole compared to the 2-pole topology.

4 Measurements and Influence on Rotor Position Control

At the prototype machine (Fig. 1) the suspension currents for horizontal rotor forces $i_{d,L}$ and for vertical rotor forces $i_{q,L}$ were measured at rotor standstill for 10 different equidistant rotor angle positions (Fig. 9). The currents are averaged over 1 s. Since the rotor position is constant, the dynamic controller properties are excluded from the measurements. The rotor is – according to the calibrated radial eddy current position sensors – in the concentric position. It is aligned with the vertical axis to avoid the influence of the gravitational force.
The measurement was conducted with the Sm_2Co_{17}-rotor (Fig. 9a) and with the NdFeB-rotor (Fig. 9b). Both rotors exhibit a sine- and cosine-shaped radial disturbing force, which is compensated by the suspension currents $i_{d,L}$ and $i_{q,L}$. Consequently, this radial force wave of amplitude $\hat{F}_{ecc}$ rotates synchronously with the rotor (time order $k = 1$).
The measured suspension force-current coefficients are $k_{F,Nd} = 1.09\ N \cdot A^{-1}$ for the NdFeB-rotor and $k_{F,Sm} = 1.01\ N \cdot A^{-1}$ for the Sm_2Co_{17}-rotor. The AC component of the suspension force is $\hat{i}_{ecc} = 3.42$ A for the NdFeB-rotor and $\hat{i}_{ecc} = 0.83$ A for the Sm_2Co_{17}-rotor. Thus, the eccentricity forces are determined to be $\hat{F}_{ecc} = 3.7$ N for the NdFeB-rotor and $\hat{F}_{ecc} = 0.83$ N for the Sm_2Co_{17}-rotor. These results agree with the simulation results at $\delta_{ecc} = 180\ \mu m$ for the NdFeB-rotor (simulated: $\hat{F}_{ecc} = 3.4$ N) and at $\delta_{ecc} = 50\ \mu m$ for the Sm_2Co_{17}-rotor (simulated: $\hat{F}_{ecc} = 0.7$ N). This eccentricity force $F_{ecc}(t)$ acts on the rotor according to Figure 8, in which the rotor plant is shown without the tilting motions and the gyroscopic effect for simplicity. The damping and oscillation behavior of the plant is determined by the controller settings, which must overcome the negative magnetic stiffness $k_{s,r}$ (Fig. 8). In contrast to rotor unbalance forces, the amplitude $\hat{F}_{ecc}$ of the magnetic eccentricity force is speed-independent. Neglecting the controller influence on the rotor movement and considering only the movement originated by $F_{ecc}(t)$, the actual amplitude $\hat{x}_{ecc}$ of the movement due to the eccentricity force decreases with rising speed $n = f_m$ according to *Newton*'s law with $\hat{x}_{ecc} \propto 1/f_m^2$ (14).

$$F_{ecc}(t) = \hat{F}_{ecc} \cdot \sin(2\pi \cdot f_m \cdot t) = m \cdot \ddot{x}_{ecc} \quad \Rightarrow \quad \hat{x}_{ecc} = \frac{\hat{F}_{ecc}}{m \cdot (2\pi \cdot f_m)^2} \tag{14}$$

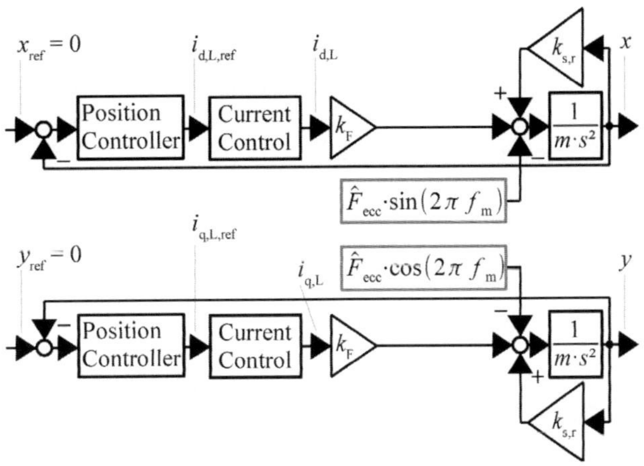

Figure 8: Simplified position controller circuit with eccentricity force F_{ecc}, acting on the rotor plant (neglect of rotor tilting motions and gyroscopic effect)

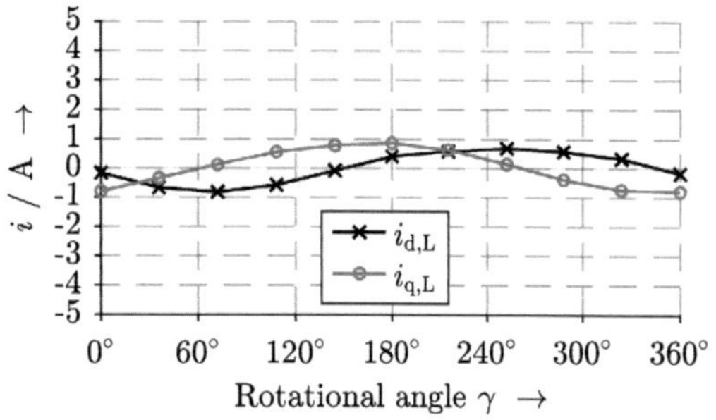

(a) Sm_2Co_{17}

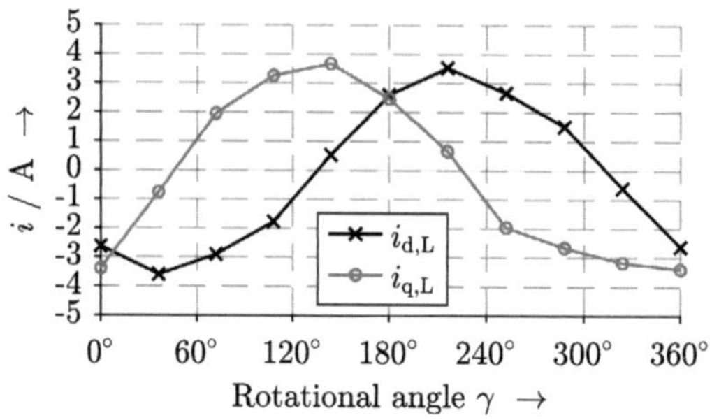

(b) NdFeB

Figure 9: Measured currents $i_{d,L}$, related to a bearing force in x-direction, and $i_{q,L}$, related to a bearing force in y-direction at concentric vertical rotor position for a varying rotation angle γ at rotor standstill ($i_{d,L}$, $i_{q,L}$ expected to be zero for no magnetic eccentricity)

Fig. 10a shows the actual x-position of the NdFeB-rotor during a speed-up from $n = f_m = 0$ to $n = f_m = 240$ Hz. Also the red line indicates the theoretical magnetic eccentricity originated rotor movement amplitude $\hat{x}_{ecc}$ according to (14). The rotor's rigid body eigenfrequencies for the chosen controller stiffness are at $f_{cig,1} = 48$ Hz and $f_{cig,2} = 66$ Hz. These eigenfrequencies do not appear in the position signal with increased oscillation amplitudes due to the good damping behavior of the position controller. The rotor orbit decreases from $\hat{x} = 52$ μm at $f_m = 5$ Hz to $\hat{x} = 26$ μm at $f_m = 150$ Hz. This decrease is owing to mechanical "damping" property described in (14). For $f_m > 155$ Hz a "force-free" rotation [11, 12] is enabled by notching out f_m in the position controller feedback signal. By this, the rotor orbit is even decreased due to the poor position controller's disturbance reaction in case of high-frequent disturbances.
Fig. 10b shows the suspension current $i_{d,L}$ for the NdFeB-rotor during a speed-up from $n = f_m = 0$ to $n = f_m = 240$ Hz. The current takes high values up to $\hat{i}_{d,L} = 3$ A at standstill due to magnetic eccentricity forces (Fig. 9b). The current amplitude even increases up to $\hat{i}_{d,L} = 4$ A at $f_m = 150$ Hz, since the position controller's disturbance reaction to the rotational-frequent forces worsens for higher frequencies. Thus, the rotation frequency f_m is notched out of the actual position sensors for $f_m > 155$ Hz. Consequently, the AC component in the suspension current $i_{d,L}$ diminishes.
The effectiveness of the notch filter can also be seen by comparing Fig. 10c (operation at $f_m = 20$ Hz) with Fig. 10d (operation at $f_m = 200$ Hz). At low speed – and, thus, a big eccentricity-originated rotor movement $\hat{x}_{ecc}$ – the position controller counteracts the rotor movement with current nearly in phase opposition to the position signal. At high speed, the rotor orbit is inherently small, since the magnetic eccentricity-originated rotor movement diminishes. Additionally, the position controller shows no reaction to the rotational-frequent rotor oscillation, which occurs due to rotor unbalance.
The investigation shows that the magnetic eccentricity force F_{ecc} does not influence the bearingless machine's operation at speeds $n > 50$ Hz. Even though it is not crucial for high-speed operation, the suspension current amplitudes at low speed can be crucial. This criticality is based on the inverter's current limit. In Fig. 10a, 10b the torque-producing current was only $i_{q,D} = 0.3$ A and, thus, by factor 10 smaller than the suspension current.
For high machine dynamics, the speed-up process may be carried out with high torque-producing current values $i_{q,D}$. Also, the torque producing current leads to radial disturbing force orthogonal to the controller's reference force vector [13], which worsens the controller's

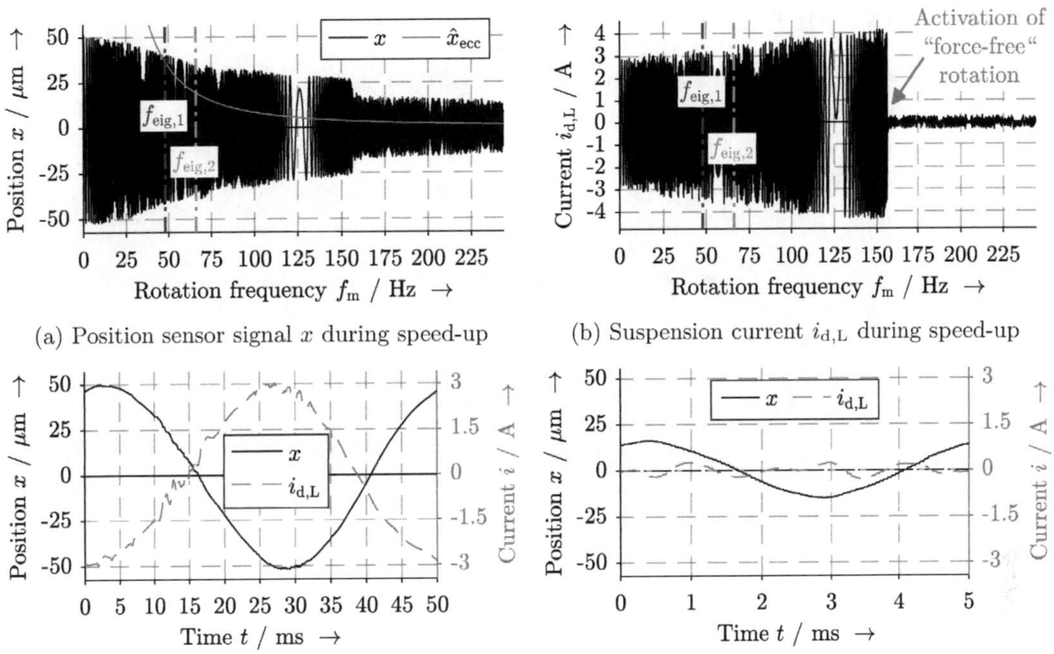

(a) Position sensor signal x during speed-up

(b) Suspension current $i_{d,L}$ during speed-up

(c) Suspension current $i_{d,L}$ and position x at rotation with $f = 20$ Hz

(d) Suspension current $i_{d,L}$ and position x at rotation with $f = 200$ Hz

Figure 10: Measured suspension current and position signal with the NdFeB-rotor at operation with rotation frequency $f < 240$ Hz

reaction to the magnetic eccentricity forces. Therefore, the inverter's current limit may be reached very fast if the magnetic eccentricity force is big and the machine is accelerated fast. For the prototype machine the inverter's current limit is at $i_{max} = 16$ A (momentary value), which is reached if the torque-producing reference current at zero-speed is $i_{q,D,ref} > 10$ A. Even though the reference currents are below the inverter's current limit, the over-current protection can trip due to ripple in the actual current, caused by the pulse width modulation. For the considered prototype, values of $i_{q,D,ref} > 10$ A are only possible at speeds $n = f_m > 150$ Hz.

5 Conclusion

Disturbing radial rotor forces in bearingless PM machines are discussed which have the origin in an undesired half-moon shaped air gap between the permanent magnet and the rotor shaft. The occurring radial force wave rotates with rotor frequency independent of the pole count. Its amplitude increases for PMs with high remanence values and for machines with a large bore diameter. Even for very small eccentricity air gaps of $\delta_{ecc} < 200$ μm the occurring force amplitude can take considerable values of several N. Measurements on a prototype with a NdFeB-rotor have shown that an eccentricity air gap of $\delta_{ecc} = 180$ μm leads to high suspension currents, which can reach the inverter's current limit. The resulting rotor orbit is very big at low speed but decreases strongly with speed ($\propto 1/n^2$) due to the rotor mass inertia, since the eccentricity force amplitude is independent of the rotor speed. Thus, at high-speed operation magnetic eccentricity forces are not crucial.

6 Acknowledgments

Funded by the Deutsche Forschungsgemeinschaft (DFG, German Research Foundation) – 437667923, BI 701/22-1 (Gefördert durch die Deutsche Forschungsgemeinschaft (DFG) – 437667923, BI 701/22-1). Supported by KEBA Industrial Automation Germany GmbH.

References

[1] Vacuumschmelze GmbH & Co. KG, "Rare Earth Permanent Magnets Vacodym Vacomax," 2014. [Online]. Available: https://vacuumschmelze.de/shared/quickNav/Downloads

[2] S. Ruoho, M. Haavisto, E. Takala, T. Santa-Nokki, and M. Paju, "Temperature Dependence of Resistivity of Sintered Rare-Earth Permanent-Magnet Materials," *IEEE Transactions on Magnetics*, vol. 46, no. 1, pp. 15–20, Jan. 2010. [Online]. Available: http://ieeexplore.ieee.org/document/5357499/

[3] H.-D. Stölting, E. Kallenbach, and W. Amrhein, Eds., *Handbook of fractional-horsepower drives.* Berlin, Heidelberg, New York: Springer-Verlag, 2008.

[4] Y. Fu, M. Takemoto, S. Ogasawara, and K. Orikawa, "Investigation of Operational Characteristics and Efficiency Enhancement of an Ultra-High-Speed Bearingless Motor at 100,000 r/min," *IEEE Transactions on Industry Applications*, vol. 56, no. 4, pp. 3571–3583, Apr. 2020.

[5] Z. Liu, A. Chiba, Y. Irino, and Y. Nakazawa, "Optimum Pole Number Combination of a Buried Permanent Magnet Bearingless Motor and Test Results at an Output of 60 kW With a Speed of 37000 r/min," *IEEE Open Journal of Industry Applications*, vol. 1, pp. 33–41, Feb. 2020.

[6] A. Borisavljevic, *Limits, Modeling and Design of High-Speed Permanent Magnet Machines*, ser. Springer Theses. Berlin, Heidelberg: Springer-Verlag, 2013.

[7] F. Rezaee-Alàm, B. Rezaeealam, and J. Faiz, "Unbalanced Magnetic Force Analysis in Eccentric Surface Permanent-Magnet Motors Using an Improved Conformal Mapping Method," *IEEE Transactions on Energy Conversion*, vol. 32, no. 1, pp. 146–154, 2017.

[8] U. Kim and D. Lieu, "Magnetic field calculation in permanent magnet motors with rotor eccentricity: with slotting effect considered," *IEEE Transactions on Magnetics*, vol. 34, no. 4, pp. 2253–2266, Jul. 1998. [Online]. Available: http://ieeexplore.ieee.org/document/703863/

[9] G. Messager, "Self-bearing permanent magnet synchronous machine configurations and control for high-speed applications," Ph.D. Thesis, Technical University of Darmstadt, Institute for Electrical Energy Conversion, Darmstadt, Germany, Sep. 2019.

[10] N. Erd, R. Koster, and A. Binder, "Computational Analysis of Air Gap Field in Electrical Machines by Fourier Coefficient Matrices," in *2020 International Conference on Electrical Machines (ICEM).* Gothenburg, Sweden: IEEE, Aug. 2020, pp. 2479–2485. [Online]. Available: https://ieeexplore.ieee.org/document/9270692/

[11] G. Schweitzer and E. H. Maslen, Eds., *Magnetic Bearings: Theory, Design, and Application to Rotating Machinery.* Berlin, Dordrecht, Heidelberg, New York: Springer-Verlag, 2009.

[12] A. Chiba, T. Fukao, O. Ichikawa, M. Oshima, M. Takemoto, and D. Dorrell, *Magnetic bearings and bearingless drives.* Amsterdam, London: Elsevier/Newnes, 2005, oCLC: 179729376.

[13] G. Messager and A. Binder, "Derivation of forces and force interferences in a double conical high-speed bearingless permanent magnet synchronous motor," in *International Electric Machines and Drives Conference (IEMDC)*, Miami, USA, May 2017, pp. 1–8.

Identification of Various Frequency Response Functions for Levitating Rotor System using Active Magnetic Bearings

Michael Kreutz[1,2], Johannes Maierhofer[1], Thomas Thümmel[1], and Daniel J. Rixen[1]

[1] Chair of Applied Mechanics, Technical University of Munich, Garching, Germany
[2] m.kreutz@tum.de

Abstract

Frequency response functions (FRFs) of rotor systems can be used as indicator functions for condition monitoring. Component-wise FRFs are of high interest to locate errors in the fault case. To enable continuous monitoring, measurements should be taken during operation.

This contribution shows methods of using active magnetic bearings (AMBs) for determining different FRFs of a rotor system. The AMBs are used as combined sensors and actuators and at the same time as support for the rotor. Two different types of FRFs, namely for the free rotor and the supported/bounded system dynamics can be determined from a single experiment. This procedure does not need any change in the assembly, because the AMBs are simultaneously used as bearing and exciter. Using the ground reaction force of the magnetic bearings gives the free rotor's FRF. The supported system's FRF is determined by considering the excitation force only.

As show-case, an academic rotor test rig is used with and without rotation to show the general feasibility of the method. To evaluate and interpret the results of the experiments, a numerical model of the rotor using finite-element formulations is used.

1 Introduction

Condition monitoring compares a rotor system's actual behavior with a desired or normal behavior using indicators to detect faults. In [16], multiple rotor faults are described. Indicator functions can be for example frequency response functions (FRFs) or modal parameters. Mechanical FRFs $\boldsymbol{H}(\omega)$ are defined as the linear relationship between harmonic forces $\boldsymbol{F}(\omega)$ and harmonic responses $\boldsymbol{X}(\omega)$ (displacement, velocity, acceleration) which depend on the excitation frequency ω, cf. [2].

$$\boldsymbol{X}(\omega) = \boldsymbol{H}(\omega)\,\boldsymbol{F}(\omega) \tag{1}$$

In a simulation model, it is easy to obtain the FRFs. For a symmetric rotor, considering gyroscopic effects, the FRFs for a rotational speed Ω can be obtained by

$$\boldsymbol{H}(\omega) = \left(-\omega^2\boldsymbol{M} + i\omega\left(\boldsymbol{C} + \boldsymbol{G}(\Omega)\right) + \boldsymbol{K}\right)^{-1} \tag{2}$$

with the system's mass, damping, gyroscopic and stiffness matrix $\boldsymbol{M}, \boldsymbol{C}, \boldsymbol{G}, \boldsymbol{K}$ respectively.

For an experimental determination, there exist a number of algorithms, based on fast Fourier transforms, that estimate the FRF from a set of given input (force) to output (response) time-signals. Determining the FRFs is the first part of experimental modal analysis (EMA). Measuring the FRFs of rotating systems is a special challenge.

Strategies to measure the FRFs of rotating systems have been proposed in the BRITE/-MARS project [4], also given in related items [3]. In these publications, active magnetic bearings (AMBs) are used as excitation system additional to existing support in ball bearings. In [6], an approach for combining the AMB's control system with a MIMO identification procedure to obtain a state-space model through the system's FRFs is given for systems with real poles. In [5], a method to combine the magnetic bearing's support with an additional excitation is given. This is similar to the method applied in this contribution. Here, we want to use a simple implementation of standard MIMO excitation signal, as e.g. burst random excitation, which can be found in text books about EMA, for example in [2, p. 323ff]. We also want to give an intuitive interpretation of the results, to understand how AMBs can be used to get different FRFs of one system in a single experiment.

In [12], an approach of using AMBs to perform EMA of the supported rotor system using an external commercial modal analysis system was shown for the levitating, but non-rotating rotor.

In order to measure the FRFs of a system, typically, its operation must be stopped. The system must be instrumented by exciters and sensors. The system is excited and from the resulting time-data, it is possible to determine the FRFs. Stopping the system's operation and performing the tests takes a lot of time and effort, so this is not feasible for condition monitoring. In this contribution, the AMBs are used as combined exciters and sensors, while still being used as bearings. This enables to measure the FRF of the rotor system during operation, without the need to stop its operation. The excitation force and the total force in the bearing is estimated by a linear current-force relationship. With this, it is possible to determine the FRFs of the supported system and of the free rotor in a single measurement run. Thus, using FRFs for a continuous condition monitoring of the rotor system is enabled. The availability of the supported system's FRF and free rotor's FRF is especially useful for localizing faults in the rotor system as the bearing effects can be separated from the rotor effects in the different FRFs.

Here, the test rig's control system is used for measuring the necessary data, without the need for an additional modal analysis system. The excitation is applied via the bearings, using an additional excitation current that is added to the AMB's controller current. The combination of the excitation and bearing task in the magnetic bearing enables to measure FRFs of different theoretical system configurations in a single experiment but by considering different force signals. Namely: Considering only the additional excitation force gives the supported rotor system's FRFs. Considering the total force, comprised of contributions from the controller and from the additional excitation, gives the free rotor's FRF.

In this contribution, the approach will be applied to an academic test rig with and without rotation. First, the test rig and a simulation model of the test rig is explained. Then results of applying the method to a non-rotating system are given and interpreted. The method is then also applied to a rotating system, which shows its feasibility. In the end, a short summary of the contribution is given.

2 Test Rig and Simulation

The magnetic bearing test rig at the Chair of Applied Mechanics of the Technical University of Munich will be used as an example of a levitating rotor in magnetic bearings. In fig. 1, the test rig with its components is given. The test rig is also described in [8, 10]. The rotor is supported in active magnetic bearings and can be driven by a servo motor which is coupled to the rotor via a double cardan joint. A mass disc and two bearing journals are attached to the rotor. Displacements are measured in planes on the front and back plate of each magnetic bearing via eddy current sensors. Measuring the acceleration by adding acceleration sensors to the rotor is not feasible, because the system should be able to rotate. Compared to the test

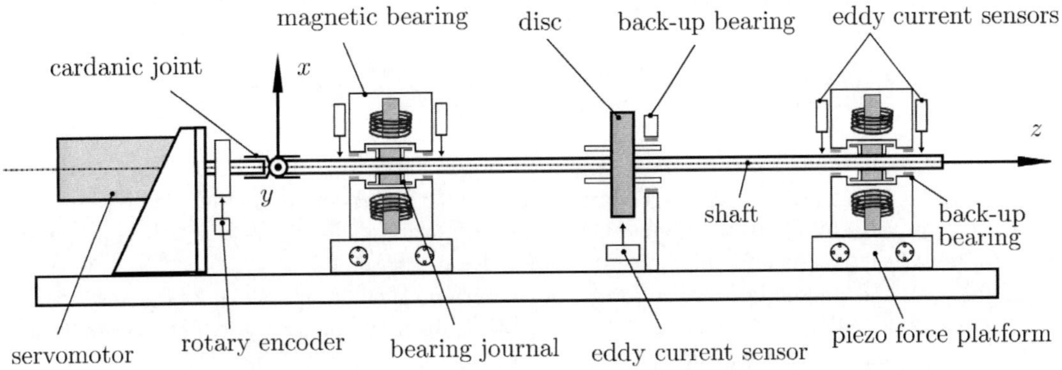

Figure 1: Schematic view of the magnetic bearing test rig, see also [10].

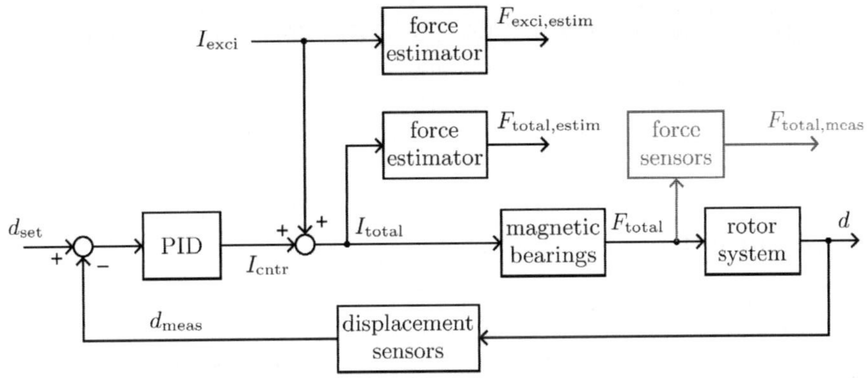

Figure 2: Control scheme including additional excitation.

rig configuration in [10], the test rig has been improved by applying additional force-torque sensors (piezo force platform Kistler 9129AA) underneath the magnetic bearings. They give an accurate reading of the magnetic bearing ground force, which corresponds to the total force of the bearing. It is used later for a comparison with the results using a force estimation for the total force.

The position of the rotor in the magnetic bearings is controlled by a PID controller, which uses the position error with the measured displacement d_{meas} as input signal and a controller current I_{cntr} as output. The PID controllers are implemented separately for each direction. A control scheme of the system is given in fig. 2. Only small vibrations inside the bearing are assumed, so that the magnetic force law is not influenced by the position in this example. An additional excitation current I_{exci} can be applied to the magnetic bearings by adding it to the controller current. The force resulting from the excitation current I_{exci} cannot be measured directly. The measured force $F_{total,meas}$ contains both contributions from the controller current I_{cntr} and the excitation current I_{exci}, which cannot be easily separated:

$$F_{meas} = f(I_{cntr} + I_{exci}, \dots). \tag{3}$$

An ideal behavior of the force sensors is assumed ($F_{total} \approx F_{total,meas}$). In order to be able to separate the forces, we assume a linear relationship between the applied current and the resulting force.

$$F = k_i I, \qquad \text{with } k_i = \frac{\sqrt{2}}{4} \cdot 50\,\frac{\mathrm{N}}{\mathrm{A}} \approx 18\,\frac{\mathrm{N}}{\mathrm{A}} \tag{4}$$

The factor k_i was determined at the chair in a similar test rig to [11]. The bias current is $I_b = 2.5\,\mathrm{A}$. The magnetic bearings are actuated in a differential driving mode, e.g. [14, p.125]. In eq. (4), the current I is split in current between opposing coils, where one coil is actuated with $I_{\text{coil},1} = I_b + I_{\text{total}}/2$ and the opposing coil with $I_{\text{coil},2} = I_b - I_{\text{total}}/2$.

In general, the magnetic bearing force law is non-linear, but in special cases, e.g. for a low current and small displacements, a linear relationship may be assumed. The non-linearity between displacement, current and force for this magnetic bearing was investigated in [11].

The excitation and total force are estimated from eq. (4)

$$F_{\text{exci,estim}} = k_i I_{\text{exci}}, \qquad F_{\text{total,estim}} = k_i I_{\text{total}}. \tag{5}$$

They should be used to calculate the FRFs of the assembled system and the free rotor in this contribution. The force $F_{\text{total,meas}}$ is directly measured by the force platforms. The results using the measured force are used as comparison to the results using the estimated total force. This is possible in the academic test rig. In an industrial application adding force measurement adds a lot of cost and complexity and is often not possible due to space restrictions.

2.1 Measurement Setup

In the experiments, FRF-measurement is performed by using the active magnetic bearings as exciters. The experiment uses burst random excitation in x-direction for both AMBs, which consists of 0.5 s excitation using band-limited noise (limited to 250 Hz, amplitude 1.6 A). This is followed by 1.548 s pause. For information about burst random excitation, see [2, p. 338]. As independent burst random excitations are applied on both AMBs, this is a multiple input multiple output (MIMO) identification procedure. It is important for MIMO identification, that the excitation signals are uncorrelated. The measurement duration is 204.8 s with a sampling rate of 10 kHz.

The resulting FRFs are calculated from the time data of force and displacement signals by using algorithms included in ABRAVIBE [1], which is a companion software for the text book [2]. The standard H_1-estimator is used, because it is widely spread and easily applicable to MIMO identification. A Hanning window (2.048 s long) with 67 % overlap is used.

2.2 Simulation

To help the interpretation of the experimental results, a numerical model of the test rig is used. The numerical model is built using the open-source rotor simulation toolbox AMrotor, cf. [7, 9]. A model of the system has been built in [8]. The model of the free rotor has been validated using a commercial modal analysis system with impact excitation of the free rotor, disassembled from the test rig. The magnetic bearings are represented as simple linear spring-damper elements with $k_{\text{AMB}} = 1.3 \times 10^5\,\mathrm{N/m}$ and $d_{\text{AMB}} = 90\,\mathrm{N\,s/m}$. The magnetic bearing forces are assumed to be applied to the middle point in each AMB. The free system is the rotor without added spring-damper-element. The assembled system is the rotor including the spring-damper element as basic model of the magnetic bearings.

3 Non-Rotating System

The experimental results for the non-rotating system are obtained in a single measurement and are validated by comparison with simulation results.

3.1 FRF of Rotor in AMBs Using Additional Excitation Force

In a first step, only the additional excitation force on the AMBs is considered as an input to the FRF calculation algorithm. The excitation force is estimated from the linear static relationship in eq. (5).

Figure 3 shows a simple model of the resulting system. As only the additional force is considered, the bearings forces are considered as internal forces and thus are included in the dynamic FRF. The abbreviation *supp* is used for the *supported* system.

$$\boldsymbol{X}(\omega) = \boldsymbol{H}_{\text{supp}}(\omega)\ \boldsymbol{F}_{\text{exci}}(\omega) \tag{6}$$

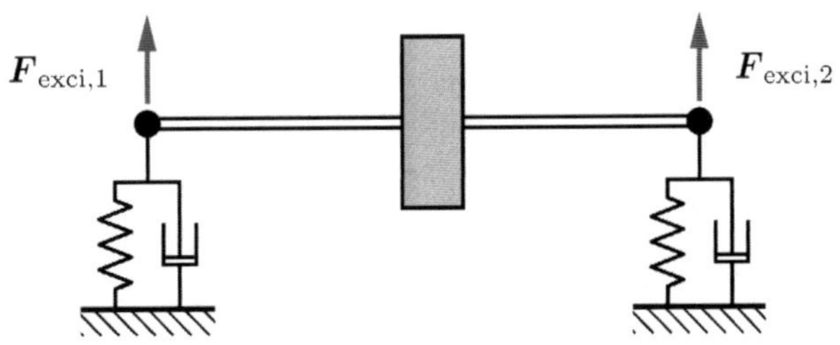

Figure 3: Model of the identified system using F_{exci} as input signal.

Figure 4 shows the experimental results for the FRF using only the additional excitation force and compares them with the results from the numerical simulation for the rotor in elastic bearings. It shows the FRF between excitation at the right AMB and the response at the rightmost eddy current sensor. The excitation and response positions are very close, so a behavior similar to a driving point FRF is expected. The experimental and simulation results are clearly similar. The coherence of the experimental FRF is shown in orange. It starts to fall at around 200 Hz, because of using band-limited noise with a cut-off frequency of 250 Hz as part of the burst random excitation.

The first peak at 16.6 Hz agrees very well in simulation and experiment. The peaks at around 60 Hz are due to the magnetic bearing influence. For these, the simulated eigenfrequencies are lower, than in the experiment. This can be explained by an error in the spring-damper-parameters, that represent the AMBs in the simulation. For the peaks above 100 Hz the simulation deviates from the experiment. They are hardly influenced by the spring-damper parameters in the simulation, but are rather influenced by the stiffness, damping and mass of the rotor itself. Especially the correct damping in the rotor must be determined more accurately. A more accurate model of the rotor should improve the results, e.g. by model updating procedures.

There is an offset in the amplitude between the experimental and simulation results. This is due to an error in the current-force-factor k_i. Additional deformation is brought to the test rig by the force platform itself.

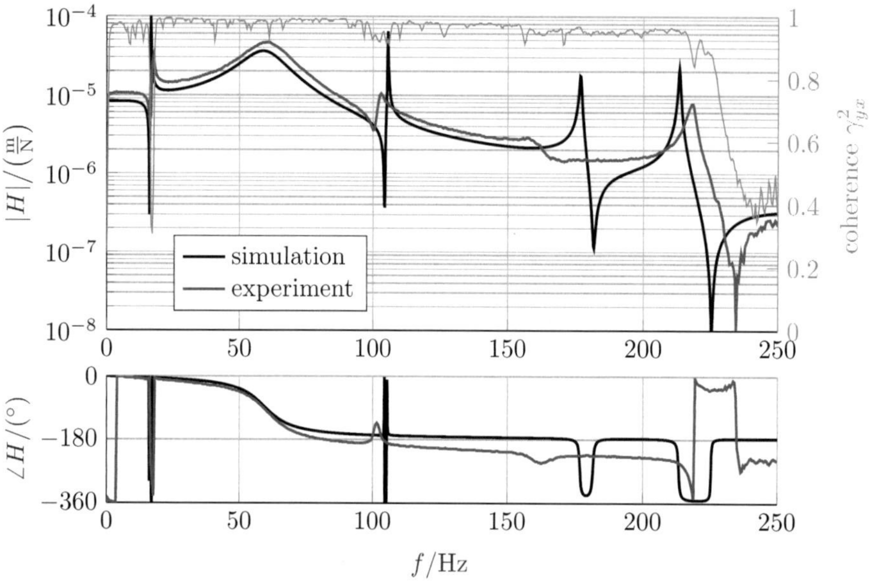

Figure 4: FRF H_{supp} of the rotor supported in magnetic bearings (using only excitation force), non-rotating.

The phase of the FRF shows a similar behavior in simulation and experiment. As expected, the phase falls −180° for resonances and rises 180° at anti-resonances. The phase of the experimental FRF falls down linearly. This is due to effects in the current amplifier, mainly time-delay. These effects are not compensated in the static force estimation (using k_i), so they show up here.

Both curves show anti-resonances close to the resonance frequencies. This means, that the points for excitation or response measurement are very close to vibration nodes. This is due to considering the points at the AMB. The AMB's controller tries to minimize the displacement, because it acts as a bearing. Naturally, the vibration nodes will be close to the bearing. The close vicinity of the excitation and measurement points to the bearing is a disadvantage of the method and means a bad observability and controllability of these points. The stiffer the AMBs act, the worse the problem gets.

This also means, that the resulting FRFs are very sensitive to the exact positioning of the sensors, as they are close to the vibration nodes. Small mistakes in the position have a rather large influence on the measured FRFs. So, an inherent problem of the method is, that the excitation position is very close to the vibration nodes.

This experiment showed the measurement of the supported rotor FRF, using a force estimation using only the additional excitation current.

3.2 FRF of Free Rotor Using Total Force

In this section the total force of the AMB is considered for the FRF calculation algorithm. Figure 5 shows the rotor that is cut free from the bearings. The total force is estimated from eq. (5) and measured by the force platform. It is the force that acts between the ground and the magnetic bearings. It includes the additional excitation force and the controller force. The controller force can be seen as the reaction force of the bearings. It is now assumed to be part of the external excitation and the bearings are therefore no longer part of the characterized system. So only the rotor without the bearings remains, i.e. the free rotor.

$$\boldsymbol{X}(\omega) = \boldsymbol{H}_{\text{free}}(\omega)\ \boldsymbol{F}_{\text{total}}(\omega) \tag{7}$$

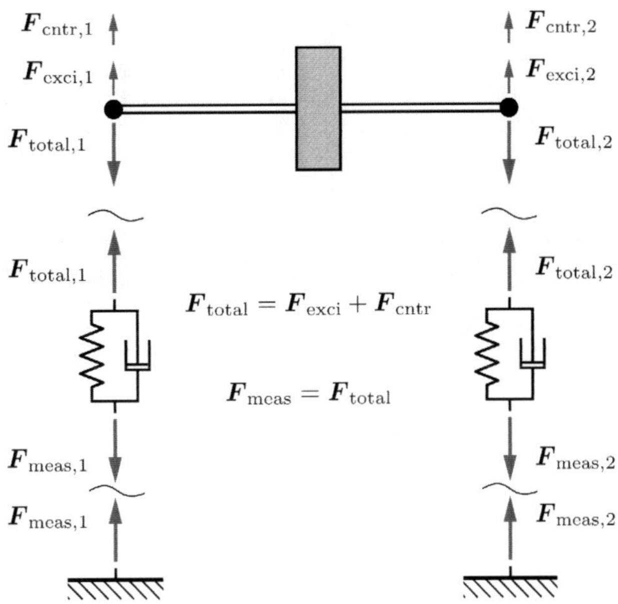

Figure 5: Model of the identified system using the total force F_{total} as input signal.

For this example the estimated total force should be used and compared with the results from the measured total force. There are several methods to determine the total force. In [5], some possibilities are given, that are shortly repeated here:

- Measuring the force between the AMB and the ground. It is suggested to consider the inertia of the AMB. Here, we neglect this influence, as the force platform is assumed to be rigid.
- Measuring the magnetic flux density B using a Hall-sensor in the AMB and estimating the force using the area of the electromagnet and the magnetic permeability constant. This method enables the magnetic bearing to measure the total force, without using additional force sensors, making the AMB a self-sensing active component.
- Estimating the force from a model. This can be a model, which has been parameterized experimentally or which used theoretical knowledge about the AMBs from an electromagnetic simulation model. Here, the force is estimated using a linear static current-force relationship, eq. (5).

Figure 6 shows the results for the experimental FRF using the total force between the right magnetic bearing and the rightmost eddy current sensor. It shows the amplitude and angle of H_{free} and its coherence for the experimental results. This experimental FRF is obtained by using different signals from the same experiment as in previous section 3.1. The curves

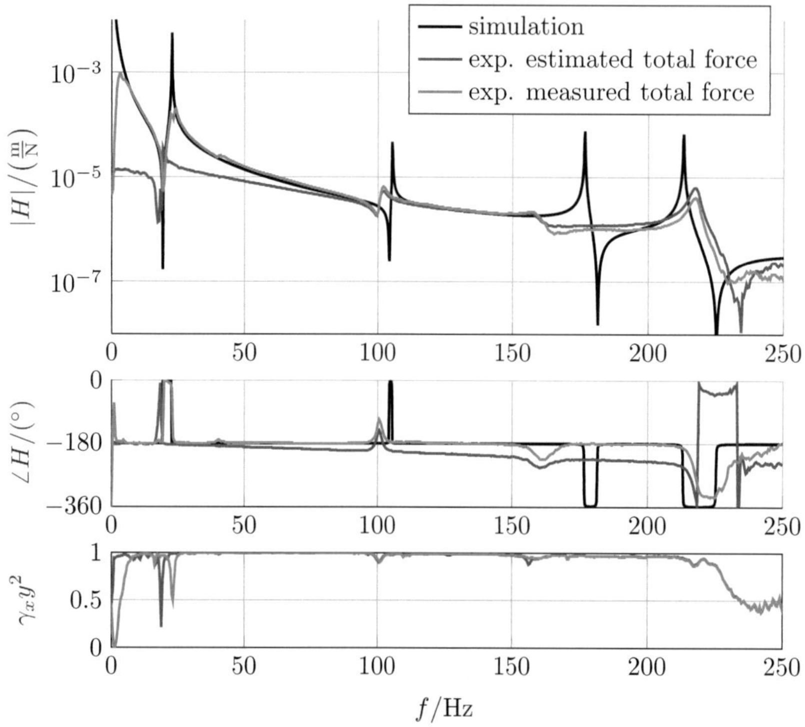

Figure 6: FRF H_{free} of the free rotor (using the total force), non-rotating, with coherence.

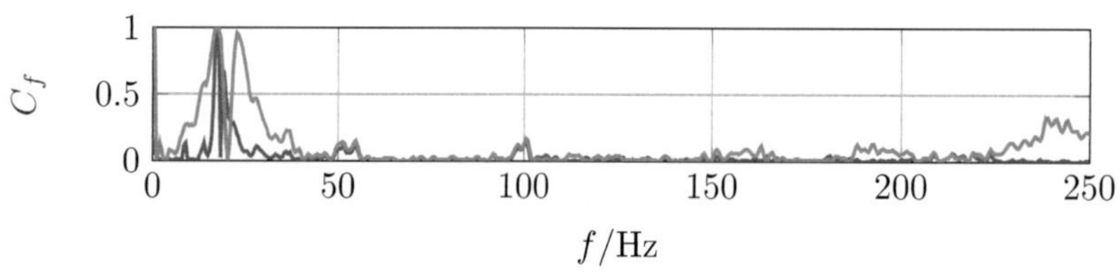

Figure 7: Ordinary coherence between the two input signals (total force).

exp. measured total force use the force that is directly measured by the force platforms F_{meas}. The curves *exp. estimated force* use the force that is estimated from the total electric current $F_{\text{est,total}} = k_i \cdot I_{\text{total}}$. It is compared to the simulation results for the free rotor system.

For the FRF of a free system, one expects to see the effect of the rigid body modes, i.e. the amplitude starts at infinity, $|H(f \to 0)| \to \infty$. This can be clearly seen in the simulation results. In the experimental results, this is less obvious. For the measured force FRF the amplitude agrees after about 6 Hz. For lower frequencies the amplitude does not match. In order to go to infinity, the FFT of the force must go to zero, which is not possible in practice, as the force signal is then primarily dominated by noise. The amplitude above 6 Hz and outside of the resonance points agrees very well, using the force platforms.

The FRF from the estimated force approaches the correct amplitudes only at approx. 100 Hz. Before that, there is a distinct error, that is due to the insufficient force estimation for high currents, which are present at lower frequencies. The solution using the estimated force deviates from the measured force. The real current I_{total} is not measured directly. But it is assumed that the behavior of the electrical system is ideal, i.e. a desired current is perfectly executed. In reality, this does not hold, which can be seen by the deviation of the FRFs in the frequency region up until 90 Hz, where large currents and thus forces are requested. The system could be improved by adding a system for current measurement. Current measurement systems are cheaper and more easily available, than the applied force platforms. Implementing the measurement method in a practical application would only use a force estimation by a measured current, as this saves costs compared to a direct force measurement.

As in the previous section, the simulation model agrees very well with the experimental result for the measured total force in the first peak and is stiffer than the experimental system for eigenfrequencies over 100 Hz.

At the position of the peaks, the amplitudes do not agree, because of inaccuracies in the rotor model (damping) and also because the free rotor has vibration nodes close to the measurement position, which makes the system sensitive to small positioning errors. In contrast to the previous section, this is not an inherent problem of the method, but rather the case for this special rotor system, which has concentrated mass at the bearing journal while the rest of the rotor is rather slim.

Compared to the previous section, there is no eigenfrequency at around 60 Hz, which was found from identification of the rotor with supports in the previous section 3.1.

The phase of the FRFs show a similar behavior between simulation and experiment.

The coherence has small dips at resonance points.This is especially relevant for the first peak. In fig. 7 the ordinary coherence (correlation) between the input signals is shown. The controller force is dominant when there is large displacement that has to be controlled. At resonance frequencies of the supported system the displacement is large and thus strongly influences the total force and thus its correlation. A high correlation of the input signals can cause problems in the FRF calculation due to the poor conditioning of the problem, cf. [2, p. 327].

Another effect that can negatively influence the results is internal feedback in the measured system. This can lead to biased FRFs, cf. [3]. In this magnetic bearings, the displacement within the magnetic bearing directly influences the magnetic force. This influence was neglected in this contribution.

3.3 Interim Conclusion

It has been shown, that it is possible to measure the FRF of the supported rotor system and of the free rotor in a single measurement run. For a better agreement of experimental and simulation results, the numerical model should be updated. Particularly, the damping is not well represented. The supported system's FRFs were experimentally determined by estimating the excitation force from the current. The results agree well. The free rotor's FRFs were determined by considering the total force. The experimental results using the force estimation do not represent well the amplitude for small frequencies. There, the estimation is insufficient, because of high currents. However, the overall behavior agrees well with simulation and experimental results using measured force.

The results for the FRFs using only force estimation showed differences in amplitude compared to simulation results or to measured force FRFs. In continuous condition monitoring the FRFs (indicator function) of a rotating machine are continuously compared to previous FRFs. So, only differences in the FRFs are important, while the exact amplitude is of less importance, which implies the suitability of the presented FRF aquisition using estimated forces.

4 Rotating System – FRF of Free Rotor

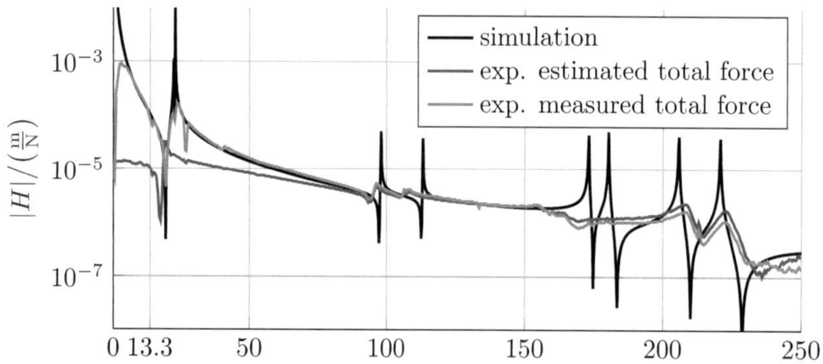

Figure 8: FRF amplitude $|H_{\text{free}}|$ of the free rotor (using total force) for 800 rpm

In this section, an example of applying the measurement method to obtain the free rotor's FRF on the rotating system is shortly displayed.

As given in eq. (2), in simulation the FRF of a symmetric rotor at a given speed can be easily obtained from the system matrices. However, experimental measurement of the FRFs for the rotating system are hard to get. Especially the FRF of the free rotor cannot be obtained by usual EMA techniques, like e.g. impact testing.

Using the technique shown in this contribution, the rotor is excited in its magnetic bearings and the total force is used for FRF-estimation. Figure 8 shows the determined FRFs in simulation and from the experiment for a rotational speed of 800 rpm $\approx 13.3/\text{s}$. Two results from the experiment are shown.

The results are similar to fig. 6. Due to the rotational speed, gyroscopic effects occur, which split the eigenfrequencies. This can be seen in occuring double peaks for the tilting modes. At the frequency of the rotation, which is marked in the figure, the FRF with the measured force is has a small dip. At frequency 2Ω, there is a disturbance from the rotational speed, which is greater in magnitude. So, the FRFs of the rotating system can be determined accurately with little disturbance from the rotation.

Figure 9 gives a summary of the experimentally determined FRFs. Each sub-figure shows the results from one measurement run, for 0 rpm and 800 rpm respectively. This clearly shows the difference between the supported system's and free rotor's FRFs, where the rigid body modes are apparent at small frequencies for the free system and the distinct peak at around 60 Hz only appears in the supported system. By comparing the left and right parts of fig. 9, the splitting of eigenfrequencies with rotation can be seen.

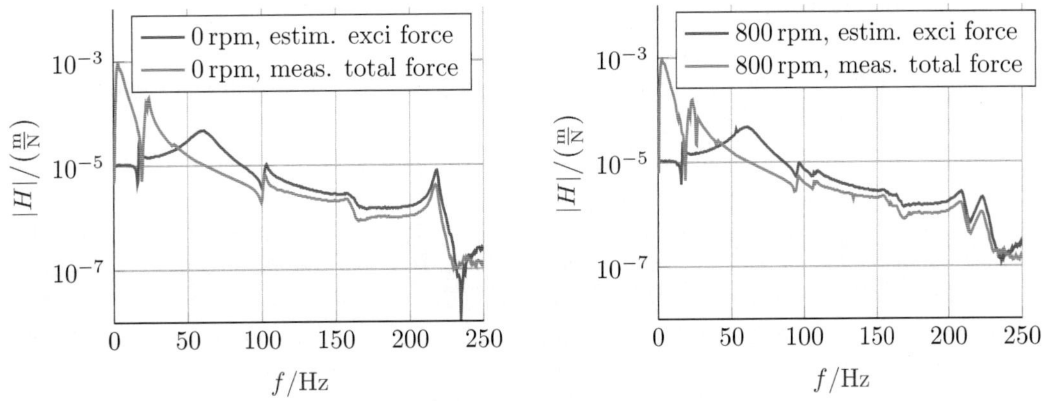

Figure 9: Comparison of the experimentally determined FRFs.

5 Summary

This contribution shows, that it is possible to measure the free FRF of a rotor and the supported system's FRF in a single measurement run using AMBs with multiple input signals. Measurement of the FRF of a rotating rotor without bearing influence could not be done using usual impact hammer FRF measurements.

With the force estimator, used in this contribution, the FRF amplitude cannot be accurately determined for low frequencies. For condition monitoring, the correct amplitude is not as critical as a continuous evaluation of the results. Thus, it could be shown, that FRFs using estimated forces are suitable for condition monitoring. The results are validated by comparison to a more accurate force measurement.

The availability of FRF of the free and the supported rotor during operation can further improve condition monitoring. It can be used to locate a possible fault. A fault in a bearing would show in the supported system's FRF, but not in the free rotor's FRF. A fault in a rotating component can be localized, if it can be observed in both types of FRFs.

The procedure for measuring FRFs of the rotating system can be investigated further as special problems arise for rotating systems. There exist a number of publications regarding suitable excitation signals for MIMO systems, like e.g. [13]. Most of them are not specific to rotating systems, but are considering feedback in general.

Free rotor FRFs can also be used to extend frequency-based substructuring, cf. [15], to rotating systems, which can for example be applied in experimental characterization of rotor system components (e.g. seals, journal bearings). This requires high-quality FRFs, i.e. force measurement or an improved force-model.

6 References

[1] Anders Brandt. *ABRAVIBE - A MATLAB toolbox for noise and vibration analysis.* www.abravibe.com. 2019. URL: `http://www.abravibe.com` (visited on 03/22/2019).

[2] Anders Brandt. *Noise and Vibration Analysis.* Chichester, UK: John Wiley & Sons, Ltd, Feb. 2011. ISBN: 9780470978160. DOI: 10.1002/9780470978160.

[3] I. Bucher and D. J. Ewins. "Modal analysis and testing of rotating structures". In: *Philosophical Transactions of the Royal Society A: Mathematical, Physical and Engineering Sciences* 359.1778 (2001), pp. 61–96. ISSN: 1364503X. DOI: 10.1098/rsta.2000.0714.

[4] D. J. Ewins et al. *MARS - Modal Analysis of Rotating Structures: Development of Validated Structural Dynamic Modelling and Testing Techniques for Vibration Predictions in Rotating Machinery.* Tech. rep. 1996.

[5] P. Förch et al. "Modale Analyse an rotierenden Maschinen mittels Magnetlager". In: *Schwingungen in rotierenden Maschinen III.* Ed. by H. Irretier and R. Nordmann. Springer, 1995, pp. 245–254. DOI: 10.1007/978-3-322-83807-0_24.

[6] Conrad Gähler, Manuel Mohler, and Raoul Herzog. "Multivariable identification of active magnetic bearing systems". In: *JSME International Journal, Series C: Mechanical Systems, Machine Elements and Manufacturing* 40.4 (1997), pp. 548–592.

[7] *GitHub: AMrotor - A MATLAB toolbox for the Simulation of Rotating Machinery.* 2021. URL: https://github.com/AppliedMechanics/AMrotor (visited on 05/20/2021).

[8] Michael Kreutz et al. "Modaler Modellabgleich eines Rotors in Magnetlagern". In: *Sechste IFToMM D-A-CH Konferenz.* Lienz, Österreich, 2020. DOI: 10.17185/duepublico/71192.

[9] J Maierhofer et al. "AMrotor - A MATLAB ® Toolbox for the Simulation of Rotating Machinery". In: *12th International Conference on Vibrations in Rotating Machinery.* Ed. by Institute of Mechanical Engineers. Liverpool: CRC Press, Oct. 2020. DOI: 10.1201/9781003132639.

[10] Johannes Maierhofer et al. "Modellbasiertes Monitoring an magnetgelagerten Rotoren". In: *SIRM 2017* (2017).

[11] Johannes Maierhofer et al. "Progress in Calibrating Active Magnetic Bearings with Numerical and Experimental Approaches". In: *Mechanisms and Machine Science.* Vol. 63. 2018, pp. 249–261. ISBN: 9783319992723. DOI: 10.1007/978-3-319-99272-3_18.

[12] Johannes Maierhofer et al. "Using the Dynamics of Active Magnetic Bearings to perform an experimental Modal Analysis of a Rotor System". In: *SIRM 2019.* February. 2019. DOI: 10.13140/RG.2.2.26743.16809.

[13] Steven Pauwels et al. "A new MIMO sine testing technique for accelerated, high quality FRF measurements". In: *Conference Proceedings of the Society for Experimental Mechanics Series* (2006). ISSN: 21915644.

[14] Gerhard Schweitzer. "Applications and research topics for active magnetic bearings". In: *IUTAM-Symposium on Emerging Trends in Rotor Dynamics.* Vol. 25. 2009. ISBN: 9789400700192. DOI: 10.1007/978-94-007-0020-8-23.

[15] Maarten V. van der Seijs, Dennis de Klerk, and Daniel J. Rixen. "General framework for transfer path analysis: History, theory and classification of techniques". In: *Mechanical Systems and Signal Processing* 68-69 (2016), pp. 217–244. ISSN: 10961216. DOI: 10.1016/j.ymssp.2015.08.004. URL: http://dx.doi.org/10.1016/j.ymssp.2015.08.004.

[16] Thomas Thuemmel et al. "Rotor Orbits at Operation Speed and Model-Based Diagnosis of Multiple Errors". In: *Proceedings of the 10th International Conference on Rotor Dynamics – IFToMM.* Springer International Publishing, 2019, pp. 222–237. ISBN: 9783319992686. DOI: 10.1007/978-3-319-99268-6_16. URL: http://link.springer.com/10.1007/978-3-319-99268-6%7B%5C_%7D16.

Summary of a Bearingless Motor with Passive Electrodynamic Axial Suspension

Guilherme Cavalcante Rubio[1], Yusuke Fujii[1], and Akira Chiba[1]
[1] Tokyo Institute of Technology, Meguro, Tokyo, Japan
cavalcante.g.aa@m.titech.ac.jp
fujii@ee.e.titech.ac.jp
chiba@ee.e.titech.ac.jp

1 Introduction

Either active control or passive stabilization can be used to achieve non-contact magnetic suspension in magnetic bearings and bearingless motors. As the rotational speed is increased, the *d*-axis current becomes harder to regulate, thus motivating the research of full passive suspension. Permanent magnet bearings and reluctance forces cannot stabilize all Degrees Of Freedom (DOF) by themselves, therefore, complementary passive methods are required to achieve full passive suspension in bearingless motors.

Among several passive suspension techniques, electrodynamic suspension is chosen for this machine, because the same coils that generate motor torque can be used for passive stabilization and it does not require superconductors with a dedicated cooling system.

Passive electrodynamic suspension in rotating machines has been already explored in several structures. It was applied to both passive electrodynamic bearings [1] and bearingless motors [2]–[4]. This paper reports an improvement of a previously reported bearingless motor with passive axial suspension through figure-eight coils previously presented in an IEEE IAS cosponsored conference paper [4]. It suffered from low axial force due to the low radial flux density. This electrodynamic suspension principle on axial direction is similar to previously published machines [2], [3]. In this paper, magnetic flux density and, consequently, axial electrodynamic force at low rotational speed are significantly increased when compared to the previous model [4]. A prototype machine is built to confirm the axial force characteristics.

2 Suspension Principle

Fig. 1 show a planar view of the figure-eight coil [2] A_1. The rotor has surface-mounted permanent magnets that generate a sinusoidal magnetic field. This magnetic field is represented by an array of vectors entering and leaving the coil plane. The permanent magnet rotates at a constant angular speed ω in radians per second. When the rotor is displaced from the axial center, there is an imbalance between fluxes linked by the upper and lower coils. Consequently, a suspension current i_{za1} circulates inside the figure-eight coil. Through the interaction of this current and the rotor magnetic field, stabilizing electrodynamic forces are generated in the stator coils, as indicated by blue arrows in Fig. 1.

Fig. 2 shows the equivalent electrical circuit for the figure-eight coil A_1. The upper and lower coils are assumed to have identical shape, identical number of turns n, identical resistance R, and identical self-inductance L. Defining the number of figure-eight coils in the stator c, the number of pole pairs p, and the constant K_λ, equivalent to the derivative of flux linkage in the upper and lower coils in terms of axial displacement z, the total axial electrodynamic force on the stator is derived as

$$F_{ze\,T} = \frac{c\,{K_\lambda}^2\,p^2\,\omega^2 L\,z}{(R^2 + p^2\omega^2 L^2)}. \tag{1}$$

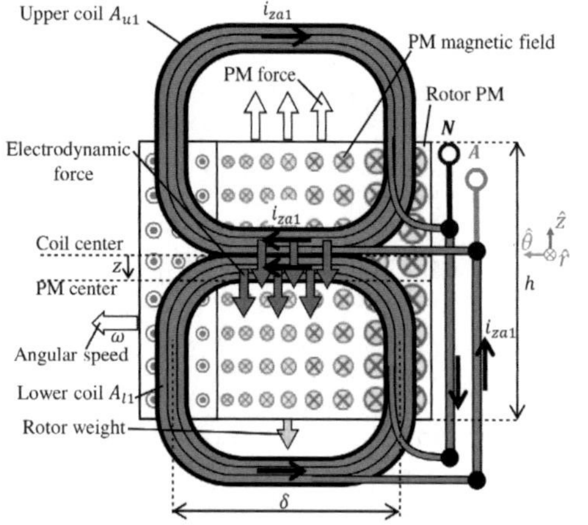

Fig. 1. Planar view of figure-eight coil A_1

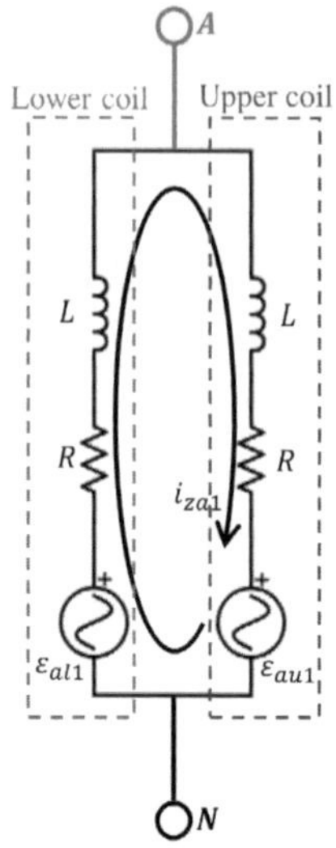

Fig. 2. Equivalent circuit of figure-eight coil A_1

3 Improvements of the Motor Structure

Figs. 4(a)-(d) show the initial structure Motor 1 [4] and the improved structures that derived from the initial machine. Motor 1 has a two-pole permanent magnet rotor and six figure-eight coils evenly distributed in the stator. To avoid unstable radial attraction forces, there are no ferromagnetic materials in the stator. The air gap flux density is very low, resulting in low axial electrodynamic force. Structural improvements were done to increase the air gap flux density and suspension force, so that the minimum rotational speed for full passive suspension is decreased.

Fig. 4(b) shows the structure of Motor 2. To increase radial flux density, two rotating back-iron were attached to the shaft. The outer and inner rotating parts are mechanically connected by an aluminum frame. The stationary coils are now surrounded by rotating parts. The leakage flux is reduced and a path with high magnetic permeability is created. Because there are no ferromagnetic parts in the stator, unstable attraction force is not generated. The number of poles was changed from two to four, so electrical frequency was doubled. The PM material was changed from neodymium N40 to N50.

Fig. 4(c) shows the improved structure Motor 3. Two permanent magnet rings are added to the rotor of Motor 2. One is on the upper side and another is on the lower side. Both have polarization opposite to that of the original central permanent magnet. Flux-linkage variation with displacement z becomes twice that of Motor 2. Owing to the increased ratio of magnetic flux linkage variation over displacement and to the force generated in the upper and lower coils ends, it is found that the electrodynamic force in Motor 3 is approximately four times higher than that of Motor 2.

Fig. 4(d) shows the final improved structure Motor 4. The radial component of the magnetic flux density, and, consequently, the magnetic flux linkage are increased by attaching additional permanent magnets to the outer shell. The magnetization direction is aligned with the PM on the inner rotor. The total volume of permanent magnets attached to the rotor is increased. Not only the peak radial flux density is increased, but it is also more uniform along the air gap between both permanent magnet surfaces than what is observed in Motor 3.

Fig. 5 shows a comparison of axial electrodynamic forces for $z = 1$ mm. These forces are calculated through 3D FEA among Motors 1, 2, 3, and 4 for rotational speeds up to 50 krpm. The axial force increases as the rotational speed is increased. At 45 krpm, Motor 4 achieved an axial force approximately 77 times higher than that of the initial structure, Motor 1.

To validate the equation for total axial electrodynamic force, the experimental value of K_λ is substituted in (1). The results of equation (1) for $z = -0.49$ mm are compared to both the FEA and experimental results for a range of rotational speeds up to 2400 rpm in steps of 300 rpm. The

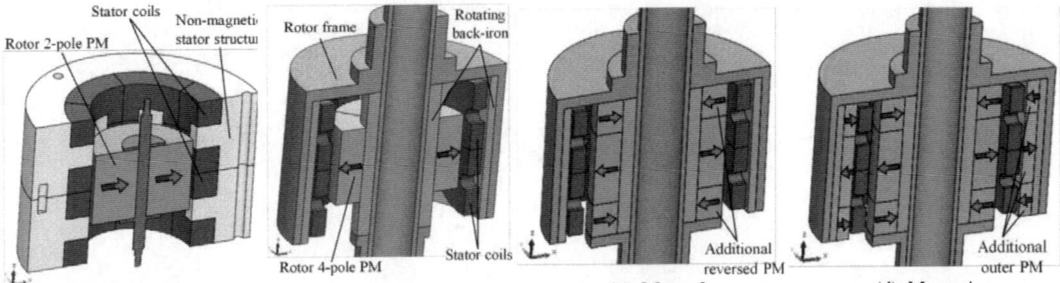

(a) Motor 1 (b) Motor 2 (c) Motor 3 (d) Motor 4

Fig. 4. Improvements upon the initial structure Motor 1. Each improved motor has higher axial electrodynamic stiffness than the previous one.

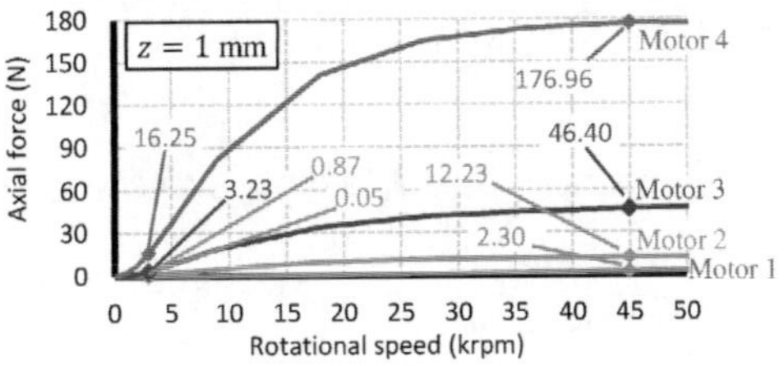

Fig. 5. Improved axial force calculated through FEA

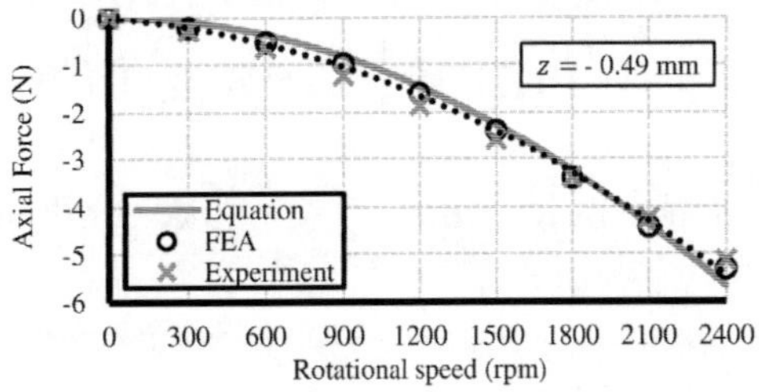

Fig. 6. Experimental axial force measured for Motor 4

axial force increases with rotational speed. At 2400 rpm, FEA and measured force were -5.31 N and -5.14 N, respectively. Discrepancy is -3.2%. The divergence observed between the equation, FEA and experimental values occurs due to uncertainties in K_λ, self-inductance, and resistance. The resistance value is dominant in this range of rotational speeds, since the coil inductive reactance is about 20% of the resistance at 2400 rpm. The constant K_λ affects force at all rotational speeds.

4 Conclusion

This paper shows the structure improvement of a bearingless motor with electrodynamic suspension in the axial direction. The motor was improved in three steps from the initial structure. The electrodynamic axial force of Motor 4 measured through the experiment was about 180 times higher than the one calculated for Motor 1 through FEA at 2400 rpm. A prototype was built for Motor 4, the machine with highest axial stiffness. The electrodynamic axial stiffness was measured for a range of rotational speeds and fixed axial displacement z.

References

[1] J. Sandtner and H. Bleuler, "Electrodynamic passive magnetic bearing with planar halbach arrays," in *Proceedings of the 9th international symposium on magnetic bearings*, 2004, pp. 3–6.

[2] H. Matsue, "Electromagnetic property of bearingless motor based on electro-magnetic induction," *Railway Technical Research Institute Report*, vol. 25, no. 3, pp. 47–52, 2011.

[3] J. Van Verdeghem and B. Dehez, "Fully Passively Levitated Self-Bearing Machines with Combined Windings," in *2020 IEEE Energy Conversion Congress and Exposition (ECCE)*, Oct. 2020, pp. 254–261, doi: 10.1109/ECCE44975.2020.9235412.

[4] G. Cavalcante Rubio and A. Chiba, "Design and analysis of a bearingless motor with passive axial suspension through null-flux coils," in *2019 IEEE International Electric Machines and Drives Conference, IEMDC 2019*, 2019, pp. 779–786, doi: 10.1109/IEMDC.2019.8785077.

Proposal of Surface-Rotating Ball for Wind-Tunnel Using Magnetic Suspension

Takeshi Mizuno[1], Yonosuke Jibiki[1], Yuji Ishino[1], and Masaya Takasaki[1]

[1] Saitama University, Saitama, Japan
mizar@mech.saimtama-u.ac.jp

Abstract

A wind-tunnel system for a spinning ball has been proposed to measure aerodynamic forces acting on the body. In the proposed system, the ball is suspended and rotated by electromagnets. The forces are measured from the control signal for suspension. However, due to the non-uniformity of the magnetic fields, rotation around a horizontal axis suffered from drag torque caused by eddy currents induced in the surface of the ball. It led to high energy consumption in the electromagnets. In this work, a surface-rotation ball with an adjustable rotating axis is proposed and developed. The ball is suspended stably with various-attitude surface rotation.

1 Introduction

Magnetic suspension provides an ideal way of supporting a model for wind tunnel tests because there is no support interference problem arising with mechanical model-support (Boyden, Britcher, & Tcheng, 1985). The forces and moments to support the model are generated by electromagnets arranged outside the test section. In addition, aerodynamic forces acting on the model are estimated from the control current of the electromagnets.

Aerodynamics around a spinning body such as golf ball is still an intriguing topic in both academic and industrial fields. Although simulation-based analysis has been making very rapid progress, the details of the dynamics have not been clarified sufficiently. The main difficulty is that the target phenomenon is a complex mix of macro-scale and micro-scale dynamics. Therefore, ideal wind tunnel tests are required for more precise and reproducible observation. However, conventional wind tunnels using magnetic suspension were not designed to test a spinning body so that they lacked the function of rotating the body (Boyden, Britcher, & Tcheng, 1985) (Sawada, Suenaga, Suzuki, & Ikeda, 1994).

We have proposed a wind-tunnel system for spinning body to measure hydrodynamic forces acting on the body (Mizuno, Sakai, Ishino, & Takasaki, 2012) (Mizuno, Sakai, Takasaki, & Ishino, 2017). In the proposed system, the body is suspended and rotated by electromagnets. The forces acting on the body are measured from the control signal for suspension. An experimental apparatus with a 60×60 mm wind tunnel was fabricated which has eight electromagnets around a wind-tunnel and an optical displacement sensor for detecting the three-dimensional positions of the body. Stable suspension, three-dimensional positioning and rotation of the body were achieved (Mizuno, Sakai, Takasaki, & Ishino, 2017).

However, due to the non-uniformity of the magnetic fields, rotation around a horizontal axis suffered from drag torque caused by eddy currents induced in the surface of the ball. It led to high energy consumption in the electromagnets. It is obviously preferable that rotation around an arbitrary axis can be achieved.

In this work, a surface-rotating ball is proposed, designed and fabricated. In this ball, the surface spherical shell rotates while the enfolded structure including magnetic cores, a motor, its driver circuit and a battery does not rotate. Therefore, no drag torque due to induced eddy currents is generated. It is experimentally demonstrated that the ball is suspended stably with surface rotation.

2 Wind Tunnel for Spinning Body Using Magnetic Suspension

Figure 1 shows a photograph of the developed wind tunnel using magnetic suspension (Mizuno, Sakai, Takasaki, & Ishino, 2017). The size of the total system is $360 \times 1200 \times 440$ mm, approximately. It consists of a magnetic suspension mechanism, a three-axis displacement sensor, a controller and a wind-tunnel.

Figure 1: Picture of wind tunnel for spinning ball using magnetic suspension

2.1 Magnetic suspension mechanism

A schematic drawing of the mechanism for magnetic suspension is shown in Figure 2 where the nearest electromagnet is removed for a better view of the inside poles. The size is $244 \times 244 \times 296$ mm. The suspension system has eight electromagnets for controlling the three-dimensional position of a sphere body made of ferromagnetic material. Each electromagnet has five 300-turn coils. The impedance of the electromagnet can be adjusted by the connections of the coils. The distance of the poles are kept long enough for a 60×60 mm wind tunnel to be inserted.

2.2 Suspension object

The object was a steel ball with a diameter of 19mm and a mass of 28.4g. There are machined dimples on the surface to simulate a golf ball.

2.3 Sensor

The operation of an optical sensor detecting the three-dimensional positions of the object is illustrated in Figure 3. It combines a pair of unit with a LED (light source), four collecting lens and four phototransistors. It can detect the three-dimensional positions of the body in the full-differential mode. When the output of the phototransistor k ($k = 1, \cdots, 8$) is denoted by S_k, the displacement in each direction is estimated from

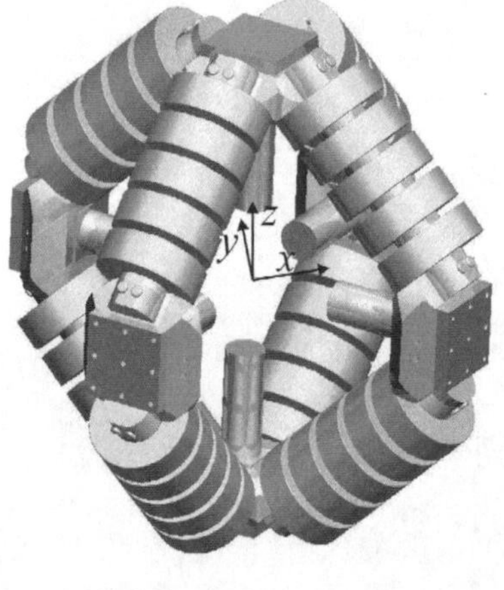

Figure 2: Schematic drawing of magnetic suspension mechanism

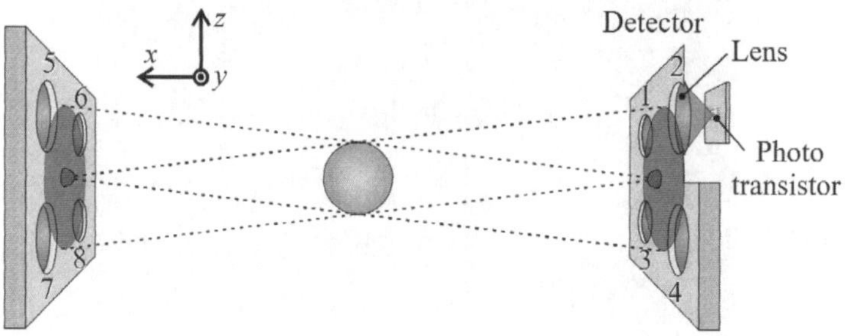

Figure 3: Operation of position sensor

$$\begin{cases} x-axis:(S_1+S_2+S_3+S_4)-(S_5+S_6+S_7+S_8) \\ y-axis:(S_2+S_4+S_5+S_7)-(S_1+S_3+S_6+S_8) \\ z-axis:(S_1+S_2+S_5+S_6)-(S_3+S_4+S_7+S_8) \end{cases} \quad (1)$$

This sensor is expected to be superior in linearity and resolution, especially as to the *x*-direction, to the sensor used in the previous work (Mizuno, Furutachi, Ishino, & Takasaki, 2010) because the latter operates in a single mode in this direction.

2.4 Levitation and rotation

Stable suspension was achieved by applying PID control. Figure 4 shows the motions of the floator in the wind of 8 m/s in the *y*-direction. It is observed that the position of the body is kept stably whereas vibratory motions were induced by a wind in the first apparatus (Mizuno, Furutachi, Ishino, & Takasaki, 2010).

The rotation of the body is realized by superimposing two-phase AC signals on the control signals for the *x*- and. *y*-directions. The rotation about the vertical axis (*z*-axis) was achieved. The relation between the excitation frequency and the rotational speed was studied experimentally. The results are summarized in Figure 5. The body is driven up to 3000rpm. Since the principle of rotation is same as that of induction motor, the slip is observed in higher frequency regions.

The rotation around a horizon axis was expected to be achieved similarly. However, due to the non-uniformity of the magnetic fields, such rotation was prevented by drag torque caused by eddy currents induced in the surface of the ball. To achieve rotation around an arbitrary axis, therefore, surface rotating ball is proposed in this work.

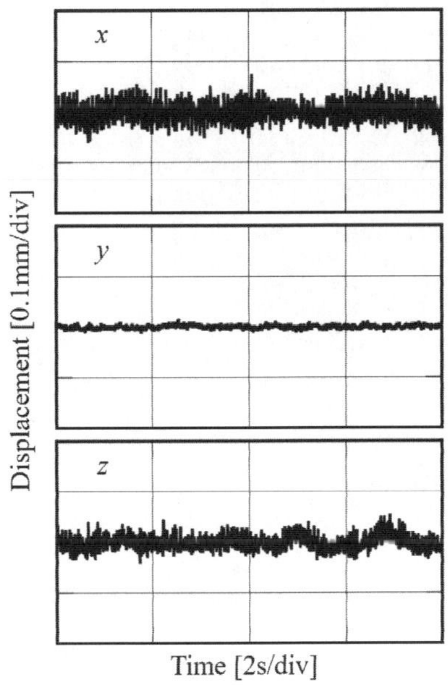

Figure 4: Steady-state displacements of the object in a wind of 6 m/s.

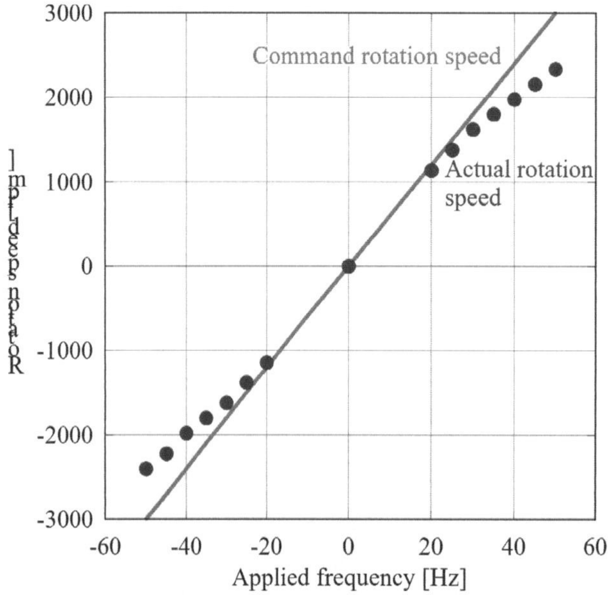

Figure 5: Relation between the rotational speed and the frequency of two-phase AC signal.

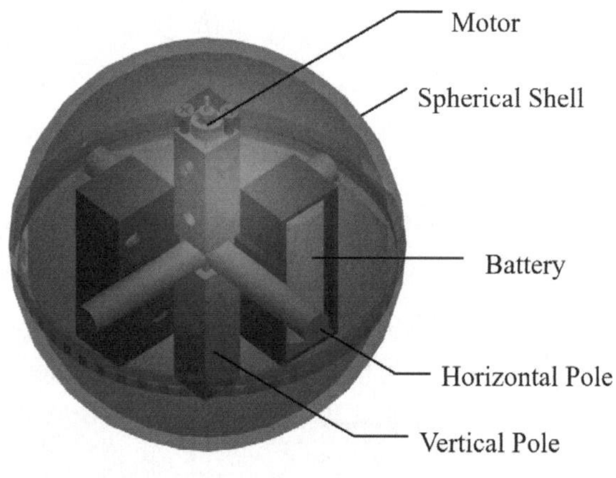

3 Surface Rotating Ball

3.1 Basic concept

A schematic view of surface-rotating ball is shown by Figure 6. It has a ferromagnetic structure for magnetic suspension inside and a rotating nonconductive spherical shell enfolding the structure. A motor is fixed at the center frame of the structure to rotate the sphere shell. In wind-tunnel testing, the enfolding spherical shell rotates while the enfolded structure maintains its position and attitude. Thereby, no eddy-current is induced even though the spherical shell is rotating so that no drag torque due to eddy-current is produced.

3.2 Fabricated ball

Figure 7 shows a picture of the fabricated ball. The inside structure has six magnetic poles which face the poles of the magnetic suspension system shown by Figure 8. They are made of SS400. The vertical (central) magnetic core with two poles is a rectangular column with square bases of 9×9 mm. The horizontal magnetic core is a cylinder with circular bases of $\phi 6$ mm.

The outside shell is a sphere with a diameter of 75 mm, which corresponds to the size of a baseball. It is made of nonmagnetic and nonconductive material (nylon). A motor for rotating shell, its driver circuit and a battery are included in the ball. The spherical shell consists of two semi-spheres that can be divided in installing the enfolded components and then assembled.

4 Realization of Rotation during Levitation

The magnetic suspension mechanism shown by Figure 2 is used to suspend the fabricated ball. Because the distance between the poles facing each other is 90 mm, it is too small to conduct wind-

Figure 7: Picture of fabricated ball

Figure 8: Picture of suspended ball

tunnel tests. In this work, however, suspension tests are carried out to examine the operation of the fabricated ball.

Stable suspension is achieved by applying PID control. Figure 8 shows a picture of the ball during levitation. Figure 5 shows the displacements of the ball. It is observed that the position of the ball is kept stably.

Then the rotation of the spherical shell was started. Figure 10 shows the displacement of the ball. The rotational speed is approximately 20 rpm that is expected from the period of the whirling motion of the ball. When the rational speed was increased, the suspension failed because the whirling motion

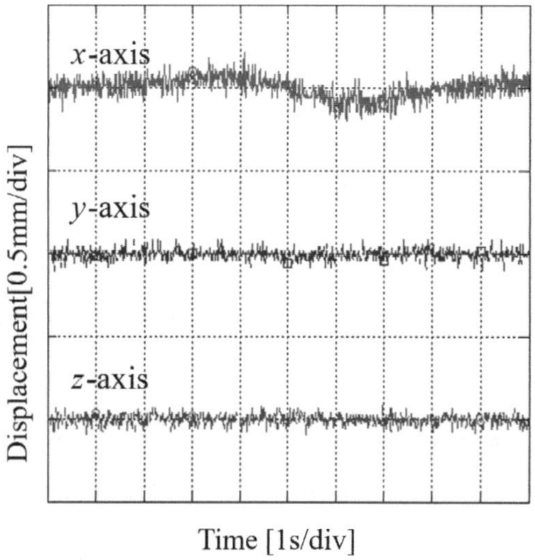

Figure 9: Displacement of ball during levitation

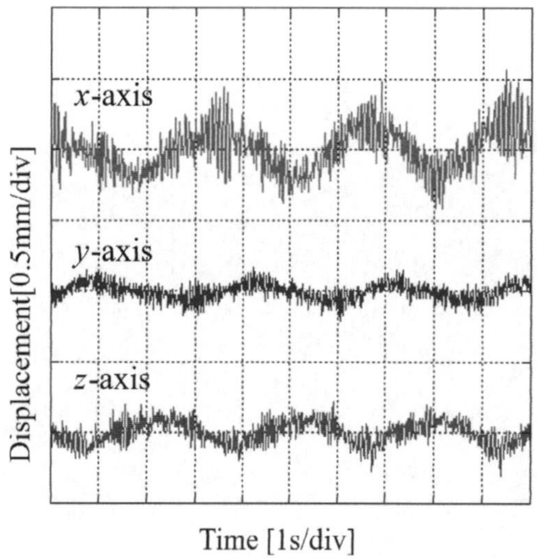

Figure 10: Displacement of ball during rotation

became too large. It is due to the unbalance of the rotating part. It can be improved by mechanically balancing or by introducing unbalance compensation (Mizuno, Analysis on the Fundamental Properties of Active Magnetic Bearing Control Systems by a Transfer Function Approach, 2011).

5 Conclusions

A surface-rotating ball was proposed, designed and fabricated. In this ball, the surface spherical shell rotates while the enfolded structure including magnetic cores, a motor, its driver circuit and a battery does not rotate. Therefore, no drag torque due to induced eddy currents is generated. It was experimentally demonstrated that the ball was suspended stably with surface rotation.

References

Boyden, P. R., Britcher, P. C., & Tcheng, P. (1985). *Status of Wind Tunnel Magnetic Suspension Research,*. SAE Technical Paper Series 851898, pp.1-9.

Mizuno, T. (2011). Analysis on the Fundamental Properties of Active Magnetic Bearing Control Systems by a Transfer Function Approach. *JSME International Journal, Series C, 44*(2), 367-373.

Mizuno, T., Sakai, Y., Ishino, Y., & Takasaki, M. (2012). Fabrication of a New Wind Tunnel for Spinning Body Using Magnetic Suspension. *the 13th International Symposium on Magnetic Bearings (ISMB13)*, (p. 23).

Mizuno, T., Sakai, Y., Takasaki, M., & Ishino, Y. (2017). Development of a wind-tunnel for spinning body using magnetic suspension. *Transactions of the Japan Society of Mechanical Engineers (in Japanese), 83*(849), 16-0048. doi:10.1299/transjsme.16-00482

Sawada, H., Suenaga, H., Suzuki, T., & Ikeda, N. (1994). Status of MSBS study at NAL. *the 2nd International Symposium on Magnetic Suspension Technology, 1*, pp. 275-289.

Optimization of the Guiding Stability of a Horizontal Axis HTS ZFC Radial Levitation Bearing

António J. Arsénio, F. Ferreira da Silva, João F. P. Fernandes, and P. J. Costa Branco
[1] Instituto de Engenharia Mecânica (IDMEC), Instituto Superior Técnico (IST), Universidade de Lisboa, Portugal
antoniojcosta@tecnico.ulisboa.pt, pbranco@tecnico.ulisboa.pt

Abstract

This document presents a study on the optimization of the 3D geometry of a horizontal axis radial levitation superconducting magnetic bearing with zero-field cooled (ZFC) high-temperature superconductor (HTS) bulks in the stator and radially magnetized permanent magnet (PM) rings in the rotor. A previous study on optimizing the 3D geometry for minimization of volume or cost considering variable dimensions of components and variable spacing was done. This initial optimization concerned only the maximization of the levitation force with a given restriction for the minimum levitation force.

Although the used geometry promotes the creation of guidance with ZFC, guiding forces depend on the spacings between PM rings in the rotor and between the rings of HTS bulks in the stator. This new optimization study aims to find the optimum spacings that maximize the guiding force for specific HTS bulk and PM ring dimensions while maintaining the minimum required levitation force. All the optimizations are based on using the non-dominated sorting genetic algorithm (NSGA-II) over 3D finite element analysis (FEA). A simplified electromagnetic model of equivalent relative permeability, calibrated with experimental measurement of magnetic levitation forces, is used on 3D FEA to reduce significantly numerical processing and optimization time.

Specifically optimized geometries for the experimental prototype were tested to validate optimization results. The experimental prototype is made of a stator of high-density polyurethane walls containing chambers where yttrium boron copper oxide (YBCO) bulks are immersed in liquid nitrogen (LN2). To hold the neodymium iron boron (NdFeB) PM rings, the rotor structure is made of polylactic acid plastic printed by a 3D computer numeric control (CNC) printer. An analysis of stable and unstable geometries domains depending on the spacings between HTS rings and PM rings is also done for two HTS bulk sizes.

1 Introduction

The studied horizontal axis radial levitation high-temperature superconductor (HTS) magnetic bearing comprises a stator with two cryostat chambers housing two discontinuous rings of zero-field cooled HTS bulks and a permanent magnet (PM) rotor with three radially magnetized PM rings with alternating polarization. Each HTS ring is positioned in the zone between adjacent PM rings.

With a uniform distribution of HTS bulks around the stator chambers, the obtained levitation force is zero with no vertical deviation of the rotor. Due to its weight, it is not possible to keep the rotor levitating at the center position. To maximize the available net force (difference between levitation force and the rotor weight), making possible the experimental measurement of levitation forces to validate electromagnetic model parameters, a topology with only 6 HTSs on the bottom of the stator was considered in experimental measurements. Figure 1(a) shows a geometry of the studied bearing, Figure 1(b) an exploded 3D view of the experimental prototype design, and Figure 1(c) a 3D perspective of its assembly.

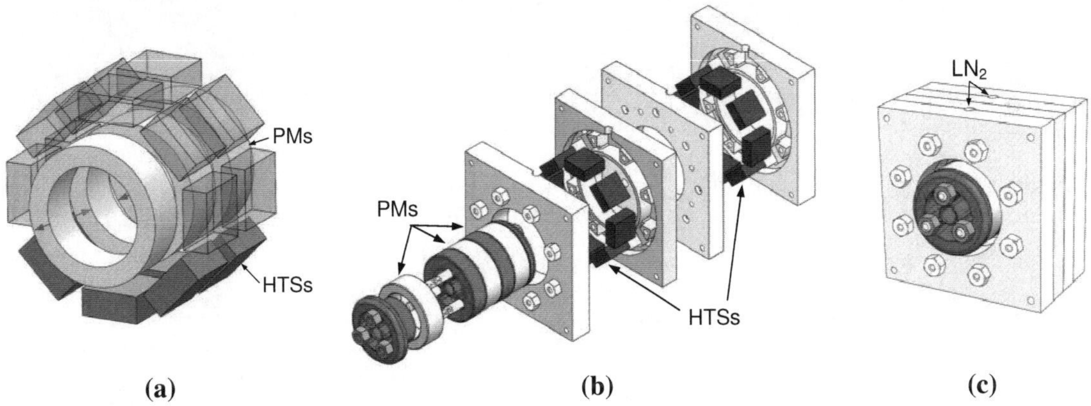

Figure 1: (a) Geometry of the studied bearing, (b) 3D exploded view design, and (c) assembly design of the experimental prototype.

In the experimental prototypes built and tested, the stator has two chambers of rigid high-density polyurethane walls, each one housing one discontinuous ring of HTS bulks. The stator walls were built by a computer numeric control (CNC) milling machine. The rotor structure in polylactic acid plastic (PLA) was printed by a 3D computer-aided design (CAD) printer (Arsénio, A.J., 2016).

Initially, a first stator named Stator I, for housing $33 \times 33 \times 14$ mm^3 size bulks was built. This stator and two rotors with 5 mm and 20mm spacings between PM rings, named respectively as rotors D5 and D20, were used to measure magnetic levitation and guidance forces and validate electromagnetic model parameters (Arsénio, A.J., 2018). Studies on the LN_2 consumption and YBCO bulk temperature evolution for different thermal processes in the initial ZFC and operation of the experimental bearing prototype were presented in (Arsénio, A.J., 2018) and (Arsénio, A.J., 2021). Work on optimizing the superconducting linear magnetic bearing of a maglev vehicle was presented (Quéval, L., 2016). A study on multi-objective multi-constraint optimization of the 3D geometry to minimize the bearing cost or volume was also performed (Fernandes, J.F.P, 2020). Optimizations using the non-dominated sorting genetic decision algorithm NSGA-II over 3D finite element analysis (FEA) were applied to obtain the optimized levitation forces. An equivalent relative permeability model was adopted for the 3D FEA to reduce the time consumption on the numerical processing significantly. This initial optimization study was based on maximizing levitation forces, not looking to guarantee and maximize the guiding stability. The optimizations are considered a restriction to guarantee a minimum levitation force equal to the one obtained with the geometry defined by 6 HTS bulks at the bottom of Stator I and rotor D5. Figure 2 compares this restriction geometry with the resulting cost and volume optimized geometries.

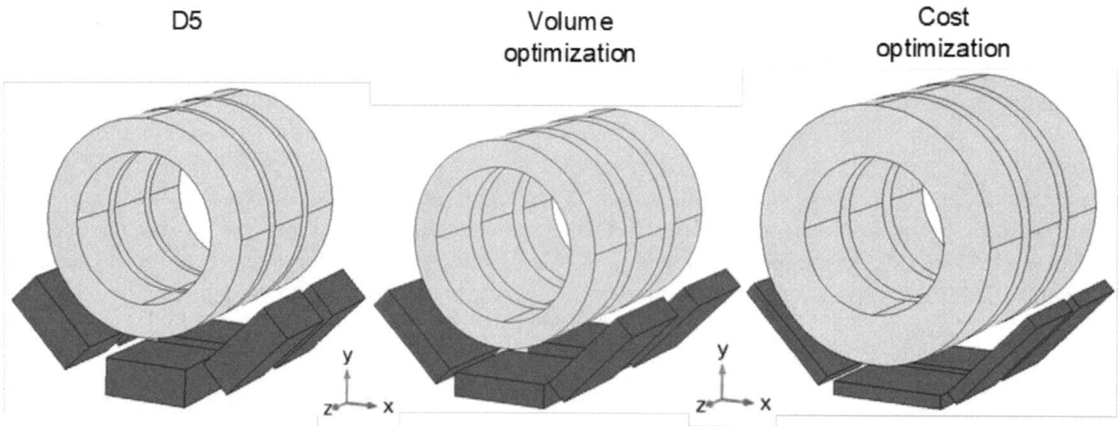

Figure 2: (a) Geometry with rotor D5 and 6 HTS bulks of $\mathbf{33 \times 33 \times 14\ mm^3}$, (b) Volume, and (c) Cost optimized geometries.

Another stator named Stator II for housing bulks with dimensions $40 \times 40 \times 10\ \mathrm{mm}^3$, close to the bulk dimensions in the volume optimized geometry was built. Figure 3 shows the chamber profiles of Stator I and Stator II.

Figure 3: Chamber profiles of (a) Stator I for $33 \times 33 \times 14\ \mathrm{mm}^3$ size bulks, and (b) Stator II for $40 \times 40 \times 10\ \mathrm{mm}^3$ size bulks.

This document presents a study on optimizing the spacings between HTS bulk rings in the stator and between PM rings in the rotor, to maximize the guiding force and so the guiding stability with a restriction on the minimum levitation force that should be guaranteed. Also, the determination of the domains in which the combination of rotor PM rings spacings and stator HTS ring spacings guarantee guiding stability is performed. This is done for the two bulk-size stators.

The purpose of using ZFC is to minimize the magnetization energy and Joule losses in the bulks, increasing their lifetime with the benefit of increasing levitation forces. Because with ZFC, there is no flux pinning as with FC, guiding stability is guaranteed by geometries with a specific arrangement of PMs and spatial distribution of HTSs. Thus, it is of extreme importance to determine which geometries present some guiding stability and find the optimum spacings between HTS rings and between PM rings to maximize the guiding stability for a given restriction on the minimum levitation force be guaranteed.

2 Electromagnetic models and parameters

The E-J model described below should be used for a detailed prediction and characterization of the induced current distribution in HTS bulks. This model implies the resolution of non-linear and partial differential equations in the superconductor domain, requiring a lot of numerical processing in the FEA. A simplified model with a calibrated value of equivalent relative permeability was used to significantly reduce the FEA processing time, especially during the optimization of the bearing 3D geometry.

2.1 E-J model

When after ZFC, a magnetic field $\mathbf{H}_a$ is applied to a superconductor bulk, an electric field **E** is induced from the variation of the penetrating magnetic field **H**, according to Faraday's law.

$$\nabla \times \mathbf{E} = -\mu_0 \frac{\mathrm{d}\mathbf{H}}{\mathrm{dt}} \tag{1}$$

According to the model in (Hong. Z., 2007), type II superconductors present a non-linear electric conductivity characteristic given by the power-law (2),

$$E = E_0 \left(\frac{J}{J_c}\right)^n \tag{2}$$

where E and J are the magnitudes respectively of induced electric field and current density, J_c the critical current density and E_0 is the electrical field at which the current density reaches J_c. Exponent n is a positive integer higher than 1. The induced current density **J** creates, by Ampere's law, a magnetization field **M**.

$$\mathbf{J} = \nabla \times \mathbf{M} \tag{3}$$

The penetrating magnetic field is given by the sum of the applied magnetic field and the magnetization field.

$$\mathbf{H} = \mathbf{H}_a + \mathbf{M} \tag{4}$$

According to the Kim-Anderson model in (Yamamoto, K., 1993) and (Fujishiro, H., 2010), the critical-current density J_c depends on the absolute value of the penetrating magnetic-flux density $|B|$,

$$J_c(B) = J_{c0} \frac{B_0}{B_0 + |B|} \tag{5}$$

where J_{c0} is the zero-field critical current density that depends on the temperature T (Koo, J., 2004), and B_0 the magnitude of the penetrating magnetic flux density for which the critical current density is half the zero-field critical current density. For the case of yttrium barium copper oxide (YBCO), $E_0 = 1 \times 10^{-4}$ Vm^{-1}, $B_0 = 0.1$ T and $n = 21$ (Zhang, M, 2012). Appropriate values of the parameter J_{c0} can be validated by comparing the magnetic forces predicted by FEA using the E-J model with the ones obtained by experimental measurement.

2.2 Equivalent relative permeability model

The relative permeability μ_r is defined by the relation between the magnitude of the penetrating field H and the magnitude of the applied magnetic field H_a, being calculated by (6).

$$\mu_r = \frac{H}{H_a} = 1 + \frac{M}{H_a} = 1 + \chi \tag{6}$$

where χ is the magnetic susceptibility. Perfect diamagnetism is traduced by χ equal to -1 and μ_r equal to 0.

In this model, an average value of relative permeability $\overline{\mu_r}$ designated by the equivalent relative permeability is considered. A methodology to compute the value of $\overline{\mu_r}$ for the total bulk volume was initially proposed in (Arsénio, A. J., 2016) and (Fernandes, J.F.P, 2020). Based on the first methodology, a more generic methodology to compute values of $\overline{\mu_r}$ in several partitions of the bulk was then proposed in (Inês, P., 2021).

Appropriate values of $\overline{\mu_r}$ can be validated by comparing the magnetic forces predicted by FEA with the ones obtained by experimental measurement. The use of this model significantly reduces the FEA numerical processing.

3 Validation of electromagnetic model parameters

Magnetic levitation and guidance forces were measured experimentally and determined by 3D FEA for two geometries respectively with rotors D20 and D5 and with 6 YBCO bulks of volume $33 \times 33 \times 14\ \text{mm}^3$ on the bottom of Stator I. In these two geometries, the spacing between the two YBCO bulk rings was 10 mm. Figures 4(a) and 4(b) show the dimensioned 3D perspective design (half part used in FEA simulations) of these two geometries tested experimentally.

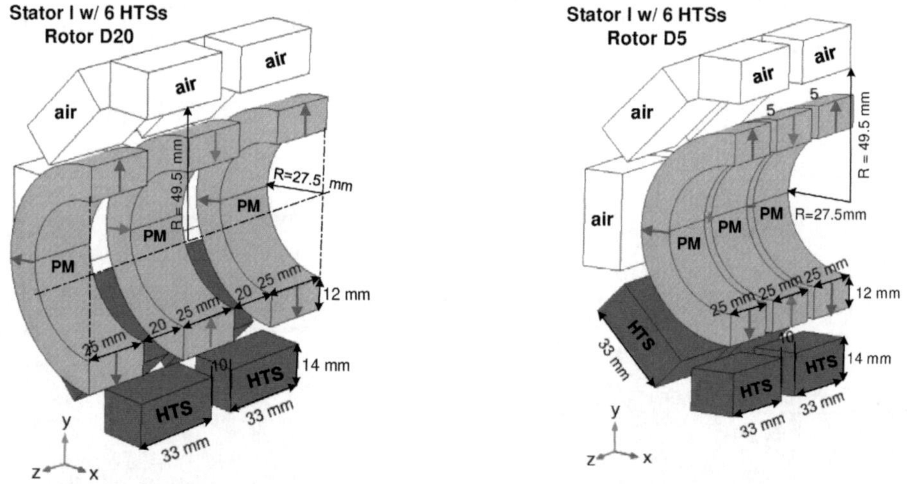

(a) **(b)**

Figure 4: Dimensioned 3D perspective of the half part (used in FEA simulations) of the tested geometries with 6 HTS bulks on the bottom of stator I and with (a) rotor D20 and (b) rotor D5.

Figures 5(a) and 5(b) show respectively transversal and longitudinal views of the magnetic flux and current density distributions for the geometry with rotor D20. Figures 6(a) and 6(b) refer to the geometry with rotor D5. The results were obtained using the *E-J* model with $J_{c0} = 8 \times 10^7\ \text{Am}^{-2}$.

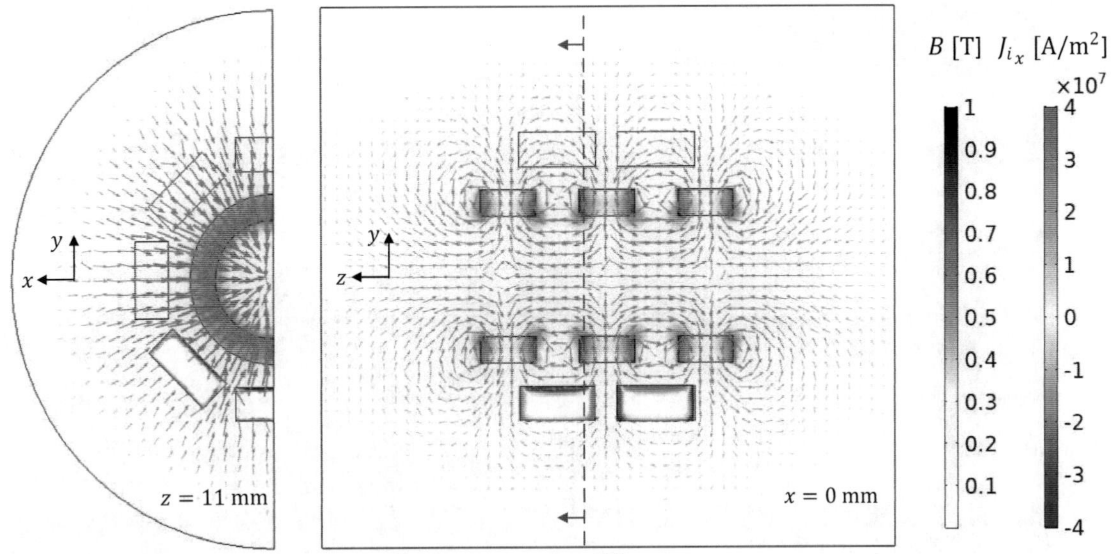

Figure 5: (a) Transversal and (b) longitudinal views of the magnetic flux and current density distributions for the geometry with 6 HTSs bulks on the bottom of Stator I and rotor D20.

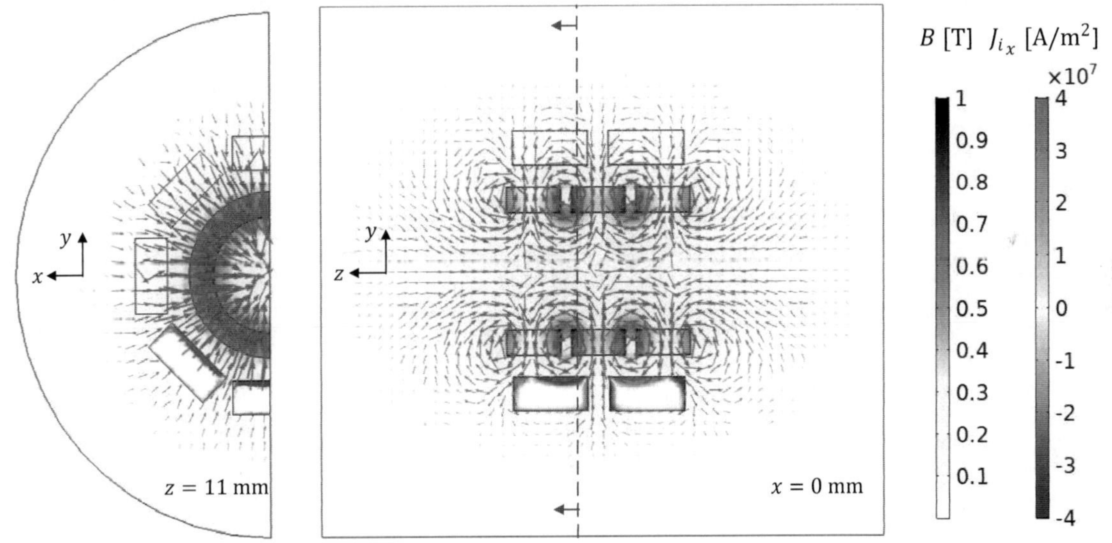

Figure 6: (a) Transversal and (b) longitudinal views of the magnetic flux and current density distributions for the geometry with 6 HTSs bulks on the bottom of Stator I and rotor D5.

Experimental measurements and FEA simulations were performed to verify the dependence of the levitation force with the rotor vertical deviation and the dependence of the guidance force with the rotor axial deviation keeping the rotor axis on the central position.

Figures 7(a) and 7(b) show pictures respectively of rotors D20 and D5, indicating the gravity force F_g of their weights. Figure 7(c) shows a photo of Stator I, with a fixed spacing of 10 mm between the two rings of HTS bulks. Figures 7(d) e 7(e) show the setups to measure the levitation force with the vertical deviation and the guidance force with the axial deviation.

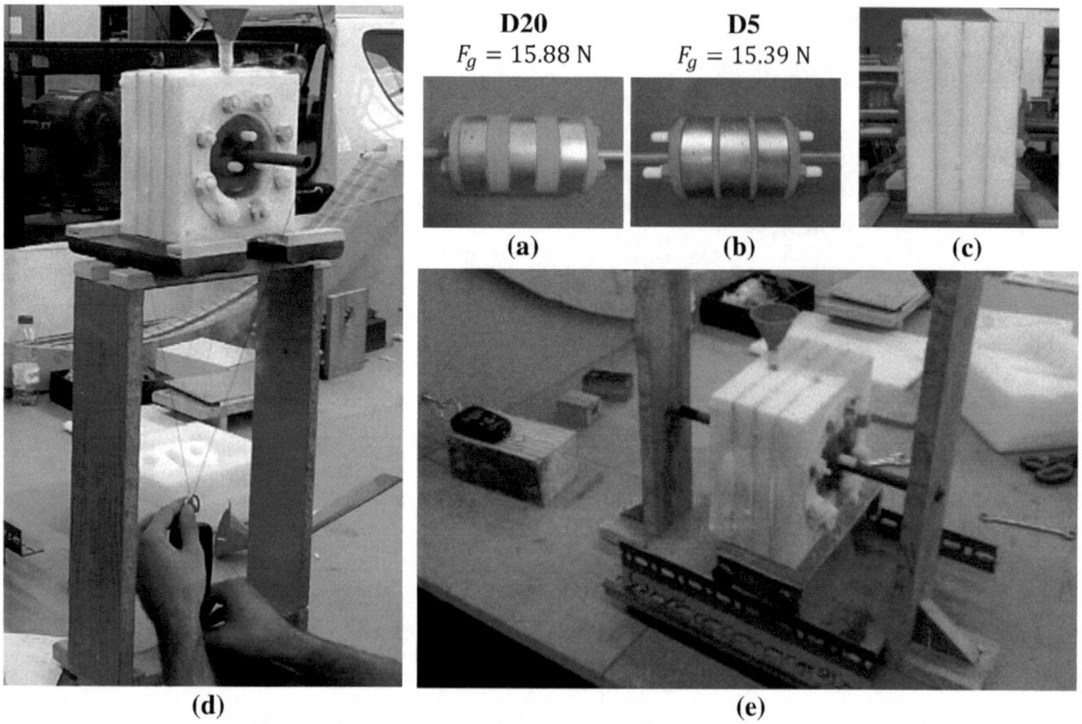

Figure 7: (a) Rotor D20, (b) Rotor D5, (c) Stator I with a spacing of 10 mm between HTS rings, and experimental setups to measure (d) levitation and (e) guidance forces.

The characteristics of dependence of the levitation force with the vertical displacement obtained by FEA using the equivalent permeability model and the E-J model and by experimental measurement respectively with Rotor D20 and Rotor D5 are shown in Figures 8(a) and 8(b). For each rotor deviation, 3D FEA simulations were performed between 1.5 hour and 6 hours with the *E-J* model and less than 30 s with the equivalent permeability model.

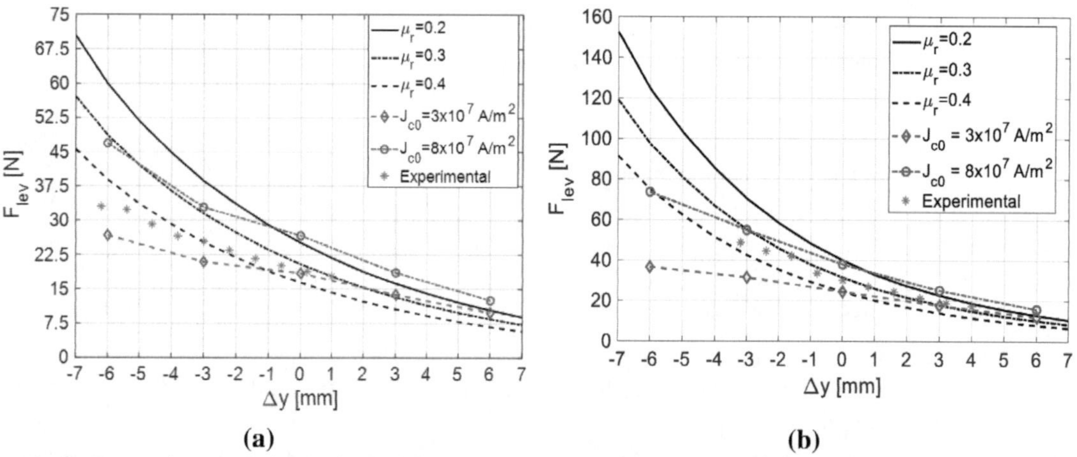

Figure 8: Dependence of the levitation force with vertical displacement by FEA using the equivalent permeability model and the E-J model and experimentally with (a) Rotor D20 and (b) Rotor D5.

As one may verify, the experimental values follow the monotony of the characteristics with the *E-J* model between $J_{c0} = 3 \times 10^7$ Am^{-2} and $J_{c0} = 8 \times 10^7$ Am^{-2}. For the second case with rotor D5, these are closer to the characteristic with $J_{c0} = 8 \times 10^7$ Am^{-2}. With Rotor D5 in the center position (no vertical deviation), the levitation force with $J_{c0} = 8 \times 10^7$ Am^{-2} is between the ones predicted with $\overline{\mu_r} = 0.2$ and $\overline{\mu_r} = 0.3$, close to the one that would be predicted with $\overline{\mu_r} = 0.25$. For a lower distance g_a between the HTSs to the PMs, the higher is the value of equivalent relative permeability for which the predicted levitation forces approximates experimental values. This is because the closer the PMs are to the HTSs, the higher the penetrating field.

Figures 9(a) and 9(b) show the characteristics of dependence of the levitation force with the vertical displacement obtained by FEA using the equivalent permeability model and the *E-J* model and experimental measurement, respectively with Rotor D20 and Rotor D5.

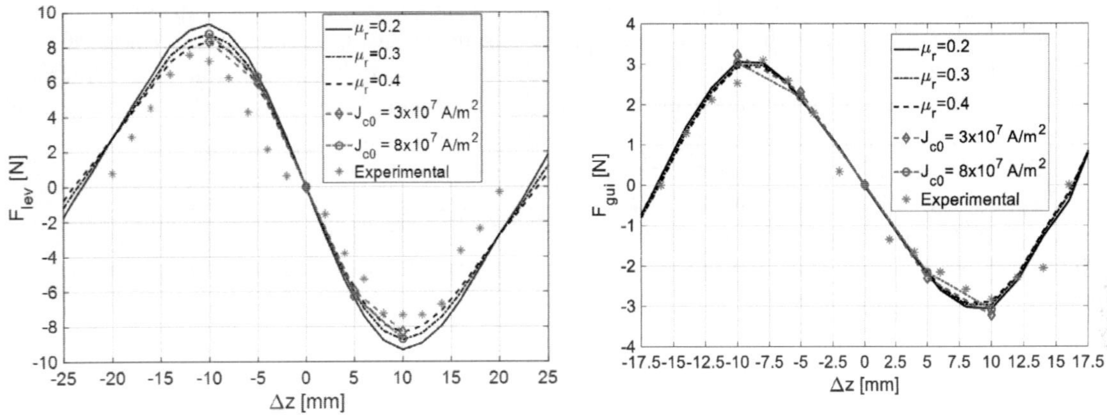

(a) **(b)**

Figure 9: Dependence of the guidance force with axial displacement by FEA using the equivalent permeability model and the E-J model and experimentally with (a) Rotor D20 and (b) Rotor D5.

As one may verify, the sensitivity of the guidance force to the value of $\overline{\mu_r}$ is much less than the one verified for the case of the levitation force. With the rotor at the center position, the applied magnetic field did not change notably in the range of axial deviations applied.

Table I resumes the values of levitation force F_{lev} with the rotor at the center position, predicted by FEA and obtained by experimental measurement. The last column presents the net force F_n (difference between levitation force F_{lev} and the gravity force F_g of the rotor weight) relatively to the experimental measurement.

Table I: Experimental and FEA values of F_{lev} with Rotor D20 and D5 at the center position.

$\Delta x = \Delta_y = 0$ mm $\Delta z = 0$ mm	F_{lev} [N]							F_n [N] Exp.
	$\mu_r =$ 0.2	$\mu_r =$ 0.25	$\mu_r =$ 0.3	$\mu_r =$ 0.4	$J_{c0} =$ 3×10^7 Am^{-2}	$J_{c0} =$ 8×10^7 Am^{-2}	Exp.	
Rotor D20 $F_g = 15.88$ N $g_a = 10.5$ mm	25.15	22.72	**20.47**	16.40	**18.47**	23.73	**19.19**	3.31
Rotor D5 $F_g = 15.39$ N $g_a = 10.5$ mm	37.62	**31.39**	26.17	21.44	22.42	**32.31**	**30.26**	14.87

With the rotors at the center position, the levitation force predicted with $J_{c0} = 8 \times 10^7$ Am^{-2} is close to the one predicted with $\overline{\mu_r} = 0.25$. The experimental are between the ones with $J_{c0} = 3 \times 10^7$ Am^{-2} and $J_{c0} = 8 \times 10^7$ Am^{-2}, being closer to the one with $J_{c0} = 8 \times 10^7$ Am^{-2} on the case of Rotor D5. With a stator made of a material with better thermal insulation than the used polyurethane foams, the average temperature of bulks would decrease, and levitation forces would get closer to the ones predicted using the *E-J* model with $J_{c0} = 8 \times 10^7$ Am^{-2}.

4 Optimization of the HTS and PM ring spacings to maximize the guiding stability

A previous study to optimize the volume and cost of the 3D geometry with a given constrain on the minimum net levitation force was performed. Optimizations were performed using the non-dominated sorting genetic algorithm NSGA-II over 3D FEA results. In this first study, the component dimensions and spacings were all considered as decision variables. The dimensions of the bulks used in Stator II are close to the ones indicated by the volume optimization.

This study aims to optimize the spacing between the two rings of HTS bulks in the stator and the spacings between PM rings in the rotor for given bulk and PM ring sizes, maintaining the minimum required levitation forces. Bulk sizes with volumes $33 \times 33 \times 14$ mm^3 (Stator I) and $40 \times 40 \times 10$ mm^3 (Stator II) are considered in this study. The considered PM ring dimensions are equal to the ones used in the built experimental bearing prototypes.

4.1 Optimization methodology

Optimizations were performed for the geometries with 6 HTS bulks on the bottom of the stator. This is done to reduce the amount of numerical processing, being expected that the results for geometries with more bulks would be similar. Also, the optimizations were performed with no vertical deviation of the rotor. Several axial deviations were performed for each genetic code (geometry solution) to find the maximum guiding force. Because of the existing symmetry, only $3/16$ of the complete geometry was simulated to reduce the amount of numerical processing. Figures 10(a) and 10(b) show the 3D perspective respectively of the geometry and its simulated partition for the case with 6 bulks of volume $33 \times 33 \times 14$ mm^3 at the bottom of Stator I.

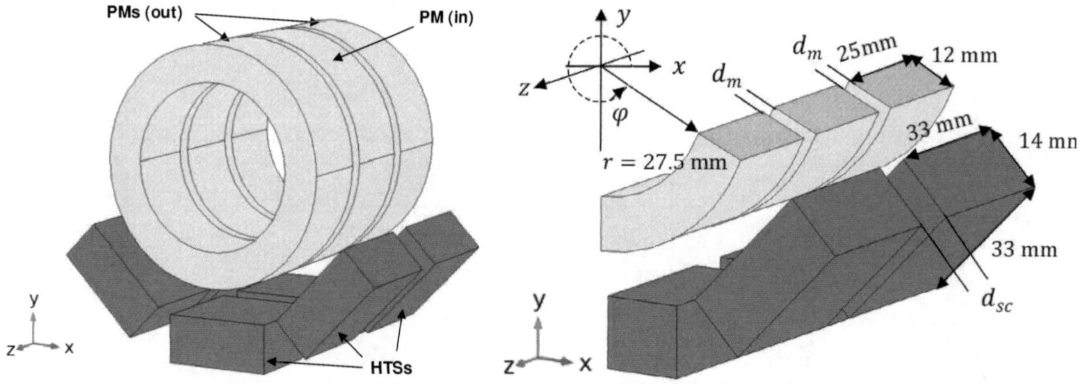

Figure 10: A 3D perspective of the (a) geometry and (b) its simulated partition for the case of Stator I.
Figures 11(a) and 11(b) the same for the case with 6 bulks of volume $40 \times 40 \times 10$ mm^3 at the bottom of Stator II.

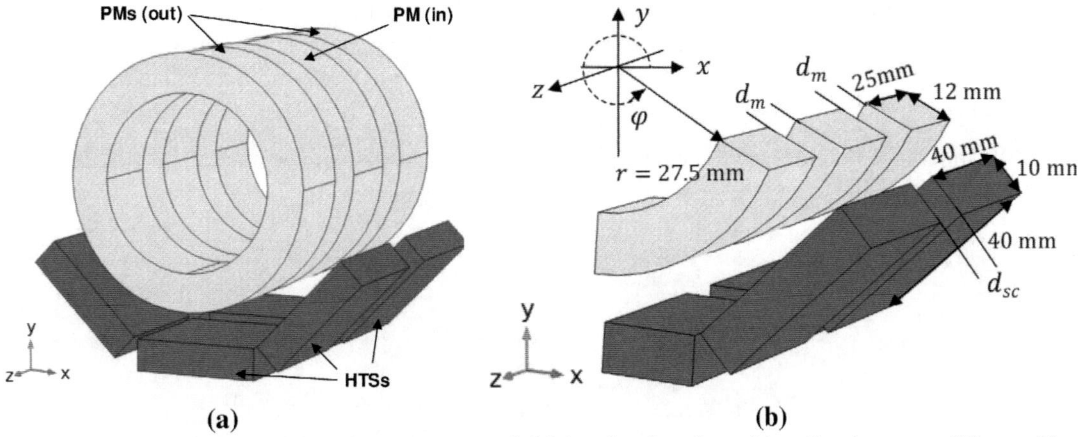

Figure 11: 3D perspective of the (a) geometry and (b) its simulated partition for the case of Stator II.

Optimizations were performed using the equivalent permeability model and not the *E-J* model to significantly reduce the numerical processing time required in the 3D FEA. With better thermal insulation conditions than the ones of the existing experimental prototypes, levitation forces would be closer to the ones predicted using the E-J model with $J_{c0} = 8 \times 10^7$ Am^{-2}. According to Table I resuming the levitation force values with the rotor axis at the center position, the values predicted with $\overline{\mu_r} = 0.25$ are close to the ones predicted with $J_{c0} = 8 \times 10^7$ Am^{-2}. For these reasons, 3D FEA optimization simulations run with $\overline{\mu_r} = 0.25$.

The optimizations in this study were also performed running the NSGA-II over 3D FEA results with $\overline{\mu_r} = 0.25$. As decision variables, it was chosen the distance d_{sc} between the rings of HTS bulks in the stator and the distance d_m between the PM rings in the rotor. The optimizations run with a population size of 150 and for 50 generations. To find the maximum guiding force and corresponding axial deviation, 10 simulations are performed with different axial deviations of the rotor for each genetic geometry.

4.2 Optimization results

Figures 12 shows the Pareto's fronts obtained for the case with 6 bulks of volume $33 \times 33 \times 14$ mm^3 at the bottom of Stator I.

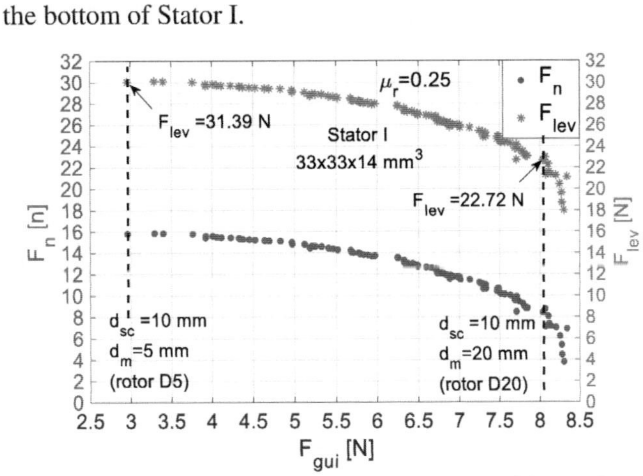

Figure 12: Pareto's front for the case of 6 HTS bulks at the bottom of Stator I.

As expected, the maximum levitation force is verified for the genetic code with $d_m = 5$ mm (rotor D5) and $d_{sc} = 10$ mm, corresponding to one of the two geometries with the levitation and guidance force characteristics presented in section 3. The two vertical dashed lines correspond to the two geometries presented in section 3.

Figure 13 shows the Pareto's front obtained for the case with 6 bulks of volume $40 \times 40 \times 10$ mm^3 at the bottom of Stator II.

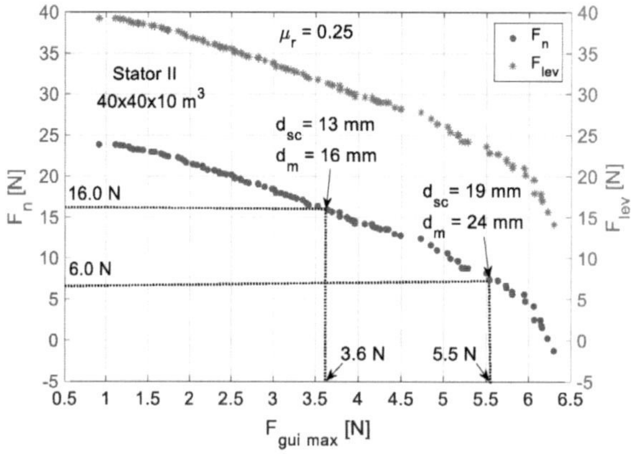

Figure 13: Pareto's front for the case of 6 HTS bulks at the bottom of Stator II.

The geometry with bulks of volume $40 \times 40 \times 10$ mm^3, with $d_m = 16$ mm and $d_{sc} = 13$ mm, guarantees the same minimum levitation force as the geometry with bulks of volume $33 \times 33 \times 14$ mm^3, with $d_m = 5$ mm and $d_{sc} = 10$ mm for which the levitation and guidance force characteristics were presented in section 3. Figure 14 shows the distribution of magnetic flux and current densities obtained by 3D FEA using *E-J* model with $J_{c0} = 8 \times 10^7$ Am^{-2}, for such geometry presenting guiding stability.

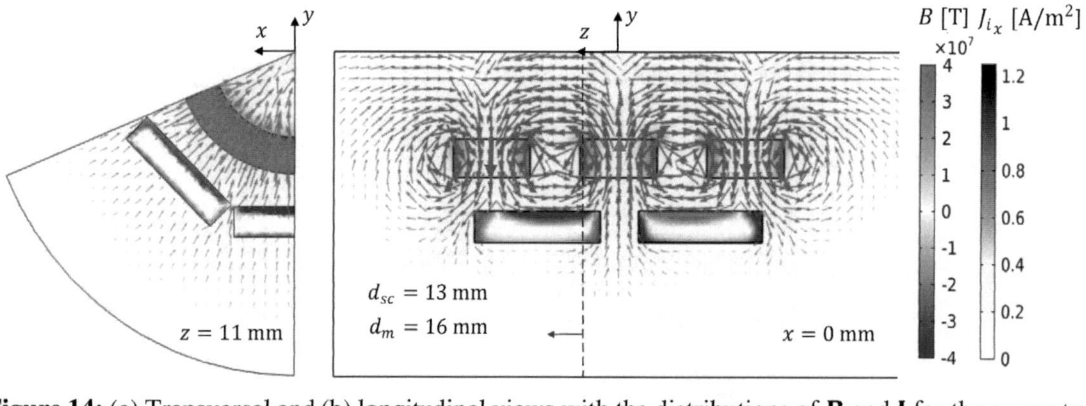

Figure 14: (a) Transversal and (b) longitudinal views with the distributions of **B** and **J** for the geometry with bulks of volume $40 \times 40 \times 10$ mm^3, $d_m = 16$ mm and $d_{sc} = 13$ mm.

With this geometry, the maximum guidance force is $F_{gui\ max} = 3.6$ N while for the geometry with bulks of volume $33 \times 33 \times 14$ mm^3, $d_m = 5$ mm and $d_{sc} = 10$ mm, the maximum guidance force was $F_{gui\ max} = 3.1$ N.

To validate optimization results with Stator II, several experimental measurements for several combinations of distances d_m and d_{sc} were performed with this stator. Plastic bars of 3 mm thick were inserted between the two rings of HTS bulks in Stator II to control d_{sc}. Plastic washers of 2.5 mm thick were inserted between the PM rings to control d_m.

Experimental measurements were performed for the case with $d_m = 17.5$ mm and $d_{sc} = 13$ mm. In this case, five washers of 2.5 mm thick were inserted between the PM rings in addition to the original 5 mm disks of Rotor D5, as shown in Figure 15(a). One plastic bar of 3 mm thick was inserted between the two chambers of Stator II, in addition to their minimum distance of 10 mm, as shown in Figure 15(b). Figure 15(c) shows the measurement of F_n and F_{lev} with no vertical deviation of the rotor from the central position. Figure 15(d) shows the measurement of $F_{gui\,max}$.

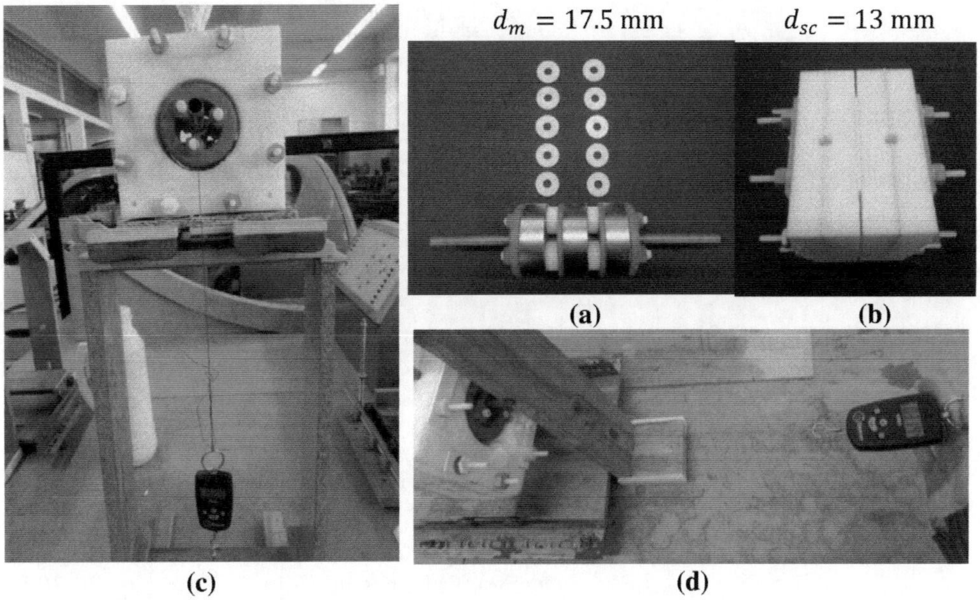

Figure 15: (a) Rotor for $d_m = 17.5$ mm, (b) Stator II for $d_{sc} = 13$ mm, (c) measurement of F_n and F_{lev} with no vertical deviation of the rotor, and (d) measurement of $F_{gui\,max}$.

Figures 16(a) and 16(b) show the F_{lev} and F_{gui} characteristics obtained by 3D FEA with $\overline{\mu_r} = 0.25$ and $J_{c0} = 8 \times 10^7$ Am^{-2}, plotting the experimental measurements of F_{lev} with $\Delta_y = 0$ and of $F_{gui\,max}$. This for the combinations with $d_{sc} = 13$ mm, and $d_m = 15$ mm or $d_m = 17.5$ mm.

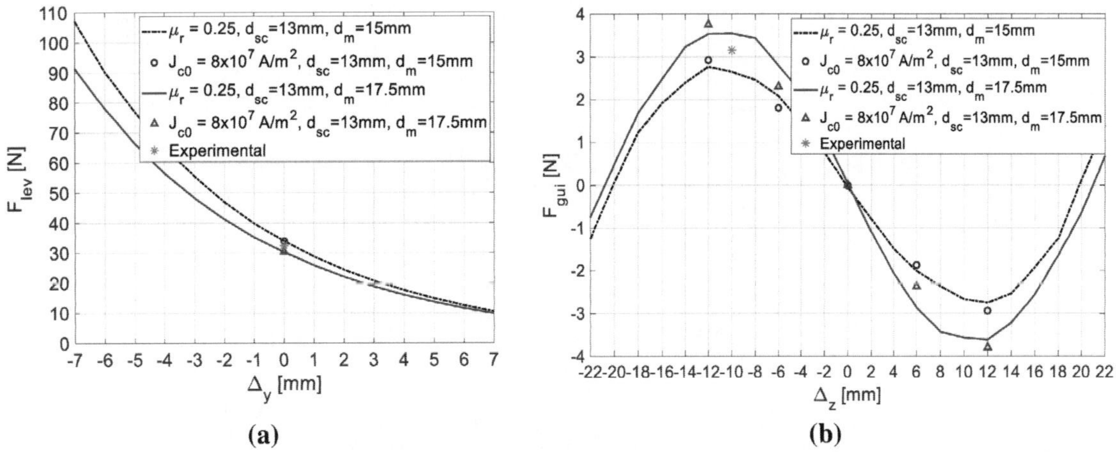

Figure 16: 3D FEA and experimental results for (a) F_{lev}, and (b) F_{gui} with $d_{sc} = 13$ mm, and $d_m = 15$ mm or $d_m = 17.5$ mm.

The geometry with bulks of volume $40 \times 40 \times 10$ mm^3, with $d_m = 24$ mm and $d_{sc} = 19$ mm, guarantees the same minimum levitation force as the geometry with bulks of volume $33 \times 33 \times 14$ mm^3, $d_m = 20$ mm and $d_{sc} = 10$ mm for which levitation and guidance force characteristics were presented in section 3. Figure 17 shows the distribution of magnetic flux and current densities obtained by 3D FEA using *E-J* model with $J_{c0} = 8 \times 10^7$ Am^{-2}, for such geometry presenting guiding stability.

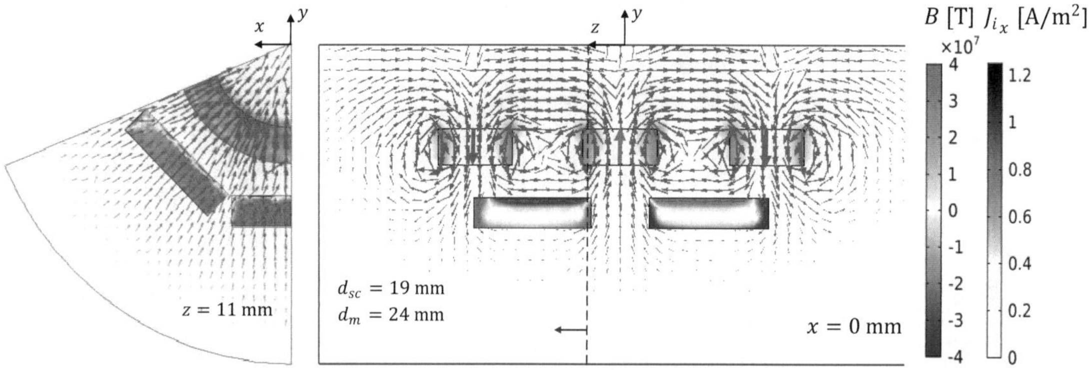

Figure 17: (a) Transversal and (b) longitudinal views with the distributions of **B** and **J** for the geometry with bulks of volume $40 \times 40 \times 10$ mm^3, $d_m = 24$ mm and $d_{sc} = 19$ mm.

With this geometry, the maximum guidance force is $F_{gui\,max} = 5.5$ N while for the geometry with bulks of volume $33 \times 33 \times 14$ mm^3, $d_m = 20$ mm and $d_{sc} = 10$ mm, the maximum guidance force was $F_{gui\,max} = 7.24$ N.

Experimental measurements were performed with $d_m = 25$ mm and $d_{sc} = 19$ mm. In this case, *two* washers of 2.5 mm thick were inserted between the PM rings in addition to the original 20 mm disks of Rotor D20, as shown in Figure 18(a). Three plastic bars of 3 mm thick were inserted between the two chambers of Stator II, in addition to the minimum distance of 10 mm, as in Figure 18(b).

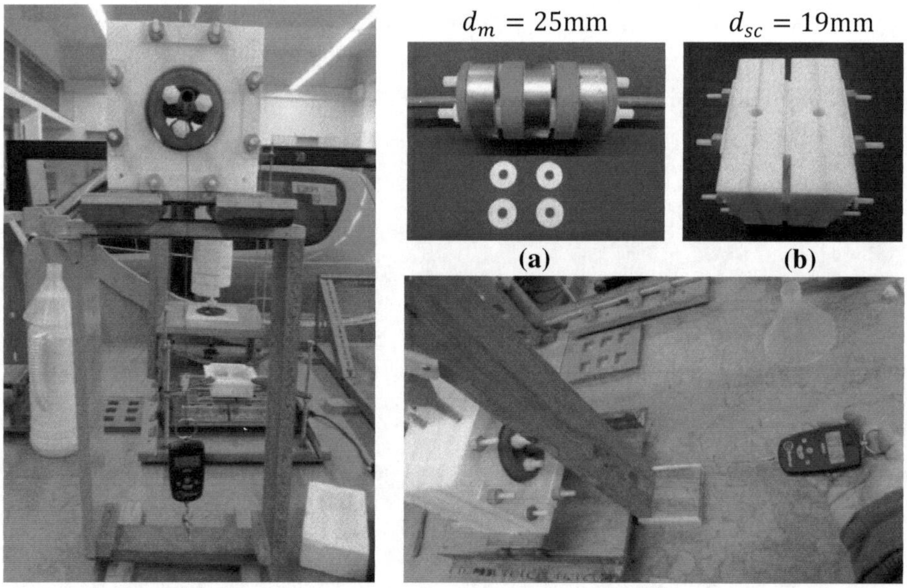

Figure 18: (a) Rotor for $d_m = 25$ mm, (b) Stator II for $d_{sc} = 19$ mm, (c) measurement of F_n and F_{lev} with no vertical deviation of the rotor, and (d) measurement of $F_{gui\,max}$.

Figures 19(a) and 19(b) show the F_{lev} and F_{gui} characteristics obtained by 3D FEA with $\overline{\mu_r} = 0.25$ and $J_{c0} = 8 \times 10^7$ Am^{-2}, plotting the experimental measurements of F_{lev} with $\Delta_y = 0$ and of $F_{gui\,max}$. This for the combinations with $d_{sc} = 19$ mm, and $d_m = 22.5$ mm or $d_m = 25$ mm.

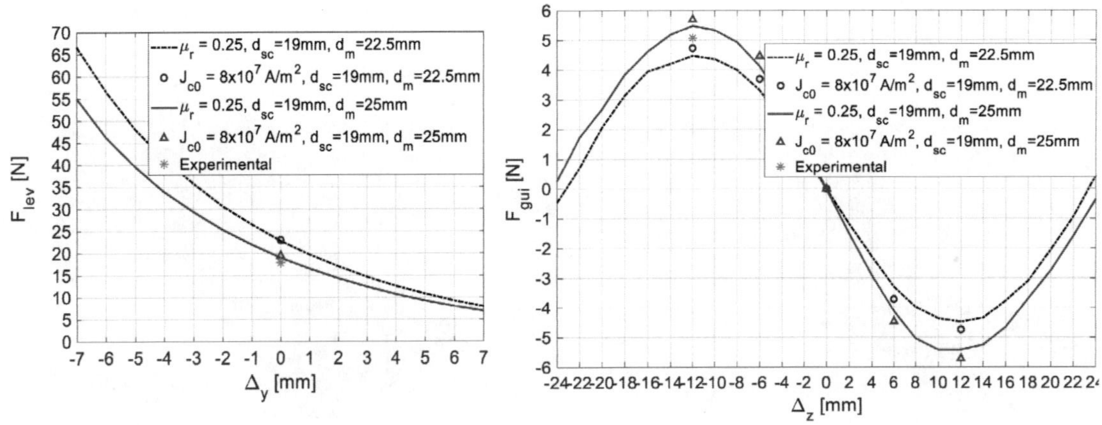

Figure 19: 3D FEA and experimental results for (a) F_{lev}, and (b) F_{gui} with $d_{sc} = 19$ mm, and $d_m = 22.5$ mm or $d_m = 25$ mm.

The value of $F_{gui\,max}$ is higher for the geometry with bulks of volume $33 \times 33 \times 14$ mm^3, spacings $d_m = 20$ mm and $d_{sc} = 10$ mm, than for the geometry with bulks of volume $40 \times 40 \times 10$ mm^3, spacings $d_m = 24$ mm and $d_{sc} = 19$ mm.

5 Identification of guiding stability zones

3D FEA simulations using $\overline{\mu_r} = 0.25$ run for several spacings d_{sc} from 10 mm to 25 mm with steps of 3 mm equal to the thick of the plastic bars inserted between the two rings of HTS bulks in Stator II, and for several spacings d_m from 5 mm to 25 mm with steps of 2.5 mm equal to the thick of the washers inserted between PM rings. Figures 20(a) and 20(b) show the surfaces of F_{lev} respectively for the cases of Stator I and Stator II.

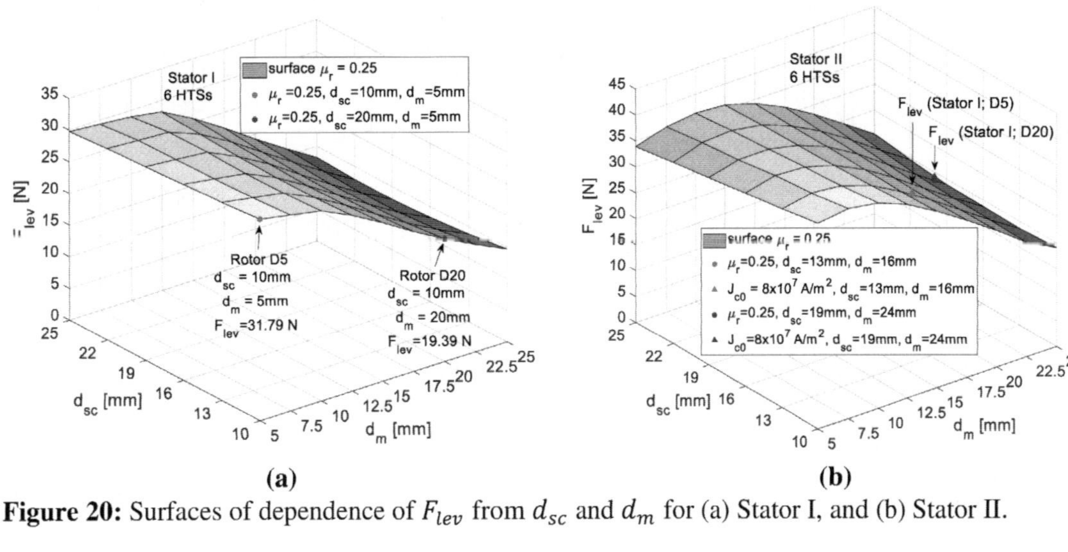

(a) (b)

Figure 20: Surfaces of dependence of F_{lev} from d_{sc} and d_m for (a) Stator I, and (b) Stator II.

As one may verify the values of F_{lev} are generally higher for the case of Stator II, including bulks of volume $40 \times 40 \times 10$ mm^3.

Figures 21(a) and 21(b) show the surfaces of $F_{gui\,max}$ respectively for Stator I and Stator II.

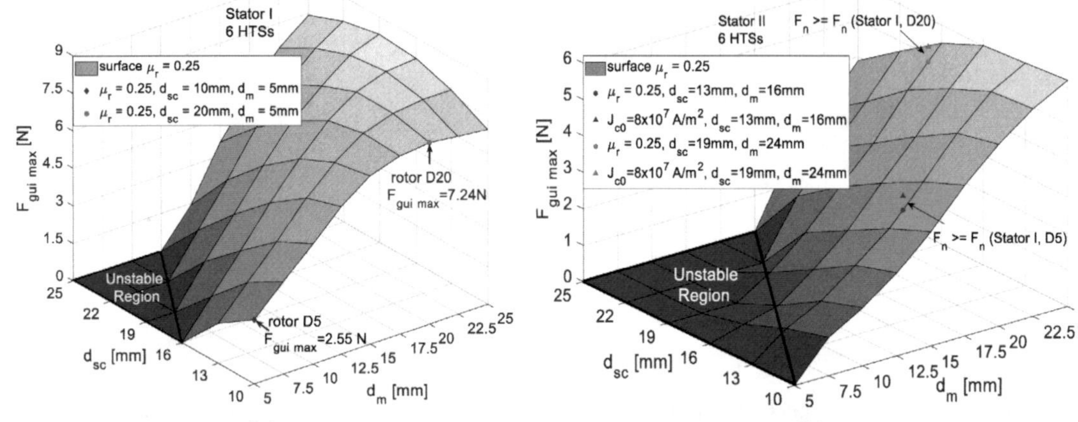

Figure 21: Surfaces of dependence of $F_{gui\,max}$ from d_{sc} and d_m for (a) Stator I, and (b) Stator II.

As one may verify with bulks of volume $33 \times 33 \times 14$ mm^3 (Stator I) the values of $F_{gui\,max}$ are generally higher than with bulks of volume $40 \times 40 \times 10$ mm^3 (Stator II). The region of no guiding stability is also wider for the last case (Stator II).

Figure 22 shows the distribution of magnetic flux and current densities obtained by 3D FEA using E-J model with $J_{c0} = 8 \times 10^7$ Am^{-2}, for one geometry on the limit of the region of no guiding stability with $d_{sc} = 19$ mm and $d_m = 10$ mm.

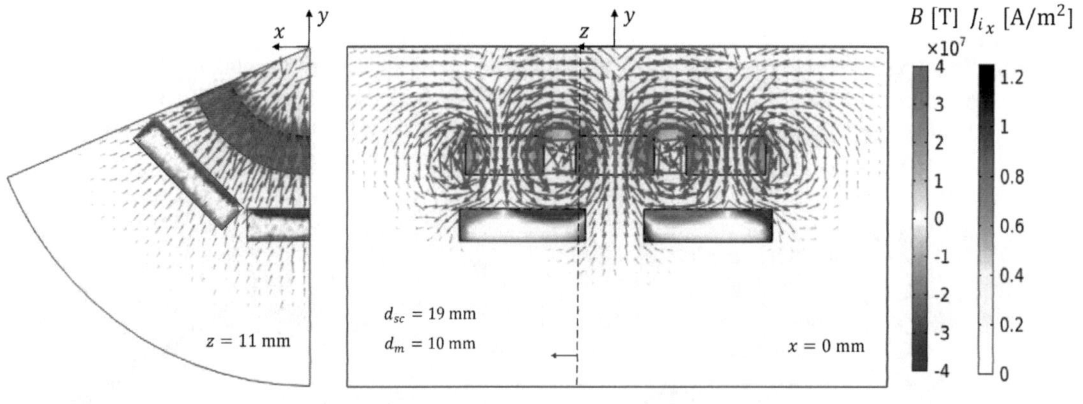

Figure 22: (a) Transversal and (b) longitudinal views with the distributions of **B** and **J** for the geometry with bulks of volume $40 \times 40 \times 10$ mm^3, $d_m = 10$ mm and $d_{sc} = 19$ mm.

As one may verify for this geometry, the bulk extremities are almost coincident with the extremities of the two side PM rings. This generally happens for the geometries on the limit line of guiding stability. Thus, the condition that should be considered for an initial check of the existence of guiding stability is the one expressed by (7)

$$\frac{d_{sc}}{2} + w_{sc} \leq d_m + 3\frac{w_m}{2} \tag{7}$$

from which may be deduced the minimum limit for d_m given by (8)

$$d_m \geq \frac{d_{sc}}{2} - 3\frac{w_m}{2} + w_{sc} \tag{8}$$

Table II presents minimum values of d_m for several values of d_{sc}, on the cases of Stator I and Stator II, calculated using (8).

Table II: Minimum values of d_m (8) for several values of d_{sc}, on the cases of Stator I and Stator II.

	Minimum d_m [mm]	
d_{sc} [mm]	Stator I	Stator II
10	0.5	7.5
13	2	9
16	3.5	10.5
19	5	12
22	6.5	13.5
25	8	15

The values in Table II follow with good approximation the lines on the limit of the guiding stability regions on Figures 21(a) and 21(b).

6 Conclusion

This study aims to present methodologies to determine geometries that maximize the guiding stability of a horizontal axis radial levitation bearing with HTS bulks cooled by ZFC. The guiding stability on the cases of two possible Stators with two different bulk sizes ($33 \times 33 \times 14$ mm^{3}and $40 \times 40 \times 10$ mm^3) for bearing experimental prototypes was studied. The regions of combinations of spacings where there is guiding stability were determined for the two Stator cases. An expression that may be used to identify the limit of the stability region was identified at first sight. The region of combinations of the spacing d_{sc} between the two HTS bulk rings and spacing d_m between PM rings, for which the guiding stability is lower for the case of $40 \times 40 \times 10$ mm^3 size bulks in Stator II than for the case of $33 \times 33 \times 14$ mm^3 size bulks in Stator I. The maximum guiding force $F_{gui\,max}$ for each combination of spacings is generally also higher on the second case. The optimization of spacings performed for the two Stator cases resulted in Pareto's curves from where one can obtain the maximum possible value of $F_{gui\,max}$, for a given restriction on the minimum levitation force F_{lev} that should be guaranteed.

References

Arsénio, A. J., Carvalho, M. V., Cardeira, C., Branco, P.J. C., and Melício, R. (2016) Conception of a YBCO superconducting ZFC-magnetic bearing virtual prototype. IEEE International Power Electronics and Motion Control Conference (PEMC), 25-28 September 2016.

Arsénio, A. J., Roque, M., Cardeira, C., Branco, P. J. C., and Melicio, R. (2018). Prototype of a Zero-Field-Cooled YBCO Bearing with Continuous Ring Permanent Magnets. IEEE Transactions on Applied Superconductivity, 28(4).

Fernandes, J. F. P., Costa, A.J. Arsénio and Arnaud, J. (2020). Optimization of a Horizontal Axis HTS ZFC Levitation Bearing Using Genetic Decision Algorithms Over Finite Element Results. IEEE Transactions on Applied Superconductivity, 30(2).

Quéval, L., Sotelo, G. G., Kharmiz, Y., Dias, D. H. N., Sass F., Zermeño, V. M. R. and Gottkehaskamp, R. (2016). Optimization of the Superconducting Linear Magnetic Bearing of a Maglev Vehicle. IEEE Transactions on Applied Superconductivity, 26(3).

Arsénio, A.J., Melicio, R., Cardeira, C., and Branco, P.J. Costa (2018). The Critic Liquid-gas Phase Transition between Liquid Nitrogen and YBCO HTS bulks: from FEM Modeling to Its Experimental Study for ZFC Levitation Devices. IEEE Transactions on Applied Superconductivity, 28(4).

Arsénio, Antonio J. Costa, and Branco, Paulo J. Costa (2021). Thermo-Hydraulic Analysis of a Horizontal HTS ZFC Levitating Bearing Concerning Its Autonomy Safety Service Time. IEEE Transactions on applied superconductivity 31, (5).

Hong, Z., Campbell, H. M. and Coombs T.A. (2007). Computer Modeling of Magnetisation in High Temperature Superconductors. IEEE Transactions on applied superconductivity, 17(2), (pp. 3761-3764).

Zhang M., and Coombs T.A. (2012). 3D modeling of high-Tc superconductors by finite element software. Superconductor Science and Technology, vol. 25.

Koo, J. H. and Cho G. (2004). Magnetic field dependence on transition temperature Tc in cuprate superconductors. Solid State Communications, 129, (pp. 191-193).

Yamamoto, K., Mazaki, H. and Yasuoka, H. (1993). Magnetization of type-II superconductors in the Kim-Anderson model. Physical Review B, 47(2).

Fujishiro, H., and Naito, T. (2010). Simulation of temperature and magnetic field distribution in superconducting bulk during pulsed field magnetization. Superconductor Science and Technology, 23, 105021.

Peixoto, Inês S. P., Ferreira da Silva, F., Fernandes, João F. P., and Branco, P. J. da Costa (2021). 3D Equivalent Space-Varying Permeability Model of HTS Bulks for Computation of Electromagnetic Forces. IEEE Transactions on applied superconductivity 31, (5).

Investigations on a new slotless homopolar hybrid active magnetic bearing

Guillaume Colinet and Bruno Dehez

Institute of Mechanics, Materials and Civil Engineering (IMMC)
Université catholique de Louvain (UCLouvain)

Abstract

Homopolar hybrid active magnetic bearings generate very low power losses thanks to the absence of bias currents and the homopolar distribution of the magnetic flux density in the airgap. However, the slots in which the winding of these bearings are inserted create fluctuations of the magnetic flux density which are at the origin of iron losses that can be non-negligible at high speeds. This paper investigates a new slotless topology of homopolar hybrid active magnetic bearings, which can totally remove these losses. An electromagnetic model based on a magnetic circuit is proposed and validated through finite element simulations. It is then used to optimize and compare the performances of this original bearing topology with an existing slotted hybrid homopolar bearing.

1 Introduction

Active magnetic bearings (AMB) generate a contact-free guiding of a rotor by actively controlling the current flowing in an armature winding. Compared with other bearings, AMB have the advantages to create very low friction as well as to operate without lubrication and mechanical wear. They are therefore widely used in industrial applications, such as in turbomachinery, from smallturbo molecular pumps to large turbogenerators, in energy storing flywheels, in momentum flywheels or even in ventricular pumps [3] [4]. The vast majority of AMB used in industry relies on a heteropolar bias magnetic flux density created by a DC component of current, superimposed to the control currents, and have thus the drawback of generating Joule losses. This is the reason why hybrid active magnetic bearing (H-AMB) topologies using permanent magnets have been proposed to solve this problem [1] [5]. Among those H-AMB topologies, those having 8 poles are widely used [9] [8]. However, for those topologies, the bias magnetic flux density in the airgap alternate between a positive and a negative value and therefore generates iron losses in the rotor during its rotation. Alternative heteropolar topologies, that have less number of poles, have been proposed to reduce these losses [12]. Beyond heteropolar topologies, homopolar hybrid AMB (HH-AMB) produces less iron losses because the polarity of the bias magnetic flux density does not vary with the position of the rotor [11] [6]. However, there are still some fluctuations of the magnetic flux density due to the slots within which the armature winding is placed. To reduce further those fluctuations, slotless hybrid homopolar AMB (SHH-AMB) have been proposed [13] [7]. Those kind of magnetic bearings drastically reduce and even theoretically remove rotor core losses due to the bias magnetic flux density, which makes it attractive for applications in the vacuum or at very high-speed. However, in the proposed topologies, the PM are placed on the rotor in a surface-mounted configuration, which makes the mechanical design of those bearings more difficult. In this context, this paper explores a new slotless homopolar hybrid topology whose PM are placed in the stator, but more singularly, inserted between two ring-shaped iron poles. This feature fundamentally changes the way the magnetic flux density is redistributed in the airgap when the rotor is off-centered. The principle of force generation is therefore different which could open the way to a potential increase in the performances of the AMB.

This paper has the following structure. Section II gives a description and a parametrization of the proposed SHH-AMB. Section III provides an electromagnetic model of the AMB that allows to link the dimensional parameters of the SSH-AMB to performance criteria while section IV presents a comparison of this model with a FEM model. In section V, a parametric analysis is performed in order to study the influence of the parameters of the SHH-AMB over the two performance criteria. Finally, the results of a bi-objective optimization combining the two performance criteria as well as a comparison of these latter with an existing slotted HH-AMB proposed in [6], are presented in section VI.

2 Magnetic bearing description and parametrization

The AMB considered in this study is shown in Figure 1. The stator is made of a PM ring inserted between two ring-shaped iron poles and a concentrated winding located in the airgap between the stator iron poles and the rotor.

The PM is polarized in the axial direction and creates a bias magnetic flux that it is guided radially inward in the airgap on the right iron pole and radially outward in the airgap on the left iron pole, as highlighted with the white arrows in Figure 1. It is thus a homopolar topology, as the magnetic flux density produced in the airgap by the PM is supposed to be perfectly

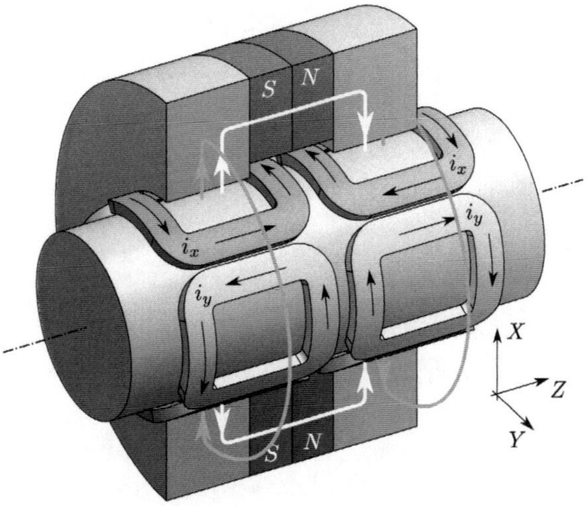

Figure 1: 3D view of the AMB.

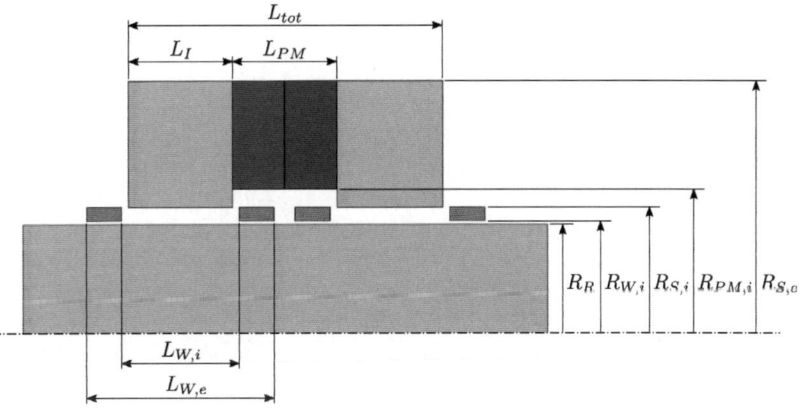

Figure 2: Parametrization of the AMB.

constant and independant of the angular position, when the rotor is centered radially.
The winding is made of four concentrated coils for each of the two radial axes. Two of them are placed at one iron pole with their magnetic axis pointing in the same direction as the axis under consideration, and the other two at the other iron pole but with their magnetic axis pointing in the opposite direction. The force generation relies in the unbalance of the magnetic pressure in the airgap arising from the superposition of the bias flux and the control flux created by the winding. Figure 1 highlights with the green arrows the control magnetic flux resulting in a force aligned with the positive X-axis. On the left iron pole, the control magnetic flux is directed such that it increases the magnetic pressure at the top and decreases it at the bottom, creating therefore an upward force on the rotor. On the right iron pole, the bias magnetic flux and the control magnetic flux are directed in the opposite direction, in order to also increase the magnetic pressure at the top and decrease it at the bottom.

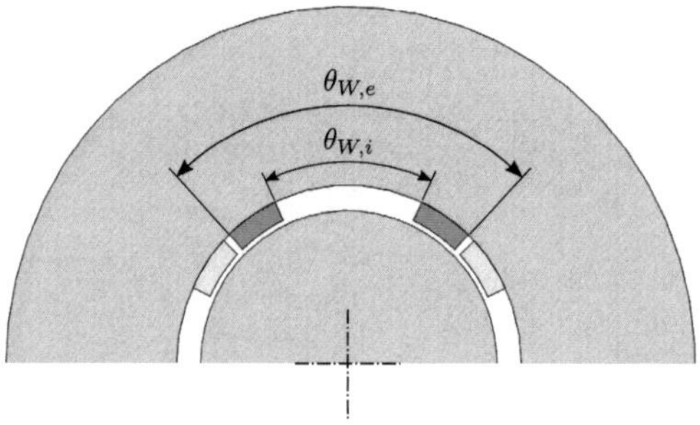

Figure 3: Parametrization of the AMB.

3 Modeling

The performances of an AMB are related to the force that is produced on the rotor. The resulting force that acts on the rotor has different components. The first component is the electrodynamic force, that arises from the interaction between the current i_x and i_y flowing respectively in the coils for the X-axis and the Y-axis and the magnetic flux of the PM that is linked by those coils. This force is related to the current through the the current stiffness k_i. By controlling the current in the winding, this force stabilizes the rotor in its centered position. The second component is the detent force, due to the interaction between the PM and the rotor iron. This force is related to the position x or y of the rotor with respect to the centered position through the position stiffness k_e. This force appears only when the rotor is off-centered, but it is directed in the same direction as the one of the off-centering, giving rise to instabilities. The third component is the reluctant force. It is due to the variation of the inductance of the winding with the rotor position and it is proportionnal to the square of the current.

In the centered position, because of symmetry, this reluctant force and its derivative relative to the rotor position is nul. For the same reason, the potential corss coupling between the two radial axes is non-existent. In consequence, in the centered position, the force acting on the rotor is given by:

$$\begin{cases} F_x = -k_e\, x + k_i\, i_x \\ F_y = -k_e\, y + k_i\, i_y \end{cases} \tag{1}$$

In this section, a model of the magnetic flux density due to the PM is first provided. From this latter, the current stiffness and the position stiffness are derived. Finally, as the force is linked to the current that flows in the winding, a model of the electrical resistance is also provided, to evaluate the Joule losses.

3.1 Airgap magnetic flux density

The magnetic flux density in the airgap, due to the PM, is obtained in two steps. The first consists in determining the magnetic potential difference between the stator and the rotor by solving the magnetic circuit shown in Figure 4. This latter assumes an infinite magnetic permeability of both the stator iron poles and of the rotor. The PM ring can be reprensented by its Northon representation with a flux source :

$$\phi_{PM} = B_r \cdot \pi(R_{S,e}^2 - R_{PM,i}^2) \tag{2}$$

in parallel with its internal permeance:

$$P_{PM} = \frac{L_{PM}}{\mu_0 \cdot (R_{S,e}^2 - R_{PM,i}^2)} \tag{3}$$

where B_r is the remanent flux density, $R_{S,e}$, $R_{PM,i}$ and L_{PM} are respectively the external radius of the stator, the internal radius of the PM and the length of the PM and μ_0 is the magnetic permeability of the vacuum.
The leakage magnetic flux going between the left and right outer radial surfaces of the stator iron parts, is taken into account by the permeance P_{PMfe}. According to [2], this permeance can be approximated using the following expression:

$$P_{PMfe} = 2\,\mu_0\,R_{S,e}\,ln\left(1 + \frac{2\,\pi \cdot R_{S,e}\,L_I + \pi\,L_I^2}{2\,L_{PM}\,R_{S,e}}\right) \tag{4}$$

where L_I is the axial length of one stator iron pole. The leakage magnetic flux going between the inner lateral surfaces of the stator iron poles. It is modeled by the permeance P_{PMfi}, whose expression is given by:

$$P_{PMfi} = \frac{\mu_0\,\pi\,(R_{PM,i}^2 - R_{S,i}^2)}{L_{PM}} \tag{5}$$

where $R_{S,i}$ is the internal radius of the stator. The flux going through the airgap is taken into account by the permeance P_g, that depends on the rotor off-centering x. Considering that the magnetic flux is purely radial, this permeance is derived from:

$$P_g = \int_0^{2\pi} dP_g \tag{6}$$

where dP_g is the permeance of a radial flux tube located at angle θ relative to the X-axis and having an infinitesimal angular opening $d\theta$. This latter is given by:

$$dP_g = \frac{\mu_0\,L_I}{\ln\left(\frac{R_{S,i} - x\cos\theta}{R_R}\right)}\,d\theta \tag{7}$$

where R_R is the external radius of the rotor. This function is not analytically integrable as it is, but considering that the off-centering x is small relative to the radii $R_{S,i}$ and R_R, on the one hand, and to the nominal airgap thickness $g = R_{S,i} - R_R$, on the other hand, it can be approximated first by developing the logarithm function in Taylor series limited to the first order so as to be rewritten:

$$dP_g = \frac{\mu_0\,L_I}{\ln\left(\frac{R_{S,i}}{R_R}\right) - \frac{x}{R_{S,i}}\cos\theta}\,d\theta \tag{8}$$

Next, as this function is a periodic function of θ, it can be expanded in Fourier series. Limited to the second order and its coefficients being themselves approximated but analytically, it results in:

$$dP_g = \frac{\mu_0 L_I}{\ln^3\left(\frac{R_{S,i}}{R_R}\right)}\left[\ln^2\left(\frac{R_{S,i}}{R_R}\right) - \left(\frac{x}{R_{S,i}}\right)\ln\left(\frac{R_{S,i}}{R_R}\right)\cos\theta + \left(\frac{x}{R_{S,i}}\right)^2\cos 2\theta\right]d\theta. \tag{9}$$

Using (6) and (9), the analytical expression of P_g finally reduces to:

$$P_g = \frac{2\,\pi\,\mu_0\,L_I}{\ln\left(\frac{R_{S,i}}{R_R}\right)}. \tag{10}$$

The fringing flux between the lateral surfaces of the stator iron parts and the rotor is taken into account by the permeances P_{gf}. According to [2], it can be expressed as follows:

$$P_{gf} = 4\cdot\mu_0\,R_R\,ln\left(1+\frac{y_{pgo2max}\,(2\,R_R+y_{pgo2max})}{y_{pgo2min}\,(2\,R_R+y_{pgo2min}}\right) \tag{11}$$

considering $y_{pgo2max} = R_{S,e} - R_{S,i}$ and $y_{pgo2min} = R_{S,i} - R_R$.

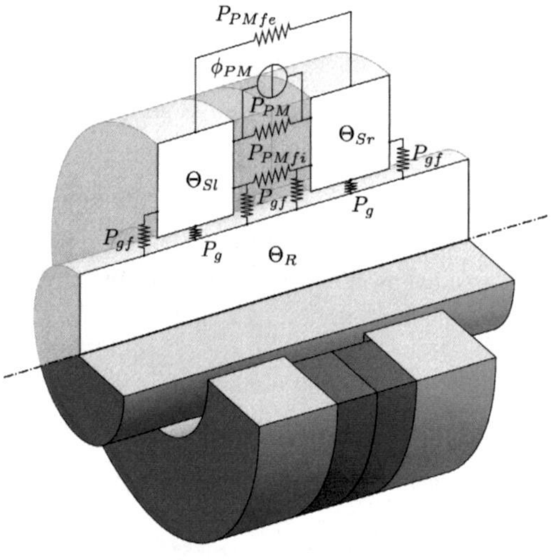

Figure 4: Magnetic circuit considering fringing effets.

Based on the equivalent circuit, the magnetic potential difference between the stator and the rotor $\Theta_{S-R} = \Theta_{Sl} - \Theta_R$ is directly given by:

$$\Theta_{S-R} = \frac{2\Phi_{PM}}{2\,P_{PMfe}+2\,P_{PM}+2\,P_{PMfi}+P_g+2\,P_{gf}}. \tag{12}$$

The second step consists in determining the magnetic flux density in the airgap considering that the flux $d\Phi_g$ going through a radial flux tube having an infinitesimal angular opening $d\theta$ is given by:

$$d\Phi_g = \Theta_{S-R}\,dP_g. \tag{13}$$

As this flux is also the product of the normal section and the magnetic flux density, this latter is finally given by:

$$B_g = \frac{\mu_0\,\Theta_{S-R}}{r\ln^3\left(\frac{R_{S,i}}{R_R}\right)}\left[\ln^2\left(\frac{R_{S,i}}{R_R}\right) - \left(\frac{x}{R_R}\right)\ln\left(\frac{R_{S,i}}{R_R}\right)\cos\theta + \left(\frac{x}{R_R}\right)^2\cos 2\,\theta\right]. \tag{14}$$

3.2 Current stiffness

The electrodynamic force is related to the current through the current stiffness k_i. This latter can be expressed as the partial derivative relative to the off-centering x of the magnetic flux Ψ_x linked by the winding:

$$k_i = \frac{\delta \Psi_x}{\delta x}. \tag{15}$$

This flux can be obtained by summing the contributions of the fluxes $d\Psi_x$ linked by coils of infinitesimal section $dS_c = r\,dr\,d\theta$, whose $\frac{N_W}{S_W} dS_c$ go and back conductors are located respectively in (r, θ) and $(r, -\theta)$. Considering the four coils constituting the X-axis winding, the current stiffness is given by:

$$k_i = \frac{4\,\mu_0 L_I \Theta_{S-R}}{R_R \ln^2\left(\frac{R_{S,i}}{R_R}\right)} \frac{N_W}{S_W} \left(R_{S,i}^2 - R_{W,i}^2\right) \left(\cos\left(\theta_{W,e}/2\right) - \cos\left(\theta_{W,i}/2\right)\right) \tag{16}$$

where N_W and S_W are the number of turns and the section of one coil respectively.

3.3 Position stiffness

The position stiffness k_e is directly related to the detent force through:

$$k_e = \frac{\partial F_{detent}}{\partial x}. \tag{17}$$

This force can be derived by integrating the X-axis component of the magnetic pressure on the rotor surface facing the stator iron poles:

$$F_{detent} = 2\,L_I \int_0^{2\,\pi} \frac{B_g^2}{2\,\mu_0} \cos\theta\, R_R\, d\theta. \tag{18}$$

Using (14) and (18), (17) leads to:

$$k_e = 2\,\mu_0 \frac{\pi\, L_I\, \theta_{S-R}^2}{R_R^2 \ln^3\left(\frac{R_{S,i}}{R_R}\right)}. \tag{19}$$

3.4 Electrical resistance

The total resistance of one winding is given by:

$$R = 4\,\rho\, N_w \frac{L_{loop}}{S_c} \tag{20}$$

where ρ is the electrical resistivity, L_{loop}, the mean length of a loop, is given by:

$$L_{loop} = 2\,L_I + \left(\frac{R_{S,i} + R_{W,i}}{2}\right) \left[\frac{\pi}{2} \left(\theta_{W,e} - \theta_{W,i}\right) + 2\,\theta_{W,i}\right] \tag{21}$$

and S_c, the cross section of a conductor, is given by:

$$S_c = \eta \frac{S_W}{N_W} = \eta \frac{\left(\theta_{W,e} - \theta_{W,i}\right)\left(R_{S,i}^2 - R_{W,i}^2\right)}{4 N_W} \tag{22}$$

where η is the fill factor.

4 FEM validation

In order to validate the proposed model, its predictions are compared with those obtained from 3D FEM simulations. These latter have been performed with the software COMSOL Multiphysics ®. The parameters of the AMB used for this validation are given in Table 1. They corresponds to the parameters of the slotted HH-AMB proposed in [6], except for the parameters R_R, $R_{W,i}$, $\theta_{W,i}$ and $L_{W,e}$ that have been obtained by maximizing the bearing constant defined in (23).

Parameter	Symbol	Value	Unit
Total length of stator	L_{tot}	21	mm
Length of PM	L_{PM}	5	mm
Length of stator pole	L_I	8	mm
Internal length of winding	$L_{W,i}$	8	mm
External length of winding	$L_{W,e}$	12.748	mm
Radius of rotor	R_R	15.89	mm
Internal radius of winding	$R_{W,i}$	16.09	mm
Internal radius of stator	$R_{S,i}$	16.2	mm
Internal radius of PM	$R_{PM,i}$	33	mm
External radius of stator	$R_{S,e}$	35	mm
Internal angle of winding	$\theta_{W,i}$	1.2767	rad
External angle of winding	$\theta_{W,e}$	$\frac{\pi}{2}$	rad
Number of loops of a coil	N_W	95	/
Fill factor	η	0.4	/

Table 1: Parameters.

The evolution of the magnetic flux density in the airgap with the angular position, for a centered and off-decentered positions along the X-axis is shown in Figure 5. It can be seen that the analytical predictions overestimate the FEM ones by approximately 3%. This offset can be explained by the fact that the model does not consider the leakage flux between the PM ring and the rotor.

A comparison on the current stiffness shows that the analytical predictions underestimates the current stiffness by maximum 4%. Even if the magnetic flux density is overestimated by the analytical model, this underestimation can be explained by the fact the FEM model takes into account the flux linked the end-windings whereas the analytical model not.

A comparison on the position stiffness shows that the analytical model overestimates the FEM model by approximately 7%, independently of the off-centering amplitude. This difference can be explained by the higher magnetic flux density predicted by the analytical model.

The evaluation of this component of the force using the FEM model has shown that it represents less than 1% of the total force and therefore can be neglected.

5 Parametric analysis

In this section, the influence of the bearing parameters on its performances is studied. Those performances depend on the electrodynamic force, that should be as large as possible, and on the detent force, that should be as low as possible. Altough the latter can be evaluated using the position stiffness, the first one depends both on the current and on the current stiffness. For

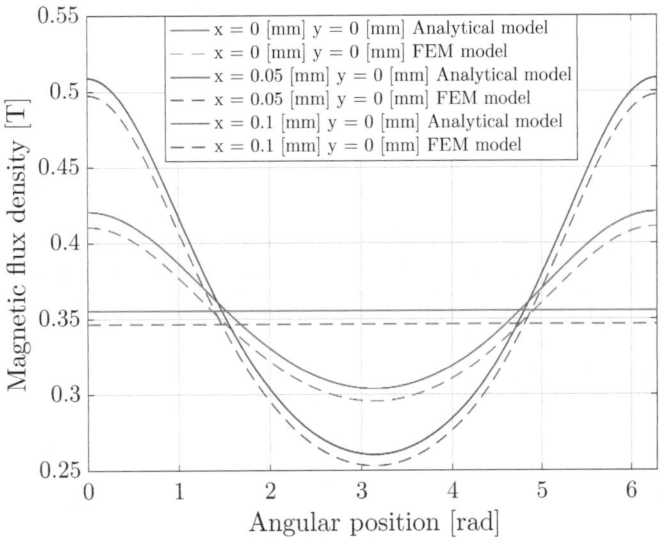

Figure 5: Magnetic flux density in the airgap.

the same current stiffness, an increase in the current would increase the electrodynamic force but this has to be limited because of the Joule losses and the temperature increase it generates. Therefore, an interesting criterion is the bearing constant defined as [10]:

$$k_b = \frac{k_i}{\sqrt{R}} \tag{23}$$

Indeed, mutliplying the numerator and the denominator by the current, the bearing constant reveals to be proportional to the ratio between the electrodynamic force and the square root of the Joule losses.

Based on these two performance criteria, a parametric analysis is performed. In a first step, the influence of the winding thickness and of the winding internal angle is studied, considering the parameters in Table 1. Then, keeping the best values for those two parameters, the influence of the other parameters is studied.

5.1 Winding thickness

The variation of the current stiffness, the electrical resistance, the bearing constant and the position stiffness with the winding thickness is shown in Figure 6. The current stiffness decreases when the thickness increases because it directly impacts the gap between the stator and rotor, and therefore reduces the airgap magnetic flux density. The number of loops beeing constant, the flux linked by the winding, as well as the resulting current stiffness have to follow the same evolution. Regarding the electrical resistance, it decreases with the winding thickness because, for a constant number of loops, the wire cross section increases with the winding section.

It can be seen that this effect has a much bigger impact for very small values of the thickness. Concerning the bearing constant, it reaches a maximum as the decrease in resistance predominates over the decrease in current stiffness for small thicknesses, whereas it is the opposite

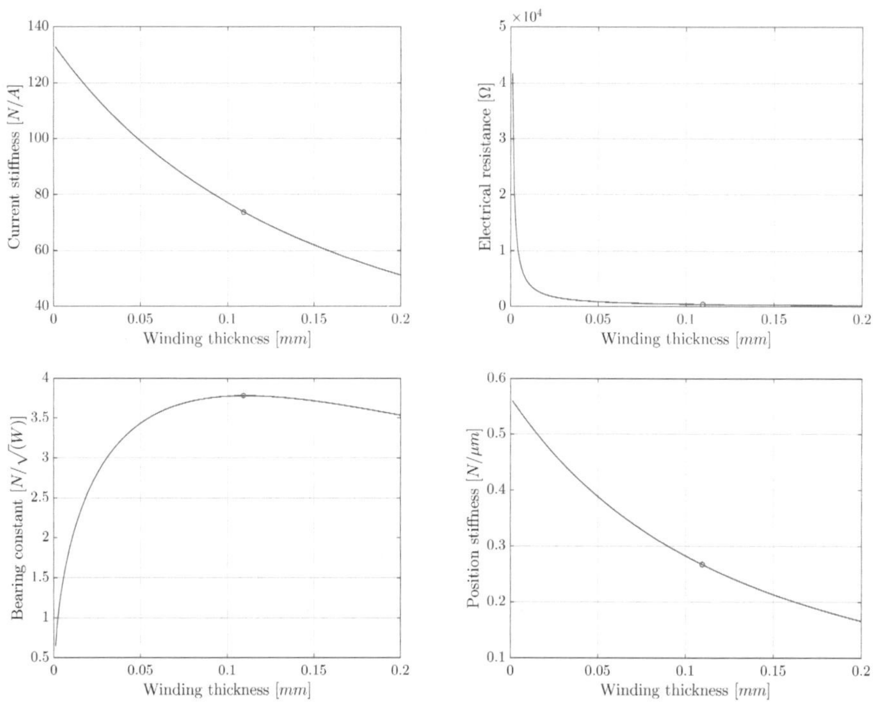

Figure 6: Analysis on winding thickness, ∘ corresponds to parameters in Table 1.

for larger thicknesses. The evolution of the position stiffness with the winding thickness quite logically looks close to the evolution of the current stiffness.

5.2 Winding internal angle

As shown in Figure 7, the current stiffness increases with the winding internal angle because each winding loop links a higher magnetic flux. Concerning the resistance, it increases slightly for small value and then increases fastly when this angle becomes close to the external angle of the winding. As shown by the evolution of the bearing constant, the increase of the resistance has more impact than the increase of the current stiffness and therefore the bearing constant is decreasing monotonically. The position stiffness is not influenced by the winding internal angle as this latter parameter does not modify the value of the magnetic flux density in the airgap.

5.3 Stator internal radius

As presented in Figure 8, a higher radius leads to a higher current stiffness. It can be explained by the fact that the winding covers a higher area and therefore links a higher magnetic flux, the airgap magnetic flux density remaining almost the same because the gap between the stator and the rotor is fixed. Concerning the resistance, it decreases with the value of the radius because the section of the winding increases. The effect on the current stiffness and the resistance goes in a way to increase the bearing constant. The radius has only a small impact on the position stiffness.

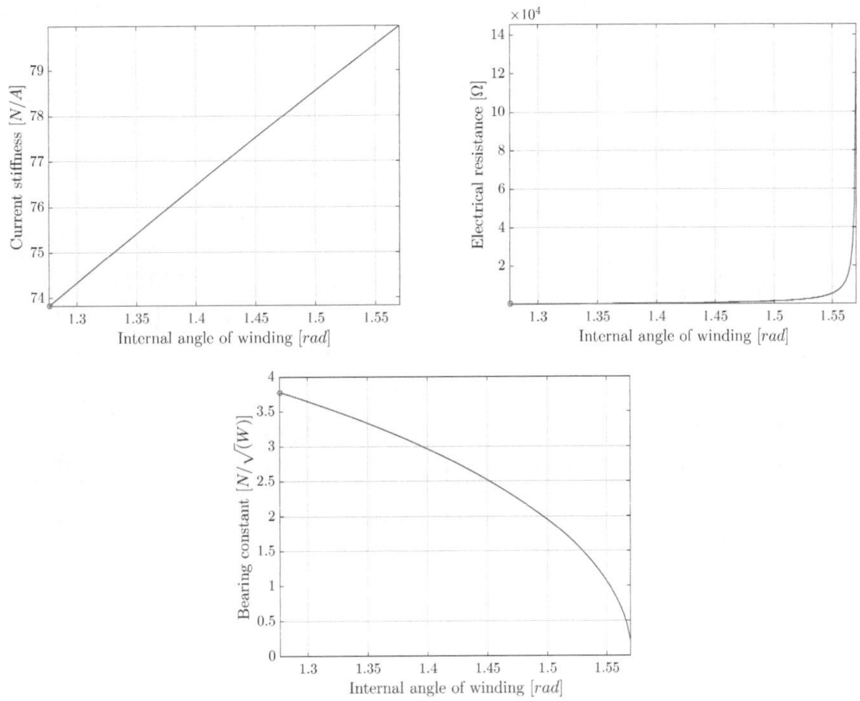

Figure 7: Analysis on winding internal angle, ∘ corresponds to parameters in Table 1.

5.4 Permanent magnet internal radius

The internal radius of the PM has a direct impact on the value of the magnetic flux density in the airgap. It can be seen in Figure 9 that the bearing constant, which has exactly the same shape as the current stiffness because the resistance does not vary, and the position stiffness depend almost linearly on the radius. However an increase of a given factor over the current stiffness is counterbalanced by an increase of the square of this factor in the position stiffness. Obviously, the electrical resistance is not affected by the PM internal radius.

5.5 Permanent magnet length

As the total length of the bearing $L_{tot} = 2L_I + L_{PM}$ remains constant in the analysis, the length of the PM is studied through the following parameter:

$$\alpha - \frac{L_{PM}}{L_{tot}} \tag{24}$$

As illustrated in Figure 10, the current stiffness reaches an optimum that can be explained by the respective evolutions with the magnet length of both the magnetic flux density in the airgap and the coil area. For increasing α, the airgap magnetic flux density increases as the leakage fluxes between the two stator ferromagnetic parts decrease on the one hand, while the coil area decreases on the other hand. For small values of α the airgap magnetic flux density increases faster than the surface of winding decreases, resulting in an increase of the current stiffness, while it is the opposite for high value of α.

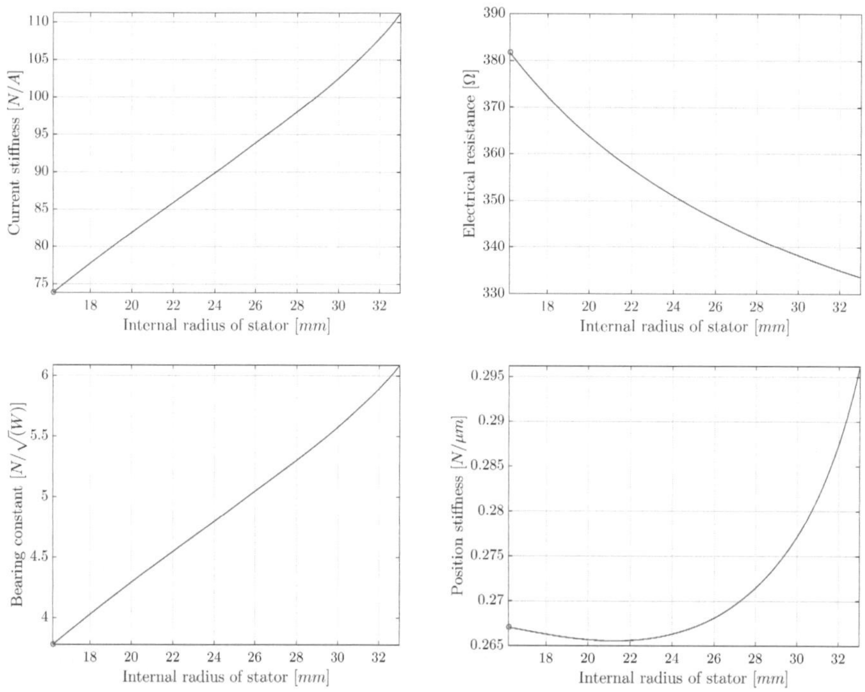

Figure 8: Analysis on stator internal radius, ∘ corresponds to parameters in Table 1.

The resistance decreases with α as it depends linearly on the stator pole length L_I. The position stiffness reaches a maximum but for higher α because of the square dependency of this stiffness on the magnetic flux density.

6 Optimization and comparison

A bi-objective optimization, considering the bearing constant k_b and the position stiffness k_e as two independant objectives is performed. The optimization parameters are R_R, $R_{S,i}$, $R_{PM,i}$ $\theta_{W,i}$, $\theta_{W,e}$ and α. The other parameters, more specifically $R_{S,e}$ and L_{tot}, are fixed to the ones in Table 1, . The optimization is performed using a sequential quadratic programming (SQP) algorithm. The results of the optimization process is shown in Figure 11 and consists of several optimizations applied on a single objective function combining with variable weights the two objectives. The Pareto front consists of the best compromise and is found between the extremas $(k_b, k_e) = (0, 0)$ and $(k_b, k_e) = (28.9, 40.5)$. In Figure 11, the geometry of two bearings from the Pareto front are shown. It can be seen that the bearing ∘ has a lower position stiffness due to the higher winding thickness and higher internal radius of PM. On the contrary, the bearing ⋄ reach a higher bearing constant because of the very low winding thickness and the lower internal radius of PM.
Comparing these results with those of the slotted homopolar AMB presented in [6] for which $(k_b, k_e) = (51.23, 0.57)$, shows that it is apparently not possible to reach the same value of the bearing constant. Furthermore, for the same position stiffness, the bearing constant is approximately five times lower.

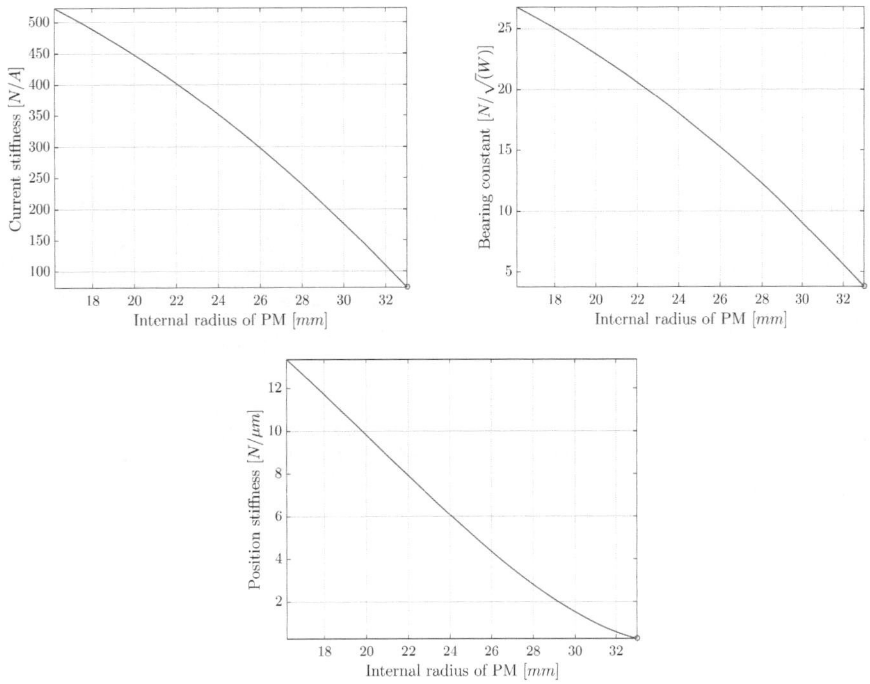

Figure 9: Analysis on PM internal radius, ○ corresponds to parameters in Table 1.

7 Conclusion

In this paper, a new slotless HH-AMB that theoretically removes iron losses due to the bias magnetic flux density has has been studied. To that end an electromagnetic model that assumes small off-centering amplitude has been presented and validated by a FEM model. From that model, two performance criteria have been derived, the bearing constant, which reflects the force capability of the bearing through the current stiffness and the winding electrical resistance, and the position stiffness. On that basis, a parametric analysis has been carried out to understand the influence of the different dimensional parameters of the proposed bearing on its performances. Then, a Pareto front was obtained via an optimisation process. This revealed that the proposed bearing seems not able to reach the performances of slotted HH-AMB in terms of bearing constant. . However, these results have been obtained by imposing the same total length of the stator and the same external radius of the PM. Keeping the same volume of the AMB while varying those two parameters might give better results. In addition, these two HH-AMB should also be compared considering the iron losses, since the slotless topology should be better on this aspect.

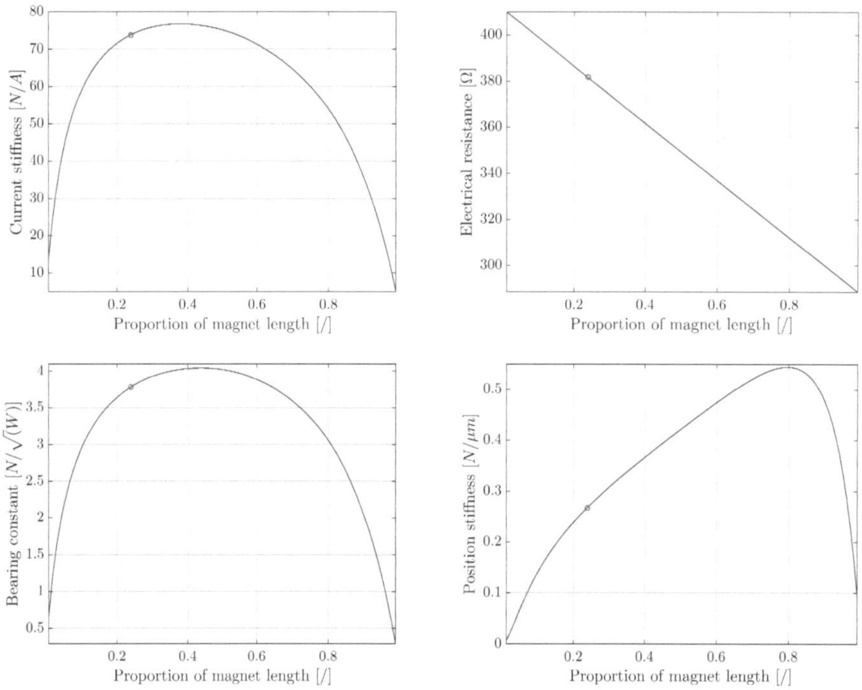

Figure 10: Analysis on PM length, ○ corresponds to parameters in Table 1.

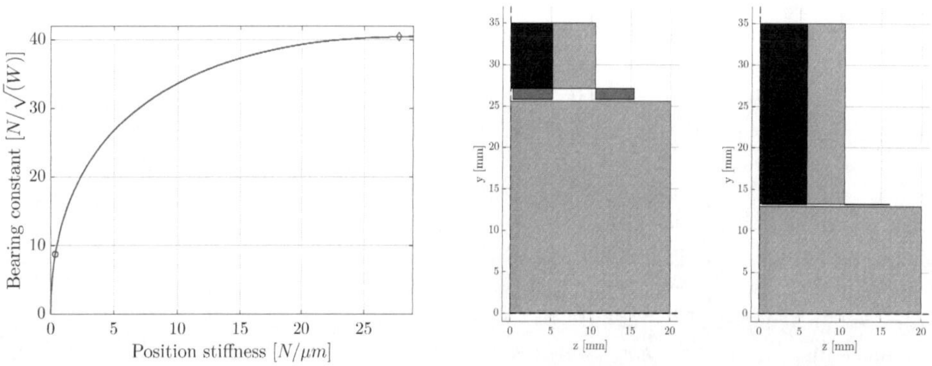

Pareto Front a: Bearing ○ b: Bearing ◇

Figure 11: Pareto front and geometry of 2 bearings.

References

[1] E. H. Allaire A. Studer and C. K. Sortore. Low power magnetic bearing design for high speed rotating machinery. *Int. Symposium on Magnetic Suspension Technology , Hampton, VA, USA, 1991*, 1991.

[2] M. A. Batdorff and J. H. Lumkes. High-fidelity magnetic equivalent circuit model for an axisym-

metric electromagnetic actuator. *IEEE Transactions on Magnetics*, 45(8):3064–3072, 2009.

[3] Alfons Traxler Gerhard Schweitzer, Hannes Bleuler. Active magnetic bearing foundation, performance and application. Yu Lie, Yuan Suijun, translation. Beijing: New Times Press. 1997.

[4] Eric H.Maslen Gerhard Schweitzer. Magnetic bearings:theory, design, and application to rotating machinery. Xu Yang, Zhang Kai, Zhao Lei translation. Beijing: Mechanical Industry Press,2012.

[5] L. Ji, L. Xu, and C. Jin. Research on a low power consumption six-pole heteropolar hybrid magnetic bearing. *IEEE Transactions on Magnetics*, 49(8):4918–4926, 2013.

[6] F. Jiancheng, W. Xi, W. Tong, T. Enqiong, and F. Yahong. Homopolar 2-pole radial permanent-magnet biased magnetic bearing with low rotating loss. *IEEE Transactions on Magnetics*, 48(8):2293–2303, 2012.

[7] Maurus Kaufmann, Arda Tüysüz, Johann W. Kolar, and Christof Zwyssig. High-speed magnetically levitated reaction wheels for small satellites. In *2016 International Symposium on Power Electronics, Electrical Drives, Automation and Motion (SPEEDAM)*, pages 28–33, 2016.

[8] Kenny and A. B. Palazzolo. Single plane rad ial, magnetic bearings biased with poles containing permanent magnets. *J. Mech. Des.*, 125:178–185, Mar. 2003.

[9] H. Koyanagi Okada and K. Kakihara. Miracb earing: New concept of miracle magnetic bearings. *Int. Symp. on Magnetic Bearings , Lexington, Kentucky, USA*, Aug. 2004.

[10] Patricio Peralta, Guzmán Borque-Gallego, Yoann Lapijover, and Yves Perriard. Experimental electromechanical characterization of slotted and slotless miniature bearingless drives. In *2021 IEEE International Electric Machines Drives Conference (IEMDC)*, pages 1–8, 2021.

[11] M. A. Pichot and M. D. Driga. Loss reduction strategies in design of magnetic bearing actuators for vehicle applications. *IEEE Transactions on Magnetics*, 41(1):492–496, 2005.

[12] R. Zhu, W. Xu, C. Ye, and J. Zhu. Novel heteropolar radial hybrid magnetic bearing with low rotor core loss. *IEEE Transactions on Magnetics*, 53(11):1–5, 2017.

[13] Christof Zwyssig, Thomas Baumgartner, and Johann W. Kolar. High-speed magnetically levitated reaction wheel demonstrator. In *2014 International Power Electronics Conference (IPEC-Hiroshima 2014 - ECCE ASIA)*, pages 1707–1714, 2014.

Space Vector Modulation for High Dynamic Current Control of a Self-Sensing Active Magnetic Bearing

Dominik Wimmer, Markus Hutterer, and Manfred Schrödl

Institute of Energy Systems and Electrical Drives
Technische Universität Wien
Gußhausstraße 25, 1040 Vienna, Austria
dominik.wimmer@tuwien.ac.at

Abstract

The control loop of active magnetic bearings requires feedback of the rotor position to enable a stable levitation of the rotor. In contrast to conventional position sensor-based systems, self-sensing control can lead to a far more simple design of the system. The focus of this study lies on a high dynamic control of a self-sensing three-phase radial active magnetic bearing. The self-sensing operation is based on the detection of coil inductance deviations of an eccentrically levitating rotor. The inductance of the actuator coils is obtained by current slope measurements in the phases of the bearing. For this reason, special voltage pulse patterns are applied to the coils, which are defined by the half-bridges of a three-phase inverter. The required voltage pulse patterns are defined by a space vector modulation in a digital signal processor. In this work, a variant of space vector modulation is presented, which allows a high dynamic of the current controller under consideration of self-sensing aspects. Finally, the designed modulation strategy was investigated with simulations and measurements on a prototype of a radial active magnetic bearing.

1 Introduction

Active magnetic bearings (AMBs) enable a stable levitation of a rotor without mechanical contact. Therefore, high rotor speeds can be reached without the need of lubricants. For a stable levitation of the rotor, a position information of the rotor is required to close the control loop. Conventional AMB systems use position sensors to determine the rotor position. Design criteria of active magnetic bearings often include requirements like system costs, construction space and the reliability of the system. Self-sensing methods can help to achieve the design goals by a reduction of the number of sensors in a system. The self-sensing control of active magnetic bearings is a field of research for many years [1]. Therefore, various control approaches have been established, which can be divided in observer-based and high frequency perturbation methods [2]. Observer-based methods rely on a design of an estimator for the air gap length by voltage and current measurements. Therefore, a detailed mathematical description of the system is required to achieve adequate estimator results as shown in [3]. High frequency perturbation methods rely on an estimation of the air gap length by the inductance of the actor. In many AMB configurations, the air gap length influences the inductance of the actuator coils. Hence, inductance changes can be detected by the current response of high frequency (HF) voltage pulses. Previous studies in this field used switching amplifiers with hysteresis [4], [5], HF voltage injection [6], [7] or an analysis of the current ripple by the pulse width modulation (PWM) in time or frequency domain [8]–[13].
In this study, a modified version of the so-called INFORM (Indirect Flux detection by Online Reluctance Measurement) method is used for the detection of the rotor position. The method

was originally designed for the rotor angle estimation of permanent magnet synchronous motors [14]. In previous studies, the INFORM method was adapted to three-phase active magnetic bearings, enabling a self-sensing control of the rotor position [15], [16]. The method is based on current slope measurements to detect inductance deviations of an eccentrically levitating rotor. Therefore, special voltage pulse patterns are applied to the actuator coils by the switches of a three-phase inverter. The voltage pulse patterns are defined by a space vector modulation, which is an essential element of the proposed self-sensing control method. On the one hand, the space vector modulation has to provide the desired output voltage for the control of the coil currents. On the other hand, the space vector modulation must ensure the possibility of current slope measurements by the fulfillment of timing requirements of the pulse pattern.
The following article deals with the design of a space vector modulation, which enables a high bandwidth of the current controller as well as a suitability for self-sensing control. In previous work, control current deviations were observed by a switchover of the space vector modulation strategy, which may lead to limitations in high precision control [17]. For that reason, special attention is paid to a precise control of the coil currents. In the following, the self-sensing principle is briefly outlined and requirements as well as limitations of space vector modulation (SVM) are discussed. Considering the design rules, a novel variant of SVM is introduced and implemented in a digital signal processor (DSP). Measurements on a prototype show the achievable current controller performance compared to a previous SVM strategy.

2 Self-Sensing Principle

The self-sensing principle is briefly outlined to introduce the requirements of self-sensing space vector modulation. The bearing configuration of Fig. 1 shows a three-phase radial AMB with six poles. The bias flux of the homopolar design is realized by permanent magnets. For the use of a conventional three-phase inverter, the coils are connected in a double wye configuration. The

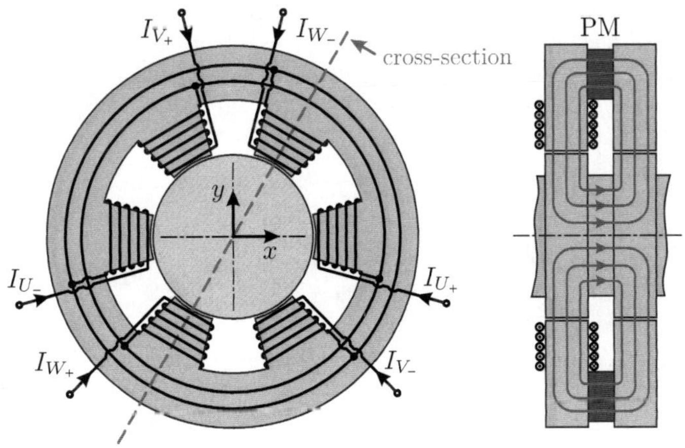

Figure 1: Structural design of the six pole radial homopolar AMB with a cross-section of the shaft. The bias flux (indicated by blue lines) is generated by permanent magnets (PM).

detection of the rotor position is based on the evaluation of the position dependent inductance of the actuator coils. Therefore, voltage pulses are applied to the coils, which are defined by the half bridges of a three-phase inverter. In simplification, the inductance can be determined by a measurement of the current slope $\mathrm{d}I/\mathrm{d}t$ due to a defined voltage pulse.

$$U \approx L \cdot \frac{\mathrm{d}I}{\mathrm{d}t} \rightarrow L \approx U \cdot \left(\frac{\mathrm{d}I}{\mathrm{d}t}\right)^{-1} \tag{1}$$

Concerning the rotor position detection, the inductance variation of opposing coils plays an important role. For this reason, an approach with a differential evaluation of opposing coil currents was chosen to cancel out inductance offsets. The determination of the inductance variation is based on a differential current slope measurement of the actuator coils. In this configuration, the differential current slope is measured by a transformer with an open-circuited sense winding as shown in Fig. 2. The transformer builds the connection between two opposing

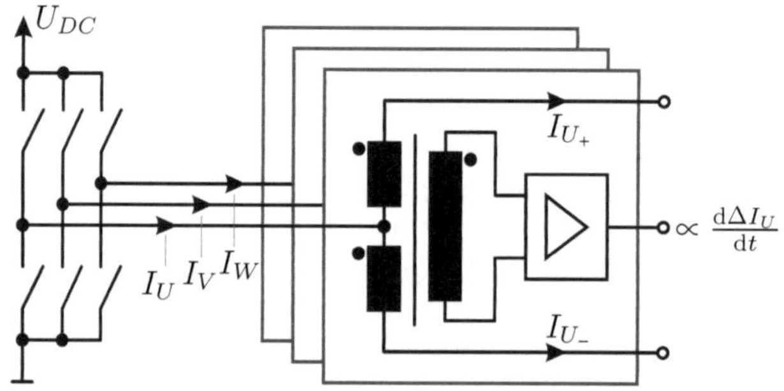

Figure 2: Illustration of the three-phase differential current slope measurement.

coils of the bearing and a phase of the inverter. Therefore, the differential design of the transformer enables the measurement of the differential current slope $\mathrm{d}\Delta I/\mathrm{d}t$ with $\Delta I = I_+ - I_-$ [18]. Thus, the rotor position can be calculated with the $\mathrm{d}\Delta I/\mathrm{d}t$ information under the consideration of the reluctance model of the three-phase magnetic circuit [16]. Concerning the settling time of the transformer, a minimal pulse width of the applied voltage pulses is required to allow a stable measurement of the current slope $\mathrm{d}\Delta I/\mathrm{d}t$. Figure 3 shows an exemplary waveform of the differential current ΔI_U and the output voltage of the transformer, which is proportional

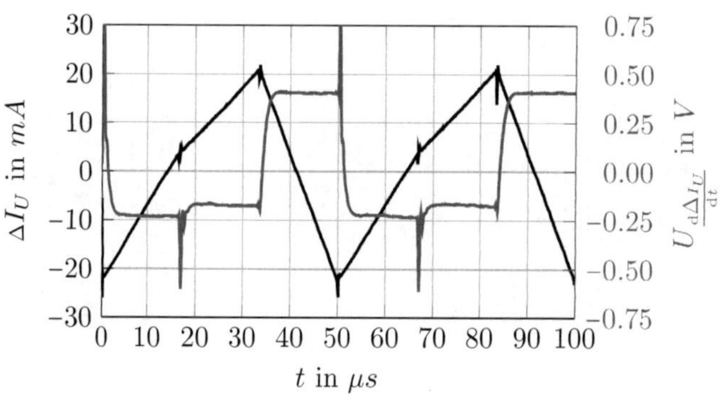

Figure 3: Current slope detection: The output voltage of the transformer $U_{\mathrm{d}\Delta I_U/\mathrm{d}t}$ is proportional to the differential current slope of the AMB phase U. ($f_{PWM} = 20\,kHz$)

to the derivative of the signal. It can be seen that the output voltage is nearly constant after the settling time of the transformer. Thus, it can be concluded that the presented procedure requires a minimum voltage pulse width to enable a proper determination of the current slope information, which must be guaranteed by the space vector modulation.

3 Space Vector Modulation

The three-phase design of the AMB enables the usage of a space vector modulation, which is responsible for applying the respective control voltage to the AMB coils. Figure 4 shows an overview of the interaction between current controller, space vector modulation and the

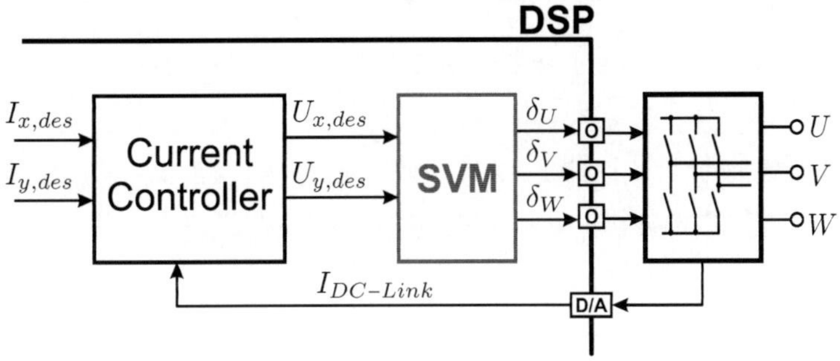

Figure 4: Structural overview of the current control loop: The SVM defines the duty cycles of the half-bridges correspondingly to the desired phase voltages of the current controller. The current feedback is realized by a measurement of the DC-Link current $I_{DC-Link}$.

half-bridges of the inverter. For achieving the force requirements of a superimposed position controller, the respective control currents ($I_{x,des}$, $I_{y,des}$) must be regulated in the AMB coils. Therefore, the current controller demands the desired coil voltages ($U_{x,des}$, $U_{y,des}$), which are the inputs of the SVM.

$$\boldsymbol{P} = \begin{bmatrix} U_{x,des} \\ U_{y,des} \end{bmatrix} \tag{2}$$

The SVM calculates the duty cycles δ_U, δ_V, δ_W correspondingly to the desired coil voltage space vector $\boldsymbol{P}$, whereby the timing of the switches of the three half-bridges is defined. The topology of the three-phase inverter enables the definition of eight fundamental voltage space vectors by the switches of the half-bridges as shown in Table 1. If the voltage transition during the

Table 1: Voltage space vector definition of a three-phase inverter

Phase	**Switch**	**Voltage space vector**							
		U_+	U_-	V_+	V_-	W_+	W_-	Z_+	Z_-
U	HS	1	0	0	1	0	1	1	0
U	LS	0	1	1	0	1	0	0	1
V	HS	0	1	1	0	0	1	1	0
V	LS	1	0	0	1	1	0	0	1
W	HS	0	1	0	1	1	0	1	0
W	LS	1	0	1	0	0	1	0	1

HS = High-Side switch, LS = Low-Side switch; 1 = closed, 0 = open switching action is neglected, the voltage pulses have only discrete amplitudes in time domain. However, the mean value over a PWM period correlates with the desired voltage space vector $\boldsymbol{P}$, which can be assembled by a linear combination of the fundamental voltage space vectors $\boldsymbol{U}_+$, $\boldsymbol{U}_-$, $\boldsymbol{V}_+$, $\boldsymbol{V}_-$, $\boldsymbol{W}_+$, $\boldsymbol{W}_-$ and the zero space vectors $\boldsymbol{Z}_+$, $\boldsymbol{Z}_-$.

$$\begin{aligned} \boldsymbol{P} = & c_{U_+}\boldsymbol{U}_+ + c_{V_+}\boldsymbol{V}_+ + c_{W_+}\boldsymbol{W}_+ + c_{Z_+}\boldsymbol{Z}_+ + \\ & c_{U_-}\boldsymbol{U}_- + c_{V_-}\boldsymbol{V}_- + c_{W_-}\boldsymbol{W}_- + c_{Z_-}\boldsymbol{Z}_- \end{aligned} \tag{3}$$

The coefficients c_{U_+}, c_{U_-}, c_{V_+}, c_{V_-}, c_{W_+}, c_{W_-}, c_{Z_+}, c_{Z_-} define the amplitude of the corresponding fundamental space vectors. In this context, the amplitude of a voltage space vector is equal to the duration of the voltage pulse in the PWM pattern. The choice of the coefficients provides a degree of freedom, which can be used for the fulfillment of the following criteria.

3.1 Design Objectives

1. For achieving a high bandwidth of the current controller, a high maximum modulation amplitude $\|\boldsymbol{P}\|_2$ is desired. Furthermore, it is advantageous if the achievable modulation amplitude is independent of the angle of the voltage space vector, allowing the same dynamic in all space directions.

2. For self-sensing control a minimum pulse width $t_{\mathrm{d}I/\mathrm{d}t}$ is required, to allow a current slope measurement as depicted in Fig. 3. For reasons of symmetry, at least one fundamental voltage space vector of each phase shall obtain the minimum pulse width $t_{\mathrm{d}I/\mathrm{d}t}$.

3. The waveform of the current ripple is affected by the timing sequence of the voltages pulse pattern. Abrupt transitions between modulation patterns can lead to transient deviations in the control current [17]. For enabling a precise current control, the timing sequence must not change for infinitesimal changes of the desired voltage space vector $\boldsymbol{P}$. Therefore, the SVM should be designed in a way that the coefficients of (3) are determined by a continuous function of $\boldsymbol{P}$.

3.2 Design Procedure

The main question of the design of the SVM is the determination of the coefficients from (3), while taking the described design objectives into account. For the following considerations, the convention of Fig. 5 is used for the definition of the angles φ and Φ. The angle φ is defined between the space vector $\boldsymbol{U}_+$ and the desired voltage space vector $\boldsymbol{P}$. Moreover, the sector angle Φ is defined as

$$\Phi = \frac{\pi}{6} \underbrace{\left\lfloor \frac{\varphi}{\frac{\pi}{6}} \right\rfloor}_{\in\{0,1,\dots,11\}} , \quad 0 \le \varphi < 2\pi \tag{4}$$

with the floor function $\lfloor\,\rfloor$. Dividing the set of sector angles into even and odd sectors

$$\Phi_{even} = \left\{ 2k\frac{\pi}{6} \middle| \ k = 0..5 \right\} \tag{5}$$

$$\Phi_{odd} = \left\{ (2k+1)\frac{\pi}{6} \middle| \ k = 0..5 \right\} \tag{6}$$

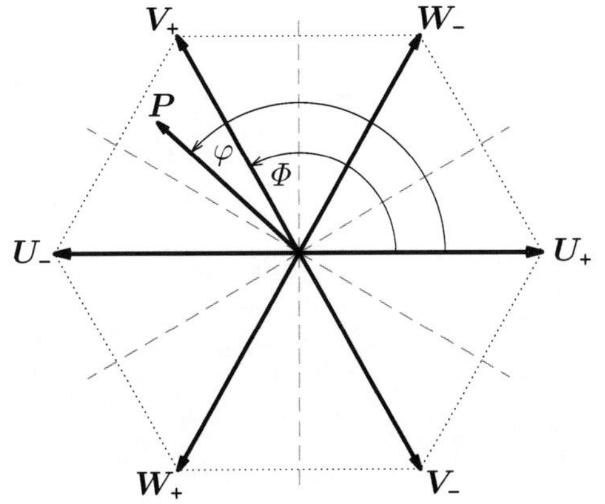

Figure 5: Illustration of the fundamental voltage space vectors and the angles φ, Φ for a desired voltage space vector $\boldsymbol{P}$.

enables the definition of the coordinate transformation

$$\overline{\boldsymbol{P}} = \begin{bmatrix} \cos(-\vartheta) & -\sin(-\vartheta) \\ \sin(-\vartheta) & \cos(-\vartheta) \end{bmatrix} \cdot \boldsymbol{P} \tag{7}$$

with

$$\vartheta = \begin{cases} \Phi & \text{for} \quad \Phi \in \Phi_{even} \\ \\ 2\varphi - \Phi - \frac{\pi}{6} & \text{for} \quad \Phi \in \Phi_{odd} \,. \end{cases} \tag{8}$$

Equation (7) transforms the space vector $\boldsymbol{P}$ in a generic coordinate system and has the property that the space vector $\boldsymbol{P}$ can always be transformed to the gray section of Fig. 6. Thus, the

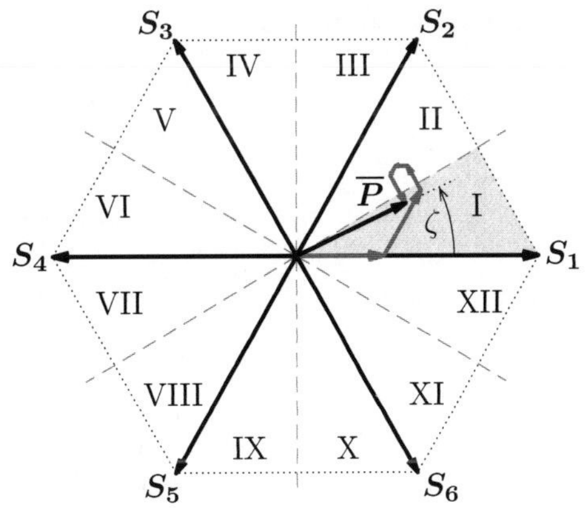

Figure 6: The voltage space vector $\overline{\boldsymbol{P}}$ represents $\boldsymbol{P}$ in a generic coordinate system after the coordinate transformation (7).

resulting space vector $\overline{\boldsymbol{P}}$ can be represented by a linear combination of the generic space vectors $\boldsymbol{S_1}, \ldots, \boldsymbol{S_6}$. The correlation of the generic space vectors to the fundamental space vectors will be discussed later on. The generic coordinate system enables a simplification of the problem (3), because the whole modulation range can be reduced to the gray area of Fig. 6. After solving the problem in the generic coordinate system, the solution can be transformed back to the respective sector, which is given by Φ. For further considerations, the angle ζ

$$\zeta = \varphi - \vartheta, \quad \zeta \in \left[0, \frac{\pi}{6}\right] \tag{9}$$

is defined as the angle between the space vector $\boldsymbol{S_1}$ and $\overline{\boldsymbol{P}}$ (Fig. 6) and is normalized to an angle of $\pi/6$

$$\hat{\zeta} = \frac{6}{\pi}\zeta \tag{10}$$

for reasons of simplicity. For solving the problem, the coefficients of the space vectors $\boldsymbol{S_1}, \ldots, \boldsymbol{S_6}$ must be determined with regard to the design objectives. In order to achieve a continuous behavior of the voltage pulse pattern, special attention is paid to a sector switchover due to a change of the sector angle Φ. Thus, the following boundaries of the modulation range in Fig. 6 must be considered:

$$B_1 = \left\{\overline{\boldsymbol{P}} \;\middle|\; \left\|\overline{\boldsymbol{P}}\right\|_2 > 0,\ \hat{\zeta} = 1\right\} \tag{11}$$

$$B_2 = \left\{\overline{\boldsymbol{P}} \;\middle|\; \left\|\overline{\boldsymbol{P}}\right\|_2 > 0,\ \hat{\zeta} = 0\right\} \tag{12}$$

$$B_3 = \left\{\overline{\boldsymbol{P}} \;\middle|\; \left\|\overline{\boldsymbol{P}}\right\|_2 = 0,\ \hat{\zeta} \in [0, 1]\right\} \tag{13}$$

The boundary B_1 (B_2) represents a switchover from sector I to sector II (XII), while the boundary B_3 enables a switchover to an arbitrary sector by a transmission via the origin of the coordinate system. Equation (14) shows the designed modulation strategy for the representation of a space vector $\overline{\boldsymbol{P}}$, which is composed as follows:

$$\begin{aligned}\overline{\boldsymbol{P}} = a_1\,\boldsymbol{S_1} + a_2\,\boldsymbol{S_2} + \left(t_{min} + \left(t_{\mathrm{d}I/\mathrm{d}t} - t_{min}\right)\hat{\zeta}\right)\boldsymbol{S_3} +\\ t_{min}\,\boldsymbol{S_4} + t_{min}\,\boldsymbol{S_5} + t_{\mathrm{d}I/\mathrm{d}t}\,\boldsymbol{S_6}\end{aligned} \tag{14}$$

In principle, the space vector $\overline{\boldsymbol{P}}$ could be represented just by the space vectors $\boldsymbol{S_1}$ and $\boldsymbol{S_2}$. To enable a current slope measurement in each phase within one PWM period, at least one coefficient of each phase must obtain a minimum amplitude of $t_{\mathrm{d}I/\mathrm{d}t}$. Hence, the space vectors $\boldsymbol{S_1}$, $\boldsymbol{S_2}$ and $\boldsymbol{S_6}$ are chosen for the current slope measurement, because the linear combination of the vectors already shows in a similar direction to $\overline{\boldsymbol{P}}$, meeting the demand of a high modulation amplitude. Consequently, the space vectors $\boldsymbol{S_1}$, $\boldsymbol{S_2}$ and $\boldsymbol{S_6}$ are desired with an amplitude of at least $t_{\mathrm{d}I/\mathrm{d}t}$.

In the first step, the coefficient of $\boldsymbol{S_6}$ is set to $t_{\mathrm{d}I/\mathrm{d}t}$ and it is assumed that the coefficients a_1 and a_2 obtain an amplitude greater than $t_{\mathrm{d}I/\mathrm{d}t}$. The validity of this assumption will be proven later on. The coefficients of non-required space vectors ($\boldsymbol{S_4}$, $\boldsymbol{S_5}$) are set to t_{min} to ensure a minimum pulse width. The minimum pulse width t_{min} avoids very short voltage pulses, which may cause problems for semiconductor switches. Concerning a transition between sectors, the space vector $\boldsymbol{S_3}$ plays an important role for a jerk-free sector switchover. An evaluation of (14) at the boundary B_1 indicates

$$\overline{\boldsymbol{P}}_{B_1} = a_1\,\boldsymbol{S_1} + a_2\,\boldsymbol{S_2} + t_{min}\left(\boldsymbol{S_4} + \boldsymbol{S_5}\right) + t_{\mathrm{d}I/\mathrm{d}t}\left(\boldsymbol{S_3} + \boldsymbol{S_6}\right) \tag{15}$$

that pairs of space vectors $\{\boldsymbol{S_1}, \boldsymbol{S_2}\}$, $\{\boldsymbol{S_4}, \boldsymbol{S_5}\}$, $\{\boldsymbol{S_3}, \boldsymbol{S_6}\}$ have the same coefficients ($a_1 = a_2$ by constraint of boundary). Due to the symmetric arrangement of the coefficients, a continuous switchover from sector I to sector II is possible by swapping the generic space vectors of the pairs ($\boldsymbol{S_1} \rightarrow \boldsymbol{S_2}$, $\boldsymbol{S_2} \rightarrow \boldsymbol{S_1}$, ...). Concerning the boundary B_2,

$$\overline{\boldsymbol{P}}_{B_2} = a_1\, \boldsymbol{S_1} + a_2\, \boldsymbol{S_2} + t_{min}\, (\boldsymbol{S_3} + \boldsymbol{S_4} + \boldsymbol{S_5}) + t_{\mathrm{d}I/\mathrm{d}t}\, \boldsymbol{S_6} \tag{16}$$

the pairs of space vectors $\{\boldsymbol{S_2}, \boldsymbol{S_6}\}$ and $\{\boldsymbol{S_3}, \boldsymbol{S_5}\}$ have the same coefficients ($a_2 = t_{\mathrm{d}I/\mathrm{d}t}$ by constraint of boundary). Therefore, also in this case, a continuous switchover from sector I to sector XII is possible by swapping the generic space vectors of the outlined space vector pairs. The relevance of the boundary B_3 will be shown after solving (14). The coefficients a_1 and a_2 can be directly calculated by converting (14) to the following form:

$$\begin{bmatrix} a_1 \\ a_2 \end{bmatrix} = \begin{bmatrix} \boldsymbol{S_1} & \boldsymbol{S_2} \end{bmatrix}^{-1} \cdot \left(\overline{\boldsymbol{P}} - \left(t_{min} + \left(t_{\mathrm{d}I/\mathrm{d}t} - t_{min}\right) \hat{\zeta}\right) \boldsymbol{S_3} - t_{min}\, \boldsymbol{S_4} - t_{min}\, \boldsymbol{S_5} - t_{\mathrm{d}I/\mathrm{d}t}\, \boldsymbol{S_6}\right) \tag{17}$$

Thus, the space vector $\overline{\boldsymbol{P}}$ can be represented by (18)

$$\overline{\boldsymbol{P}} = \bar{c}_1\, \boldsymbol{S_1} + \bar{c}_2\, \boldsymbol{S_2} + \bar{c}_3\, \boldsymbol{S_3} + \bar{c}_4\, \boldsymbol{S_4} + \bar{c}_5\, \boldsymbol{S_5} + \bar{c}_6\, \boldsymbol{S_6} \tag{18}$$

with the coefficient vector

$$\bar{\mathbf{c}} = \left[\bar{c}_1\ \bar{c}_2\ \bar{c}_3\ \bar{c}_4\ \bar{c}_5\ \bar{c}_6\right]^T. \tag{19}$$

The coefficients of (19) must fulfill the constraint that the sum

$$\sum_{i=1}^{6} \bar{c}_i = 1 \tag{20}$$

is equal to 1. This constraint procures that the whole PWM period is filled by the voltage space vectors, which is a consequence of a constant PWM switching frequency. For that reason, the remaining time

$$t_{rem} = 1 - \left(a_1 + a_2 + t_{\mathrm{d}I/\mathrm{d}t} + 3t_{min} + \left(t_{\mathrm{d}I/\mathrm{d}t} - t_{min}\right) \hat{\zeta}\right) \tag{21}$$

is distributed equally among all generic voltage space vectors, to allow a maximum measurement window for $\mathrm{d}\Delta I/\mathrm{d}t$ measurements. Thus, the coefficients of $\bar{\mathbf{c}}$ are given by a coefficient comparison between (14) and (18),

$$\bar{\mathbf{c}} = \begin{bmatrix} a_1 \\ a_2 \\ t_{min} + \left(t_{\mathrm{d}I/\mathrm{d}t} - t_{min}\right) \hat{\zeta} \\ t_{min} \\ t_{min} \\ t_{\mathrm{d}I/\mathrm{d}t} \end{bmatrix} + \frac{t_{rem}}{6} \begin{bmatrix} 1 \\ 1 \\ 1 \\ 1 \\ 1 \\ 1 \end{bmatrix} \tag{22}$$

with an extension by the value of $t_{rem}/6$. The additional amplitude of $t_{rem}/6$ does not influence the representation of $\overline{\boldsymbol{P}}$ because it is equally distributed to all space directions, which results in a composed zero space vector. Figure 7 shows the admissible modulation area, which is located inside the red boundaries. The minimum and maximum symmetrical modulation amplitude (R_{min}, R_{max}) define the symmetrical modulation area, which is indicated by the gray area. The limitation of the admissible modulation area to the symmetrical modulation area provokes that the current controller obtains the same dynamic in all space directions. It can be seen that the symmetrical modulation area does not include the origin of the modulation area for the chosen parameters $t_{\mathrm{d}I/\mathrm{d}t} = 0.14$ and $t_{min} = 0.02$ (parameters are normalized to PWM period).

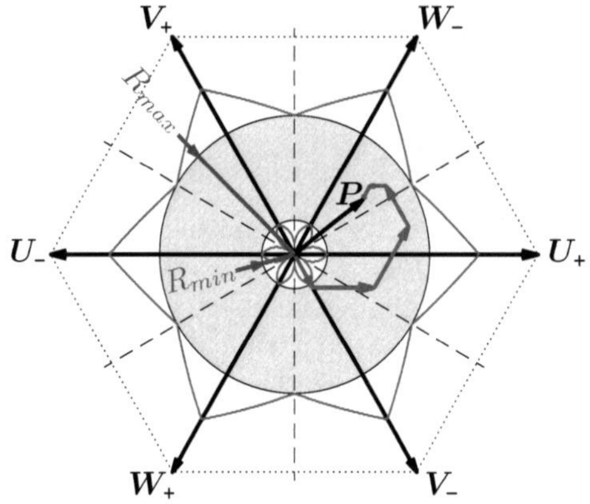

Figure 7: SVM by Equ. (14): Illustration of the possible symmetrical modulation range (gray area) with the minimum and maximum modulation boundaries (red). ($t_{\mathrm{d}I/\mathrm{dt}} = 0.14, t_{min} = 0.02$)

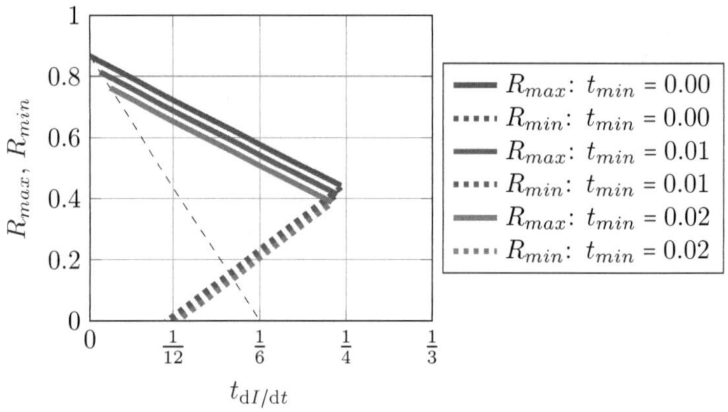

Figure 8: Comparison of the symmetrical modulation amplitude for the designed space vector modulation.

Figure 8 shows the symmetrical modulation amplitudes as a function of $t_{\mathrm{d}I/\mathrm{dt}}$ for fixed values of t_{min}. It can be seen that the maximum symmetrical modulation amplitude R_{max} decreases with a higher value of $t_{\mathrm{d}I/\mathrm{dt}}$. At a certain value of $t_{\mathrm{d}I/\mathrm{dt}}$, also a minimum symmetrical modulation amplitude R_{min} occurs, which is an issue for small amplitudes of $\overline{\boldsymbol{P}}$.
For that reason, the calculation approach of (14) is extended for dealing with small modulation amplitudes. Equation (23) shows an alternative set of coefficients for the representation of $\overline{\boldsymbol{P}}$

$$\hat{c} = \frac{1}{6} \cdot \boldsymbol{T}^{6\times1} + \frac{1}{3} \left[\boldsymbol{S_1}\ \boldsymbol{S_2}\ \boldsymbol{S_3}\ \boldsymbol{S_4}\ \boldsymbol{S_5}\ \boldsymbol{S_6}\right]^T \cdot \overline{\boldsymbol{P}} \tag{23}$$

with a 6×1 matrix $\boldsymbol{T}^{6\times1}$ with all entries chosen as 1. The calculation (23) has the property that each coefficient has an offset of 1/6, which is superposed by the inner product $\overline{\boldsymbol{P}}$ and the corresponding space vector $\boldsymbol{S}$. Thus, all coefficients have the amplitude of 1/6 if the amplitude of $\overline{\boldsymbol{P}}$ is zero ($\|\overline{\boldsymbol{P}}\|_2 = 0$). This characteristic is appropriate for a continuous sector switchover at the boundary B_3 (13), because with an equal amplitude the mapping between the coefficients (23) and the fundamental space vectors can be done in any order (Equ. (23) also fulfills the boundaries B_1 and B_2). For the design of a SVM without restrictions for small modulation amplitudes as shown in Fig. 8, the coefficients of (19) and (23) are combined

$$\boldsymbol{c} = \alpha\,\bar{\boldsymbol{c}} + (1-\alpha)\,\hat{\boldsymbol{c}} \tag{24}$$

by a weighting factor α. The weighting factor α

$$\alpha = \frac{\|\boldsymbol{P}\|_2}{R_{max}} \tag{25}$$

is amplitude dependent by $\|\boldsymbol{P}\|_2$ and normalized to R_{max}. Therefore, the SVM of (23) is dominant for low amplitudes and the modulation strategy (19) is dominant for large amplitudes of $\|\boldsymbol{P}\|_2$. Regarding the transition between two modulation strategies, it must be checked that all coefficients obtain at least an amplitude of t_{min} within the modulation range. Furthermore, at least one coefficient of each phase direction must allow a $\mathrm{d}\Delta I/\mathrm{d}t$ measurement by an amplitude of at least $t_{\mathrm{d}I/\mathrm{dt}}$. This circumstance was proven with a numerical simulation over the whole admissible modulation area. With the coefficients of (24), the desired voltage space vector $\overline{\boldsymbol{P}}$ can be composed as

$$\overline{\boldsymbol{P}} = c_1\,\boldsymbol{S_1} + c_2\,\boldsymbol{S_2} + c_3\,\boldsymbol{S_3} + c_4\,\boldsymbol{S_4} + c_5\,\boldsymbol{S_5} + c_6\,\boldsymbol{S_6}. \tag{26}$$

This kind of SVM will be stated as 6-Active HMR SVM (High Modulation Range) in the following considerations. The term "6-Active" refers to the number of active space vectors within a PWM period. Figure 9 shows the possible modulation range for the 6-Active HMR

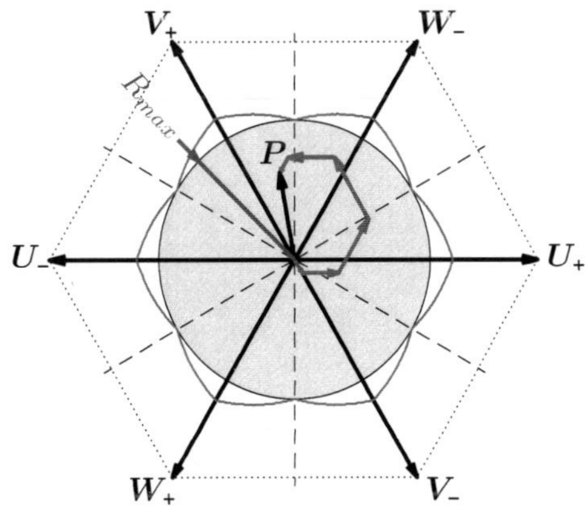

Figure 9: 6-Active HMR SVM: Illustration of the possible symmetrical modulation range (gray area) with the corresponding maximum modulation boundary (red). The origin of the coordinate system is included in the admissible modulation range. ($t_{\mathrm{d}I/\mathrm{dt}} = 0.14$, $t_{min} = 0.02$)

SVM. In contrast to Fig. 7, the admissible modulation range also includes the origin of the coordinate system. Considering the possible modulation range, Fig. 10 depicts the maximum symmetrical modulation range as a function of $t_{\mathrm{d}I/\mathrm{dt}}$ for the modulation strategy (26).

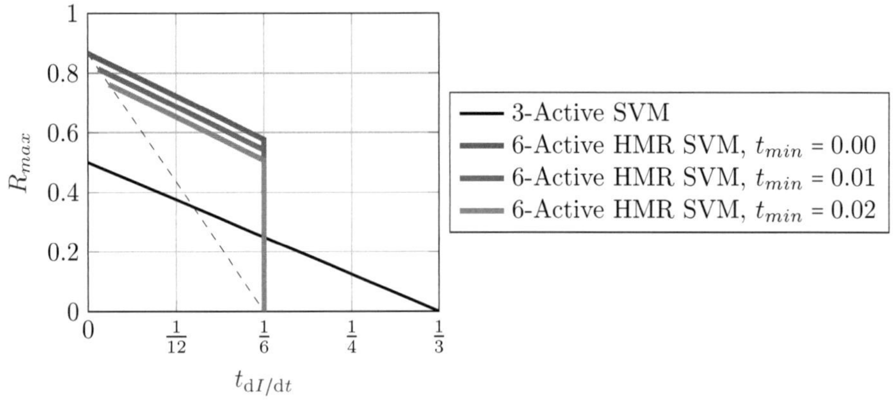

Figure 10: Comparison of the symmetrical modulation amplitude for the 6-Active HMR SVM.

In contrast to Fig. 8, the 6-Active HMR SVM has the advantage of R_{min} = 0 in the admissible range of $t_{\mathrm{d}I/\mathrm{d}t}$. For reasons of comparison, the characteristic of the 3-Active SVM is also shown, which was used in previous studies for self-sensing control [16]. Concerning the modulation amplitude, the 6-Active HMR SVM enables a significant higher value of R_{max} compared to the 3-Active SVM, which allows a higher dynamic in current control.

Equation (26) shows the coefficients of the SVM in the generic coordinate system, which was introduced with the coordinate transformation (7). However, the coefficients of the fundamental space vectors $\boldsymbol{U}_+$, $\boldsymbol{U}_-$, $\boldsymbol{V}_+$, $\boldsymbol{V}_-$, $\boldsymbol{W}_+$, $\boldsymbol{W}_-$ have to be determined for an implementation of the SVM in a DSP. The coefficients of (3) are determined by a back transformation of the generic space vector coefficients. Therefore, the coefficients of the fundamental space vectors c_{U_+}, c_{U_-}, c_{V_+}, c_{V_-}, c_{W_+}, c_{W_-}, c_{Z_+}, c_{Z_-} are determined by a sector angle dependent mapping of the coefficients of (26) by means of Table 2. Regarding self-sensing, at least three current slope measurements are desired within one PWM period. Figure 11 shows the number of achievable current slope measurements within one PWM period for different modulation regions. It can be seen that the inner regions of the modulation area allow more than three current slope measurements. The additional information could be used for increasing the quality of the self-sensing position signal.

Table 2: Coefficient Mapping

Coeff.	**Sector angle Φ**											
	$0\frac{\pi}{6}$	$1\frac{\pi}{6}$	$2\frac{\pi}{6}$	$3\frac{\pi}{6}$	$4\frac{\pi}{6}$	$5\frac{\pi}{6}$	$6\frac{\pi}{6}$	$7\frac{\pi}{6}$	$8\frac{\pi}{6}$	$9\frac{\pi}{6}$	$10\frac{\pi}{6}$	$11\frac{\pi}{6}$
c_{U_+}	c_1	c_2	c_6	c_3	c_5	c_4	c_4	c_5	c_3	c_6	c_2	c_1
c_{W_-}	c_2	c_1	c_1	c_2	c_6	c_3	c_5	c_4	c_4	c_5	c_3	c_6
c_{V_+}	c_3	c_6	c_2	c_1	c_1	c_2	c_6	c_3	c_5	c_4	c_4	c_5
c_{U_-}	c_4	c_5	c_3	c_6	c_2	c_1	c_1	c_2	c_6	c_3	c_5	c_4
c_{W_+}	c_5	c_4	c_4	c_5	c_3	c_6	c_2	c_1	c_1	c_2	c_6	c_3
c_{V_-}	c_6	c_3	c_5	c_4	c_4	c_5	c_3	c_6	c_2	c_1	c_1	c_2
c_{Z_+}	0	0	0	0	0	0	0	0	0	0	0	0
c_{Z_-}	0	0	0	0	0	0	0	0	0	0	0	0

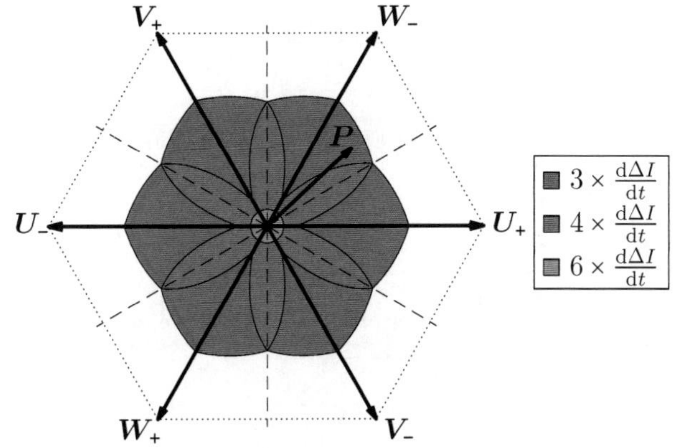

Figure 11: 6-Active HMR SVM: Visualization of the number of possible $\mathrm{d}\Delta I/\mathrm{d}t$ measurements for different modulation regions. ($t_{\mathrm{d}I/\mathrm{d}t} = 0.14$, $t_{min} = 0.02$)

4 Measurements and Results

The following section contains measurement results of the introduced space vector modulation strategy. Therefore, the SVM was implemented on a three-phase inverter with a TMS320F28335 digital signal processor from Texas Instruments. The current controller is based on the control structure as shown in Fig. 4. In this configuration, only the DC-Link current was measured and the phase currents (I_U, I_V, I_W) were calculated and transformed in a Cartesian coordinate system (I_x, I_y) by the Clarke transformation. The control law of the control currents I_x, I_y was based on non-coupled proportional integral (PI) controllers. In the following, all measurements were performed on a radial homopolar AMB as shown in Fig. 1. The six-pole AMB was driven by a three-phase inverter with a DC-Link voltage of $U_{DC} = 18\,V$. The parameters of the SVM were chosen as $f_{PWM} = 20\,kHz$, $t_{\mathrm{d}I/\mathrm{d}t} = 0.14$ and $t_{min} = 0.02$. This configuration avoids voltage pulses shorter than $1\,\mu s$ and provides at least $7\,\mu s$ duration for current slope measurements.

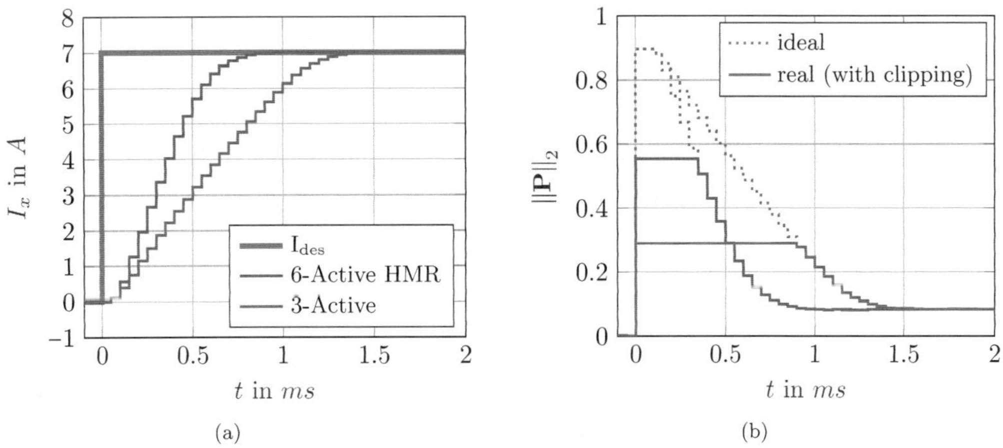

Figure 12: (**a**) Measured step response of the current controller for different modulation strategies. (**b**) Corresponding amplitude of the desired voltage space vector $\boldsymbol{P}$.

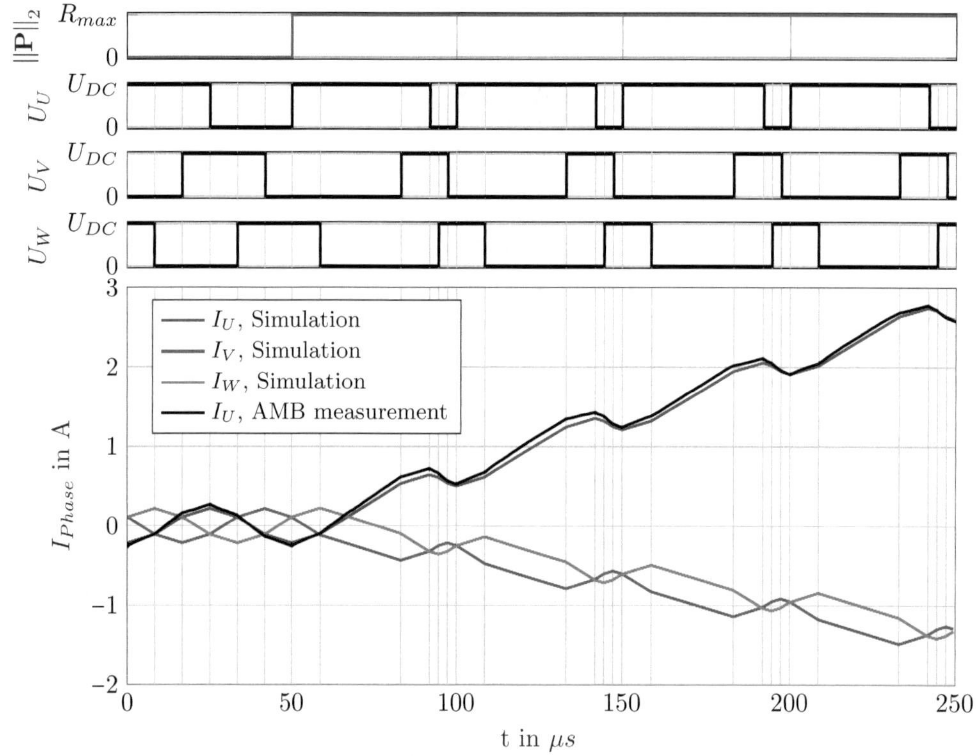

Figure 13: Modulation pattern of the 6-Active HMR SVM: Illustration of the phase voltages and currents for a step in the amplitude of $\boldsymbol{P}$ from zero to R_{max} ($\varphi = 0$).

Figure 13 shows the operation of the designed SVM in time domain with the respective phase voltages and currents. The desired voltage space vector $\boldsymbol{P}$ was chosen as a step from zero to the maximal symmetrical amplitude R_{max} in direction of $\boldsymbol{U}$+ ($\varphi = 0$). The phase voltages U_U, U_V and U_W depict the timing of the voltage pulse pattern corresponding to $\boldsymbol{P}$. It can be seen that always six space vectors are active within one PWM period. In case of $||\boldsymbol{P}||_2 = 0$, all space vectors obtain the same amplitude, which results in zero mean value of the phase voltages in contrast to the case of $||\boldsymbol{P}||_2 = R_{max}$. The bottom plot shows a comparison between a simulation and a measurement on the AMB prototype for the phase U. Although measurement and simulation show good consistency, a remaining deviation can be seen because of parameter uncertainties and non-linear AMB material effects. Regarding self-sensing operation, the SVM enables at least one $\mathrm{d}\Delta I/\mathrm{d}t$ measurement per phase in each PWM period (Fig. 3), which can be used for self-sensing position detection.

An adequate bandwidth of the current controller is important to follow the requirements of superimposed force and position controllers. Figure 12 shows a measurement of the current controller step response for different SVMs. The 3-Active SVM has already been investigated in previous work [16] and is therefore used as reference. It can be seen that the 6-Active HMR SVM achieves a fast current response without overshoot. A comparison of the maximum space vector amplitude $\left\|\overline{\boldsymbol{P}}\right\|_2$ indicates that the 6-Active HMR modulation has about 90% more modulation amplitude than the 3-Active SVM for the chosen parameters $t_{\mathrm{d}I/\mathrm{d}t}$ and t_{min}. A bigger modulation amplitude can be used to increase the dynamic of the current controller or to reduce the DC-link voltage in order to minimize power losses in the AMB without losing control bandwidth.

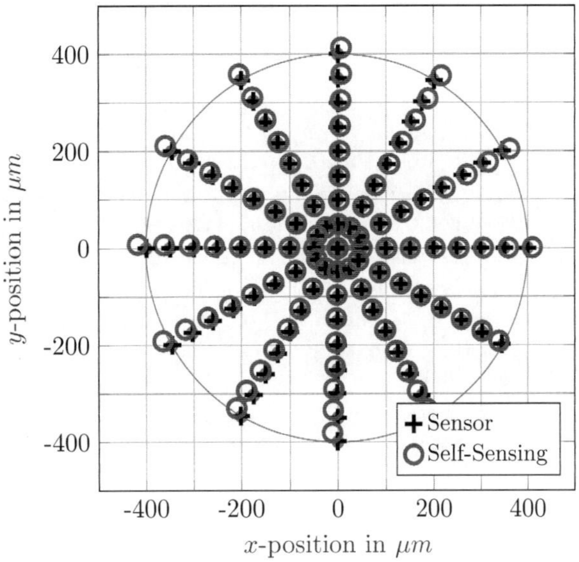

Figure 14: Comparison of the self-sensing and sensor position for different rotor setpoints.

Concerning self-sensing position control, the AMB was operated with a position estimator, which uses the $d\Delta I/dt$ information obtained by the 6-Active HMR SVM. Figure 14 shows a comparison of the self-sensing and sensor position of a levitating rotor, whereby the AMB prototype has an air gap of $800\,\mu m$. The steady-state error between the sensor and self-sensing method shows an overall low error with slight amplitude and angle deviations in the outer regions.

5 Conclusions

The focus of this study lied on the design of a space vector modulation for a three-phase self-sensing radial magnetic bearing. In contrast to conventional SVM methods, the design of the SVM in this study underlies additional restrictions to allow a self-sensing operation of the magnetic bearing. Special attention was paid to the modulation amplitude for achieving a high dynamic of a superimposed current controller. Furthermore, considerations concerning precise current control were taken into account. This work contains the principles of self-sensing space vector modulation with a detailed description of a novel self-sensing SVM strategy. The designed SVM strategy was implemented and verified on a three-phase inverter in combination with a prototype of a self-sensing magnetic bearing. Measurements on the prototype showed that the dynamic of the current controller could be increased compared to a previous modulation strategy. Finally, a measurement with different setpoints of a levitating rotor showed the steady-state performance of the self-sensing control compared to an external sensor. Further analysis on the quality of the self-sensing position detection, especially under dynamic conditions, will be part of future investigations.

References

[1] H. Bleuler, D. Vischer, G. Schweitzer, A. Traxler, and D. Zlatnik. New concepts for cost-effective magnetic bearing control. Automatica, Vol. 30, No. 5, pp. 871–876, May 1994.

[2] E. H. Maslen. Self–sensing for active magnetic bearings: overview and status. Proc. 10th International Symposium on Magnetic Bearings (ISMB10), Martigny, Switzerland, 21–23 Aug. 2006.

[3] T. Glück, W. Kemmetmüller, C. Tump, and A. Kugi. A novel robust position estimator for self-sensing magnetic levitation systems based on least squares identification. Control Engineering Practice, Vol. 19, No. 2, pp. 146–157, Feb. 2011.

[4] T. Mizuno, T. Ishiim, and K. Araki. Self-sensing magnetic suspension using hysteresis amplifiers. Control Engineering Practice, Vol. 6 No. 9, pp. 1133–1140, Sept. 1998.

[5] T. Mizuno and Y. Hirasawa. Self-sensing magnetic suspension using an H-bridge type hysteresis amplifier operating in two quadrants. Proc. 28th IEEE IECON, pp. 1818–1823, Nov. 2002.

[6] J.-S. Yim, J.-H. Kim, S.-K. Sul, H.-J. Ahn, and D.-C. Han. Sensorless Position Control of Active Magnetic Bearings Based on High Frequency Signal Injection method. Proc. 18th Annu. IEEE APEC, Vol. 1, pp. 83–88, Feb. 2003.

[7] P. Garcia, J. M. Guerrero, F. Briz, and D. D. Reigosa. Sensorless Control of Three-Pole Active Magnetic Bearings Using Saliency-tracking Based Methods. IEEE Transactions on Industry Applications, Vol. 46, No. 4, pp. 1476–1484, Jul.-Aug. 2010.

[8] W. Gruber, M. Pichler, M. Rothböck, and W. Amrhein. Self-Sensing Active Magnetic Bearing Using 2-Level PWM Current Ripple Demodulation. Proc. 7th International Conference on Sensing Technology (ICST), Wellington, New Zealand, Dec. 2013.

[9] J. Wang and A. Binder. Position estimation for self-sensing magnetic bearings based on the current slope due to the switching amplifier. Eur. Power Electr. and Drives (EPE), Vol 26., No. 4, pp. 15–141, 2016.

[10] A.C, Niemann, G. Van Schoor, and C.P. Du Rand. Self-Sensing Active Magnetic Bearing Based on a Direct Current Measurement Approach. MDPI, Sensors, Vol. 13, No. 9, 2013.

[11] Y. Jiang, K. Wang, M. Sun, and J. Xie. Displacement Self-Sensing Method for AMB-Rotor Systems Using Current Ripple Demodulations Combined With PWM Command Signals. IEEE Sensors Journal, Vol. 19, No. 14, Jul. 2019.

[12] E. O. Ranfta, G. van Schoor, and C. P. du Rand. An Integrated Self-Sensing Approach for Active Magnetic Bearings. SAIEE Africa Research Journal, Vol. 102, No. 4, Dec. 2011.

[13] Y. Jiang, X. Ma, and Y. Fan. Rotor Displacement Self-Sensing Approach for Permanent Magnet Biased Magnetic Bearings Using Double-Axis PWM Demodulation. IEEE Sensors Journal, Vol. 18, No. 19, Oct. 2018.

[14] M. Schrödl. Sensorless control of AC machines at low speed and standstill based on the "INFORM" method. IAS '96. Conference Record of the 1996 IEEE Industry Applications Conference Thirty-First IAS Annual Meeting, San Diego, CA, USA, 6-10 Oct. 1996.

[15] M. Hofer, M. Hutterer, T. Nenning, and M. Schrödl. Improved Sensorless Control of a Modular Three Phase Radial Active Magnetic Bearing. Proc. 14th International Symposium on Magnetic Bearings (ISMB14), Linz, Austria, 11–14 Aug. 2014.

[16] T. Nenning, M. Hofer, M. Hutterer, and M. Schrödl. Setup with two Self-Sensing Magnetic Bearings using Differential 3-Active INFORM. Proc. of the 14th International Symposium on Magnetic Bearings (ISMB14), Linz, Austria, 11–14 Aug. 2014.

[17] D. Wimmer, M. Hutterer, M. Hofer, and M. Schrödl. Space Vector Modulation Strategies for Self-Sensing Three-Phase Radial Active Magnetic Bearings. MDPI Actuators, Vol. 8, No. 2, 2019.

[18] D. Wimmer, M. Hutterer, and M. Schrödl. Design and Analysis of a PCB Integrated Differential Current Slope Sensor with Ferrite Support for High dV/dt Operations. PCIM Europe digital days 2021, Nuremberg, Germany, 3–7 May 2021.

Building a Bench with Magnetic Bearings for Study of Rotor Dynamics

Jefferson S. Coelho[1] and Fernando Augusto N. C. Pinto[2]

[1] Federal University of Amazonas, Amazonas, Brazil
jeffersoncoelho@ufam.edu.br
[2] Federal University of Rio de Janeiro, Rio de Janeiro, Brazil.
fcpinto@ufrj.br

Abstract

Along with the rise of great technological advancements came new products and processes, demanding an increase in quality and production growth. This work presents a study on Active Magnetic Bearings, a no contact, friction-free device that supports a rotor through magnetic forces. Thus, a few of its advantages include a low maintenance cost and the possibility of working at high rotation speeds. A disadvantage is the need for a control loop. Therefore, a PID controller is implemented on an embedded electronic FPGA-based platform reconfigurable for levitation, stabilization and the analysis of the rotor's dynamic behavior. To achieve the proposed objective, a prototype of the system was developed to carry out measurements and experimental analyses and to observe the advantages and disadvantages offered by the system. The results obtained show that the rotor maintains itself within the acceptable displacement zone around the bearings' geometric center. Therefore, the controller managed to stabilize the system so its dynamic behavior does not exceed the limit established in this project.

Contents

1 Introduction

Rotating machines have tendency to work in increasingly severe conditions, with high spin speeds and high load values applied, even maintaining the high standard demand in productivity and reliability [17]. Therefore, there is need for modifications and improvements in the machine designs so that they can meet the new performance standards and with high efficiency. An example in modern industry is compressors [2]. These are used in applications that require pressurization of gases or liquids, for example, in the petrochemical industry. In oil exploration in deep or ultra-deep waters (pre-salt layer) where high pressures occur, compressors are not easily accessible for diagnosis and frequent maintenance. dog. Therefore, it becomes necessary equipment with a high technological level that can work for long periods of time without maintenance, efficiently and with great reliability. Rotating machines such as pumps, electric motors and generators, compressors, steam turbines, gas, are basically formed by three main components: the rotor, the bearings and the support structures [6]. The bearings are responsible for supporting loads and also guiding the axes even in the presence of external forces, such as, for example, vibrations due to rotor unbalance. In this scenario, the bearings are presented as important parts, which inuence the dynamic performance of the rotor, life and reliability of the machine.

Recent developments in electronics have increased the need for advanced electronic control systems, due to the increased capacity and speed of digital platforms such as Digital Signal Processors (DSP), microcontrollers, FPGA (Field Programmable Gate Array) high-performance, Application-specific integrated circuits (ASICs), among others. Thus, active magnetic bearings (AMBs) have gained a great ally, as they are systems where they need a control loop for rotor stabilization, as they are unstable in nature. With that, its use became more attractive and serving as an alternative to replace the current conventional bearings in modern machines due to the advantage of the complete elimination of the mechanical contact between the rotating parts and the fixed parts . Its operating principle is based on the use of forces from magnetic fields, produced by electromagnets, to maintain to maintain relative position of a rotating assembly (rotor) to a stationary component (stator), providing high rotation speeds, absence of lubrication, longer service life and lower maintenance cost [19]. For Yoon [20], these are highly desirable features for compressors. Other advantages that we can mention are in relation to the monitoring and diagnosis of systems during the machine operation [10], this can be designed for analysis and interpretation of measurement data that are collected through sensors arranged in the bearings, making it possible to modify the dynamic characteristics, such as stiffness and damping [11]. Some application areas with active magnetic bearings can be found in [14], such as: turbomachinery [1], energy storage system (Flywhells) [7], Biomedical [8], Oil and Gas [5, 16].

One of the greatest challenges in the use of AMBs is to control and stabilize the rotor, so that it can support loads and guarantee a good dynamic behavior against high rotation speeds and disturbances that can occur. For this, new control strategies that can meet these needs must be developed and tested. The main aim of this study is to investigate dynamic behavior of the magnetic bearing system, evaluating the characteristics of the response for cases where the rotor is subject to a static (no rotation), transient, dynamic (with rotation) and with perturbations.

2 Active Magnetic Bearing System Design

The test rig consists of a shaft/rotor supported by two radial AMBs, polypropylene plate to limit the radial movement of the rotor and also as protection of the stator or rotor surface due to eventual failures, and support axial to avoid movement in the longitudinal direction. The AMB system is composed of position sensors, power amplifiers, on-board controller responsible for signal processing, and a drive system through a DC motor (Direct Current), whose speed control is done through a PWM signal (Pulse Width Modulation) and measured with the aid of a digital tachometer. These elements can be identified in Figure 1.

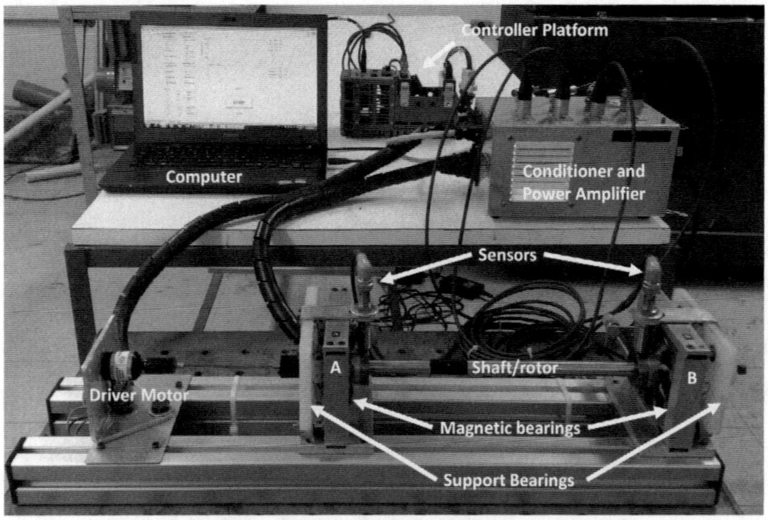

Figure 1: Assembled AMB test rig.

Table 1 the main constructive parameters of the AMB system are presented.

Parameter	Value	Unit
Number of coil turns per pair	260	–
Resistance of each coil pair	$1,2$	Ω
Base current in the coils	$1,0$	A
Rotor/Bearing Clearance (gap)	$1,0$	mm
Rotor Clearance/Support	$0,5$	mm
Shaft length	400	mm
Shaft/rotor assembly mass	$1,0$	kg
Distance between bearings	315	mm

Table 1: AMB System Parameters.

The assembly of the bearings was built on a structural aluminum profile in order to keep the components centralized and aligned, however, an assembly that meets these conditions is almost impossible. One of the advantages that was observed in the prototype is that the magnetic bearings have mechanical characteristics where it is possible to adjust the control for proper functioning. The AMB system has a platform where they were allocated the power sources of the sensors and power system. This platform is internally composed of a $12V/20A$, a font of $24V$, four power controllers, circuit for conditioning the sensor signal and serial port for connection with controller platform (Compact-RIO).

2.1 Magnetic Bearing

The constructive aspects of the prototype with its structure and materials used in the mechanical design of the magnetic bearing are presented. The Electromagnet shows Figure 2 two It is composed of a stator (ferromagnetic core) and electrical coils.

The stator has a heteropolar configuration that is the most used, where the magnetic field travels radially. The sheet material used was non-oriented grain silicon steel with thicknesses of $0.5mm$ due to its magnetic properties such as high permeability and mainly high magnetic saturation. These characteristics were obtained from the BH magnetization curve described by [15].

Figure 2: Stator and coils set.

The coil is made of a copper conductor covered by a thin insulating enamel layer, gauge AWG22 ($0.33mm^2$), wound with 130 turns per tooth, totaling 260 per pair of coils, providing a maximum current of $10A$.

2.2 Rotor

The rotor consists of a set of blocks that will be fitted to the shaft, these are composed of three discs. The first disc is made of carbon steel and has the function of supporting the other discs and serving as a surface for reading the sensors. The second disc is made of aluminum, whose function is to move the sensors away from the coils actuation site and prevent it from interfering with the reading, in addition to creating a necessary physical spacing When mounting the sensor with the bearing. Finally, the last disc is made up of a set of plates of the same material used in the stator, nonoriented grain silicon steel, with a good surface finish, concentricity and uniform diameter to ensure a good alignment with the bearing stator. The main shaft is made of stainless steel due to its mechanical resistance property, its length is $40mm$, with $14.28mm$ of diameter. Figure 3 shows the complete rotor.

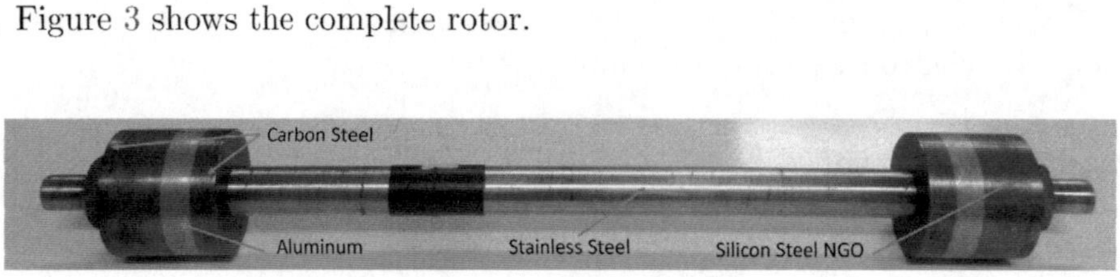

Figure 3: Shaft-Rotor set.

2.3 Sensors

The position sensors used are of the inductive type and provide an analog output signal in voltage that varies linearly between $0V$ and $10V$ proportional to the distance measured in the air gap, with a measuring range of $0.2mm$ to $2mm$ and noise of approximately $3\mu m$. Its operation is more compatible with what is desired and its measurement is the least influenced by external agents [3]. A total of four sensors were mounted in the radial directions of the two bearings, with two sensors per bearing (one in the horizontal direction and one vertically perpendicular to each other) to measure the position without mechanical contact with the shaft.

The bearing has a nominal air gap of $1.0mm$ radial with the rotor at the operating point. In this way, the rotor can move radially in a range from 0 to 2.00 mm. However, the maximum movement clearance is given by the support bearing, whose function is to avoid contact between the rotor and the bearing in case of malfunction. In this case, the clearance between the rotor and support is $0.5mm$, therefore, the radial displacement region of the rotor starting from the operating point is from 0 to $1.0mm$. During the installation of the sensors, a displacement of $1.1mm$ was considered between the sensor and rotor from the operating point. Thus, the operating range of the sensors varies from 0.5 to $1.6mm$. Figure 4 shows the mounted sensors on the bearing and rotor.

Figure 4: Inductive position sensor.

2.4 Electronic Power System

The purpose of the electronic power system is to transform the weak control output signal, into voltage, into current that flows through the bearing coils, and thus provide the necessary force to suspend the rotor. The system used is a board Monster Moto Shield gives Sparkfun Electronics which has a pair of VNH2SP30 full bridge controllers capable of controlling high currents independently (Figure 5). The integrated circuit board of the controller has an internally switched amplifier with a full H-bridge, which can supply up to $14A$ of direct current and PWM frequency of up to $20kHz$. Although the system has a complete H-bridge in the magnetic bearing, only half a bridge was used, as it is not necessary to invert the direction that the current flows through the coil.

Figure 5: Electronic Power System [18].

For more precise control of the bearing current, the PWM technique was used, where the output of the control signal determines the pulse width, or duty cycle, and is directly related to the voltage average. This signal will be transferred to the digital controller with a PWM frequency of $10kHz$ at the output and applied to the input of the power amplifier (bridge H) used to control the current value in the coils. In total, four devices were used, each of which controls a shaft with two bearing directions.

2.5 Acquisition and Control System

The controller platform used is the compactRIO 9030 from National Instruments. It features 1,33GHz dualcore Intel Atom processorGHz, 4 GB non-volatile storage, 1 DDR3 memoryGB, and 4 compartments for input and output modules (analog or digital). The controller operates with an embedded processing system running LabVIEW Real-Time for control, data acquisition and analysis. The platform chassis is where the input and output modules are assembled and allows communication between these modules and the controller, it integrates a matrix of logic circuits with reconfigured port arrangements (FPGA). This technology provides the reliability of a hardware dedicated and deterministic performance of closed-loop controls at extremely fast rates exceeding $1MHz$ [13].

The control loop implementation was developed with the high-level graphical programming tool based on data flow in LabVIEW where it has function blocks for developing FPGA-based control systems. The programming is done using fixed-point data type, as it can optimize the FPGA resources, and with simple precision floating point. The logic used to implement the PID control is presented in the block diagram in 6. Many control solutions have been used over time, but the PID controller is the most used in the industry today [4]. The block diagram reads the sensor signal inputs, in voltage, is then converted to millimeters, and sent to the PID controller where it compares it to a reference value, then is treated by the law of control. Its control signal will be added and subtracted to a constant value (base value) causing the direction to act in a differential way guaranteeing equal forces in both directions for the coils, these signals of operating mode outputs are sent to a duty cycle input (PWM) on the power system.

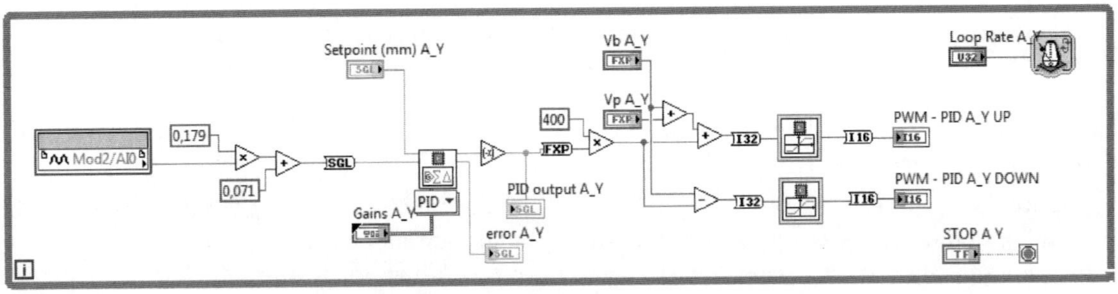

Figure 6: FPGA block diagram for the implementation of the PID controller.

The controller was implemented in FPGA with a sampling frequency of $10kHz$ ($100\mu s$) that can be changed. This value was chosen based on the orientations of Schweitzer [14], where they usually use rates between 5 to $10kHz$ for magnetic bearing systems in a digital control implementation.

2.6 Drive System

The bearings are driven by a DC motor from the manufacturer JDrones. Control is done through a PWM signal generated by an embedded device from National Instrument to vary its angular velocity. Figure 7 shows the drive system used. The bearings can be driven by other types of motors according to the application need.

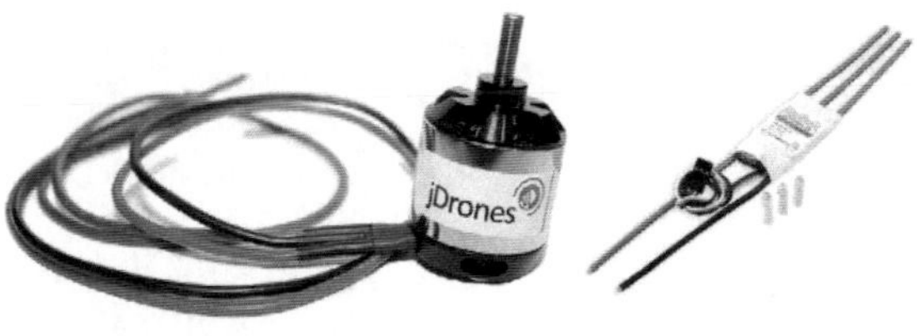

Figure 7: DC motor and speed controller [9].

3 Results

This section aims to present and discuss the experimental results obtained in the prototype of the AMS system. Tests were carried out with different situations, evaluating the following response characteristics: static behavior (without rotation), transient, dynamic (with rotation) and perturbations. The graph of the orbit is presented with the trajectory of the movement of the geometric center of the rotor in the plane xy defined by the position sensors. The orbit provides important information as it represents the real path of the rotor's geometric center during lateral vibrations [12].

3.1 System response during run-up

Initially, the response of the system with the rotor at rest supported by the support bearings was analyzed, and then the system was started. For mechanical purposes, this is important to verify if the system will effectively manage to reach stability within a time interval and the level of disturbance that can occur in the contact of the support bearings with the rotor. In general, analyzing the dynamic behavior of the system during startup means studying the transient response.

Figures 8a and 8a show the changes in rotor displacement in the vertical direction at the time the system is activated. The system stabilized maintaining an oscillation around $0.005mm$ $(5\mu m)$ of amplitude. It was observed the rotor is not taken directly to the zero position, taking approximately 2s to reach the steady-state, which is the system's accommodation time. This effect is due to the control system and is related to the integrative term, which has as its main function to eliminate the steady-state error and keep the rotor in the center of the magnetic bearings.

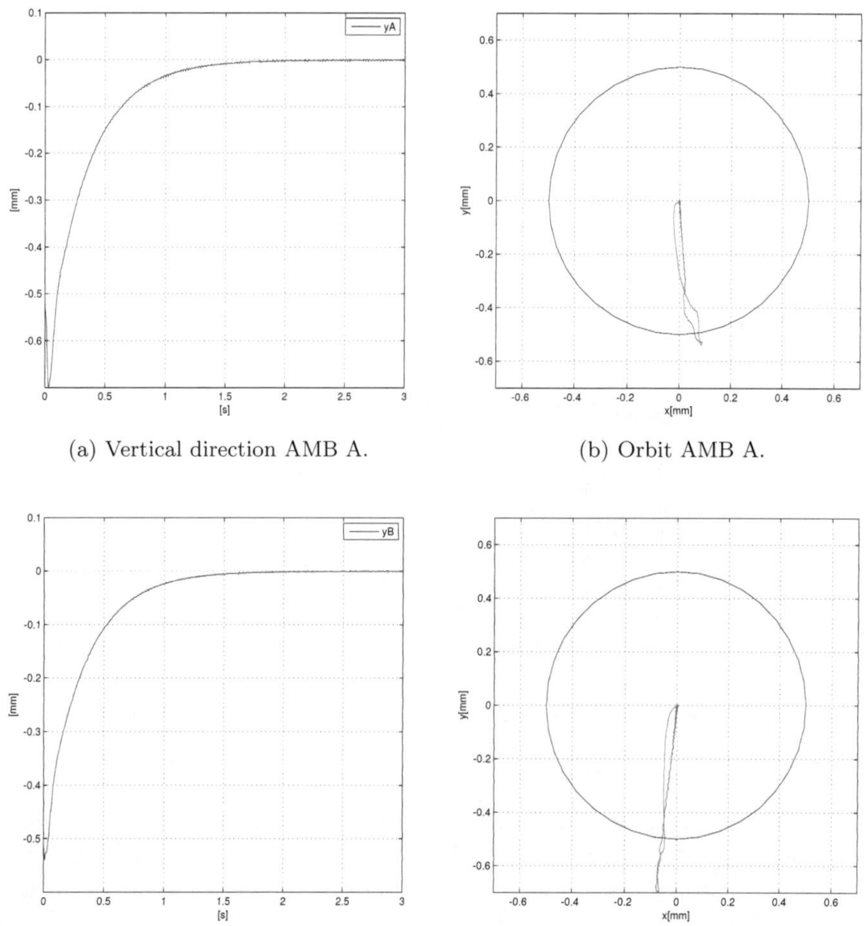

(a) Vertical direction AMB A. (b) Orbit AMB A.

(c) Vertical direction AMB B. (d) Orbit AMB B.

Figure 8: Transient response of the rotor during run-up.

Figures 8b and 8d the orbits of rotor movement within the bearings are shows. It is verified that the rotor has asymmetrical movements during start caused by the irregular initial position of the rotor inside the support bearing. This could be due to the non-ideal manufacture of the bearing, a possible eccentricity with the stator or also misalignment of the bearings. It is also observed that the displacements are greater than the radial clearance. This may have happened because of the loss of the displacement sensors calibration adjustment due to the rotor vibration in the support bearings during the start, caused by the fact that the sensors are fixed in the support structure and these have been made of elastic material. The results show that even with a non-ideal geometry, the system performed satisfactorily, keeping it stable.

3.2 System response in steady-state

The analysis of the dynamic behavior of the system for an operation with zero rotation speed will be presented. Therefore, the maximum displacement of the shaft in relation to the central position of the bearings will be seen.

Figures 9a and 9b present the graphs of the orbits of the axis put into operation for the system response is measured steady-state. It is observed that the rotor remains most of the time in a range of $0.004mm$ with respect to the zero points, but not being completely static. This was expected, since the system has a decentralized control structure, that is, a controller for each bearing direction acting independently to keep the rotor position stable. However, the vibration produced by the vertical control also influences the horizontal direction. Even with these small displacements, the system presents a satisfactory response indicating an operation with an adequate performance of the control system.

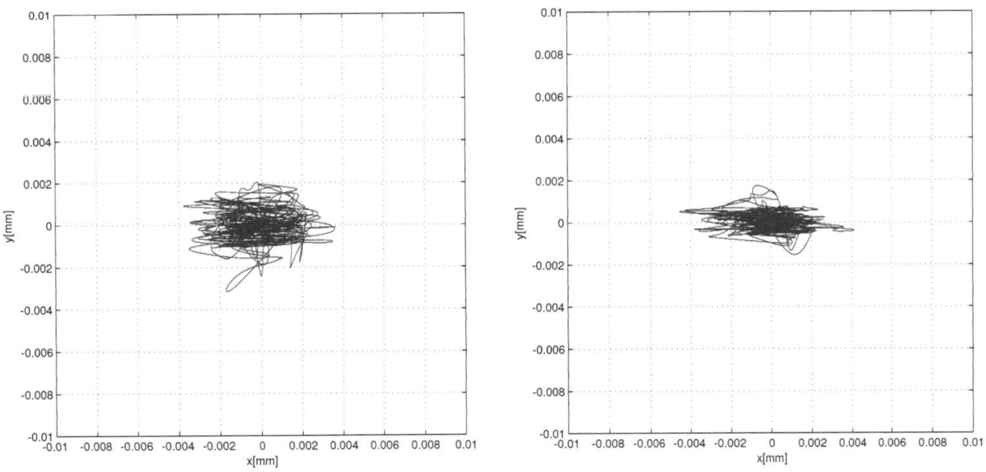

(a) AMB A. (b) AMB B.

Figure 9: Measured rotor orbits (in mm) at radial AMB.

3.3 Rotating system response

The analysis of the dynamic behavior of the system will be done with the rotor rotating at rotation speeds of 2000 and 3000 RPM. The system is driven by a DC motor coupled at the end of the shaft close to bearing A. Initially, the analysis will be made for the case of the system without unbalancing excitation, only considering the rotor's own weight. The values in the orbit graphs indicate the maximum and minimum limits given by the bearing support structure with a radial clearance of $0.5mm$.

Figure 10a shows the measured orbit in the bearings for a speed of 2000 RPM. An increase in amplitude is noticed in relation to the static system case. The movement of the rotor remains in a range less than $0.1mm$ ($100\mu m$) for bearings A and B. Figure 10b for a speed of 3000 RPM, it was observed there is a decrease in the vibration amplitude of the rotor in the bearing in relation to the system with a rotation speed of 2000 RPM, remaining at around $0.050mm$ ($50\mu m$) for bearings A and B. The results show that the system kept the rotor position within the limit in a stable and satisfactory way. As the system always presents a residual unbalance, with an increase in the rotation speed the amplitudes suffer alterations. In this case, it presents a decrease in the position amplitude due to the increase of rigidity related to the increase of the rotation frequency.

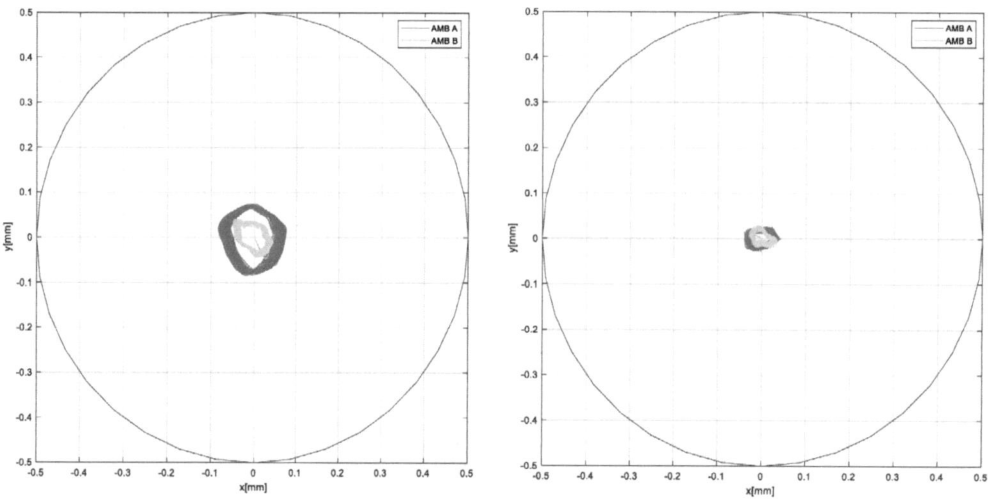

(a) 2000 RPM. (b) 3000 RPM.

Figure 10: Measured rotor orbits (in mm) at radial AMB locations at different spin speeds.

3.4 System response to unbalance excitation

Unbalance is a phenomenon characterized by a high amplitude at the peak of the rotation frequency. The behavior analysis takes into account some situations that can occur in reality. Thus, the results of the behavior of the system will be for excitation with general unbalance caused by two identical masses of $0.001kg$ positioned close to the bearings at a distance of $0.042m$ of the rotation axis of the rotor with phase angle is $120°$ as shown in Figure 11. The analysis will be done with the rotor rotating at rotation speeds of 2000 and 3960 RPM. A general unbalance is equivalent to a combination of static and dynamic effects, that is, the center of mass is outside the rotation axis and the main axis of inertia is inclined in relation to this. This combination implies a displacement of the center of mass and an inertia tensor with non-zero terms outside the main diagonal. The position of the new center of mass, known as eccentricity, given by a point mass that is outside the main rotor axis and outside the plane of the central section of the rotor [19].

Figure 11: Test rig to General Unbalanced.

In the orbit of the Figure 12a for a sping speed of 2000 RPM, it is verified that the system presents a satisfactory response, remaining in a stable way, where the displacements remain with oscillations smaller than $0.30mm$ of amplitudes for both bearings. Figure 12b shows the behavior of the system with an increase in spin speed of approximately 3960 RPM. This maximum spin speed was restricted to the mechanical coupling of the motor, which induced considerable forces on the bearing. Nevertheless, the system presented a satisfactory response within the limit determined by the support bearings, remaining stable with oscillations smaller than $0.2mm$ of amplitude in the bearings. Note that there is a decrease in amplitude concerning the previous case (Figure 12a), due to the increase in stiffness related to the increase in the rotation frequency.

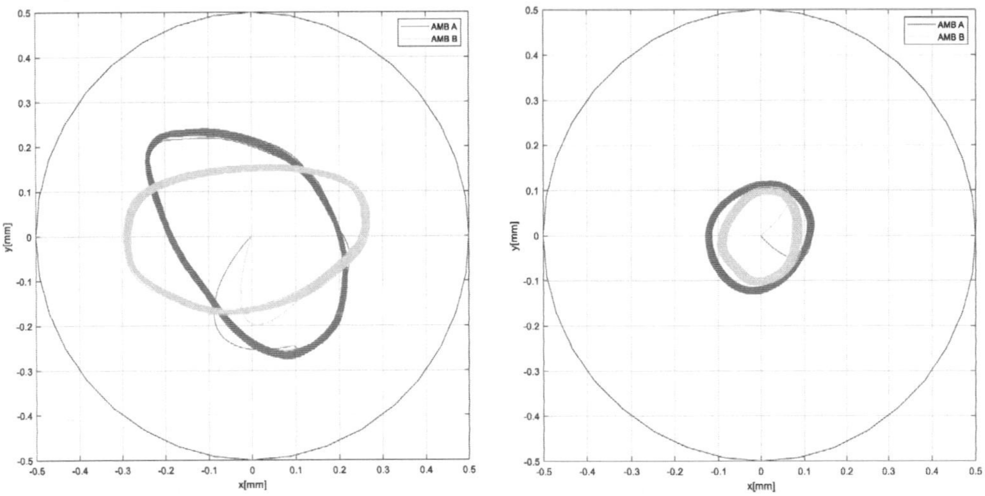

(a) 2000 RPM. (b) 3960 RPM.

Figure 12: Rotor orbit on AMBs with general unbalance.

4 Conclusions

In this study, the construction and operation of an AMB system for rotor levitation and stability are presented and discussed. The project presented satisfactory results with low vibration amplitudes and within the established limit, even without a mechanical precision in relation to alignment and centering that had little effect on the system's operation.

The experimental results were presented and discussed in several situations for the dynamic behavior of the system, such as the rotor in static, transient and rotor in steady-state driven by a DC motor. In the stationary regime, the system presented low amplitude in the measurements of the rotor positions. During the transient, the system responded satisfactorily, remaining stable. In the steady-state with motor activation, the results were adequate since there was an increase in amplitude, due to the tests that were carried out with the influence of loads. Despite this, the shaft remained within the design limits by levitating and turning.

References

[1] M. Aenis, E. Knopf, and R. Nordmann. Active magnetic bearings for the identification and fault diagnosis in turbomachinery. *Mechatronics*, 12(8):1011 – 1021, 2002.

[2] B. Aeschlimann, M. Hubatka, R. Stettler, and R. Housseini. Commissioning of off-shore gas compressor with 9-axes magnetic bearing system: Commissioning tools. In *15th International Symposium on Magnetic Bearings*, volume 9, pages 215–221, 2017.

[3] Angelo Bonfitto, Ran Gabai, Andrea Tonoli, Luis Miguel Castellanos, and Nicola Amati. Resonant inductive displacement sensor for active magnetic bearings. *Sensors and Actuators A: Physical*, 287:84–92, 2019.

[4] Rakesh P. Borase, D. K. Maghade, S. Y. Sondkar, and S. N. Pawar. A review of PID control, tuning methods and applications. *International Journal of Dynamics and Control*, (June), 2020.

[5] E.A. Costa, I. E. Chabu, and J. J Cruz. Application of active magnetic bearing for tubular linear induction motor in oil pumping. *Proceeding 11th International symposium on magnetic bearing*, August 2008.

[6] Michael I. Friswell, John E. T. Penny, Seamus D. Garvey, and Arthur W. Lees. *Dynamics of Rotating Machines*. Cambridge University Press, New York, 1 edition, 2010.

[7] Giancarlo Genta. Kinetic energy storage: an ideal application for magnetic bearings. *Proceedings of the 14th International Symposium on Magnetic Bearings (ISMB)*, August 2014.

[8] Oswaldo Horikawa, Rogério I. Yamamoto, and Isaías da Silva. Development of single axis controlled attraction type magnetic bearing and its application in ventricular assist device. *1st Brazilian Workshop on Magnetic Bearings*, Outubro 2013.

[9] JDrones. Arducopter motor. Acessado em 05/06/2020.

[10] Zbigniew Kozanecki, Małgorzata Gizelska, and Dorota Kozanecka. Monitoring and diagnostics of the rotating system with an active magnetic bearing. 198:547–552, 2 2013.

[11] S. Liu, D. Li, W. Yu, and J. Guo. Stiffness analysis of the magnetic bearing system. *Proceedings of the 9th International Symposium on Magnetic Bearings (ISMB)*, August 2004.

[12] Agnieszka Muszynska. *Rotordynamics*. CRC Press, New York, 1 edition, 2005.

[13] G. Rata, M. Rata, and C. Prodan. Analysis of the deforming regime generated by different light sources, using reconfigurable system - compactrio. pages 748–751, Oct 2014.

[14] Gerhard Schweitzer and Eric H. Maslen. *Magnetic Bearings - Theory, Design an Application to Rotating Machinery*. Springer-Verlag, Berlin, 1 edition, 2009.

[15] R. S. Siqueira. Projeto e implementação de um mancal magnético ativo com controle por modos deslizantes. Dissertação de Mestrado - UFRJ - COPPE/PEM - Programa de Pós-graduação em Engenharia Mecânica, 2013.

[16] Joaquim Siva, François Carrere, and Frédéric Ponson. Subsea gas compression station: Collaborative design using multiphysics/multiscale system simulation applied to active magnetic bearings. *1st Brazilian Workshop on Magnetic Bearings*, Outubro 2013.

[17] A Smirnov, Nikita Uzhegov, Teemu Sillanpää, Juha Pyrhönen, and Olli Pyrhönen. Design and evaluation of high-speed solid rotor induction machine supported by AMBs with a multidisciplinary tool. *15th International Symposium on Magnetic Bearings*, pages 467–473, 2016.

[18] Sparkfun. Sparkfun - monster moto shield. Acessado em 02/05/2020.

[19] Richard M. Stephan, Fernando A. N. Castro Pinto, Fernando Celso D. N. Gomes, José A. Santisteban, and Andres Ortiz Salazar. *Mancais Magnéticos - Mecatrônica sem atrito.* Editora Ciência Moderna, Rio de Janeiro, 1 edition, 2013.

[20] Se Young Yoon, Zongli Lin, and Paul E. Allaire. *Control of Surge in Centrifugal Compressors by Active Magnetic Bearings.* Springer-Verlag, London, 1 edition, 2013.

Dynamics of Vertical Magnetic Suspension Rotor and Touchdown Bearing: Touchdown Mechanism

Zilin Li[1], Congtao Wang[2], Zixi Wang[1*] and Yuming Wang[1]
[1] State Key Laboratory of Tribology, Department of Mechanical Engineering, Tsinghua University, Beijing 100084, Beijing, China;
[2] Shenyang University of Technology, Shenyang 110003, Liaoning, China.

Abstract

During touchdowns of active magnetic bearings (AMB), rotor will generate different trajectory on touchdown bearings (TDB). The full rub will cause violent collisions and damages to both parts. By modeling radial and axial contact between rotor and TDBs, the dynamic and thermodynamic touchdown model of a vertical rotor with internal TDBs were established. Numerical simulation showed that the orbit response of vertical rotor after touchdowns was forward whirl, which is divided into two states: dry-friction whirl and dry-friction whip. There's a limit speed associated with the natural frequency that enables the rotor to convert from synchronous positive precession to sub-synchronous positive precession. The contact force and temperature of inner race has increased significantly when rotor entering the sub-synchronous forward precession.

Keywords: Touchdown Bearing; Vertical Rotor; Dynamic Model; Forward whirl.

1 Introduction

Once the Active Magnetic Bearing (AMB) is powered off or overloaded, the rotor will fall and rub against the touchdown bearings (TDB). The function of TDB is to limit the maximum radial and axial displacement of the rotor, to avoid the collision and damage of the main parts of AMB. But in fact, TDB itself is a part that is prone to damage. A number of theoretical and experimental studies have found that severe damage often occurs on TDB, so fatigue failure isn't the main cause of TDB failure. At present, there's no effective life evaluation system for TDB, which has become a key factor restricting the development of AMB. (Schweitzer Kuesnacht (Switzerland)] & Maslen Charlottesville, VA (United States). Dept. Mechanical and Aerospace Engineering], 2009)

* Corresponding authors: Tel: 86 10 62794845; Email: zxwang@mail.tsinghua.edu.cn (Zixi Wang).

The initial states such as rotor imbalance (Hawkins et al., 2006), surface friction coefficient (Sun et al., 2004), material pairing (Zhu, 2010), and assembly error (Kärkkäinen et al., 2008), etc., will have a significant impact on the rotor dynamic behavior of the touchdown. In severe cases, the rotor and TDB will vibrate violently and cause damage to TDB. According to ISO 14839, three typical orbit responses (pendulum vibration, combined rub and bouncing, and full rub) can be detected in touchdown events. (ISO, 2012) Lyu Mindong established a touchdown trajectory response identification method for online monitoring and resuspension of the rotor touchdown. (Lyu et al., 2018) In different touchdown trajectories, the whirl of the rotor (shown as full rub) was prone to significantly higher contact force and heat than other states, which is one of the main reasons for the failure of TDB. (Zhu et al., 2018) Zhao Yulan established a multi-degree-of-freedom rigid rotor drop model, based on the nonlinear Hertzian contact and collision theory, which can effectively simulate the movement, force and heating of the rotor touchdown. (Zhao et al., 2015) Helfert built an AMB-rotor touchdown test bench, using a high-speed camera to capture the displacement, speed and acceleration of the rotor, inner ring, and rolling elements after the fall. (Helfert et al., 2006)

In the study of rotor dynamic, 'whirl' generally refers to the orbital movement of the eccentric mass of the rotor around the axis, mainly for the contact between the flexible rotor and the stator. Literature (Jacquet-Richardet et al., 2013) provides a literature review on this topic. According to the direction of the frictional force, the rotor presents a backward whirl state during the contact. Wilkes (J. C. Wilkes et al., 2010) divides the contact form into the following two categories:

- Dry-friction whirl. The rotor rolls on the surface of the stator without sliding, and the precession frequency is governed by the radius-to-clearance ratio at the contact position;
- Dry-friction whip. The rotor continuously slides on the surface of the stator, and the precession frequency is governed by the combined natural frequency of the connected rotor-stator system.

Although the touchdown of AMB rotor is similar to the traditional rotor rubbing, touchdown rubbing has characteristics such as mainly occurs in the inner ring of the bearing, rotatabel bearing support, and not distinguished rotor mode. Therefore, the touchdown dynamic behavior is very different from the traditional rotor rubbing.

When the vertical rotor drops, it first collides and rubs with the axial end surface of the stationary TDB inner ring. The inner ring accelerates to the rotor speed in a short time. Then the rotor will continue colliding and rolling in the radial gap. Since there is no radial constraint, the rotor will always be in full rub state. The direction of the friction force indicates that the rotor will form forward whirl during vertical rotor drops. In a vertical rotor drop test, Caprio found that the forward whirl speed of the rotor monotonously increased with the rotor speed at a low rotation speed, but the whirl speed no longer increased when the speed was higher than 2000r/min. (Caprio et al., 2004) Wilkes verified this phenomenon through a numerical model in a subsequent study, and proved that the aforementioned dry-friction whirl and dry-friction whip states also occurred in the vertical rotor forward whirl. (J. Wilkes et al., 2014) There is a combined natural frequency between two state transitions.

The current dry-friction whirl/whip research of vertical rotor touchdown still stays in the rotor trajectory, motion and related vibration frequency. Researchers generally pay more attention to the behavior of the rotor, and there is still a lack of related research on TDB. For the design of the AMB-rotor system especially TDB, more parameterized, detailed and specific calculation are needed. Therefore, this research focuses on the establishment of system touchdown dynamics and thermodynamic models to quantify the force and temperature characteristics of TDB. By improving and supplementing the existing theoretical system, it provides parameter optimization for the overall structure of the preliminary designed protective bearing. In the future, we will also verify our model through experiments.

2 Model

2.1 AMB-rotor-TDB test system

The AMB-rotor-TDB test system consists of a rotor, 2 groups of radial and 1 group of axial AMBs and 2 groups of TDBs (Figure 1). The upper TDB bears the axial and radial load of the rotor, and the lower TDB only bears the radial load of the rotor. See Table 1 for relevant structural parameters of the rotor and bearing system. Use hybrid ceramic ball bearings as TDB solution. At each side of the rotor, a pair of 71938AC angular contact ball bearings are installed, and they are installed face to face. The inner and outer rings of the bearing are made of GCr15(SAE 52100) bearing steel, and the rolling elements are made of Si_3N_4 ceramic balls. The mechanical and thermal parameters related to materials are shown in Table 2.

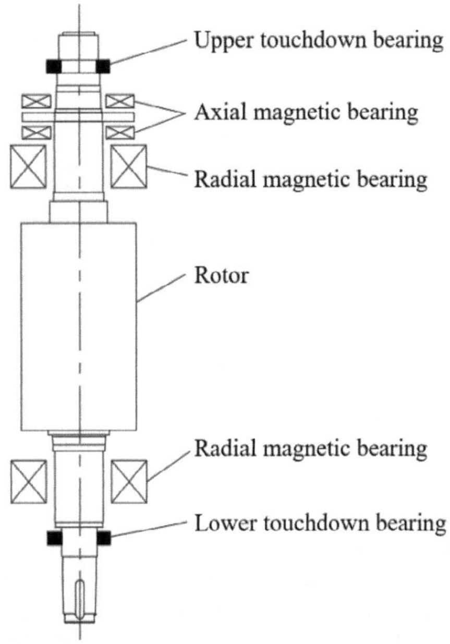

Figure 1: Schematic diagram of AMB-rotor-TDB system.

Rotor parameters	Value	TDB parameters	Value
Mass, kg	3000	Inner ring mass, kg	1.19
Length, mm	3054	Outer ring mass, kg	1.52
Momentum of Inertia, kg·m^2	130	Inner ring diameter, mm	190
Dynamic balance quality level	G6.3	Outer ring diameter, mm	260
Eccentricity, mm	0.02	Bearing width, mm	33
Radial clearance, mm	0.3	Bearing internal COF	0.0018
Axial clearance, mm	0.3	COF between rotor and inner ring	0.15

Table 1: AMB system parameters. (COF: Coefficient of friction)

Material parameters	Bearing steel GCr15	Ceramic ball Si3N4	Rotor Carbon steel
Elastic Modulus, GPa	219	310	210
Poisson's ratio	0.3	0.269	0.31
Thermal conductivity, W/(m·K)	40	3.5	50
Specific heat capacity, J/(kg·K)	450	800	480

Table 2: Material mechanics and thermal performance parameters.

2.2 Dynamic model of rotor-TDB system

When rotor dropped, considering the dynamic behavior of the upper TDB, the contact between rotor and TDB was divided into two parts: the axial end face and the radial inner face. Rotor first collided with the annular end surface of the inner ring. And then contacts and whirls in radial clearance. The contact and collision generated support force, friction force and friction torque. The system dynamics model of rotor-TDB and support base was shown in Figure 2.

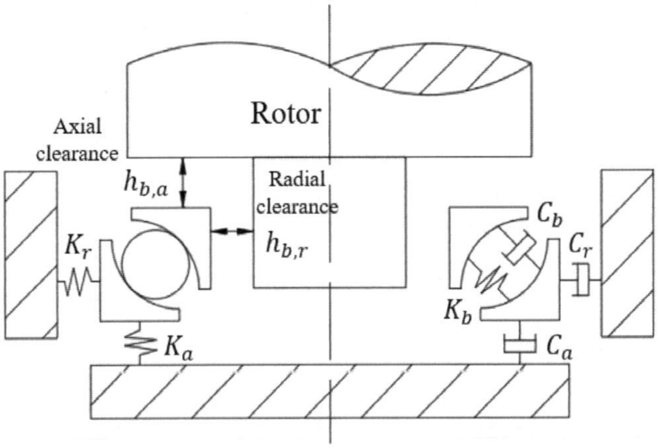

Figure 2: Rotor-TDB-support base system model.

$h_{b,a}, h_{b,r}$ represented the axial and radial drop clearances between the rotor and TDB respectively. K_b, C_b represented the stiffness and damping inside TDB. K_a, C_a represented the axial stiffness and damping between the outer ring of TDB and the support seat. K_r, C_r represented the radial stiffness and damping between the outer ring of TDB and the support seat.

The contact force between the rotor and TDB was calculated using the Hunt-Crossly collision model (Hunt & Crossley, 1975), which can calculate the impact of touchdown accurately. The calculation formula is:

$$F = \begin{cases} K\delta^n(1 + 1.5\kappa\dot{\delta}), & \delta > 0 \\ 0, & \delta \leq 0 \end{cases}$$

δ was the amount of penetration between two contact surfaces. n was the contact coefficient. The coefficient κ is distributed in the interval from 0.08 to 0.2. K was the contact stiffness.

The friction of the system was calculated using the Coulomb friction model, as shown in Figure 3. The formula for calculating the radial friction force was $F_t = \mu F_r$. For the friction of the end surface, the force and torque were obtained by integrating the contact surface. Establish the coordinate system Oij according to the relative position of the rotor and the inner ring, then the relative speed between the inner ring and the rotor at point A in Figure 3 can be expressed by the following formula:

$$\overrightarrow{v_{A,rb}}(r,\varphi) = [\dot{\rho} + (\dot{\theta}_r - \dot{\theta}_b)r\sin\varphi]\vec{i} + [(\dot{\alpha} - \dot{\theta}_r)\rho - (\dot{\theta}_r - \dot{\theta}_b)r\cos\varphi]\vec{j}$$

$\dot{\theta}_r, \dot{\theta}_b$ represented the rotation speed of the rotor and TDB; $\dot{\alpha}$ represented the rotation speed of the geometric center of the rotor around the fixed coordinate system Oxy, that was, the whirl speed; ρ represented the distance between the geometric center of the rotor and the origin of the coordinate.

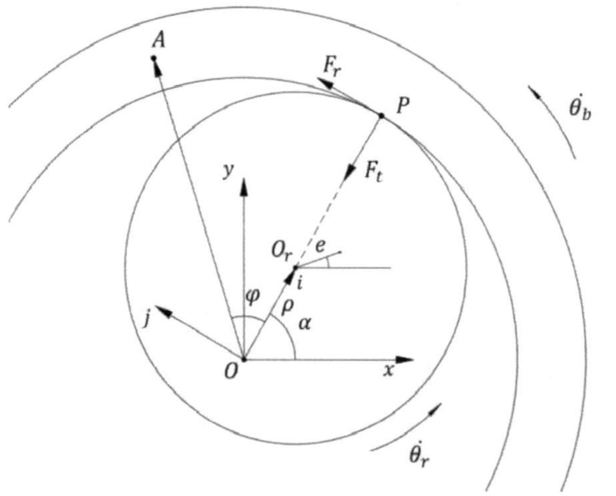

Figure 3: Model of friction force between rotor and TDB.

Approximately, the axial load can be regarded as uniformly distributed on the inner ring. Use P_a to represent the contact pressure of the end surface. The friction force on the entire surface can be calculated in two directions with the following formula:

$$F_{a,i} = -P_a\mu_f \cdot \int_{R_1}^{R_2}\int_0^{2\pi} \frac{[\dot{\rho} + (\dot{\theta}_r - \dot{\theta}_b)r\sin\varphi]\vec{i}}{\left\{[\dot{\rho} + (\dot{\theta}_r - \dot{\theta}_b)r\sin\varphi]^2 + [(\dot{\alpha} - \dot{\theta}_r)\rho - (\dot{\theta}_r - \dot{\theta}_b)r\cos\varphi]^2\right\}^{\frac{1}{2}}} rd\varphi dr$$

$$F_{a,j} = -P_a\mu_f \cdot \int_{R_1}^{R_2}\int_0^{2\pi} \frac{[(\dot{\alpha} - \dot{\theta}_r)\rho - (\dot{\theta}_r - \dot{\theta}_b)r\cos\varphi]\vec{j}}{\left\{[\dot{\rho} + (\dot{\theta}_r - \dot{\theta}_b)r\sin\varphi]^2 + [(\dot{\alpha} - \dot{\theta}_r)\rho - (\dot{\theta}_r - \dot{\theta}_b)r\cos\varphi]^2\right\}^{\frac{1}{2}}} rd\varphi dr$$

The axial friction torque can also be obtained as:

$$T_a = \frac{2F_a\mu_f(R_2^3 - R_1^3)}{3(R_2^2 - R_1^2)}$$

R_1, R_2 represented the inner and outer radius of the end face of the bearing inner ring.

According to the Lagrange equation, the dynamic equations of the rotor drop in three directions during contact are established as follows:

$$m_r\left[\ddot{x}_r - e\sin(\theta_r)\ddot{\theta}_r - e\cos(\theta_r)\dot{\theta}_r^2\right] = -F_r\cos\alpha - F_t\sin\alpha$$

$$m_r\left[\ddot{y}_r + e\cos(\theta_r)\ddot{\theta}_r - e\sin(\theta_r)\dot{\theta}_r^2\right] = -F_r\sin\alpha + F_t\cos\alpha$$

$$m_r\ddot{z} = -m_r g + F_{a,z}$$

$$m_r e[-\sin(\theta_r)\ddot{x}_r + \cos(\theta_r)\ddot{y}_r] + (m_r e^2 + J_r)\ddot{\theta}_r = F_r e\sin(\alpha - \theta_r) - F_t e\cos(\alpha - \theta_r)$$

2.3 Thermodynamic model of rotor-TDB system

During the drop process of the rotor, the main heat sources considered come from three aspects: viscous friction torque, spin friction torque and contact load friction torque. Palmgren derives the friction torque inside the rolling bearing, and Harris further derives the spin torque of the rolling element relative to the inner ring and the outer ring. (Harris & Barnsby, 1998)The contact friction torque can be directly calculated by the mechanical model. The total frictional heat generated by the final fall can be expressed as:

$$H_{total} = \sum (H_{i,e}^{j} + H_{r}^{j} + H_{a}^{j})$$

$H_{i,e}$ was the friction heat generated by the rolling element with the inner and outer rings. H_r was the friction heat generated by the radial contact surface of the rotor and TDB. H_a was the friction heat generated by the axial end surface of the rotor and the upper TDB; the angle mark j indicated each pair of bearings.

Based on the theory of heat transfer, a one-dimensional heat transfer network model was established by calculating the heat transfer impedance of each component during the touchdown process of the rotor, considering the heat conduction and heat convection of the system (Figure 4). The heat transfer between the temperature nodes was expressed by the first-order differential equation as follows:

$$mC_p\Delta T = \Delta Q$$

m represented the mass, C_p represented the specific heat capacity, ΔT represented the temperature difference between the two temperature nodes. And Q represented the heat flux, R represented the thermal resistance of each part.

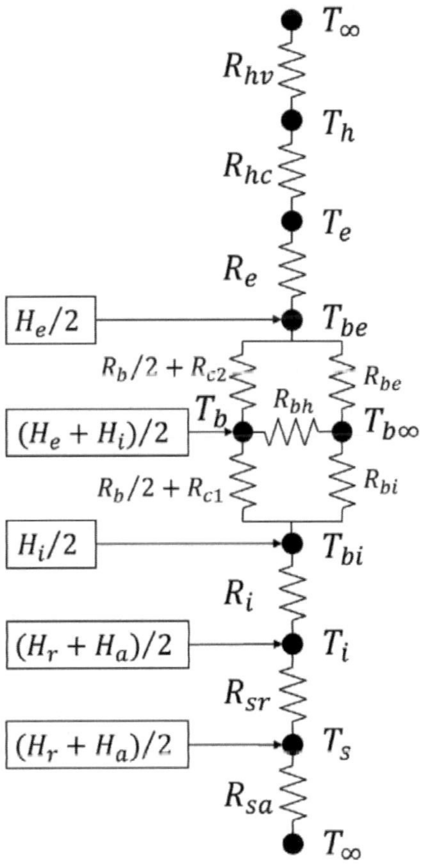

Figure 4: Schematic diagram of the heat transfer network of the rotor-TDB system.

Furthermore, each component of TDB will expand as the temperature rises. Considering the thermal expansion of the inner and outer rings and rolling elements, the thermally induced load inside TDB is:

$$F_{th} = K[\varepsilon_b + 0.5(\varepsilon_i - \varepsilon_e)\cos\alpha_0]^{3/2}$$

Among them, $\varepsilon_b, \varepsilon_i, \varepsilon_e$ respectively represented the expansion of the protective bearing rolling element, inner ring and outer ring. α_0 represented the initial contact angle.

3 Result

The model of the rotor-TDB system was established above, then the numerical simulation calculation was carried out. According to the speed target designing the test bench, two typical calculation examples were simulated. The initial speed was set to 3000 r/min and 6000 r/min. Suppose that the rotor lost AMB support and started to fall freely on the set speed at time 0. Figure 5 shows the operating conditions of the rotor touchdown event at an initial speed of 3000r/min within 1s, including axial displacement and contact force, radial touchdown trajectory, radial contact force, rotor and bearing speed, temperature of various parts, and temperature ascent rate of the inner ring.

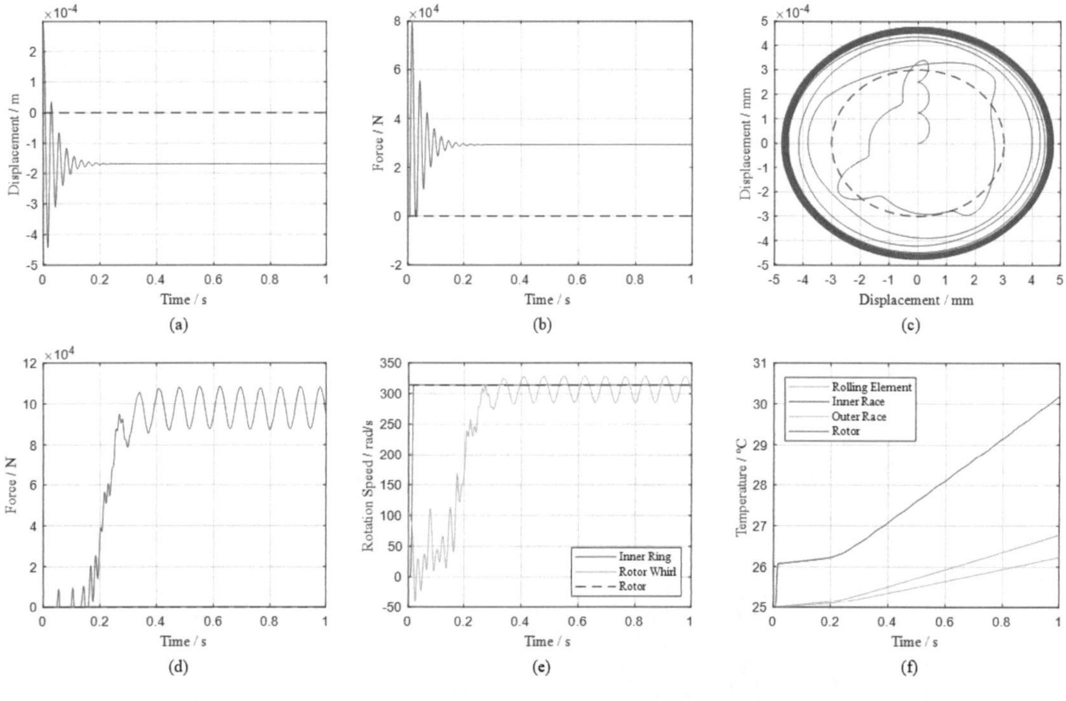

Figure 5: Numerical simulation results of rotor touchdown at 3000 r/min: (a) Rotor axial displacement; (b) Axial contact force of inner ring; (c) Rotor radial trajectory; (d) Radial contact force of inner ring; (e) Rotation speed; (f) System temperature change.

Numerical simulation results showed that, in the axial direction, the rotor reached a stable support state after several collisions and rebounds (Figure 5(a)). The maximum axial contact force occurred at the first collision (Figure 5(b)). The trajectory in the radial direction (Figure 5(c)) showed that the rotor entered a forward whirl state after several collisions, making an approximate circular motion. The radial contact force curve (Figure 5(d)) showed that the collision force was small in the first few collisions, then continued to rise after the rotor enters the forward whirl, finally fluctuated in a fixed interval. Figure 5(e) showed the change curve of the rotation speed of the rotor, the inner ring and the rotor whirl. It can be seen that during the initial axial contact between the rotor and inner ring, the inner ring quickly accelerated to near the rotor speed and maintained a constant speed. The whirl speed of the rotor rose relatively slowly. It rose to the rotor speed as the rotor entered a forward whirl, and oscillated within a certain range. At this time, the rotor and the inner ring of TDB enter the synchronous positive precession state. Figure 5(f) showed the temperature changes of the various components of the system. The inner ring had a significant temperature increase during the first contact due to the rapid acceleration and collision friction in the axial direction. The temperature of other parts increased with time. The temperature rise of the rotor is not obvious due to its large mass. After the rotor enters the equilibrium state, the rate of temperature change was also balanced.

Change the initial drop speed to 6000 r/min for touchdown simulation. The partial results obtained were shown in Figure 6. It can be seen from this calculation that the whirl radius of the rotor trajectory increased (Figure 6(c)). The radial contact force was increased by 159.5% compared to 3000 r/min (Figure 6(d)). The whirl speed of the rotor did not rise to the rotor speed, but stabilized at about 400 rad/s (Figure 6(e)). The temperature of each part has a significant increase compared to the low speed (Figure 6(f)).

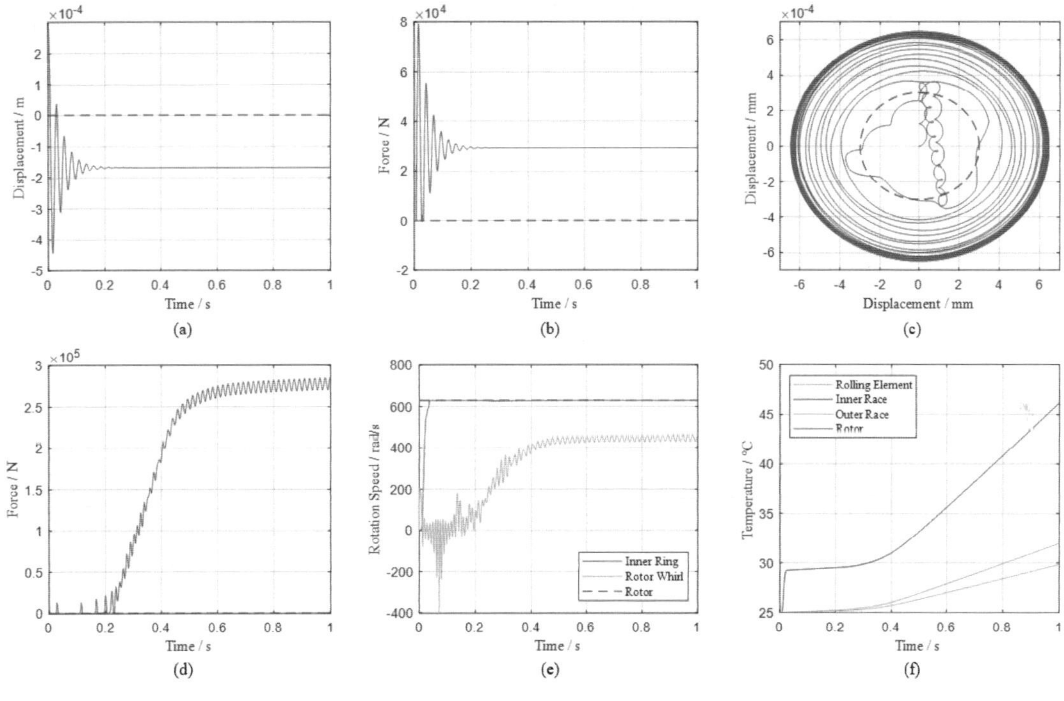

Figure 6: Numerical simulation results of rotor touchdown at 6000 r/min: (a) Rotor axial displacement; (b) Axial contact force of inner ring; (c) Rotor radial trajectory; (d) Radial contact force of inner ring; (e) Rotation speed; (f) System temperature change.

In order to study the relationship between the whirl speed and the rotor speed of vertical magnetic suspension rotor, a waterfall plot of the whirl speed distribution changing with the initial dropping speed was made, as shown in Figure 7. The X-axis represented the whirl speed, and the Y-axis represented the rotor rotation speed of the initial drop. When the rotor speed was lower than the limit speed (4000 r/min), the rotor forward whirl speed was equal to the rotor speed. The rotor and the inner ring underwent synchronous positive precession. At this time, the radial contact appeared as dry-friction whirl. When the rotor speed increased to 4000r/min or more, the rotor forward whirl speed became lower than the rotor speed. As the rotor speed increased, the difference between the rotor speed and the whirl speed increased. The rotor and the inner ring underwent sub-synchronous positive precession. At this time, the form of radial contact was dry-friction whip. There was continuous sliding friction, causing more significant energy conversion and heating processes in the system.

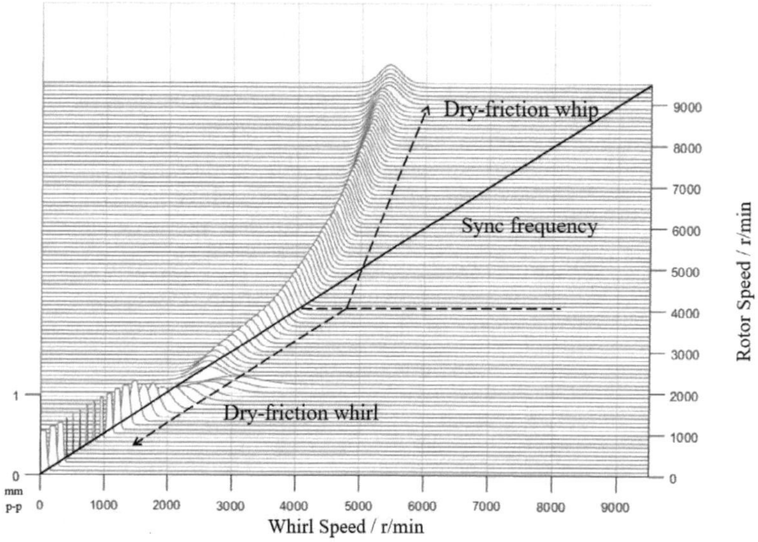

Figure 7: Waterfall plot of the rotor whirl speed at the upper TDB.

The above phenomenon was basically consistent with the rotor touchdown whirl phenomenon pointed out in the literature [14]. But it's worth noting that, literature [14] proposed that when the rotor speed exceeded the limit speed ω, the rotor whirl speed was equal to the limit speed ω. And our model observed that when the rotor speed exceeded the limit speed ω, the whirl speed was still increasing as the rotor speed increased. The differential speed between the rotor speed and the whirl speed also increased.

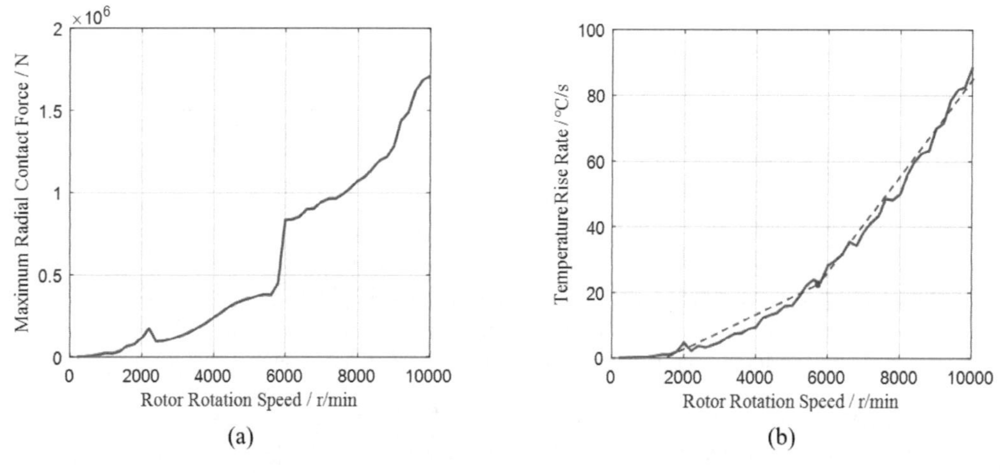

Figure 8: Change curve of (a) maximum radial contact force and (b) temperature rise rate.

Figure 8(a) plotted the change curve of the maximum radial contact force with the rotor speed when touchdown happened. It can be found that the maximum contact force increased monotonously with the increase of the speed. In addition, it's observed that around 6000 r/min, that is, when the increase in the whirl speed stopped, the contact force had a sudden increase. This was due to the rotor form conversion of the synchronous positive precession to the sub-synchronous positive precession. The result was a significant increase in contact force and friction force. Figure 8(b) showed the change curve of the temperature rise rate of the inner ring with the speed, after the rotor entered the oscillation balance. It can also be found that the temperature rise rate increased with the rise of the speed. And after the rotor enters the sub-synchronous positive precession, the temperature rise slope increased obviously.

The above phenomenon can be summarized when the vertical rotor dropped on TDB, first axial collision was the most violent at the initial stage of the rotor drop. Its collision force was the largest. The radial collision force increased continuously until it enters the forward whirl. The forward whirl was divided into two states: dry-friction whirl and dry-friction whip. The two are distinguished by the rotor speed. At low speeds, it's dry-friction whirl, and at high speeds, dry-friction whip occurred. Monitoring the parameters can find that the latter brought greater radial contact force and more severe temperature rise to the rotor, both of which will have a significant impact on the stability and reliability of TDB, and then affect the life of TDB.

4 Discussion on the behavior of rotor whirl

In order to explain the dry-friction whirl and whip phenomena found in the rotor touchdown experiment by Caprio [13] and Wilkes [14], the difference in dynamic behavior between the two is derived from a theoretical perspective. When the inner ring is stationary at the beginning of the rotor drop, the axial touchdown will produce a large friction torque. Since the rotor mass is much larger than the bearing mass, the inner ring will quickly accelerate to the rotor speed $(\dot{\theta}_r \approx \dot{\theta}_b)$. For radial contact, the tangential velocity at point P (in Figure 3) on the rotor and bearing can be expressed by the following formula:

$$V_r = \rho \cdot \dot{\alpha} + R_r \cdot \dot{\theta}_r$$
$$V_b = R_1 \cdot \dot{\theta}_b$$

Among them, R_r represents the rotor radius and R_1 represents the inner radius of the bearing inner ring, $R_1 = R_r + h_{b,r}$. When the rotor is in contact with the inner ring, there is $\rho = h_{b,r}$. Therefore, $V_r < V_b$ in the initial stage of the fall, and the friction force direction is forward, relative to the direction of rotation. The rotor has a trend of forward whirl. The rotor is driven by the tangential friction F_t, which causes the whirl speed $\dot{\alpha}$ to increase continuously. Until $\dot{\alpha} = \dot{\theta}_r = \dot{\theta}_b$, there's $V_r = V_b$. At this time, the tangential friction force F_t at point P is 0, rotor and inner ring contact with dry-friction whirl. And the entire system reaches a balanced state.

However, as mentioned in literature [14], There is a limit rotation speed ω for the acceleration of $\dot{\alpha}$. When $\dot{\theta}_r$ is less than the limit speed ω, $\dot{\alpha}$ will take $\dot{\theta}_r$ as the equilibrium speed. When $\dot{\theta}_r$ is greater than ω, $\dot{\alpha}$ will use ω as the balance speed. At this time, the system will increase the penetration ρ between the rotor and inner ring to achieve balance, which will lead to an increase in contact force.

Then, according to different contact behaviors, the motion form of the rotor and the inner ring is defined as synchronous positive precession and sub-synchronous positive precession. The former corresponds to the dry-frictional whirl between the rotor and inner ring, indicated that the rotor speed $\dot{\theta}_r$ is equal to the whirl speed $\dot{\alpha}$. The latter corresponds to dry-friction whip, which means $\dot{\alpha} < \dot{\theta}_r$.

5 Conclusion

When the magnetic suspension rotor falls, it will have severe impact and rubbing on TDB. In this paper, a dynamic analysis of the vertical rotor touchdown process was carried out. By modeling the contact of the axial end face, the inner ring radial direction and the rolling elements, a three-dimensional numerical simulation model of the dynamics and thermodynamics of the vertical rotor touchdown was established.

The movement mode of the vertical rotor after touchdown event was forward whirl. Under different working conditions, it can be divided into two states: dry-friction whirl and dry-friction whip. The former is synchronous positive precession, and the latter is sub-synchronous positive precession. There is a limit speed ω. When the rotor speed is lower than the limit speed, it falls into synchronous positive precession, the balance whirl speed is equal to the rotor speed. And when the rotor speed is higher than the limit speed, the rotor falls into sub-synchronous positive precession, the balance whirl speed is equal to the limit speed. The contact force and temperature between the rotor and TDB increased significantly since entering the sub-synchronous positive precession.

In the follow-up research, we will focus on the reasons for the sub-synchronous positive precession of the system, and investigate the influence of various parameters on the limiting speed that causes the whirl state to change. We are carrying out a drop test. In the future, the reliability of the rotor-TDB touchdown will be quantitatively evaluated from the perspective of system design.

References

Caprio, M. T., Murphy, B. T., & Herbst, J. D. (2004). Spin Commissioning and Drop Tests of a 130 kW-hr Composite Flywheel. *The 9th International Symposium on Magnetic Bearings.*

Harris, T. A., & Barnsby, R. M. (1998). Tribological Performance Prediction of Aircraft Gas Turbine Mainshaft Ball Bearings. *Tribology Transactions*, *41*(1), 60–68. https://doi.org/10.1080/10402009808983722

Hawkins, L., Filatov, A., Imani, S., & Prosser, D. (2006). Test Results and Analytical Predictions for Rotor Drop Testing of an Active Magnetic Bearing Expander/Generator. *Journal of Engineering for Gas Turbines and Power*, *129*(2), 522–529. https://doi.org/10.1115/1.2436549

Helfert, M., Ernst, M., Aeschlimann, B., & Nordmann, R. (2006). High-Speed Video Analysis of Rotor-Retainer-Bearing-Contacts Due to Failure of Active Magnetic Bearings. *10th International Symposium on Magnetic Bearings 2006*. http://tubiblio.ulb.tu-darmstadt.de/58906/

Hunt, K., & Crossley, E. (1975). Coefficient of restitution interpreted as damping in vibroimpact To cite this version : HAL Id : hal-01333795 Coefficient of Restitution Interpreted as Damping in Vibroimpact. *Journal of Applied Mechanics, American Society of Mechanical Engineers*, *42*, 440–445.

ISO. (2012). *ISO 14839-4 Mechanical Vibration - vibration of rotating machinery equipped with active magnetic bearings Part4: Technical guidelines*. https://www.iso.org/standard/50442.html

Jacquet-Richardet, G., Torkhani, M., Cartraud, P., Thouverez, F., Nouri Baranger, T., Herran, M., Gibert, C., Baguet, S., Almeida, P., & Peletan, L. (2013). Rotor to stator contacts in turbomachines. Review and application. *Mechanical Systems and Signal Processing*, *40*(2), 401–420. https://doi.org/10.1016/j.ymssp.2013.05.010

Kärkkäinen, A., Helfert, M., Aeschlimann, B., & Mikkola, A. (2008). Dynamic analysis of rotor system with misaligned retainer bearings. *Journal of Tribology*, *130*(2), 1–10. https://doi.org/10.1115/1.2908921

Lyu, M., Liu, T., Wang, Z., Yan, S., Jia, X., & Wang, Y. (2018). Orbit Response Recognition during Touchdowns by Instantaneous Frequency in Active Magnetic Bearings. *Journal of Vibration and Acoustics, Transactions of the ASME*, *140*(2), 1–11. https://doi.org/10.1115/1.4037850

Schweitzer Kuesnacht (Switzerland)], G. [Mechatronics C., & Maslen Charlottesville, VA (United States). Dept. Mechanical and Aerospace Engineering], E. H. [Virginia U. (2009). *Magnetic*

bearings. Theory, design, and application to rotating machinery. Springer, Berlin (Germany). https://doi.org/10.1007/978-3-642-00497-1
Sun, G., Palazzolo, A. B., Provenza, A., & Montague, G. (2004). Detailed ball bearing model for magnetic suspension auxiliary service. *Journal of Sound and Vibration*, *269*(3–5), 933–963. https://doi.org/10.1016/S0022-460X(03)00207-4
Wilkes, J. C., Childs, D. W., Dyck, B. J., & Phillips, S. G. (2010). The Numerical and Experimental Characteristics of Multimode Dry-Friction Whip and Whirl. *Journal of Engineering for Gas Turbines and Power*, *132*(5), 52503-1-52503–52509. https://doi.org/10.1115/1.3204658
Wilkes, J., Moore, J., Ransom, D., & Vannini, G. (2014). An improved catcher bearing model and an explanation of the forward whirl/whip phenomenon observed in active magnetic bearing transient drop experiments. *Journal of Engineering for Gas Turbines and Power*, *136*(4), 1–11. https://doi.org/10.1115/1.4025890
Zhao, Y., Yang, G., Shi, Z., & Zhao, L. (2015). Thermal Analysis and Simulation of Auxiliary Bearings and Its Application in the High Temperature Reactor-10. *Journal of Tribology*, *138*(1), 011102. https://doi.org/10.1115/1.4031003
Zhu, Z. Y., Mo, J. L., Wang, D. W., Zhao, J., Zhu, M. H., & Zhou, Z. R. (2018). Study on the Correlation Between Dynamical Behavior and Friction/Wear Mechanism Under the Effect of Grooves. *Journal of Materials Engineering and Performance*, *27*(6), 2875–2884. https://doi.org/10.1007/s11665-018-3373-z
ZHU C. (2010). Effect of backup bearing impact surface pairs on rotor dropping transient response after active magnetic bearing failure. *Journal of Vibration Engineering,* 23(05): 475-479. https://doi.org/10.16385/j.cnki.issn.1004-4523.2010.05.007

A Novel Inductive Sensor Input Circuit with Improved Tolerance to Non-Concentricity with Touch-Down Bearings

Richard Jayawant [1], Andrea Masala [1], Roy Leung [1], Nigel Davies [1]
[1] Waukesha Magnetic Bearings, Unit J, Downlands Bus. Pk. Lyons Way, Worthing, W. Sussex, UK.
rjayawant@doverprecision.com

Abstract

When using inductive sensors, non-concentricity between the sensor and touch-down bearing stators of an active magnetic bearing (AMB) system can result in the sensor signal being “offset” within the touch-down bearing clearance. The resulting signal will have a larger amplitude at the extremes of motion compared to a well-centered system. A novel signal conditioning circuit is presented which allows the signal to be “re-centered” prior to demodulation. The impact of this circuit on required machining tolerances and sensor noise levels is discussed. We also present an implementation that is well-suited to automated and remote commissioning and the techniques used to set up the sensor sub-system during such a commissioning.

1 Conventional Inductive Sensor Signal Conditioning

Inductive sensors have common usage in AMB systems due to their non-contacting measurement capability, simple and robust construction, acceptable bandwidth and their low conductor count. Inductive sensors have been described by Traxler and Maslen, (Traxler, 2009) but we will recap some of the key elements.

Inductive sensors are typically built into a sensor ring, which is fitted around a laminated target on the shaft and are wired to form a bridge arrangement in each of the 2 axes of the radial bearing. Each axis has as a minimum an opposed pair of sensing elements, where each element is a variable inductor incorporating the rotor/stator air gap, such that variations in gap result in variations in inductance. The sensor is then driven with a balanced AC drive signal which may be either sinusoidal or a square wave. Waukesha prefer sinusoidal drive due to the linear behavior of such systems. The drive frequency is selected to be as high as possible whilst still working within the “inductive” frequency range of the sensor elements. If the drive frequency is set too high then the elements are affected by eddy currents and resonance with cable capacitance.

The radial sensor in figure 1 has single wound coils giving a single ended signal relative to the center point of the drive signals, Alternative dual wound arrangements are possible which provide a differential output from the sensor ring, giving improved noise immunity. In either case, the resulting output from the sensor ring is a waveform whose amplitude is a function of the displacement of the rotor away from the center of the sensor ring and the phase of the waveform relative to the drive waveform is nominally either 0° or 180° depending on the direction of the displacement. When the rotor is centered in the sensor ring the output should be zero.

The schematic of a typical axial sensor arrangement is shown in figure 2.

Because the output of the bridge is phase sensitive, phase sensitive de-modulation of the sensor signal is required. This may be performed either using analogue electronics or digitally. Waukesha use convolution techniques as shown in Figure 3 to perform the phase sensitive demodulation. Jayawant and Dabbs (Jayawant R. a., 2013) describe a digital phase sensitive demodulation circuit.

Whichever method is chosen, the signal must be within the working range of the de-modulation circuit

Within this whole arrangement, there are several sources of noise and other non-ideal behavior, which can be seen in Figure 4:

1. Common mode noise. For the single wound sensor arrangement, each output signal is referenced relative to the mid-point of the balanced drive waveform (which is derived within the AMB controller). Consequently, any common mode noise which couples into the sensor signal may result in a measurement error. However, the de-modulation circuit will have a limited bandwidth (typically $1/10^{th}$ of the carrier frequency) and so out of band noise will not impact on the resulting measured signal. The dual wound sensors are immune to common mode noise and so are better suited to systems with long cable runs.

2. Bandwidth. The bandwidth of the sensor is limited by the bandwidth of the de-modulation circuit, which in turn is limited by the carrier frequency, since we will normally want to limit the amount of carrier frequency "breaking through" into the output waveform. As previously noted, eddy current and resonance effects limit the carrier frequency.

3. Electronic noise. For analogue de-modulation and the analogue front end of digital demodulation circuits, the circuits will have electronic noise added into the waveform at each stage. The larger the waveform we are working with, the lower will be the impact of the noise. However, we need to ensure the signal remains within the working voltage of all circuits in order to avoid clipping and the resulting non-linear behavior.

4. Quantization noise. For digital demodulation, the waveform will be sampled through an A to D converter (ADC). This will result in quantization noise. We want to have as many "counts" as possible across the working range of the sensor. This may be achieved by either having an ADC with a higher resolution or by optimizing the gains in the sensor system so that the resulting signal corresponds with the range of the ADC when the rotor is traversed across working range of the sensor. Use of a high resolution ADC may not be available for some Digital Signal Processors (DSPs) and will incur extra cost.

5. Other sources of noise can be attributed to the electro-mechanical assembly of the sensing elements and rotor system. Differences in positioning and inductance of the sensor elements can result with signal distortion and increased level of sensors noise. Additional sources of noise, identified during rotation, are attributed to uneven surface of the rotor lamination target in the rotor. These geometric defects manifest as rotor runout, produced during the machining. Operation of the rotor, or localized rotor lamination defects, light dents or scratches, produced during assembly or operation. These type of noise effects can be mitigated by the mechanical configuration of the sensor elements, increasing the number of the sensing elements, or digitally with filtering algorithms embedded in the control structure.

A typical AMB system will have an associated touchdown bearing system (TDB) which will limit the rotor motion in the event of failure of the AMB. The motion of the rotor is limited by the clearance within the TDB together with the compliance of the TDB. It is noted that the clearance of the AMB magnets and sensors is typically much larger than the TDB clearance.

An AMB system will normally be set up so that the "zero" positions are when the rotor is centered in the TDBs.

It is imperative that the sensor is able to provide a valid signal across the whole range of motion (clearance plus compliance) and a small margin is typically added to allow for thermal effects, wear in the TDB and rotor and housing deflection. For Waukesha this results in a design guideline that the sensor working range should be 1.5 to 2 times the TDB allowable movement (clearance plus compliance).

The sensor rings and the TDBs are separate parts within the AMB / machine assembly and will have non-concentricity to each other. The novel sensor circuit described in this paper mitigates the negative impact on required sensor working range of such non-concentricity.

2 The Results of Misalignment with Touch-Down Bearings

The mechanical hardware of a magnetic bearing system typically consist of assembled components, including actuator, sensor and auxiliary bearings, assembled in stacked configuration, as represented in Figure 5. Due to the manufacturing tolerances associated to each component, some residual misalignment is always present between these components. While axial misalignment can be partially corrected with shims or spacers between the stacked components, the radial misalignment is more problematic. Tight machining tolerances, manufacturing and assembly procedures can be applied to correct radial misalignment but this results with additional cost and assembly time.

For an ideal AMB supported rotor system, the two radial sensor stators will be concentric with the two radial TDB. In this case, when the rotor is centered in the TDB, it will be centered in the sensor rings. When centered, the bridge circuit formed by the inductive sensors for the given axis will be balanced and the output of the bridge will be zero as shown in Figure 6.

Also shown in Figure 6, we see that if the rotor is displaced in the +ve direction to the TDB clearance we will get an amplitude Y. If the rotor is displaced to the TDB clearance in the –ve direction, we will get the same amplitude, but out of phase by 180°. In this case we can set up the gain at the front end of the de-modulation circuit so that (nominally) the amplitude Y corresponds to the input range of the ADC or the analogue de-modulation circuit.

Remembering that the clearance in the AMB magnets and the AMB sensors may be several times that of the TDB, we now consider the case where the radial sensors stators are not concentric with the TDB stators. In Figure 7, we show the (extreme) case where the sensors are non-concentric with the TDB by twice the TDB clearance.

When centered at the TDB, the bridge circuit formed by the inductive sensors for the given axis will no longer be balanced and the output of the bridge will have amplitude 2Y as shown in Figure 7.

Also shown in Figure 7, we see that if the rotor is displaced in the +ve direction to the TDB clearance we will get an amplitude 3Y. If the rotor is displaced to the TDB clearance in the –ve direction, we will get amplitude Y with the same phase.

In order for the sensor range to be operational over the range of TDB motion, the gain at the front end of the de-modulation circuit needs to be set so that a signal of amplitude 3Y can be correctly processed. This means the sensitivity compared to the well centered case is reduced (in this case) by a factor of 3, with a corresponding increase in quantization and other electronic noise. It should also be noted that 2/3 of the sensor range is not used.

3 Overview of the Novel Sensor Input Circuit

The circuit shown in figure 8 allows the effect of the non-concentricity between the sensor stator and the TDB stator to be compensated electronically, giving a well centered signal with respect to motion within the TDB clearance. Because the signal is well centered, it is then possible to optimize the gain to fully use the input range of the phase sensitive sensor demodulation circuit.

The way the patented (UK Patent No. GB2582286, 2021) circuit works is to apply an "offset" level of a copy of the drive waveform to each incoming sensor output signal. The offset signal will either be in phase with the drive waveform or 180° out of phase with the drive waveform. The amplitude of each offset is adjusted so that the relevant signal is well centered.

The summing amplifiers which combine each sensor signal with the relevant offset, need to be of sufficient dynamic performance (bandwidth and slew rate) and input range to be able to handle the two signals without clipping, distortion or other signal degradation. Because the bandwidth of this input circuit is likely to be many times the carrier frequency, the impact on the overall sensor system bandwidth of this circuit will be insignificant. If the gains, "gain1" and "gain2" of Figure 8 are made to be variable, then the signal can be optimized for the input range of the signal processing circuit (in Figure 8 the ADC). If the gains are combined with the summing amplifiers then any concerns regarding "clipping" of the input signal can be eliminated.

The resulting (idealized) waveforms are shown in figure 11. These show the sensor output of Figure 7 after suitable compensation, together with the waveforms at the ADC input after suitable gain adjustment. In this case, we have assumed an ADC input range of +/-8Y as shown in the figure. The offset and gain have been adjusted so the sensor has been setup with 7/8 of the ADC range corresponding to the maximum TDB motion (as an example).

The circuit shown in figure 8 shows "single ended" sensors. For a dual wound differential sensor, the differential signal would be combined together with the offset to give the compensated signal.

In all of the examples so far, the sense signals have been in phase with the drive waveform. In a real world application, the sense waveforms will be phase shifted from the drive due to resonance effects from the cable capacitance in parallel with the sensor inductance.

The circuit presented here is able to compensate the in-phase component of the waveforms. In the example in Figure 12, the sense signal is shifted by 15°. In Figure 12, we can see the resulting waveforms, together with the in-phase component of the resulting signals. The gain applied to the signal again results in a maximum amplitude of 7Y, but the useful component now has a maximum amplitude of approximately 6Y i.e. 1/7 of the ADC range is not used. This is still significantly better than the case where no correction is applied where 2/3 of the ADC range is effectively unused.

In the extreme case where the phase shift is 90°, then the circuit would not be able to compensate for TDB offset, however, phase shifts of this magnitude have not been observed within the author's Company.

Algorithms and circuits with variable phase shift on the offset have been investigated, but the algorithms required for setup become overly complex and result in excessive configuration time. Practical experience shows that this capability is also not needed.

4 Deployment of the Sensor Circuit in a New AMB Controller Designed for Low-Cost Machinery

The first deployment of the novel sensor circuit was into a compact AMB controller designed for low cost machinery. Designated Alisio™, the controller includes the fully digital demodulation circuit presented by Jayawant and Dabbs (Jayawant R. a., 2013), together with the gain and offset adjustments presented here. Figure 13 shows the controller (with its enclosure removed) and Figure 15 shows the sensor card which includes the novel sensor circuit.

The sensor card also includes signal conditioning for PT100 temperature sensors and dual speed sensors.

The controller is available in two build configurations, a low power 1KVA version and a 2KVA version. The 1 KVA version has 10 magnet drives, each with a 300V DC link and maximum continuous output current of 3A. The 2KVA version has the same DC link but with 5A maximum continuous output current. In both cases the peak current is still being refined, but is expected to be 4A and 6A respectively.

The remaining major specifications related to the position sensor sub-system are shown in table 1.

The digital demodulation circuit presented by Jayawant and Dabbs (Jayawant R. a., 2013) is shown in Figure 14. When combined with the Software controlled Offset and Gain, this allows the full sensor setup to be performed remotely under the control of the Waukesha magBOOST™ automated commissioning tools.

The first application of the Alisio™ controller was into a centrifugal compressor for a chiller application.

In this case, the mechanical components for the actuator, sensor and auxiliary bearing were stacked together and the relative centering resulting from the manufacturing tolerances of the individual components and the positioning of the axial shims. For these type of AMB systems, normally intended for high volume applications, although the axial centering of the auxiliary bearing and position sensor can be adjusted with a set of mechanical shims, the correction of the centering with an electronic system provides a faster and more economical way to adjust the axial concentricity without an additional set of shims.

Table 1 - Alisio™ Specification

Function	Specification
Controller Sample rate	15 KHz (upgrade to 20KHz planned)
Sensor sample rate	210KHz
Sensor carrier frequency	26.25KHz
Input Adjustment functions	Software controlled offset and gain using digital potentiometers.
Phase sensitive demodulation	Fully digital with fully adjustable phase
Dimensions (3A version)	292 x 239 x 86.5 mm
Cooling (3A version)	Cold plate mounted
Maximum Cold plate temperature (3A)	60 °C

5 Techniques for Automatic Configuration of the Sensor Input Circuit

The sensor input circuitry is set up by moving the rotor within the TDB clearance (and compliance), providing measurement points for varying position. So that valid measurements are obtained, it is important to ensure that the input signal does not saturate whilst still maintaining as large a signal as possible. In order to do this, the phase sensitive de-modulation circuit is used to evaluate the amplitude of the signal at each measuring point. The phase is swept through 360° and the peak response at each measurement point will give the magnitude and phase of the relevant measurement. If the response at any of the measurement points is larger than the permitted level then the gain of the input circuit is reduced until a valid measurement can be obtained at all measurement locations.

There are various techniques for moving the rotor within the TDB clearance:

1. Single axis open loop clearance check

With this technique, open loop current is driven in the magnets of a given bearing axis to pull the rotor around within the TDB clearance. This technique can be used for the initial configuration of the sensor system. This is shown schematically in figure 9.

2. Partially Levitated Clearance check

This technique uses levitation on all axes except one, with the remaining axis pulled within the TDB clearance using open loop current. This is shown in Figure 10. This technique will normally be iterated twice across the axes to obtain the optimal settings.

Now that we are able to move the rotor within the TDB clearance, the sensor configuration techniques follow.

For each axis, we start with a low gain setting and zero offset. Pulling the rotor will yield two measurement points P1 and P2. These are shown in the argand diagram of Figure 16. We then apply the offset to bring the vector of the mid-point P3 so that it is orthogonal to the vector between the 2 measurement points (also shown in figure 16). Resulting in points P1', P2' and P3'. We note that the magnitude of P1' and P2' are now equal. Once we have the correct offset applied, we can then adjust the gain to bring the amplitude of the signal at the two extreme positions (P1' and P2') to the desired value. This method is only applicable to small angles β. At angles above 30°, the length of the vectors P1' and P2' will start to diverge and at angles above 60° we may see vectors with magnitude greater than the original vectors.

The techniques described above, including:

a. Using the phase sensitive de-modulation to evaluate the amplitude and phase at each measurement point;
b. Adjustment of gain to ensure all measurements are valid;
c. Single axis open loop clearance check;
d. Partially levitated clearance check;
e. Computation of offset;
f. Computation of gain;

are all fully automated within the Waukesha magBOOST™ automated commissioning tools (described by Jayawant and Daves) (Jayawant R. a., 2013) allowing for rapid execution of this relatively complex procedure.

6 Results from prototype testing

Prototype testing with 3 machines was conducted. These were commissioned remotely (including the initial sensor setup) which was conducted using the Waukesha magBOOST™ automated commissioning tools. In particular, use was made of the single axis open loop clearance check function. Each axis had 2 measurement points, one with current in the "top" magnet flowing and one with current in the "bottom" magnet.

Table 2 - Analysis of Axial Signals Before and After Application of Offset.

Original Measurement				offset (Count)
Bot Magnet pull		Top Magnet pull		
Mag (count)	phase (°)	Mag (count)	phase (°)	
257562	180	132373	191	188474
246595	180	102635	180	174615
272130	191	100968	180	200920

Final Configuration				Increased gain which can be applied
Bot Magnet pull		Top Magnet pull		
Mag (count)	phase (°)	Mag (count)	phase (°)	
69088	180	64079	-24	3.73
71980	180	71980	0	3.43
84687	-141	99952	0	2.72

The currents were set so as to be able to lift the rotor. However, when pulling with the top magnet in a radial axis, it may not be possible to fully displace the rotor in the direction of the relevant axis due to gravity and the resulting force vector at the contact point within the TDB. This effect is not relevant to axial bearings. For the initial set-up this effect is considered not significant and would be eliminated in the subsequent partially levitated clearance check. Unfortunately, at the time of writing this paper, the results from the partially levitated clearance checks were not available.

It is noted that in these machines the controller is integrated with each machine and consequently very low cable capacitance results. Figure 17 shows the useful vector for the various axes and

Figure 18 shows the ratio of the useful vector and the actual offset vector for each of the axes, i.e. the ratio of the amplitudes and the relative phase angle between the two vectors.

We can see that all of the useful vectors lie close to the x-axis. We can also see that the offset on the axial axis in all three cases has greater magnitude than the useful signal, but is close to -180° and so is eminently correctable using the new input circuit.

We will look at what happens when we apply the optimal correction using the technique of Figure 16 to the measured axial signals.

Table 2 shows the original measured signals together with the optimal correction and the resulting signals. The table also shows the increased level of gain that can be applied prior to the ADC (compared to the case where the offset is not present).

7 Conclusions and Future Work

The use of the patented offset circuit allows for relaxed tolerancing in the mechanical build, leading to faster build times and reduced costs. The normal down side to this relaxed tolerancing (sensor offset) is compensated by the new circuit, allowing larger gain to be applied to inductive sensor signals prior to processing by demodulation circuits (either analogue or digital). This increased gain allows similar performance (with regard to quantization and other electronic noise present within the sensor demodulation circuits) compared to a tightly tolerance system.

Tests on prototype systems allowed increased gain of between 2.7 and 3.5 to be applied to the axial sensor signals prior to A to D conversion compared to the case where the offset circuit was not present.

The use of software controlled gain and offset adjustments when combined with the fully digital phase sensitive de-modulation circuit supports the fully remote commissioning model, particularly when supported by sophisticated automated commissioning tools such as the Waukesha magBOOST™ software.

Further work (ongoing and planned) includes:

- Systematic measurement of noise levels with and without the offset deployed.
- Incorporation of the offset circuit into the Zephyr™ AMB controller. This is in progress and expected to be available in Q3 2021.
- Evolution of algorithms for computation of offset when the useful vector has a large phase angle.
- Testing with longer cables where phase shifts may be more significant. This is waiting on completion of the work on the Zephyr™ controller.

References

Jayawant, R. and Dabbs, S., (2013). A Novel Digital Circuit for Position Sensor De-modulation Using Advanced FPGA, Including Field Experience with a 1.45MW Motor Driven Compressor. *Proceedings of the 1st Brazilian Workshop on Magnetic Bearings, Rio de Janeiro, 2013.* Rio de Janeiro: Available at http://www.magneticbearings.org/publications/.

Jayawant, R. and Davies, N., (2013). Integration of signal processing capability in an AMB controller to support remote and automated commissioning. *Proceedings of the 1st Brazilian Workshop on Magnetic Bearings.* Rio de Janeiro: Available at http://www.magneticbearings.org/publications/.

Leung, C. and Davies, N.,. (2021). *UK Patent No. GB2582286.*

Traxler, A. and Maslen, E.,. (2009). Hardware Components. In Schweitzer, G. and Maslen, E., *Magnetic Bearings* (pp. 101-102). Zurich: Springer.

Figures

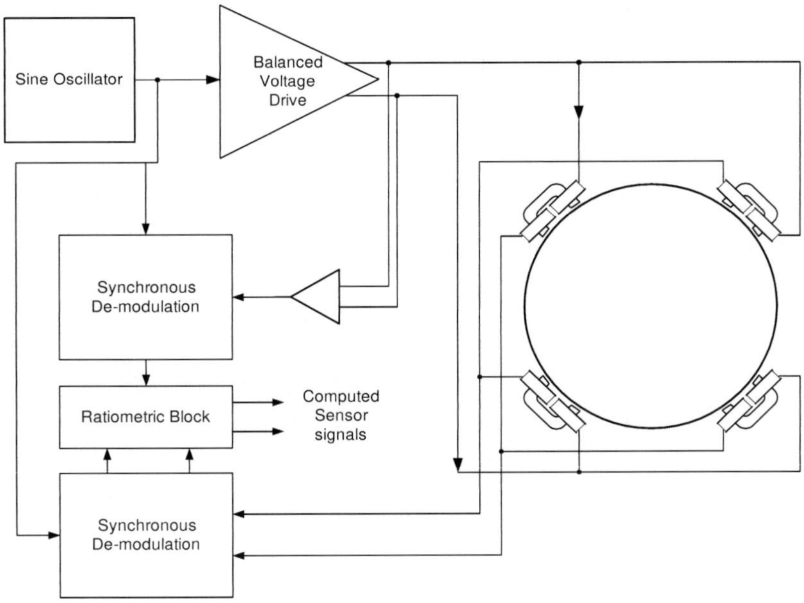

Figure 1 - Radial Inductive Sensor Schematic

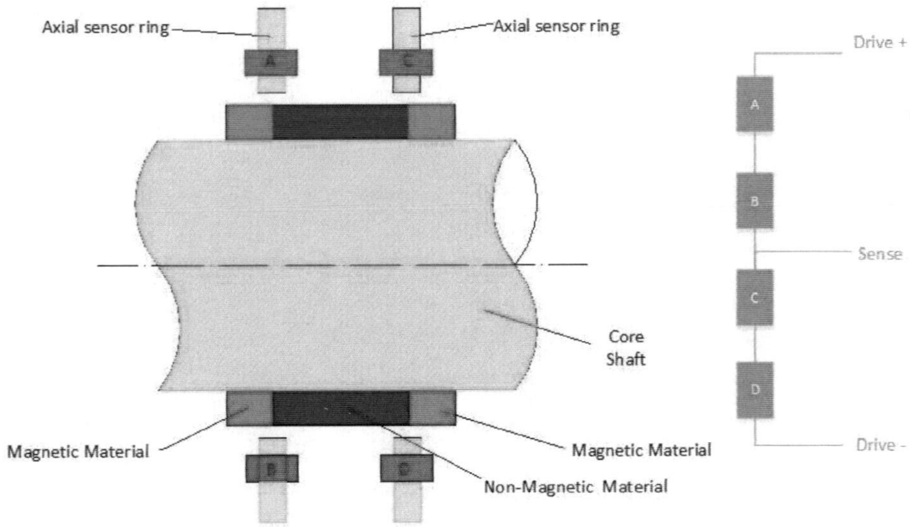

Figure 2 - Axial Inductive Sensor Schematic

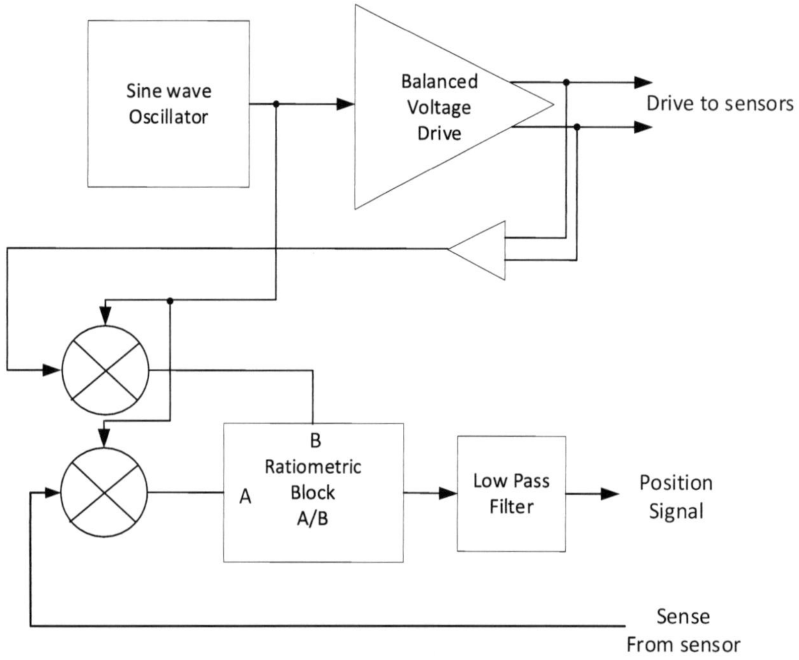

Figure 3 - Analogue Phase Sensitive Demodulation.

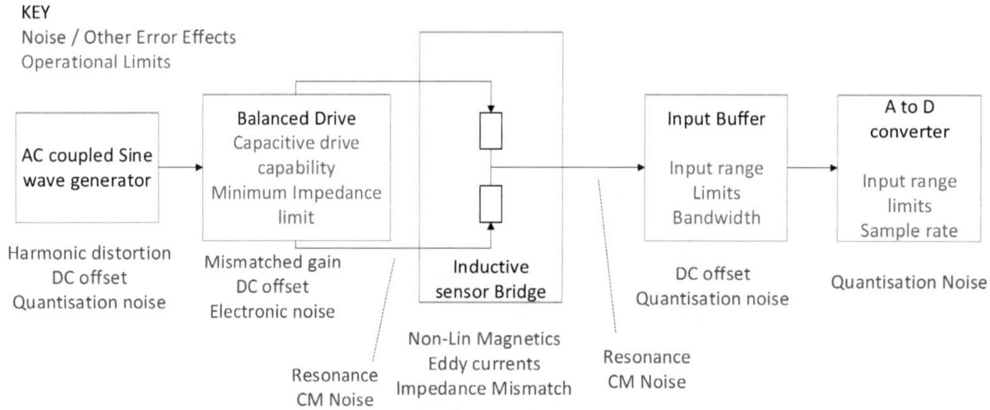

Figure 4 - Constraints within the Signal Processing Circuit and Sources of Noise.

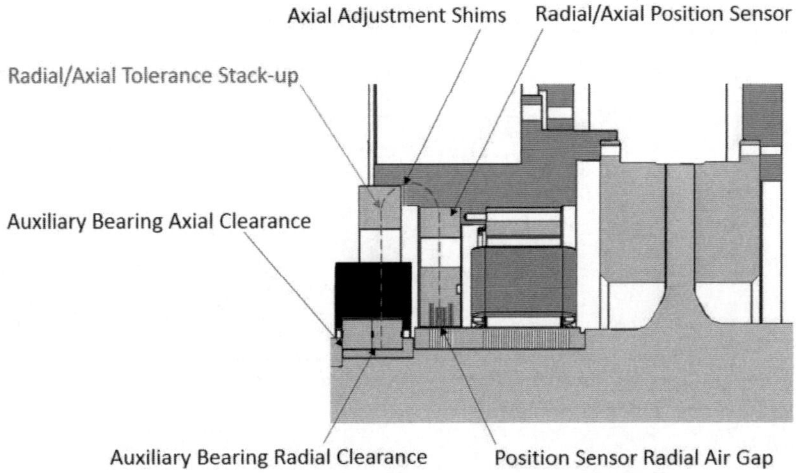

Figure 5 - Non-Concentricity Due to Non-Concentricity in the Housing or AMB Components.

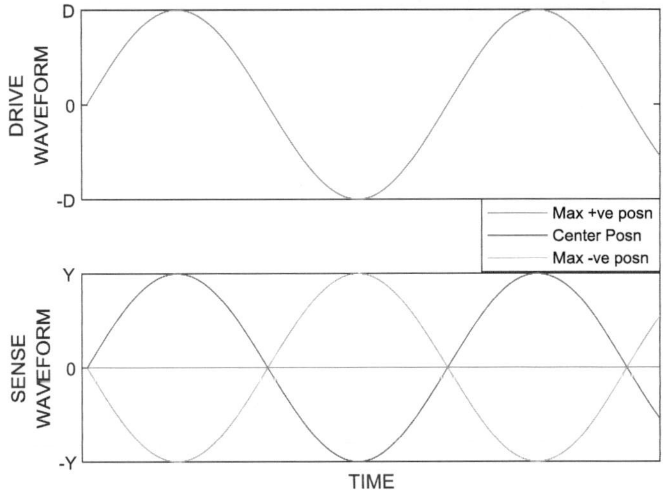

Figure 6 - Ideal Signals from a Well-Centered Sensor Ring.

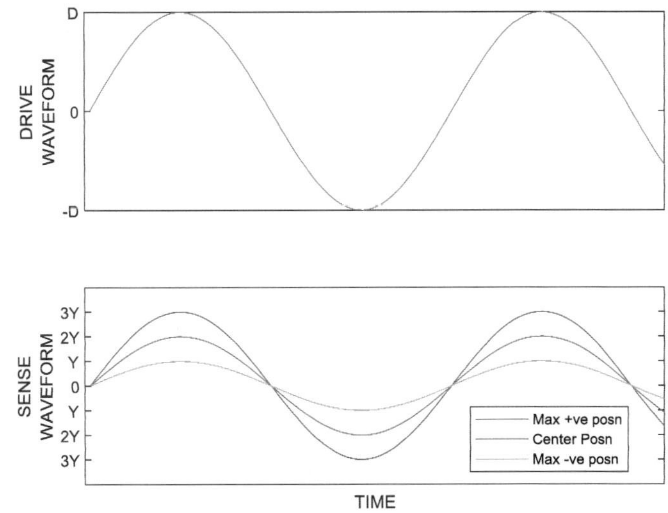

Figure 7 - Signals from a Non-Concentric Sensor Ring.

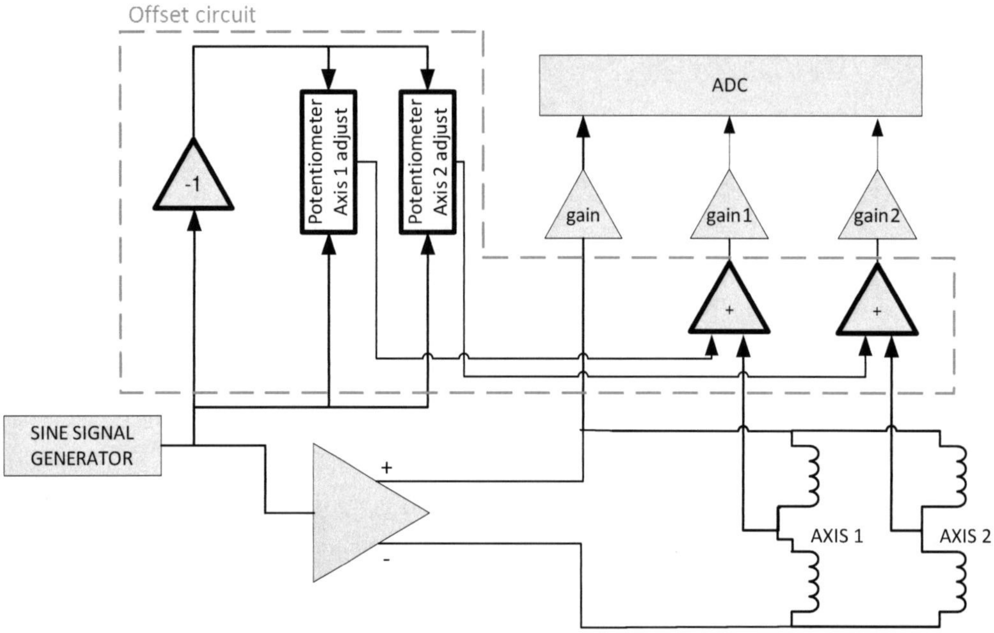

Figure 8 - . Schematic Diagram of the Sensor Input Circuit.

Figure 9 - Single Axis Open Loop Clearance Check.

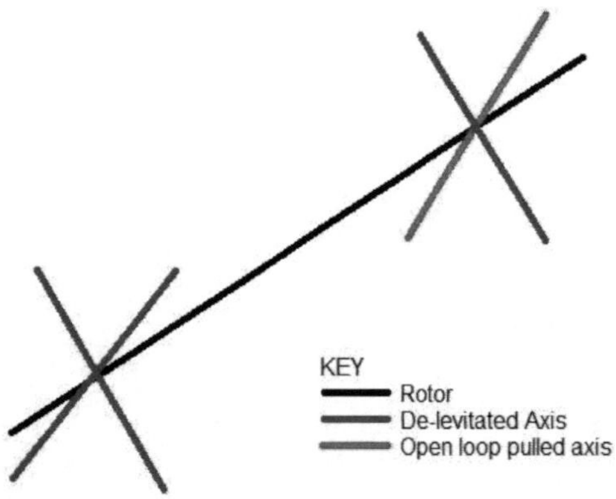

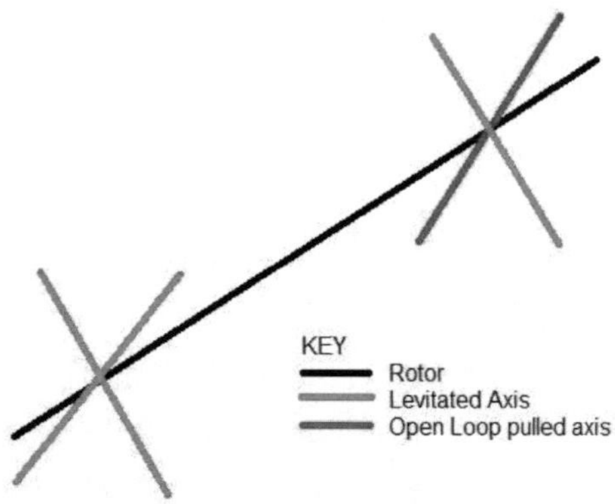

Figure 10-Partially Levitated Clearance Check.

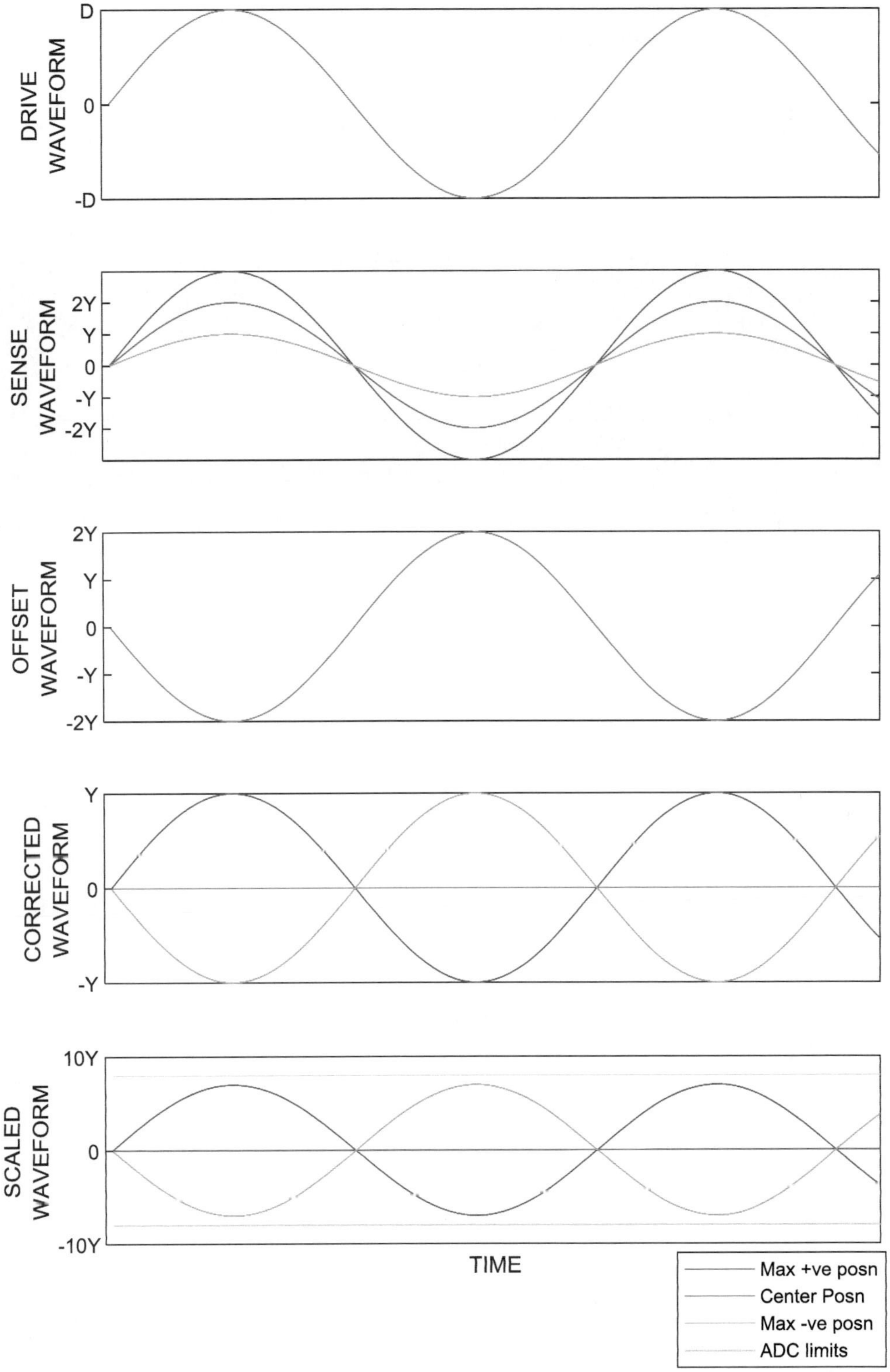

Figure 11 - Sensor Waveforms as Modified by the Sensor Input Circuit.

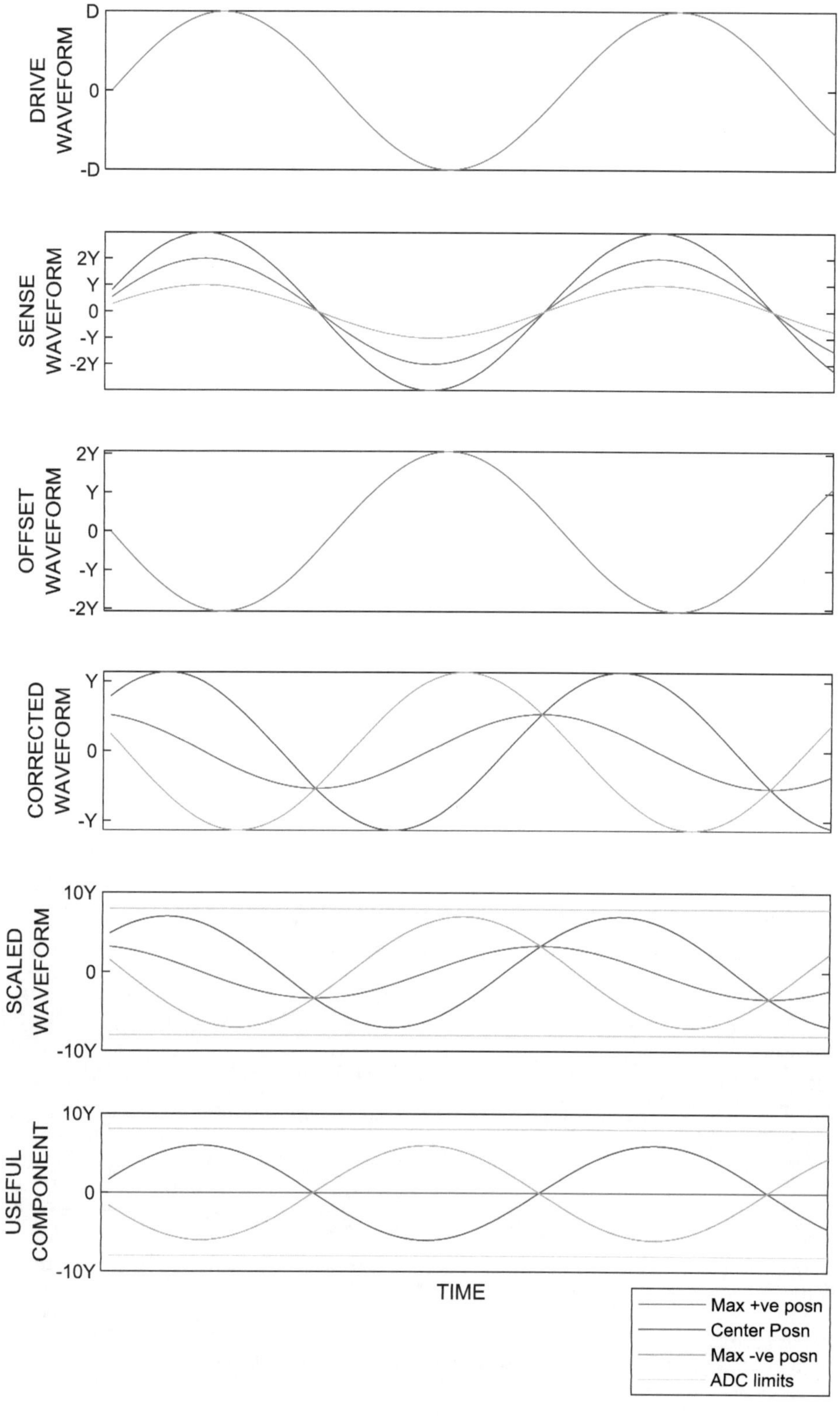

Figure 12 - Sensor Waveforms when the Sense Signal Is Out of Phase with the Drive Signals.

Figure 13 - Alisio™ Controller (enclosure removed).

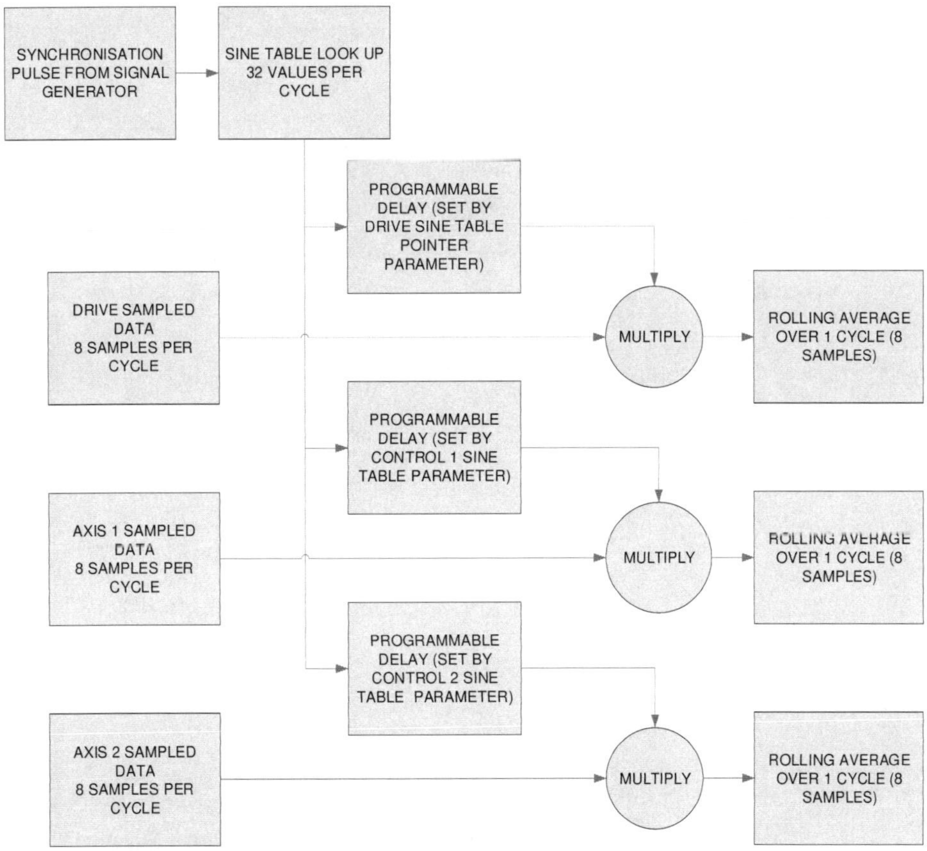

Figure 14 - Digital Phase Sensitive Demodulation.

Figure 15 - Alisio™ Sensor Card.

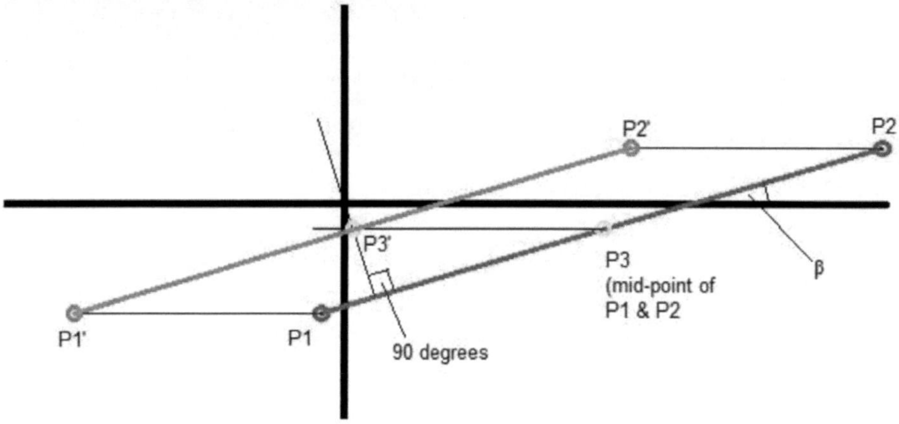

Figure 16 - Argand Diagram Showing Calculation of Offset.

Figure 17 - Argand Diagram Showing Useful Vector for the Various Axes

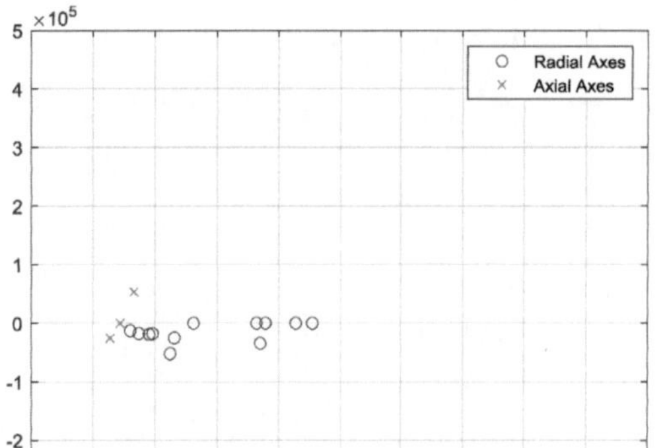

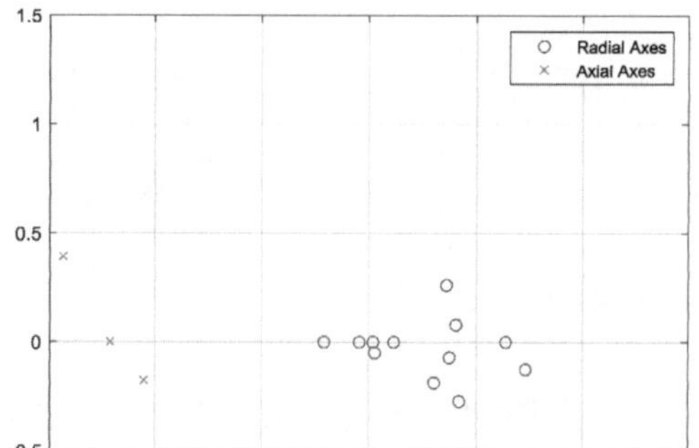

Figure 18 - Argand Diagram Showing Relative Magnitude and Phase of Actual Offset Vector and Useful Vector (Offset/Useful)

Design and Comparison of Dual-Purpose Stator Windings for Active Chatter Suppression in Milling Spindles

Dennis Guhl, Robin Liebfried, and Wilfried Hofmann

Technische Universität Dresden, Chair of Electrical Machines and Drives, Dresden, Germany
dennis.guhl@tu-dresden.de,

Abstract

The effect of chatter is one of the main obstacles to increasing metal removal rates in modern high-speed cutting (HSC). Active damping approaches using electromagnetic actuators have gained popularity in recent years but none of them found significant industrial usage. One of the major drawbacks of the existing solutions is the missing geometrical and electrical integration which inevitably leads to a lengthening of the milling spindle inherently with lower rigidity values or a loss of power. In this paper, we aim for an approach without additional actuators. By using the concept of a bearingless motor to apply the damping forces onto the rotor, we achieve a fully geometrical and electrical integrated design. In addition, the implementation of a so-called dual-purpose or combined winding allows us to freely shift the total spindle power between torque and force generation which can lead to better-dimensioned spindle motors. The main premise is to keep high industrial proximity so that no geometrical changes are to be made to the motor, but the force generation is achieved solely by rewinding the stator. The chatter damping function shall always just be an extension of the drive function so that the new machine is never worse than the original one. After a pre-study, the two most promising dual-purpose windings—dual-purpose no-voltage and multiphase—are compared against the conventional separated windings design. The main evaluation criteria are a torque characteristic equivalent to the original motor and high radial forces with a low ripple for all rotor positions. The simulations are carried out as 2D finite element simulations at an uncontrolled machine. Many references between the separated and dual-purpose winding designs can be drawn. A multiphase approach with twelve phases in total works best with minimal changes to the original winding, an identical torque, and excellent force characteristics and leads to the commission of a prototype.

Frequently used variables

F_r	radial force	y_c	coil span
$\hat{i}_k$	amplitude of current system $\boldsymbol{i}_k$	z_s	number of turns per slot
k	phase-shift parameter	α	mech. angle along the motor circumference
m	number of phases		
N	number of slots	Θ_k	MMF caused by $\boldsymbol{i}_k$
n_i	winding distribution of phase i	ν	harmonic order of the winding distribution
$\hat{n}_\nu$	ν-th order Fourier coefficient of the winding distribution	φ_{ik}	phase-angle of current system $\boldsymbol{i}_k$
p, p_s	pole pairs in the torque, suspension winding	$\varphi_{n\nu}$	phase-angle of the ν-th order winding distribution harmonic
q	number of slots per phase	φ_r	mech. rotor revolution angle
t	time	ω	electrical angular frequency

1 Introduction

In today's high-speed milling, process instabilities are one of the main causes of productivity limitations [1]. In processes with high metal removal rates, chatter effects form the basis of these instabilities. Self-excited vibrations, called chatter, deteriorate the surface quality of the workpiece and lead to a reduction of the machine tool's life, especially the cutting tool [1]. Chatter vibrations belong to the self-excited vibrations. As with forced oscillations, an external energy source is necessary, but the process itself determines the rate of energy supply. The oscillation frequency follows from the dynamic properties of the process itself and cannot simply be influenced by the exciting force as in the case of forced oscillations.

The primary reason for chatter is the regenerative effect, which results from an interaction between the tooth engagement frequency and the vibration frequency of the cutting tool itself [2]. The chip thickness varies at each tooth engagement, depending on the current cutting tool position and on its position one revolution (single edge tools) or one engagement (multi-edge tools) before. The varying chip thickness, assuming a constant chip width, is accompanied by a varying thrust, which leads to vibrations of the machine tools and results in a closed-loop effect. At too high feed rates and/or cutting depths the process becomes unstable. Chatter marks are visible on the workpiece's surface like shown in Fig. 1. In most applications such a surface cannot be used and needs reworking.

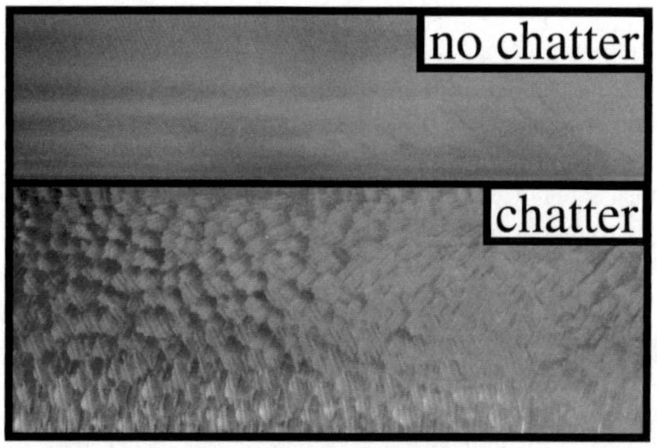

Figure 1: Chatter marks

Since the causes of chatter were discovered in the late 1950s, numerous methods have been investigated to suppress chatter. One of them are active damping techniques. These techniques are characterized by the measurement of one or more quantities characterizing the chatter vibration, the subsequent signal processing, the generation of control signals, and the actual force effect of the actuator on the process in a closed-loop control scheme. One can distinguish between series or parallel applications, depending on whether or not the actuator is subject to cutting forces [1]. There exist plenty solution approaches regarding an active structural chatter suppression [3]–[5], active tools (mostly focused on boring bars [6]) or workpiece holders [7].

This research focuses on actively damped spindle systems. Even in this subfield, two main groups of solutions can be distinguished. While some focus on the usage of active magnetic bearings [8], [9], others attempt to apply forces on the spindle axis at numerous different positions by other means. In many of the latter approaches piezoelectric actuators are used, for example, to apply forces directly on conventional contact bearings [10] or to manipulate the bearing's position [11].

In recent years electromagnetic actuators attracted attention. Abele et al. used an actuator next to the front bearing close to the milling tool [12]. Klaffert positioned the actuator in between the front bearing and the motor [13], while Bickel went for a split motor design [14]. Especially about the positioning of the actuator, far-reaching investigations were carried out. Positions between both mechanical bearings turned out to work best. But still, the volume of the additional actuator inevitably lengthens the milling spindle and therefore reduces the rigidity values. Alternatively, a corresponding reduction in the motor volume leads to a power and torque loss.

For this reason, Königsberg et al. developed a motor integrated actuator design [15]. Two main challenges need to be faced. Firstly, the integration leads to two harmonic spectrums—one for the motor, one for the actuator stator winding—which need to be decoupled. Secondly, you need to model the achievable damping force as a function of the rotor position. Regardless of the significantly increased motor power with the same volume of the spindle, a force acting over the entire active length of the motor, and a reduced manufacturing effort compared to the other solutions presented above, this design still has one serious disadvantage: It is not possible to freely shift the motor and damping power back and forth. In a first publication 40 % of the total available copper cross-sectional area is used for the damping winding, 60 % for the motor winding [15] while in the latest publication (December 2020) the research group postulates a 25/75 ratio [16]. This approach reduces the torque also in operating points where no damping is required and more torque might be desirable.

The ability to freely shift the total available electric power between torque and damping force generation depending on the need at the specific operating point would be a clear advantage. Additionally, current spindle motors normally feature a double-layer design. This allows short-pitching of the coils to reduce harmonics and therefore the torque ripple. The proposed integrated actuator design just allows a single-layer motor winding with no possibility of short-pitching.

To achieve this goal, we take a different approach. The concept of a bearingless motor—here used for damping, not for suspension—can combine the advantages like force application over the entire active length of the motor and easy manufacturing of a fully integrated actuator as in [15] with the advantage of a free power shift. The idea to use bearingless motors for vibration suppression was already proposed in 2008 by Chiba et al. [17], but this method does not seem to have been applied in connection with industrial milling spindles yet. Especially, the free power shift represents a clear advantage over all other methods mentioned above.

Structure Section 2 presents the investigation setup including the evaluation criteria, the investigated winding designs, and details about the original spindle motor. A pre-study in section 3 of two different separated winding designs forms the basis of the investigation of two different dual-purpose winding approaches in sections 4 and 5. The paper concludes with a final comparison and an outlook in section 6.

2 Investigation Setup

The main premise is to achieve high industrial proximity. Due to the conservative design approaches in mechanical engineering, no geometric changes are to be made to the motor. For easy replacement, the damping functionality is to be achieved exclusively by rewinding the stator. This goes hand in hand with many advantages. The manufacturing effort stays low and under consideration of a constant maximum current density in the stator winding, the electromagnetic design should not be a considerable challenge.

2.1 Evaluation Criteria

To follow the main premise of achieving high industrial proximity the torque curve in an exclusively motor operating mode should vary as little as possible from the original torque curve or at least should be equivalent in mean value and ripple. This implies that the new machine will never be worse than the original machine in no-damping operation. The ripple of the load torque due to the cutting edges is normally significantly higher than the machine's cogging torque.

The force's mean value resp. ripple shall be as high resp. low as possible at an operation point where the nominal current is divided equally between torque and force generation. A small force ripple ensures a uniform damping behavior over all rotor positions and simplifies a subsequent force control if it is not necessary to take a dependency of the rotor angle into account. The chosen operating point corresponds to a milling process of S275JR steel using a 16 mm four flute end mill with a feed per tooth and revolution of 0.1 mm, a radial depth of cut of 16 mm and an axial depth of cut of 5 mm [18]. The investigations are deliberately carried out on an uncontrolled machine, as it should already exhibit advantageous behavior by itself. Considerations regarding the control system will be the subject of later publications.

2.2 Investigated Winding Designs

To allow a free power shift between torque and damping force generation, winding concepts known as dual-purpose windings are necessary. These windings combine the conventional separated motor and damping windings into one. There exist several different design solutions such as [19]–[23]. Due to the premise of not changing the geometry of the original motor, but rather achieving the damping operation solely by rewinding the stator, some solutions were ruled out. Others were tested only in very distant operating areas and/or power ranges, so that we finally chose a "dual-purpose no-voltage" winding (section 4) and a multiphase one (section 5) as they seemed the most promising.

Additionally the conventional approach of a separated single-layer suspension winding with $p_s = p \pm 1$ is investigated as a reference and presented first. Many parallels can be drawn from it to the dual-purpose windings.

2.3 Motor

The investigated spindle motor is a highly utilised four-pole permanent magnet synchronous motor with surface magnets. The stator has a three phase double-layer 7/9 short pitch winding with four parallel branches per phase. Table 1 provides an overview of some of the motor specifications.

Table 1: Motor specifications

symbol	quantity	value
U_N	nominal voltage	308 V
I_N	nominal current	55 A
P_N	nominal power	25 kW
n_N	nominal speed	10000 min^{-1}
n_{max}	maximum speed	20000 min^{-1}
p	pole pairs	2

All simulations are carried out as 2D-FE simulations in ANSYS Maxwell. We consider saturation inside the electric steel M270-35A but neglect hysteresis and eddy current losses. We assume a linear magnet model with nominal remanence and coercivity values at 20 °C.

3 Separated Windings

The basic functional principle of the bearingless motor (here used for damping, not bearing) is based on the superposition of two magnetic fields with different pole pair numbers in the air gap of the motor, so that a defined torque and force are created via the Maxwell stress tensor. In bearingless PMSM (BLPMSM) the desired superposition for torque/force is between the rotor magnetic field and the respective stator one. Conventional bearingless motors contain two separate winding systems which pole pair numbers usually differ by one [24]

$$p_s = p \pm 1. \tag{1}$$

The original motor has two pole pairs $p = 2$. So the pole pair of the suspension winding can be $p_s = 1$ or $p_s = 3$, respectively.

Fig. 2 shows the simulated torque and force characteristics over one full rotor revolution for $p_s = 1$ and $p_s = 3$. As mentioned above, the original nominal current is divided equally between the two windings so

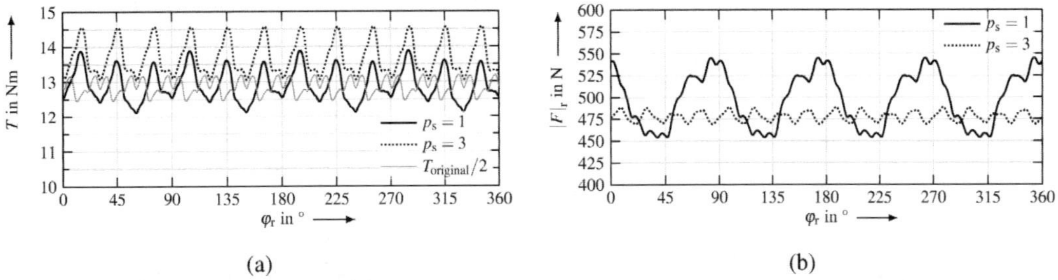

Figure 2: Simulated (a) torque and (b) force characteristics of the conventional BLPMSM with separated windings

that the maximum current density remains the same. You can see that the torque's mean value has, as expected, nearly halved compared to the original motor, whereby the torque for $p_s = 1$ is lower due to saturation effects. The ripple increases significantly because of the missing short-pitching as the double layer winding turned into a single-layer one to create space for the force winding. The influences of the slot harmonics are visible in both torque and force. The mean value of the force magnitude is slightly higher for $p_s = 1$ (494 N) than for $p_s = 3$ (478 N), although the force ripple is disproportionately higher in the first case (18.5 % resp. 4 %). The reason for this is the high saturation of the original machine. Using a six-pole damping winding, the air gap fields of the PM and the damping winding are more evenly superposed than with a two-pole configuration, so that the electrical steel in the rotor and stator is less saturated. The dominant fourth-order force harmonic ($p_s = 1$) can therefore be traced back to the four-pole rotor magnetic field, whereas the dominant 16th-order harmonic ($p_s = 3$) is based on the superposition of the harmonics of the PM and damping winding fields. The significant beat in the torque characteristic for $p_s = 1$ can also be traced back to saturation effects.

Even though $p_s = 3$ seams more suited, both designs shown above shall be used as a reference for the approaches presented next. Since you need to decide on the winding distribution before manufacturing the motor, it does not suit the required free power shift, which is the major drawback of the separated windings design.

4 Dual-purpose no-voltage winding

Khoo proposed a bearingless motor design with a combined torque and suspension winding by the use of bridge connections in [25]. This design requires only one three-phase inverter for torque generation and one isolated H-bridge for each suspension phase. The bridge connections ensure that the suspension terminals do not see any back-EMF and corresponding inverters can be dimensioned substantially smaller or have increased control dynamics compared to other combined bearingless motor designs [22]. There is also the possibility to short-circuit the suspension terminal coils to get fully passive damping.

In [26] Chiba et. al. developed a bearingless motor design with a combined torque and suspension winding for vibration suppression. Due to the middle-point-current injection, there are only two three-phase inverters necessary, one for torque and one for suspension. This idea was refined by Oishi et. al. in [27] so that the suspension terminals experience no back-EMF. Hence, both winding designs, the one by Khoo and the one by Oishi et. al., have the same two main characteristics: a combined winding for torque and force generation as well as suspension terminal connections which see no back-EMF.

This led Severson et. al. [28] to introduce the term "dual-purpose no-voltage" (DPNV) windings which sums up these two main characteristics. Followed by results of the practical implementation of such a winding design in an ac homopolar motor in [29], Severson et. al. generalized the design theory for DPNV windings in [23] and proposed practical design guidelines.

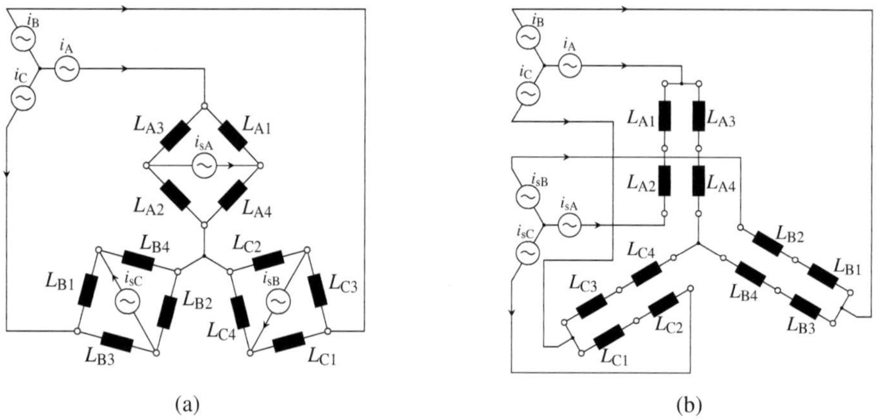

Figure 3: Bridge (a) and parallel (b) design of a DPNV winding (based on [23])

4.1 Design process

Both types of DPNV windings, the one with bridge connections [25] and the one with parallel connections [27] are depicted in Fig. 3. A comparison regarding the practical implementation of these two types is given in [29].

The parallel configuration was chosen as both three-phase inverters can be fed by isolated DC-buses. On the one hand, it counteracts any concerns regarding circulating currents, allows the conventional space vector modulation, and reduces the necessary amount of current sensors to four [29], on the other hand, it enhances the possibility to investigate the influence of the suspension inverter dynamics on the control scheme. All in all twelve switches and two isolated DC-buses are required, compared to 18 switches and four DC-buses in the bridge configuration which lead to lower hardware costs. Advantages of the bridge configuration regarding the fault redundancy were not considered in this evaluation as the motor to be manufactured will be a prototype.

The design process of a DPNV winding, starting from a conventional motor winding, is extensively covered in [23]. The following section only presents the main design steps for general understanding. For deeper insights into the design process, please refer to the denoted literature.

According to [23] there are restrictions to the relations between the number of stator slots N, the motor pole pair p and the suspension pole pair p_s. These restrictions lead to three conditions that must be fulfilled to make the conversion of a conventional winding into a DPNV winding possible.

The *first condition* checks for the symmetry of the torque winding. The number of phases m must be relatively prime to the pole pair of the torque winding as well as to the pole pair of the suspension winding:

$$p \perp m \quad \wedge \quad p_s \perp m. \tag{2}$$

The standard choice of $m = 3$ and the specification $p = 2$ allows only $p_s = 1$ for the suspension winding (see eqs. (1) and (2)).

The *second condition* ensures the fulfillment of the no-voltage requirement. For the given original stator winding with $N = 36$ slots the no-voltage requirement is already inherently fulfilled, so that we must only check for a sufficient amount of coils, which is obviously the case.

$$\frac{N}{4m} = 3 \in \mathbb{N}. \tag{3}$$

The *third condition* checks for a non-zero suspension winding factor. According to [23] the winding here represents a "type 1" winding and therefore one must check if the pole pair number of the motor winding is even

$$p = 2 \in \mathbb{N}_{\text{even}}. \tag{4}$$

When fulfilling all three conditions, what we do, it is possible to redesign the original motor winding into a DPNV one. The starting point is the star of slots of the motor winding, depicted in Fig. 4a. Therefrom, the connection star of slots is drawn as one-half of opposite voltage phasors have to be turned by 180° so that the winding becomes symmetrical and the distribution factor becomes non-zero (see Fig. 4b).

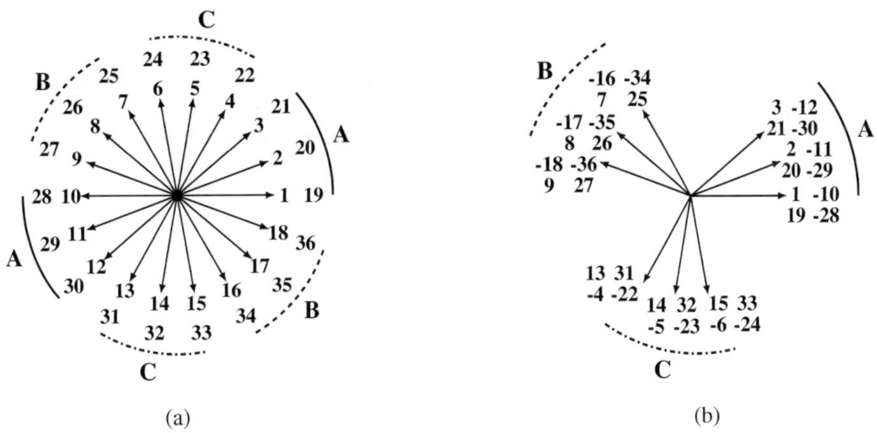

Figure 4: Star of slots (a) and connection star of slots (b) of the torque winding

For self-compensation of the back-EMF at the suspension terminals ("no-voltage"), half of the coils need to be reversed. To determine the correct coils one redraws the connection star of slots for each phase at the suspension frequency (see Fig. 5). For a "type 1" winding any 180° band would work (grey semi-circle). But not all combinations result in a symmetrical connection star of slots of the suspension winding [23]. Using the star of slots in Fig. 6a one can project the position of the connection phasors. The resulting connection star is shown in Fig. 6b. Compared to the one of the motor winding in Fig. 4b phases B and C are interchanged, which has to be taken into account when specifying the suspension currents.

To finalize the design process the coils have to be assigned to the coil groups 1-4, see Fig. 3b. The corresponding slots of the lower layer can be obtained from the chosen coil span. The following two

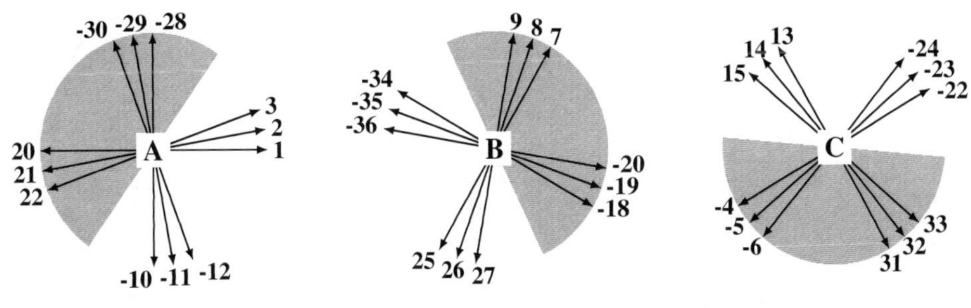

Figure 5: Connection stars of slots of the motor winding at suspension frequency (based on [23])

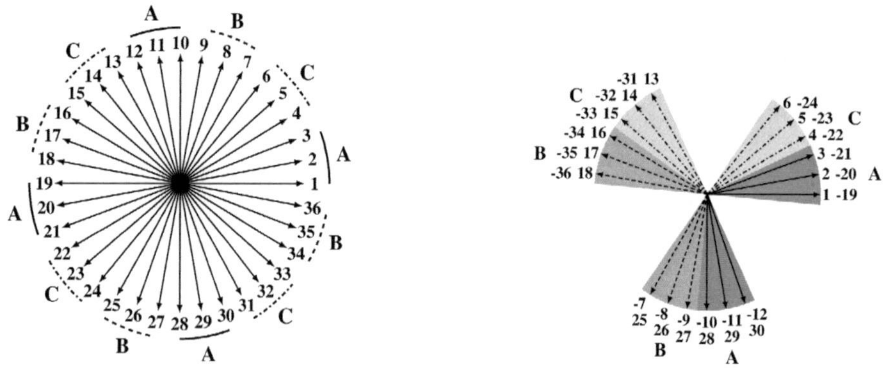

(a) (b)

Figure 6: Star of slots (a) and connection star of slots (b) of the suspension winding

paragraphs will compare a full pitch and a 7/9-short pitch configuration with each other as well as with the standard BLPMSM with $p_s = 1$ from section 3.

4.2 Simulation results

The simulated torque and force characteristics of the two DPNV designs over one full rotor revolution can be found in Fig. 7. The original nominal current is again divided equally between the torque and force generation (see section 3). One can see that the torque characteristics of the short pitch DPNV and the original motor are almost identical. This is expected as just the coil arrangement within a phase has changed but nothing else. The little deviations in uncontrolled operation result from the superposed suspension field. In an exclusive torque operation mode, the torque stays the same. The torque characteristic of the full pitch winding corresponds well to the one with separated windings (see Fig. 2). Due to the full pitch the torque mean value is higher compared to the short pitch design (13.5 Nm to 12.8 Nm) but also the ripple is significantly increased (12.4 % to 6.2 %). The characteristic beat is also clearly visible, although it is much less pronounced for the short pitched winding.

The force characteristics are very similar to the separated winding design with $p_s = 1$ as well, but the mean values have nearly halved for the full pitch winding (267 N) and have even more than halved for the short pitch one (217 N). The absolute force ripples stayed almost the same but the relative ones doubled due to the halved mean value (35.2 % resp. 45 %). An explanation therefore can be found in Fig. 8. As the suspension current distribution in the subfigure's upper parts shows, half of the suspension current is wasted in terms of MMF production to turn an original four-pole field into a two-pole field (see

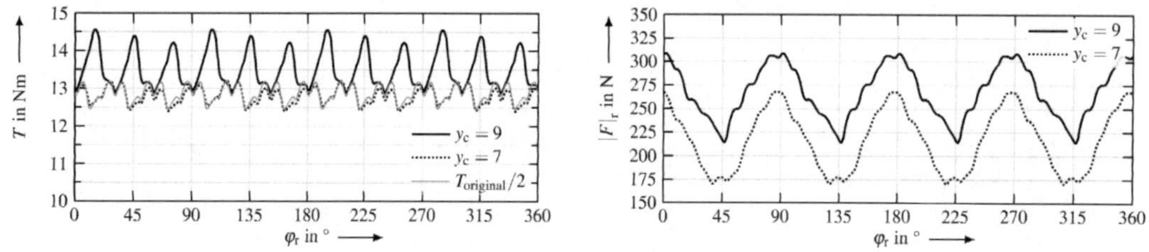

(a) (b)

Figure 7: Simulated force (a) and torque (b) characteristics of the full and short pitch DPNV bearingless motor over one rotor revolution

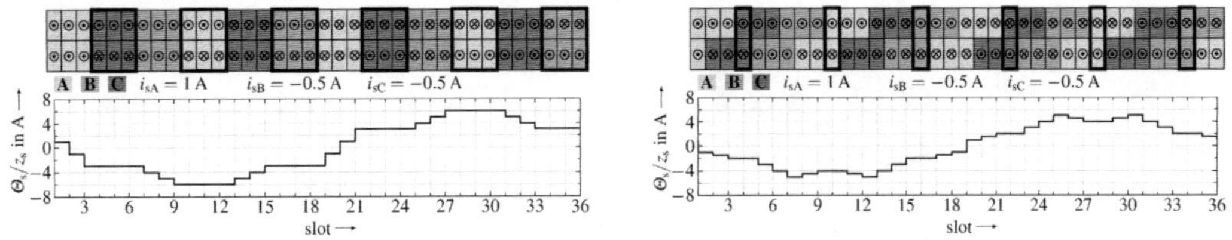

(a) $y_c = 9$ (b) $y_c = 7$

Figure 8: 1-dimensional suspension MMF of the full (a) and short pitch (b) DPNV bearingless motor

black rectangles). The suspension MMF characteristic looks even worse for the short pitch design and explains the lower mean value.

Investigations for different suspension current space vectors (rotated stepwise by 22.5°) in Fig. 9 show a similar behavior for all simulated space vector orientations. The black curves correspond to the ones in Fig. 7 and can be treated as a worst-case. Because of the low pole pair number of the suspension air gap field, the superposition with the PM field is more unevenly and results in higher force ripples (see section 3).

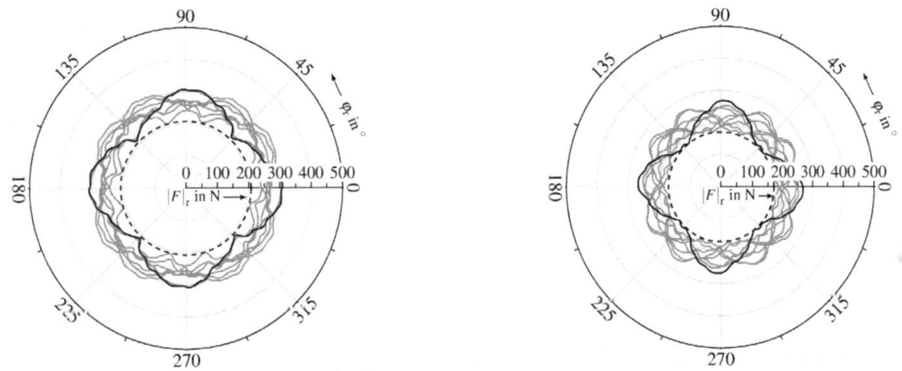

(a) $y_c = 9$ (b) $y_c = 7$

Figure 9: Simulated force and torque characteristics of the full (a) and short pitch (b) DPNV bearingless motor for different suspension current space vectors over one rotor revolution

All in all, the DPNV winding approach does not seem to suit the investigated highly utilized spindle motor if one is interested in high suspension/damping forces as half of the suspension current is just used to create a two-pole suspension field and is therefore wasted MMF wise. The suspension pole pair is fixed to $p_s = p - 1$, which leads to a worse behavior for high speeds in general [30]. This unsuitability may also count for HSC spindle motors in general as, due to the high rotational speeds, the most common numbers of pole pairs are $p = 2$ resp. $p = 3$.

5 Multiphase winding

Another realization of a dual-purpose winding is given by a research group around Kang. In various publications [22], [31], [32] they applied multiphase machine theory [33] to design a multiphase bearingless motor. The general idea is to superpose two different stator current systems and to specifically influence the fundamental and harmonic waves of the magnetic air gap field by carefully combining the three central degrees of freedom of this winding design: number of phases, phase shift between the stator currents, and coil span. The winding distribution n_i can be expressed as a Fourier series

$$\begin{aligned} n_{\mathrm{A}}(\alpha) &= \sum_{\nu=1}^{\infty} \hat{n}_\nu \cos(\nu\alpha + \varphi_{\mathrm{n}\nu}) \\ n_{\mathrm{B}}(\alpha) &= \sum_{\nu=1}^{\infty} \hat{n}_\nu \cos\left(\nu\left(\alpha - \frac{2\pi}{m}\right) + \varphi_{\mathrm{n}\nu}\right) \\ &\vdots \\ n_{\mathrm{m}}(\alpha) &= \sum_{\nu=1}^{\infty} \hat{n}_\nu \cos\left(\nu\left(\alpha - (m-1)\frac{2\pi}{m}\right) + \varphi_{\mathrm{n}\nu}\right). \end{aligned} \tag{5}$$

One can see that the phase windings are distributed equally along the circumference whereas each phase itself is in general not symmetrical (see Fig. 10). Unlike conventional symmetrical three-phase stator windings, $\hat{n}_\nu$ exists for even harmonics as well (except full pitch multiphase windings). In this context, a full pitch means a coil span of half of the stator's circumference.

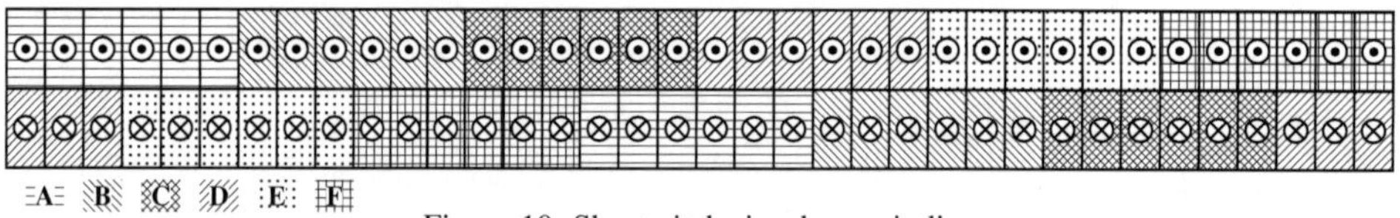

Figure 10: Short pitch six-phase winding

The phase currents are assumed to be totally sinusoidal

$$\boldsymbol{i}_k = \begin{bmatrix} i_{\mathrm{A}k} \\ i_{\mathrm{B}k} \\ \vdots \\ i_{mk} \end{bmatrix} = \hat{i}_k \begin{bmatrix} \cos(\omega t + \varphi_{\mathrm{i}k}) \\ \cos\left(\omega t + \varphi_{\mathrm{i}k} - k\frac{2\pi}{m}\right) \\ \vdots \\ \cos\left(\omega t + \varphi_{\mathrm{i}k} - (m-1)k\frac{2\pi}{m}\right) \end{bmatrix} \tag{6}$$

where k determines the current's phase shift. The resulting MMF Θ_k for $\boldsymbol{i}_k$ is:

$$\Theta_k(\alpha,t) = \sum_{\xi=1}^{m} \Bigg[\underbrace{\hat{i}_k \cos\left(\omega t + \varphi_{\mathrm{i}k} - (\xi-1)k\frac{2\pi}{m}\right)}_{\text{current}} \cdot \sum_{\nu=1}^{\infty} \underbrace{\hat{n}_\nu \cos\left(\nu\left(\alpha - (\xi-1)\frac{2\pi}{m}\right) + \varphi_{\mathrm{n}\nu}\right)}_{\text{winding}} \Bigg]. \tag{7}$$

Applying $\cos(x)\cdot\cos(y) = \frac{1}{2}(\cos(x-y)+\cos(x+y))$ one can describe the ν-th MMF harmonic caused by $\boldsymbol{i}_k$ as

$$\Theta_{k\nu} = \begin{cases} \frac{\hat{i}_k \hat{n}_\nu}{2} \cos(\omega t + \varphi_{\mathrm{i}k} + \nu\alpha + \varphi_{\mathrm{n}\nu}) & \text{for } k+\nu = l\cdot m \\ 0 & \text{for } k \pm \nu \neq l\cdot m \\ \frac{\hat{i}_k \hat{n}_\nu}{2} \cos(\omega t + \varphi_{\mathrm{i}k} - \nu\alpha + \varphi_{\mathrm{n}\nu}) & \text{for } k-\nu = l\cdot m \end{cases} \tag{8}$$

with $l \in \mathbb{Z}$ [22]. It is now possible to evaluate (8) for various combinations of number of phases and phase shifts.

As the number of slots per phase q shall be an integer, the number of phases is restricted to $m = 6,9,12,18$. Fig. 11a shows the evaluation of (8) for a six-phase motor. Each stator current system evokes forward and backward rotating or pulsating harmonics. We will focus on the upper $(m/2-1)$-rows as the bottom half is mirrored with reversed rotating directions. If m is even, the $m/2$-th and m-th row (phase shifts of π resp. 2π) just cause pulsating field waves which are useless for force and torque generation. The bottom row displays the harmonics of the magnetic field caused by the permanent magnets. Torque is generated by the interaction of a harmonic of the PM-field and a harmonic of the same order caused by the stator currents. Only the fundamental waves ($\nu = p = 2$) cause a useful torque. All interactions by higher harmonics result in alternating torque just producing torque ripple. So the intention is to minimize these interactions by diminishing the respective harmonic components n_ν of the stator MMF. The same applies to the generation of force. Every $(p(2l+1)\pm 1)$-th ($l \in \mathbb{N}$) rotating current harmonic interacts with the PM-harmonics ($p(2l+1)$, $l \in \mathbb{N}$) generating force effects whereas only the fundamental harmonics contribute a useful force component and the higher ones just enhance the force ripple. Special attention must be paid to the force-generating interactions between the harmonics of different stator current systems (e.g. $\boldsymbol{i}_{k=1}$ and $\boldsymbol{i}_{k=2}$, see black rectangles in Fig. 11a). Generally speaking, an increase of the number of phases leads to a more equalized harmonic spectrum.

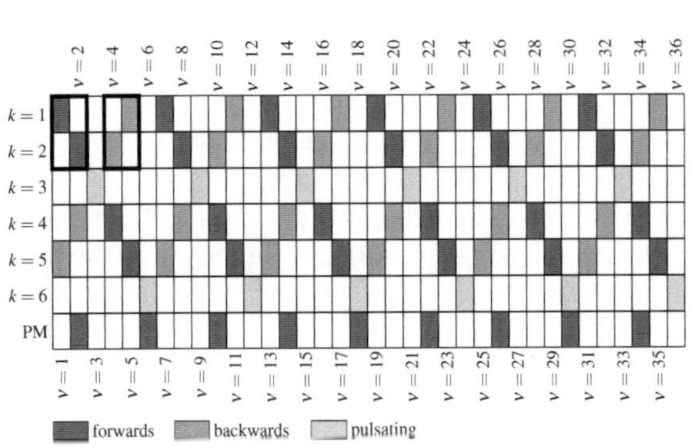

(a) $m = 6$

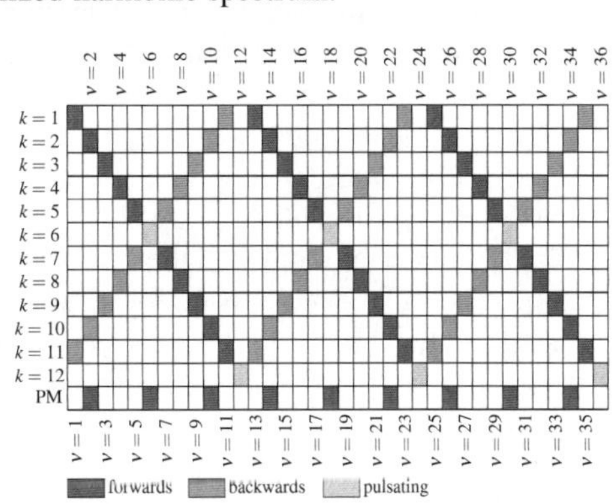

(b) $m = 12$

Figure 11: Evaluation of the MMF harmonics of a six (a) and twelve (b) phase motor (based on [22])

An analytical expression for the harmonic components of the stator MMF is therefore needed. We will assume just a radial MMF (1D-representation) caused by windings concentrated in the middle of each slot. A more detailed description as e.g. the 2-D approach in [34] is not necessary as the basic relationships already become clear in a 1D representation and are sufficient for the choice of the number of phases and the coil span.

We can decompose the winding distribution into two sums of step functions:

$$\begin{aligned} n_{\mathrm{A}}(\alpha) &= n_{\mathrm{up\,A}}(\alpha) + n_{\mathrm{dw\,A}}(\alpha) \\ &= z_{\mathrm{s}} \sum_{l=1}^{q} \mathbb{1}\left(\alpha - (l-1)\frac{2\pi}{N}\right) - z_{\mathrm{s}} \sum_{l=1}^{q} \mathbb{1}\left(\alpha - ((l-1)+y_{\mathrm{c}})\frac{2\pi}{N}\right). \end{aligned} \tag{9}$$

Combined with the Fourier expansion of the shifted step function

$$\mathbb{1}(\alpha - \varphi) = 1 - \frac{\varphi}{2\pi} + \frac{1}{\pi}\sum_{\nu=1}^{\infty}\left[-\frac{\sin(\nu\varphi)}{\nu\pi}\cos(\nu\alpha) + \frac{\cos(\nu\varphi - 1)}{\nu\pi}\sin(\nu\alpha)\right] \tag{10}$$

and the superposition of harmonic oscillations with same frequency, this leads to the Fourier coefficients in (5).

5.1 Design process

Since there are four different numbers of phases possible, a preselection had to be made. The bearingless operation relies on the superposition of two different current systems. From Fig. 11 one can see that $\boldsymbol{i}_{k=2}$ is necessary for torque generation and $\boldsymbol{i}_{k=2}$ or $\boldsymbol{i}_{k=3}$ — equivalent to $p_s = 1$ resp. $p_s = 3$ — for force generation (for $m = 6$ only $\boldsymbol{i}_1$). This superposition results in unsymmetrical phase currents so that an H-bridge is needed for each phase. A choice of $m = 18$ would therefore result in a disproportionately high need of switches. Initial studies showed an unfavourable force and torque characteristic for $m = 9$ due to an undesirable interaction between the magnetic fields' second harmonic, so that the design process focus on $m = 6$ and $m = 12$.

Fig. 12 shows the scaled Fourier coefficients of the winding distribution for six and twelve phases from eqs. (9) and (10). The vertical dashed lines indicate the chosen coil spans for $p_s = 1$ (black) and for $p_s = 3$ (grey), respectively. These choices represent a good compromise between a sufficiently high generation of torque and force (see fundamental harmonics) and also a minimization of the critical higher-order coefficients. For example for $m = 6$ the 5,7,11,13,...-th order harmonics of $\boldsymbol{i}_1$ will interact with the PM harmonics and cause undesired force ripples. The same counts for the 4,8,10,14,...-th orders harmonics of $\boldsymbol{i}_2$ in interaction with the harmonic spectrum of $\boldsymbol{i}_1$. Since lower harmonics tend to have a greater influence we focus for this example on $\nu = 1,2,4,5$. The product of $\hat{n}_1$ and $\hat{n}_2$ shall be as large as possible whereas $\hat{n}_4$ and $\hat{n}_5$ shall be as small as possible at the same time. For a coil span of nine $\hat{n}_4$ becomes zero and $\hat{n}_5$ rather small and $\hat{n}_2$ maximal, but this choice leads to a considerable loss of force generation as $\hat{n}_1$ is far from optimal. A coil span of eleven, on the other hand, provides a significant increase of $\hat{n}_1$, just a small decrease of $\hat{n}_2$ and $\hat{n}_4$ as well as $\hat{n}_5$ stays reasonably low.

Fig. 12b indicates a much higher force generation by $\boldsymbol{i}_1$ than by $\boldsymbol{i}_3$ but this only counts for a linear machine. The spindle motor is highly utilized and therefore even under normal operation considerably saturated, so that the resulting force is approximately equal. The chosen coil span for $m = 12$, $p_s = 3$ corresponds exactly to the original winding.

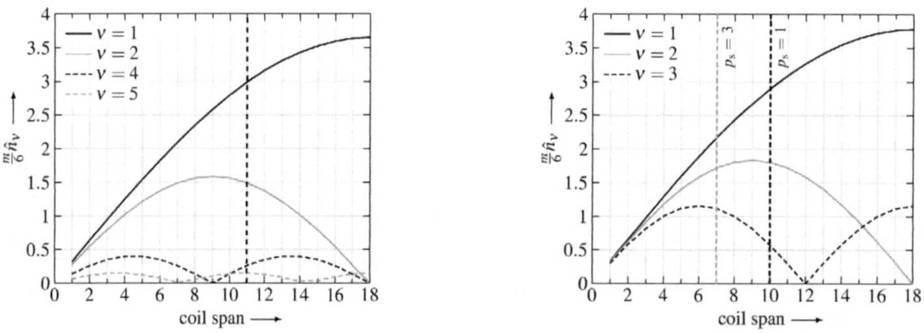

(a) $m = 6$ (b) $m = 12$

Figure 12: Scaled Fourier coefficients for six (a) and twelve (b) phase windings

5.2 Simulation results

We carried out the same simulations for the multiphase bearingless motors as we did in section 4.2. The torque and force characteristics are shown in Fig. 13. The former have a very even course, whereby the torque for $m = 6$ still shows some noticeable influences of a fourth order harmonic and the distinctive beat (see section 3). Its mean value (10.7 Nm) is, as indicated by Fig. 12, lower than the one for $m = 12$ (12.9 Nm). For the torque ripple it is, due to the influence of the fourth order harmonic and the lower mean value, just the other way round (7.6 % resp. 5.4 %). As the twelve-phase winding corresponds to the original one, the torque characteristics are nearly identical. The small deviations in uncontrolled operation are caused by the superposed suspension field. In an exclusively torque operating mode the torque will be the same.

The force characteristics show a very familiar behavior to the separated winding designs. Using a six-phase motor the force has a very dominant fourth order harmonic and a very large ripple of 25.1 %. Its mean value of 412.6 N is, compared to section 3, even a bit lower than the one of 430.8 N of the twelve-phase motor which has a ripple of just 2.8 %.The results for $m = 12$ with $p_s = 1$ are not shown here as they just represent a undesirable combination of the torque characteristic of the twelve-phase motor with $p_s = 3$ and the force characteristic of the six-phase motor.

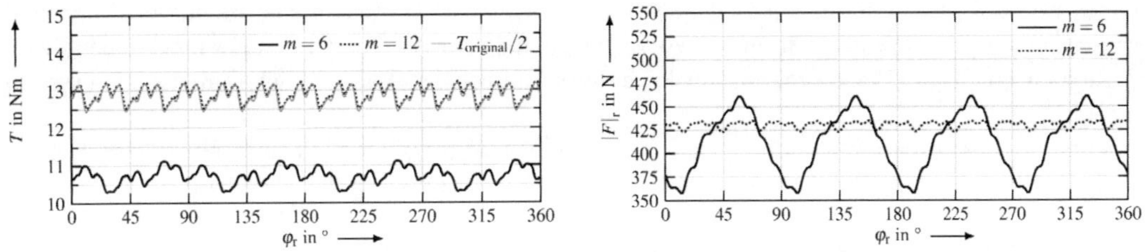

(a) (b)

Figure 13: Simulated force (a) and torque (b) characteristics of the six- and twelve-phase bearingless motor over one rotor revolution

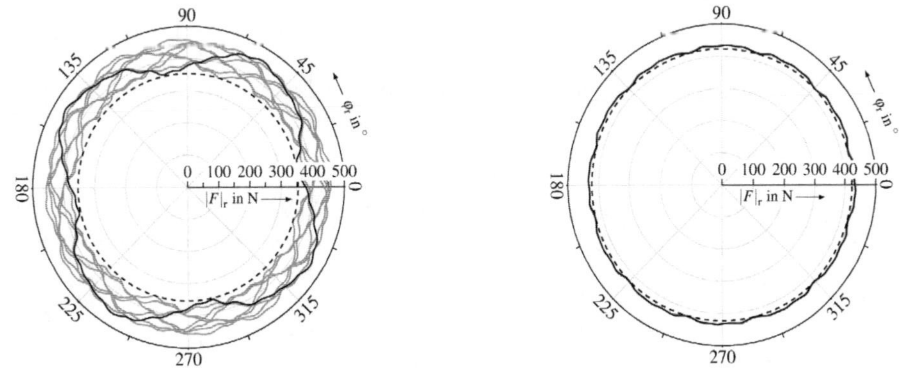

(a) $m = 6$ (b) $m = 12$

Figure 14: Simulated force and torque characteristics of the six (a) and twelve (b) phase bearingless motor for different suspension current space vectors over one rotor revolution

Fig. 14 confirms the excellent force characteristic of the twelve-phase motor. The same simulations as in Fig. 9 were carried out for the multiphase motors. The deviations in force magnitude for different suspension current space vectors are so small that they are barely depictable in this figure. The six-phase motor in Fig. 14a shows a similar behavior as the DPNV designs in Fig. 9 as all of them correspond to $p_s = 1$.

6 Conclusion

DPNV windings present in general a valid approach to turn standard motors into bearingless ones with low demand for power electronics, but they prove to be unsuitable for the spindle motor investigated here and HSC spindle motors with pole pairs of $p = 2$ and $p = 3$ in general.

The multiphase approach leads to significantly better results at the cost of high demand for power electronics. A twelve-phase motor with $p_s = 3$ worked best. The stator winding is equivalent to the original one just with the difference that the once parallelized coil groups in each phase are now supplied each by an own H-bridge. Therefore, in normal operation mode, the torque characteristic is identical to the original motor. The investigations above also show an excellent force characteristic even at an uncontrolled state.

Generally speaking, a two-pole suspension field ($p_s = p - 1$) leads to worse torque and force characteristics than a six-pole one ($p_s = p + 1$) due to the interactions of the specific harmonic spectrums and therefore a more evenly superposition of the suspension and the rotor field. This goes hand in hand with the results from [30], [35] and even ensures lower rotor eddy-current losses which were not addressed in this paper.

Based on these findings, a twelve-phase prototype was commissioned. The practical investigation and control scheme of this bearingless spindle motor is the subject of further publications.

Acknowledgments The authors would like to thank the "Sächsische Aufbaubank" (SAB) and the European Union (EU) for funding this research under the European Regional Development Fund (ERDF) with SAB application no. 100348076.

References

[1] J. Munoa, X. Beudaert, Z. Dombovari, *et al.*, "Chatter suppression techniques in metal cutting," *CIRP Annals*, vol. 65, no. 2, pp. 785–808, 2016.

[2] S. A. Tobias and W. Fishwick, "Theory of regenerative machine tool chatter," *The Engineer*, vol. 205, no. 7, pp. 199–203, 1958.

[3] A. Cowley and A. Boyle, "Active dampers for machine tools," *CIRP Annals*, vol. 18, no. 2, pp. 213–222, 1970.

[4] J. Munoa, X. Beudaert, K. Erkorkmaz, *et al.*, "Active suppression of structural chatter vibrations using machine drives and accelerometers," *CIRP Annals*, vol. 64, no. 1, pp. 385–388, 2015.

[5] E. Abele, M. Roth, C. Ehmann, *et al.*, "Aktiver Strukturdämpfer. Dimensionierung, Konstruktion und Verifikation an einem Bearbeitungszentrum," German, *wt Werkstatttechnik online*, vol. 100, pp. 105–111, 2010.

[6] H. Tanaka, F. Obata, T. Matsubara, *et al.*, "Active chatter suppression of slender boring bar using piezoelectric actuators," *JSME International Journal. Ser. C, Dynamics, control, robotics, design and manufacturing*, vol. 37, no. 3, pp. 601–606, 1994.

[7] C. Brecher, D. Manoharan, U. Ladra, *et al.*, "Chatter suppression with an active workpiece holder," *Production Engineering*, vol. 4, no. 2-3, pp. 239–245, 2010.

[8] C. R. Knospe, "Active magnetic bearings for machining applications," *Control Engineering Practice*, vol. 15, no. 3, pp. 307–313, 2007.

[9] J.-H. Kyung and C.-W. Lee, "Controller design for a magnetically suspended milling spindle based on chatter stability analysis," *JSME International Journal. Ser. C, Mechanical systems, machine elements and manufacturing*, vol. 46, no. 2, pp. 416–422, 2003.

[10] M. Ries and S. Pankoke, "Increasing the stability of the milling process by an active milling spindle," in *Proceedings of the 1st international conference on process machine interactions*, 2008, pp. 95–102.

[11] P. Albertelli, S. Elmas, M. R. Jackson, *et al.*, "Active spindle system for a rotary planing machine," *The International Journal of Advanced Manufacturing Technology*, vol. 63, no. 9-12, pp. 1021–1034, 2012.

[12] E. Abele, R. Nordmann, C. Ehmann, *et al.*, "Aktive Motorspindel stabilisiert den Prozess," German, *Werkstatt und Betrieb*, vol. 138, no. 11, pp. 22–27, 2005.

[13] T. Klaffert, *Selbstoptimierende HSC-Motorspindel*, German. Verlag Wissenschaftliche Scripten, 2006.

[14] W. Bickel, *Frässpindel mit motorintegrierter aktiver Dämpfung*, German. PZH Verlag, 2015.

[15] J. Königsberg, J. Reiners, B. Ponick, *et al.*, "Highly dynamic spindle integrated magnet actuators for chatter reduction," *International Journal of Automation Technology*, vol. 12, no. 5, pp. 669–677, 2018.

[16] B. Denkena, B. Bergmann, and F. Böhse, "Mechatronic damping systems as enablers for autonomous machine tools," *Proceedings of the Machining Innovations Conference (MIC) 2020*, pp. 166–173, 2020.

[17] A. Chiba, T. Fukao, and M. A. Rahman, "Vibration suppression of a flexible shaft with a simplified bearingless induction motor drive," *IEEE Transactions on Industry Applications*, vol. 44, no. 3, pp. 745–752, 2008.

[18] O. Kienzle, "Die Bestimmung von Kräften und Leistungen an spanenden Werkzeugen und Werkzeugmaschinen," German, *VDI-Zeitschrift*, vol. 94, no. 11–12, pp. 299–305, 1952.

[19] D. Dietz, G. Messager, and A. Binder, "1 kW/60,000 min^{-1} bearingless pm motor with combined winding for torque and rotor suspension," *IET Electric Power Applications*, vol. 12, no. 8, pp. 1090–1097, 2018.

[20] S. Kobayashi, M. Ooshima, and M. N. Uddin, "A radial position control method of bearingless motor based on d-q-axis current control," *IEEE Transactions on Industry Applications*, vol. 49, no. 4, pp. 1827–1835, 2013.

[21] W. Gruber, W. Amrhein, and M. Haslmayr, "Bearingless segment motor with five stator elements—design and optimization," *IEEE Transactions on Industry Applications*, vol. 45, no. 4, pp. 1301–1308, 2009.

[22] M. Kang, J. Huang, J.-q. Yang, *et al.*, "Analysis and experiment of a 6-phase bearingless induction motor," in *2008 International Conference on Electrical Machines and Systems (ICEMS)*, IEEE, 2008, pp. 990–994.

[23] E. L. Severson, R. Nilssen, T. Undeland, *et al.*, "Design of dual purpose no-voltage combined windings for bearingless motors," *IEEE Transactions on Industry Applications*, vol. 53, no. 5, pp. 4368–4379, 2017.

[24] A. Chiba, T. Fukao, O. Ichikawa, *et al.*, *Magnetic Bearings and Bearingless Drives*. Elsevier, 2005.

[25] W. K. S. Khoo, "Bridge configured winding for polyphase self-bearing machines," *IEEE Transactions on Magnetics*, vol. 41, no. 4, pp. 1289–1295, 2005.

[26] A. Chiba, K. Sotome, Y. Iiyama, *et al.*, "A novel middle-point-current-injection-type bearingless pm synchronous motor for vibration suppression," *IEEE Transactions on Industry Applications*, vol. 47, no. 4, pp. 1700–1706, 2011.

[27] R. Oishi, S. Horima, H. Sugimoto, *et al.*, "A novel parallel motor winding structure for bearingless motors," *IEEE Transactions on Magnetics*, vol. 49, no. 5, pp. 2287–2290, 2013.

[28] E. L. Severson, R. Nilssen, T. Undeland, *et al.*, "Dual-purpose no-voltage winding design for the bearingless ac homopolar and consequent pole motors," *IEEE Transactions on Industry Applications*, vol. 51, no. 4, pp. 2884–2895, 2015.

[29] E. L. Severson, S. Gandikota, and N. Mohan, "Practical implementation of dual-purpose no-voltage drives for bearingless motors," *IEEE Transactions on Industry Applications*, vol. 52, no. 2, pp. 1509–1518, 2016.

[30] T. Schneider, J. Petersen, and A. Binder, "Influence of pole pair combinations on high-speed bearingless permanent magnet motor," *4th IET International Conference on Power Electronics, Machines and Drives*, pp. 707–711, 2008.

[31] M. Kang, J. Huang, H.-b. Jiang, *et al.*, "Principle and simulation of a 5-phase bearingless permanent magnet-type synchronous motor," in *2008 International Conference on Electrical Machines and Systems (ICEMS)*, IEEE, 2008, pp. 1148–1152.

[32] B. Li, J. Huang, M. Kang, *et al.*, "Analysis and simulation of a 7-phase pm bearingless motor," in *2011 International Conference on Electrical Machines and Systems (ICEMS)*, IEEE, 2011, pp. 1–5.

[33] J. Huang, M. Kang, J.-q. Yang, *et al.*, "Multiphase machine theory and its applications," in *2008 International Conference on Electrical Machines and Systems (ICEMS)*, IEEE, 2008, pp. 1–7.

[34] Z. Q. Zhu and D. Howe, "Instantaneous magnetic field distribution in brushless permanent magnet dc motors. II. armature-reaction field," *IEEE Transactions on Magnetics*, vol. 29, no. 1, pp. 136–142, 1993.

[35] Z. Liu, A. Chiba, Y. Irino, *et al.*, "Optimum pole number combination of a buried permanent magnet bearingless motor and test results at an output of 60 kw with a speed of 37000 r/min," *IEEE Open Journal of Industry Applications*, vol. 1, pp. 33–41, 2020.

Operating a MagLev Train Prototype with Supercapacitors and Charge by Opportunity

Gabriel R. Pereira[1], Richard M. Stephan[2], and Felipe Costa[3]

[1] Federal University of Rio de Janeiro, Rio de Janeiro, Rio de Janeiro, Brazil
gabrielramosbr@poli.ufrj.br
[2] Federal University of Rio de Janeiro, Rio de Janeiro, Rio de Janeiro, Brazil
richard@dee.ufrj.br
[3] Federal University of Rio de Janeiro, Rio de Janeiro, Rio de Janeiro, Brazil
felipe@pee.ufrj.br

Abstract

This work presents an alternative energy supply for urban MagLev vehicles, using supercapacitors. The concept is known as "charge by opportunity". During the necessary stops for passengers′ embarkation and disembarkation, and the few seconds for the completion of this activity, supercapacitors inside the vehicle will be charged. The supplied energy is calculated to ensure the movement until the next station. The low friction of MagLev trains, the short distance between stations for public transportation inside a city, and the flat track given by elevated lines, all contribute to small energy storage requirements. This study not only describes the application, but also maps the market for supercapacitors, describing the main suppliers, prices and characteristics. In this way, an innovative power supply option for urban MagLev vehicles will be offered, which, unlike the existing options, does not visually pollute with overhead cables or extra rails. Experimental results would have been presented with tests on the prototype MagLev-Cobra, from the Federal University of Rio de Janeiro. However, due to a lack of financial support and the coronavirus pandemic the tests were canceled. MagLev-Cobra is a levitating vehicle based on the diamagnetic property of superconductors in the vicinity of rails made up of rare earth magnets. As a prototype system, the ideal test conditions are available. Furthermore, a converter DC-DC is specified to work with the supercapacitors.

Contents

5 Future Work

1 Introduction

1.1 Contextualization

The increase in urbanization in large cities and the need to preserve the environment are at the heart of current discussions about the direction of sustainability and technological innovation. Within this scenario, the issue of transport and urban mobility deserves special mention because it interferes daily in the lives of thousands of people, contributing significantly to the emission of pollutants resulting from the use of fossil fuels, as well as noise pollution. In addition, the congestion present in large cities results in both economic losses and losses to the health of the population. In order to provide options for solving these problems, MagLev-Cobra was developed.

This vehicle is an efficient, non-polluting mode of transportation, with competitive implementation and maintenance costs, and is the result of a project started in 2000 by Escola Politécnica (EP) and COPPE/UFRJ using magnetic levitation technology. The vehicle operates from the diamagnetic property presented by high-temperature critical superconductors when interacting with special permanent magnet tracks, also called rare earth magnets due to the presence of the chemical element Neodymium (Nd), used in its construction.

With its physical structure and operating principle, the vehicle uses a traction system by means of a linear induction motor, being formed by small modules that allow the MagLev to be articulated, alluding to a snake. This articulation allows curves with radii of 50 meters to be made and goes over slopes of up to 15%, permitting its implantation with greater freedom in urban environments, being able to be used in elevated magnetic levitation roads, without impacting the traffic of large cities [13].

The MagLev-Cobra travels without mechanical contact between the vehicle parts and rails, which together with the current traction system, makes it a silent vehicle. Levitation also allows the vehicle's mechanical load to be evenly distributed along the track, unlike traditional systems where all the weight is concentrated on the wheels. This feature allows the routes for using the vehicle to be constructed using less material, consequently, making them cheaper when compared to other means of transport on rails, such as the tracks of a monorail.

The project has already gone through three stages: the proof of concept in which a small scale prototype was built; the functional prototype that aimed to demonstrate technical feasibility within a controlled laboratory environment; and the operational prototype that had been operating weekly since October 2015 on the UFRJ campus, meeting the demand of the university community, as well as countless visitors. However, these operations ended with the outbreak of COVID-19. Currently, the project is in the industrialization stage, which consists of the certification of technology and the implementation of an effective transportation line containing five kilometers within the campus of the university. [2].

1.2 MagLev-Cobra Power Supply

The MagLev-Cobra prototype is powered by a DC bus present throughout the entire two-hundred-meter runway. This bus consists of three parallel bars, located on the opposite side of the vehicle entrance, so that visitors do not have access. The external busbar has 550V, the central busbar 0V and the busbar closest to the vehicle is solidly grounded. In addition,

Figure 1: AFW-11 panel.

Figure 2: MPW-100 switch.

the vehicle has two sets of collecting brushes attached to the side, responsible for maintaining contact with the DC bus throughout the operation.

This 550V DC voltage, responsible for energizing the vehicle's internal circuits, is obtained from the internal electrical network of UFRJ Technological Center, having an AC voltage of 440V and a frequency of 60Hz. The three phases of the network are connected to the high side of a 100kVA Δ-Δ three-phase transformer that has a 440V: 380V transformation ratio. The low voltage side of the transformer is connected to the CFW11-105-RB bidiretional converter bridge, responsible for rectifying the transformer voltage and supplying the DC bus.

The CFW11-105-RB converter bridge is part of the AFW-11 [14] drive panel assembly, from the manufacturer WEG, shown in figure 1, being an induction motor drive panel, equipped with a set of components that allows the regenerative conversion of energy. It is also worth mentioning that there is a breaker MPW-100 [17], shown in figure 2, located in the center of the track, making the connection between the CFW11-105-RB converter bridge and the bus External DC, that is, performing the protection of that bus.

1.3 Motivation

Man has always needed energy sources. Since these sources are not always available, there is a need to store energy for later use. This is true that living beings, by feeding themselves, conserve the energy of food in the form of chemical compounds in the body itself for later use.

Within this context, supercapacitors emerge as energy storage devices, as well as several others existing on the market. Among them, we can mention batteries that store energy in chemical form, magnetic flywheels popularly known as Flywheel Energy Storage (FES), in which energy is stored in mechanical form, and Superconducting Nagnetic Energy Storage (SMES), in which energy is stored in the form of a magnetic field. Consequently, the way in which these technologies store energy makes them completely different.

The fact that supercapacitors already store energy in electrical form means that they are able to transfer energy more quickly when compared to traditional batteries [3], making them useful in applications that require a considerable amount of energy in a short time [12].

Thus, it would be possible to replace the side bus, described in section 1.2, with supercapacitors. With the supply provided by them, there is an improvement in regards to two aspects. The first is economic in nature, there is no longer the need for collecting brushes, shown in figure

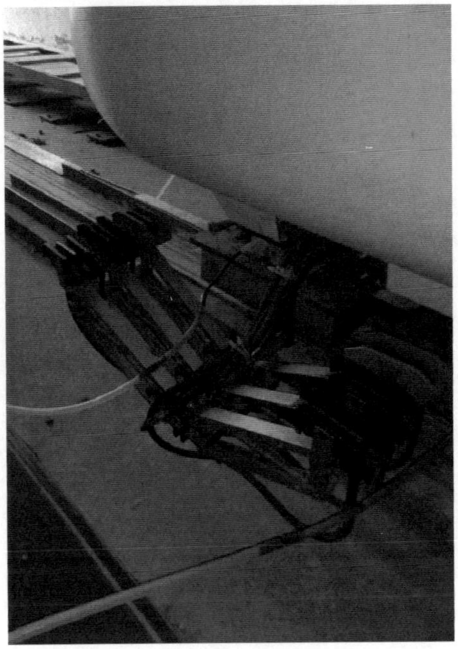

Figure 3: Collecting brushes.

Figure 4: Side bus on the track.

3, being the design element that suffers the most wear during the operation of the vehicle. The second related is to safety, since it is no longer necessary to have an energized bus on the road as illustrated in figure 4. In addition, as it is not necessary to use the side bus that represents the vehicle's only point of contact with external surfaces, it makes levitation, the main feature of MagLev, even more impressive [1].

2 Research Methods

2.1 Energy Density and Power Density

To elucidate the main properties of a supercapacitor, it is necessary to introduce two concepts: energy density and power density. The energy density represents the relationship between the stored energy and the mass (J/kg) or volume (J/m^3) of the equipment, that is, a high energy density means that the equipment can store a larger amount of power. Power density, on the other hand, provides the rate at which a certain amount of energy can be transferred by the device per unit of mass (W/kg) or volume (W/m^3).

For a capacitor of parallel flat plates, the capacitance is given by equation 1, where ϵ is the permittivity of the dielectric, A is the area of the plates and d is the distance between them. The energy stored in any capacitor is given by equation 2, where C is the capacitance and V is the terminal voltage of the device, where (V/d) represents the resistance to rupture the dielectric.

$$C = \frac{\epsilon \times A}{d} \quad (1)$$

$$E_c = \frac{1}{2} \times C \times V^2 \quad (2)$$

Using the equations 1 and 2 and considering (V/d) = 200 MV/m and ε= 8.85 x 10^{-11}F/m, we calculate the density of energy of a capacitor by equation 3. Also, it can be seen in equation 4 that the order of magnitude of this measurement for supercapacitors is around 40 times greater [12].

$$\frac{E_c}{V_o} = \frac{1}{2} \times \epsilon \times \frac{V^2}{d^2} = 1.76 \times 10^6 J/m^3 \tag{3}$$

$$\frac{E_c}{V_o} = 0.7 \times 10^8 J/m^3 \tag{4}$$

These two concepts are the main distinction between supercapacitors and batteries. While supercapacitors can provide very high power density with less energy stored, batteries can store more energy, as they have a higher energy density and a lower power density as can be seen in figure 5. These characteristics are related to the way energy is stored in each device.

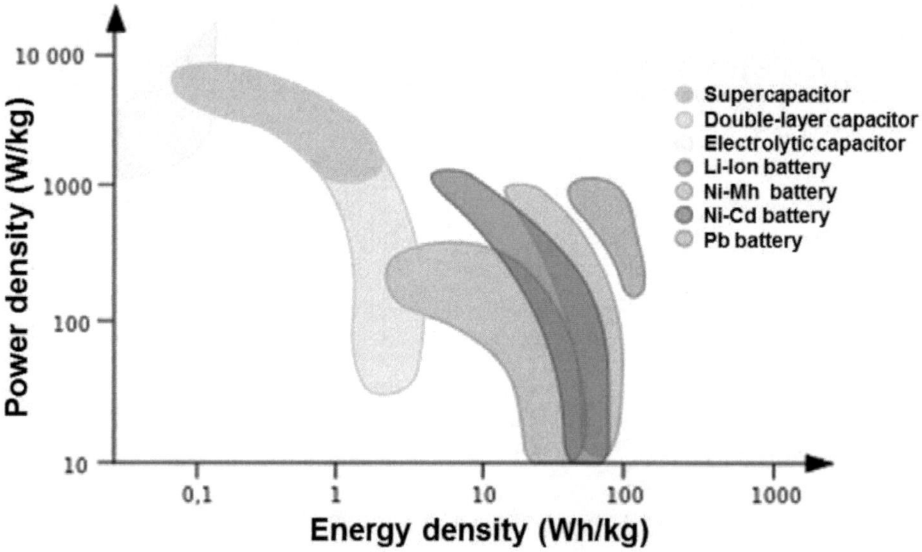

Figure 5: Ragone diagram: energy storage devices.
Source: image authorized by the author [6].

Electrolytic capacitors have a higher power density, but supply peaks in the current for only a few milliseconds. Batteries, on the other hand, have the best cost and energy storage ratio, but end up supplying energy more slowly. In addition, it is possible to notice that the supercapacitors fill the gap that existed between capacitors and batteries.

In figure 5, it can be seen that the lithium batteries also deserve to be highlighted. They differ from other batteries because they have a higher theoretical energy density which is comparable to the energy density of fossil fuels.

Because of these characteristics, supercapacitors and lithium batteries are promising devices for storing energy and will play an important role in the coming years.

It is also possible to verify that supercapacitors are passive energy storage devices with a high power density and a moderate energy density. In addition, they have other features that make their application interesting. They have a long service life (> 1000000 cycles) and an increasingly lower production cost [20].

Through figures 6a and 6b, it is also possible to explain these two concepts through an analogy with a bucket full of water with a hole in its bottom. In figure 6a, the bucket on the right has a higher rate at which a certain amount of energy can be transferred, that is, it has a higher power density when compared to the bucket on the left, this is because it has a hole with a larger diameter in the same position as the bucket on the left. In figure 6b, both buckets have the same power density. However, as the bucket on the left is at a higher height, there is a greater amount of energy stored, even if both buckets have the same size and volume of area.

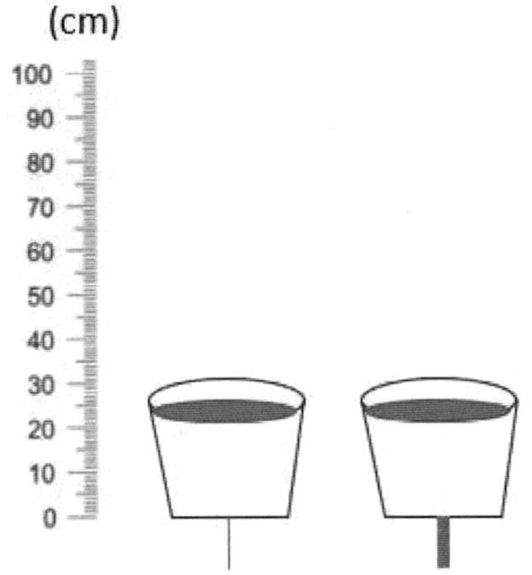

(a) Same energy density, different power density.

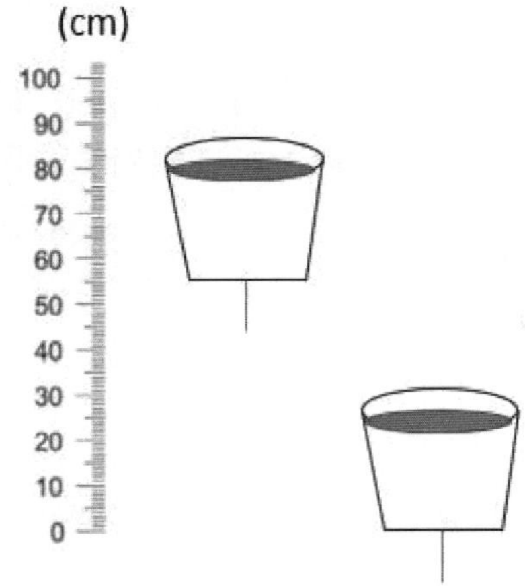

(b) Same power density, different energy density.

Figure 6: Analogy: energy density and power density.

2.2 Equivalent Circuit

It is practical to represent the supercapacitors by a simplified circuit model. Figure 7 shows the representation of the circuit, where C is the capacitance, ESR is the equivalent series resistance (which represents all ohmic losses, which depending on the application can cause a significant voltage drop, reducing the amount of energy available in the circuit, affecting its functioning, resulting from the materials used), Rp is the parallel resistance (which represents the leakage current caused by losses due to the fact that during the storage of electrical energy in the double layer, the carriers of charge inside the pores are separated by very short distances, thus irregularities can occur, leading to a small change of charge carriers and gradual discharge) and ESL is the equivalent series inductance. Therefore, like Rp $\gg$ ESR, the resistance Rp can be neglected, since it is in parallel, being considered an "opening". In addition, they behave in an ideal way, due to the porous material used in the formation of the electrodes, and as C $\gg$ ESL, it is common to simplify the circuit to that represented in figure 8 [18].

Therefore, the values of the quantities in figure 8 are the values commonly found in the datasheets of cells and modules.

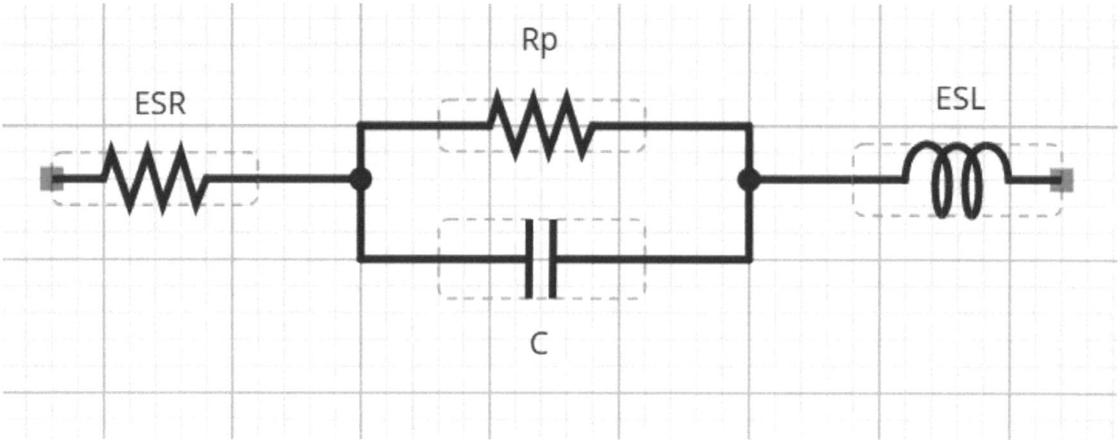

Figure 7: Circuit of a supercapacitor.

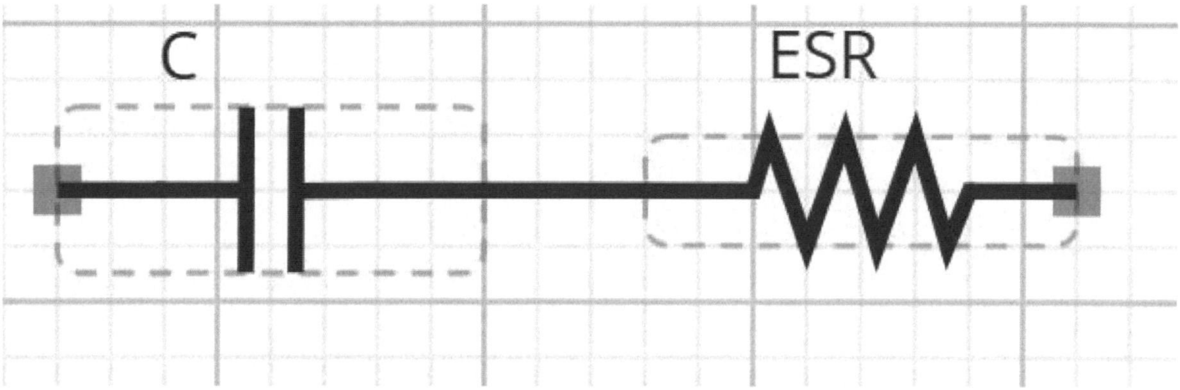

Figure 8: Simplified circuit of a supercapacitor.

Therefore, the voltage profile has resistive and capacitive components. Several models have been proposed and studied to represent the non-idealities of the device. However, the complexity of the model to be adopted depends on the specific application and the admissible error in the analysis [7].

3 Obtained Results

With the theoretical basis formed in the previous chapter, it is possible to understand and execute the project presented.

The power supply of the external 550V DC bus, which will be replaced by supercapacitors, feeds the vehicle's linear induction motor, being previously connected to the frequency inverter WEG CFW11-88A [16], whose function is to generate the three-phase voltage alternating supply of primary circuits, in addition to establishing motor speed control. It is worth mentioning that it is not part of the present work to modify this internal configuration of the vehicle, since there would be no economic or visual gain with this change, that is, it is only necessary to know the energy consumption of the load in relation to the induction motor. while traveling between stations.

3.1 Energy Consumption

According to [4], the data that will be presented in table 1 were acquired considering a constant mass of 2565kg.

Weight (kg)	Time of travel (seconds)	Energy (kWh)
2565	235	0.111

Table 1: Energy consumed in a complete trip of 2x200 = 400 meters.

3.2 Commercially Available Modules

As shown in table 1, the module must deliver at least 111Wh, as well as having a voltage that allows a boost-type DC-DC converter to be used to raise this voltage to 550V that the existing DC bus in the track provides.

Manufacturers with a sales catalog about modules were: Maxwell, Skeleton and Yunasko, the latter of which had not yet started production.

Item	Product	Voltage (V)	Capacitance (F)	Quantity	Brand
(1)	BMOD0063 P125 B08	125	63	1	Maxwell
(2)	BMOD0165 P048 C01	48	165	3	Maxwell
(3)	SMA170V53FAF	170	53	1	Skeleton
(4)	SMA51V177FAF	51	177	3	Skeleton
(5)	48V / 165F Module	48	165	3	Yunasko

Table 2: Modules analyzed for acquisition.

The data sheets required in table 2 can be found respectively on the reference sites [9], [8], [10], [11] and [19].

In addition, it was necessary to understand the situation of each company in the current scenario. Maxwell Technologies was purchased by Tesla, Inc. in 2019, an American electric vehicle and clean energy company. Since then, Maxwell's share of supercapacitors has been reduced more and more. Thus, negotiations with this supplier were interrupted, since, for future acquisitions or possible maintenance, there would be no offers. In fact, the module consulted for the first configuration had already been discontinued.

In contrast, Skeleton Technologies, founded in 2009, appears as the main supplier of modules in the current scenario, even offering technical support for any project, being the safest option. Yunasko, founded in 2010, appears as a good alternative, as it is looking for new partners in the market, however it only has a plant located in Ukraine and, therefore, is unable to expand its production.

The idea would be to use only a single module in which its voltage is raised to 550 V by means of a boost converter to be used to energize the vehicle. With all aspects considered, the configuration (3), made up of a single SMA170V53FAF module from the Skeleton manufacturer, becomes the best option for application in the MagLev-Cobra vehicle, considering it recharges after each complete trip.

However, due to its dimensions and mass, it was not possible to allocate the module inside the vehicle. Thus, it became necessary to carry out new market research and, in that time, four more manufacturers appeared with modules above 100V, they are: Eaton, SPSCAP, SECH and GTCAP.

After a new filtering, the XVM-315R9515-R [5] module from the manufacturer Eaton was chosen, which is ideal because it is more compact. The properties are described in table 3.

Product	XVM-315R9515-R
Voltage	315 V
Capacitance	5.13 F
Energy	71.1 Wh
Weight	13.5 kg
Length	0.44 m
Width	0.38 m
Height	0.10 m

Table 3: XVM-315R9515-R module features.

To meet the vehicle's energy demand, it will be necessary to use two XVM-315R9515-R modules in a series circuit. However, it will be necessary to recharge each trip, because with this arrangement we have the following stored energy.

$$E_c = \frac{1}{2} \times 2.565 \times 630^2 = 509024J = 141.40Wh \tag{5}$$

However, the maximum DC voltage on the road is 550V, so the maximum energy obtained with this arrangement would be:

$$E_c = \frac{1}{2} \times 2.565 \times 550^2 = 387956J = 107.70Wh \tag{6}$$

That would make it impossible to carry out a recharge for each complete trip.

3.3 Determination of the Cyclic Ratio of the Converter

As already mentioned, the voltage of the supercapacitors decreases with their discharge. Thus, it is necessary to have a converter that has a control in order to maintain the constant voltage required in the 550V load, that is, it varies the value of the duty cycle D with the unloading of the module.

As the required voltage on the load is equal to the value on which the supercapacitor module will be loaded, the initial value of the duty cycle D is equal to 1. When analyzing a DC converter that can raise the voltage from 180V to 550V, we have the following scenario: an amount of energy will not be used, as the module cannot be completely discharged, to keep the use of the DC-DC converter viable. This unused energy can be quantified by the following equation:

$$E_c = \frac{1}{2} \times 2.565 \times 180^2 = 41553J = 11.54Wh \tag{7}$$

In this way, it is possible to obtain the total energy during the discharge, subtracting the energy obtained with the maximum voltage from the energy obtained with the lowest voltage, calculated by equations 6 and 7.

$$E_c = 107.7 - 11.54 = 96.16Wh \tag{8}$$

In addition, using the equation of a boost converter, it is possible to determine the minimum value of the cyclic ratio D.

$$550 = \frac{1}{1-D} \times 180 \tag{9}$$

$$D = 0.67 \tag{10}$$

Therefore, the control of the boost converter should vary the cyclic ratio D between $0 \leq D \leq 0.67$.

3.4 Allocation of Technologies in MagLev-Cobra

With the modules defined, it was possible to allocate them in the vehicle. In order not to affect the operation of the MagLev-Cobra, which is already well adapted to the track and not to hinder the comfort of passengers inside, both modules were placed under the floor in such a way that they do not interfere during the operation. The converter will be in the first module of the vehicle so that it can be connected to the supercapacitors by cables that will also travel under the floor. The modules and the converter can be better viewed through figures 9 and 10, where the modules are highlighted in yellow and the converter is highlighted in red.

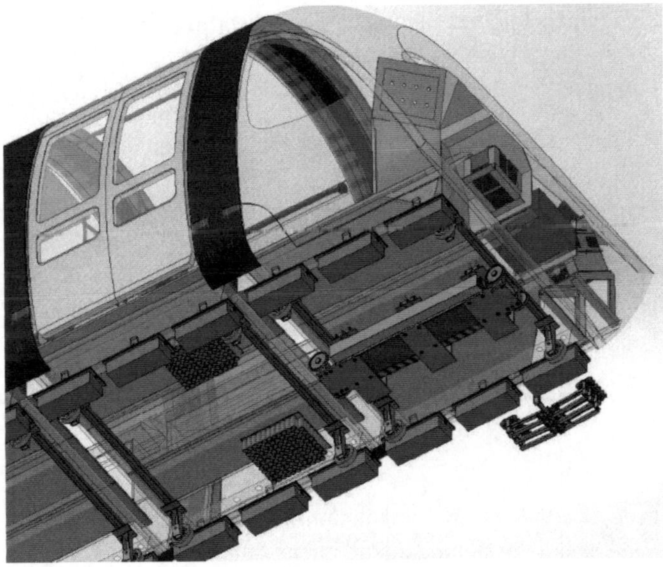

Figure 9: Bottom view of the vehicle: the modules are highlighted in yellow and the converter is highlighted in red.

3.5 Charge of Supercapacitors

We chose to use WEG CFW11-105-RB [15] converter. This converter will be responsible for rectifying the voltage of the internal electrical network of the UFRJ Technological Center, generating a voltage of 550V for the supercapacitors. It is worth mentioning that this converter has a rated output current, that is, a current limiter, solving the peak load problem.

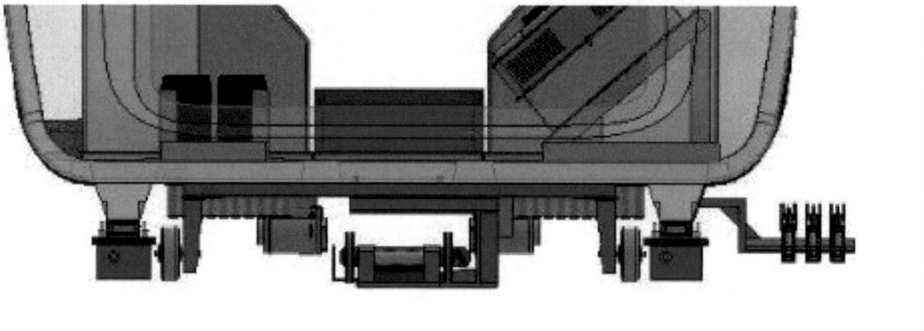

Figure 10: Front view of the vehicle: the modules are highlighted in yellow and the converter is highlighted in red.

3.6 Final System

With the necessary technologies for the application defined and properly arranged in the vehicle, the energy supply and storage system is presented as follows: two XVM-315R9515-R modules from the manufacturer Eaton in series loaded at 550V, being connected to the DC converter of boost topology that is connected to the supercapacitor bank and the WEG CFW11-88A frequency inverter. This converter has the function of keeping the voltage constant at the load, in this case, the linear induction motor, even with the unloading of the module. The final system of operation can be seen in figure 11.

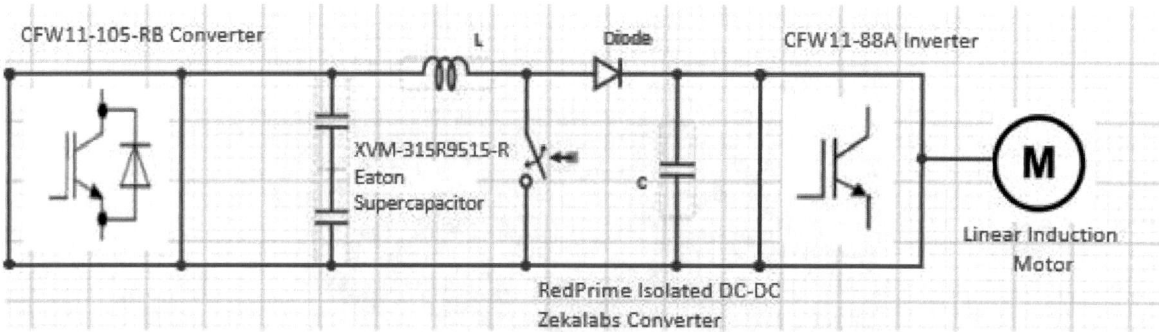

Figure 11: Electrical application diagram.

4 Concluding Summary

With this project, it was initially possible to better understand characteristics of supercapacitors, as well as some electrical models used for their representation. The complexity of the model to be adopted depends on the specific application and the admissible error in the analysis. With the mapping of the market, it was discovered that there are modules that deliver a quantity of energy capable of setting heavy vehicles in motion, such as trains and trucks. Another point highlighted was the characterization of the main suppliers of these modules in the current scenario.

In addition, the low energy consumption of the MagLev-Cobra vehicle was confirmed, which ends up allowing for other possibilities for powering the vehicle, one of which being the use of supercapacitors. Regarding the DC-DC converters, it was possible to choose the most suitable equipment to work with the module previously determined. Although, there is no specialized national manufacturer, there is international equipment with wide voltage gain as well as bidirectional and high power. With the supercapacitors module and the DC-DC converter defined, they were allocated in such a way that they did not affect the vehicle's operation or the passengers' comfort.

5 Future Work

Another project that may come from this application of supercapacitors in the vehicle is the automation of the charging process. For example, on buses that use this technology, the modules are coupled to the top of the vehicle so that when the bus arrives at a particular station for boarding and disembarking passengers, the terminals of the modules are lifted automatically , allowing contact with an external bus present at the top of the station for automatic charging of the modules.

In addition, with the creation of a longer track [2], it will be possible to use a hybrid system, that is, using batteries and supercapacitors together. Thus, a more in-depth study of this combination will be necessary.

Finally, there is also the possibility of harnessing the energy generated by decelerating the vehicle. It would be possible to absorb kinetic energy during braking in supercapacitors and release that energy during acceleration.

References

[1] J. S Carneiro and L. M Serrano. Descrição do funcionamento dos componentes do veículo de levitação maglev-cobra. *POLI UFRJ*, 2019.

[2] Comitê Técnico do Plano Diretor. Plano diretor. [online], 2011. `http://redpgv.coppe.ufrj.br/index.php/pt-BR/producao-da-rede/universidades-com-pgvs/1052-plano-diretor-ufrj-2020/file`.

[3] B. E. Conway. Electrochemical Supercapacitors. 1999.

[4] V. R De Moura. Análise energética do módulo de transporte do maglevcobra. *POLI UFRJ*, 2016.

[5] Eaton. "xvm-315 supercapacitor". [online]. `https://bit.ly/315Vsuper`.

[6] Elcap. "power vs energy density". [online]. `https://commons.wikimedia.org/wiki/File:Power_vs_energy_density.png`.

[7] André Augusto Ferreira and José Antenor Pomílio. Estado da arte sobre a aplicação de supercapacitores em eletrônica de potência. *Eletrônica de Potência*, 10(2):25 – 32, 2005.

[8] Maxwell. "bmod0165 p048 c01". [online]. `https://www.maxwell.com/images/documents/48V_ds_DuraBlue_3000685_4.pdf`.

[9] Maxwell. "ibmod0063 p125 b08". [online]. `https://www.maxwell.com/images/documents/125V_Module_datasheet.pdf`.

[10] Skeleton. "sma170v53faf". [online]. `https://bit.ly/170Vskeleton`.

[11] Skeleton. "sma51v177faf". [online]. `https://bit.ly/51vskeleton`.

[12] R. M Stephan, R de Andrade, and G. G Sotelo. Third Generation Of Flywheels: A Promising Substitute To Batteries. *Eletrônica de Potência*, pages 171–176, 2008.

[13] Richard M. Stephan, Rubens De Andrade, and Antônio Carlos Ferreira. Superconducting light rail vehicle: A transportation solution for highly populated cities. *IEEE Vehicular Technology Magazine*, 7(4):122–127, 2012.

[14] WEG. "afw 11 - acionamento com inversor de frequência modular". [online]. https://static.weg.net/medias/downloadcenter/h5f/h37/WEG-AFW11-catalogo-50026205-pt.pdf.

[15] WEG. "cfw-11 rb". [online]. https://bit.ly/2GXg4SQ.

[16] WEG. "inversor de frequência cfw-11 - manual do usuário". [online]. https://static.weg.net/medias/downloadcenter/hab/hc0/WEG-cfw11-manual-do-usuario-200-400v-mec.-a-a-d-10000062964-manual-portugues-br.pdf.

[17] WEG. "manual motor protectors - mpw series". [online]. https://www.galco.com/techdoc/weg/mpw25-3-u016_cp.pdf.

[18] Mazen Yassine and Drazen Fabris. Performance of commercially available supercapacitors. *Energies*, 10(9), 2017.

[19] Yunasko. "edlc: 48v / 165f module". [online]. https://yunasko.com/images/data/YUNASKO_Datasheet_module_48V_165F_E.pdf.

[20] Jiang Zhou, Ying Huang, Xiehong Cao, Bo Ouyang, Wenping Sun, Chaoliang Tan, Yu Zhang, Qinglang Ma, Shuquan Liang, Qingyu Yan, and Hua Zhang. Two-dimensional NiCo2O4 nanosheet-coated three-dimensional graphene networks for high-rate, long-cycle-life supercapacitors. *Nanoscale*, 7(16):7035–7039, 2015.

Monitoring of Active Magnetic Bearings

Bert-Uwe Köhler[1], Matthias Lang[2], Kristin Krenek[3]
[1] HAW Hamburg, Germany
[2] BBW Hochschule Berlin, Germany
[3]Berliner Hochschule für Technik, Germany
BertUwe.Koehler@haw-hamburg.de, 46778@bbw-hochschule.de, Kristin.Krenek@bht-berlin.de

Abstract

Two simple but efficient methods for integrity monitoring of position transducers within an active magnetic bearing are presented. Position transducers may be considered as a critical component in a control loop of the active magnetic bearing. Continuous monitoring of their integrity during operation may hence increase reliability and robustness of the active magnetic bearing and may improve quality and safety of a related industrial application. The proposed methods rely on a cross validation of sensorial information between different control axes of the active magnetic bearing and may be implemented entirely in software without additional hardware.

1 Introduction

Two simple but efficient methods for integrity monitoring of position transducers within an active magnetic bearing are presented. The methods are intended to increase reliability and robustness of active magnetic bearings used in applications where high quality and safety levels are required. In general, industrial components are typically designed to address demanding reliability and safety requirements and to provide a high maturity level of industrial equipment (see also [1]). However, it is common practice to additionally monitor particularly critical components and, in case of detecting a critical failure, to bring the machinery into a safe state.

Closed loop position control in active magnetic bearings may use signals produced by position transducers as feedback signals. As such, a transducer can be considered as a critical component wherein a transducer failure may cause damage to equipment or even injuries to operating personnel. Failure cases concerning position transducers may comprise, for example, a broken transducer cable which can normally be detected by appropriate electronic circuitry. More difficult to manage are transducer failures when a malfunction is less obvious such as a drift or a sporadic dropout which may lead to wrong position values.

The proposed methods are intended to provide a means to detect the latter kind of malfunctions. They are based on a consistency check between the signals produced by the position transducers of at least two different individually controlled active magnetic bearing control axes, wherein each of the axes has two position transducers in, with respect to a cross-sectional view of the rotor, diametrically opposite positions. Although the application of the proposed methods is particularly advantageous in radial bearings, they may be used for monitoring the correct functioning of position transducers of an axial active magnetic bearing as well. Furthermore, for an implementation, there is no need for extra hardware. Instead, the proposed methods can be implemented with a relatively simple software routine within a magnetic bearing control unit (see also [2, 3]).

2 Methods

2.1 Technical Background

A schematic cross-sectional view of an arrangement of a rotor with four position transducers as typically used in an active magnetic bearing is depicted in Figure 1. The transducers s_1, s_2, s_3 and s_4 are aligned with two orthogonal axes denoted as x axis and y axis. The center of the rotor is typically disposed at the point of intersection (x_0, y_0) of the two axes, wherein the point of intersection is also denoted as a central position. In order to keep the rotor at the central position, closed loop position control may be performed independently with respect to each of the axes (decentral control).

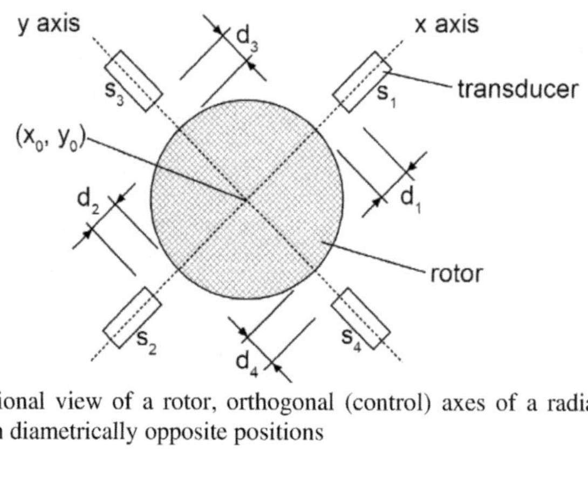

Figure 1: Cross-sectional view of a rotor, orthogonal (control) axes of a radial active magnetic bearing and transducers in diametrically opposite positions

The transducers s_1, s_2, s_3 and s_4 are configured to measure the distances d_1, d_2, d_3 and d_4 between the rotor and the respective transducers. They are usually calibrated such that, when the rotor cross-sectional center is at the central position, the distances are $d_1 = d_2 = d_3 = d_4 = d_0$, with d_0 as the nominal distance between the respective transducer and the rotor.

During normal, that is, error-free operation, the distance between each of the transducers and the rotor may vary according to

$d_1 = d_0 - \Delta_{th} - x$,
$d_2 = d_0 - \Delta_{th} + x$,
$d_3 = d_0 - \Delta_{th} - y$, and
$d_4 = d_0 - \Delta_{th} + y$,

where the term Δ_{th} represents a thermal expansion of the rotor, depending on an operating temperature and time, and where x and y denote a deflection of the rotor in a direction of the x and y axis, respectively. Furthermore, since, at least for a large rotor diameter, the deflections are small in comparison to the rotor diameter, in the above equations the deflection in the y direction is neglected for d_1 and d_2 and the deflection in the x-direction is neglected for d_3 and d_4.

For closed loop position control, it is possible to use average measured signals x_{meas} and y_{meas} of the transducer signals as feedback signals, wherein x_{meas} and y_{meas} are calculated from

$$x_{meas} = \frac{d_2 - d_1}{2} \text{ and } y_{meas} = \frac{d_4 - d_3}{2}.$$

This kind of averaging is advantageous since the thermal expansion term Δ_{th} disappears in x_{meas} and y_{meas}. Furthermore, by way of averaging, it is possible, for example, to reduce measurement noise, thereby reducing noise emissions and/or power consumption of the active magnetic bearing.

In the case where a transducer produces wrong distance values, x_{meas} and/or y_{meas} may become erroneous and the closed loop position control may then bring the rotor into a wrong or at least an undesired position. For example, if the value of d_1 exhibits an offset drift ε, that is $d_1 = d_0 - \Delta_{th} - x + \varepsilon$ then the feedback value x_{meas} becomes

$$x_{meas} = \frac{d_2 - d_1}{2} = \frac{d_0 - \Delta_{th} + x - d_0 + \Delta_{th} + x - \varepsilon}{2} = x - \frac{\varepsilon}{2}.$$

Assuming further that, by convention, the nominal (central) position in x direction is $x_0 = 0$, then the magnetic bearing closed loop control will modify the magnetic field of the magnetic bearing until $x_{meas} = 0$ such that the true position of the controlled rotor is (on average) at $x = \varepsilon/2$. That is, in the exemplary case, the rotor position will exhibit an undesired offset from the desired position.

A further implication of $x_{meas} = 0$ and closed loop position control is that (on average) $d_1 = d_2$. Due to this relation, it is not possible to detect the offset drift ε by comparing d_1 with d_2. Instead, to detect the offset drift ε, it seems to be possible to compare, for example, d_1 or d_2 with the nominal distance d_0. However, since d_1 and d_2 also depend on the radial thermal expansion term Δ_{th} it is not possible to exactly determine what change of d_1 or d_2 is due to the rotor expansion or due to the offset drift. Therefore, the drift ε cannot be reliably detected from comparing d_1 or d_2 with d_0. A possible solution to detect the offset drift would be to install a monitoring system with an additional position transducer that may monitor the rotor position independently. The disadvantage of such a monitoring system is that such an independent measurement chain may be expensive, it may require space, it may also need maintenance and calibration, and it may be defective as well.

The methods proposed in this paper are based on the observation that radial rotor expansion due to a temperature increase applies to both, x and y axis. Therefore, the expansion term applies with the same amount to all four transducer signals d_1, d_2, d_3 and d_4. Consequently, when comparing for example d_1 and d_3, there is only equality $d_1 = d_3$ if the offset drift term $\varepsilon = 0$.

Therefore, if the rotor position in the x axis direction is controlled independently from the rotor position in the y axis direction, it is possible to use position signals of the x axis transducers to verify position signals of the y axis transducers and vice versa, that is, to mutually verify the plausibility of x axis transducer signals and y axis transducer signals and hence the integrity of the transducers. Two methods, method A and method B, based on this idea are proposed in the following.

2.2 Method A

Method A directly compares mean distances (low pass filtered distances) between the rotor and the transducers of the x and the y axis. The mean distances are used for this comparison instead of the instantaneous distance values since the central position of the rotor usually moves on an orbit during operation corresponding to unbalances, rotordynamics, process forces and the forces produced by the active magnetic bearing. An example for such an orbit and $\varepsilon = 0$ is depicted in Figure 2 where, in the left diagram, an example for a trajectory of the rotor center on an orbit is depicted and, in the right diagram, corresponding signals x_{meas} and y_{meas}, calculated as indicated above, are shown.

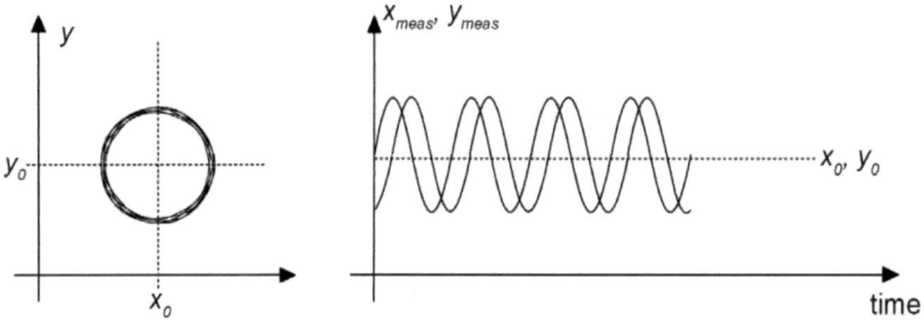

Figure 2: Left - trajectory of movement of the rotor center on an orbit; Right – movement of the rotor center in x and y direction versus time

For $\varepsilon = 0$, the mean distances between the transducers and the rotor are

$$\bar{d}_1 = \bar{d}_2 = \bar{d}_3 = \bar{d}_4 = d_0 - \Delta_{th},$$

with, as defined above, d_0 as the nominal (calibrated) distance between the respective transducer and the rotor and the radial thermal expansion term Δ_{th}. However, in the case of $\varepsilon \neq 0$ and assuming ε to be present in the transducer signal d_1, one obtains

$$\bar{d}_1 = \bar{d}_2 = d_0 - \Delta_{th} + \varepsilon/2 \neq \bar{d}_3 = \bar{d}_4 = d_0 - \Delta_{th},$$

where the equalities $\bar{d}_1 = \bar{d}_2$ and $\bar{d}_3 = \bar{d}_4$ are due to closed loop position control. So, the mean transducer signals of the different axes differ from each other in case of an offset drift error ε.

To summarize method A: at least one low pass filtered transducer signal of the x axis is compared with at least one low pass filtered transducer signal of the y axis in order to detect an offset drift error of at least one of the transducers. A substantial inequality between the compared signals indicates an offset drift error of at least one of the transducers.

Due to the necessity to filter the position signals with a low pass filter, it may be impractical to use this method for low rotational speeds of the rotor.

2.3 Method B

As method A primarily focuses on the detection of an offset drift of a transducer signal, it is also possible to combine the offset drift detection with detecting a drift of a transducer amplification.

For this, the transducer signals are added to obtain an estimate of the rotor thermal expansion values, that is

$$\Delta d_x = \frac{(d_1 - d_0) + (d_2 - d_0)}{2} = \frac{-\Delta_{th} - x - \Delta_{th} + x}{2} = -\Delta_{th}$$

and

$$\Delta d_y = \frac{(d_3 - d_0) + (d_4 - d_0)}{2} = \frac{-\Delta_{th} - y - \Delta_{th} + y}{2} = -\Delta_{th}$$

Note: the division by 2 is introduced to obtain the same scaling as for method A; it is, however, not essential for the method B.

If there is no drift with respect to an offset or to a transducer amplification then

$$\Delta d_x = \Delta d_y.$$

Both, Δd_x and Δd_y assume values corresponding to the thermal expansion of the rotor relative to the position of the transducers. In this context, one should keep in mind that also machine parts that carry the transducers may be subject to thermal expansion.

In case of an offset drift error ε and a transducer amplification error α, for example in transducer s_1, one obtains for d_1 and d_2 (assuming closed loop position control, that is, ε is visible as ε/2 in d_1 and as ε/2 in d_2)

$$d_1 = d_0 - \Delta_{th} - \alpha x + \varepsilon/2 \text{ and } d_2 = d_0 - \Delta_{th} + x + \varepsilon/2,$$

and hence for Δd_x

$$\Delta d_x = \frac{(d_1 - d_0) + (d_2 - d_0)}{2}$$
$$= \frac{-\Delta_{th} - \alpha x + \varepsilon/2 - \Delta_{th} + x + \varepsilon/2}{2} = -\Delta_{th} + \frac{\varepsilon}{2} + \frac{(1-\alpha)}{2} x$$

At the same time, if the transducers in the y axis as assumed to be error free, then, as calculated above, one obtains for the y axis $\Delta d_y = -\Delta_{th}$.

Therefore, in order to detect a transducer error, it is possible to check Δd_x and Δd_y for inequality. In the above equation, Δd_x has three terms: the first term corresponds to the thermal expansion of the rotor, the second term to an offset drift of a transducer and the third term corresponds to a drift of the transducer amplification. While the first two terms are substantially DC components, the third term comprises a frequency component which corresponds to the rotational speed of the rotor (which is DC only at stand still). For the detection of an offset drift of one of the transducers one may compare the DC components of Δd_x and Δd_y and for the detection of an amplification drift one may monitor the evolution of the AC component of Δd_x and/or Δd_y over time. It is appreciated that this kind of monitoring may be performed in time and/or in frequency domain.

Note: to simplify the analysis, the amplification error α is only applied to the rotor deflection x. In general, the transducer amplification may affect the transducer offset drift ε as well.

3 Experiments

3.1 Simulation setup

The principles of method A and method B are analyzed by way of simulation. For this it is assumed that the active magnetic bearing is operated at a sampling rate of 16 kHz. Furthermore, in this simulation, the rotor deflections x and y are considered to be AC signal (sinusoidal) components having a frequency of 10 Hz (600 rpm).

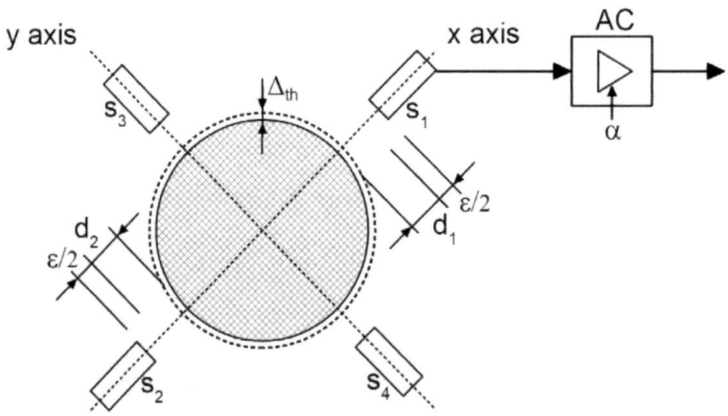

Figure 3: Simulation setup with an offset drift error ε and an AC amplification drift $\boldsymbol{\alpha}$ in the x axis and thermal rotor expansion is both x and y axis

The transducer signals d_1, d_2, d_3 and d_4 are simulated according to four setups. Each simulated transducer signal d_1, d_2, d_3 and d_4 includes four adjacent signal segments corresponding to the four setups, respectively. The setups are schematically depicted in Figure 3 and are defined as follows:

Setup I: Floating rotor without rotation. The transducer signals are given as constant values $d_1=d_2=d_3=d_4=d_0$ wherein d_0 is the calibrated nominal air gap value of the magnetic bearing and exemplarily set to $d_0=200\mu m$; the time span of the segment is $t \in [0s; 50s]$.

Setup II: Thermal rotor expansion during rotation. The thermal rotor expansion is simulated as exponential air gap reduction according to $\Delta_{th}= 40\mu m \cdot (1 - exp(-(t - 50s)/10s)$ with time span $t \in [50s; 100s]$. The transducer signals are hence given with

$$d_1 = d_0 - \Delta_{th}(\mathrm{t}) - \mathrm{x}$$
$$d_2 = d_0 - \Delta_{th}(\mathrm{t}) + \mathrm{x}$$
$$d_3 = d_0 - \Delta_{th}(\mathrm{t}) - \mathrm{y}$$
$$d_4 = d_0 - \Delta_{th}(\mathrm{t}) + \mathrm{y}$$

Setup III: Linear offset drift of transducer 1, wherein the offset drift is simulated with $\varepsilon = 50\mu m \cdot (t - 100s)/50s$ with time span $t \in [100s; 150s]$. The transducer signals are thus given with

$$d_1 = d_0 - \Delta_{th}(100\mathrm{s}) - x + \varepsilon(\mathrm{t})/2$$
$$d_2 = d_0 - \Delta_{th}(100\mathrm{s}) + x + \varepsilon(\mathrm{t})/2$$
$$d_3 = d_0 - \Delta_{th}(100\mathrm{s}) - \mathrm{y}$$
$$d_4 = d_0 - \Delta_{th}(100\mathrm{s}) + y$$

Setup IV: Linear drift of the amplification of the rotor deflection, wherein the amplification drift is simulated as $\alpha(t) = (t - 150s)/50s + 1$ with $t \epsilon [150s; 200s]$. Therefore, the transducer signals are given with

$$d_1 = d_0 - \Delta_{th}(100s) - \alpha(t) \cdot x + \varepsilon(150s)/2$$
$$d_2 = d_0 - \Delta_{th}(100s) + x + \varepsilon(150s)/2$$
$$d_3 = d_0 - \Delta_{th}(100s) - y$$
$$d_4 = d_0 - \Delta_{th}(100s) + y$$

It should be noted that an amplification error could, in a real setup, also affect the offset value of a transducer. However, in order to demonstrate the effect more clearly, the impact of the amplification drift is, in this simulation, limited to the AC signal component of the rotor deflection.

As an example, the signal d_1 of the transducer s_1 is depicted in Figure 4a where all four setups are depicted in a timely adjacent order. Figure 4b shows a low pass filtered signal of the signal of Figure 4a. For low pass filtering an 3rd order elliptic filter was used with a cutoff frequency of 1 Hz and a stop frequency at 5 Hz, pass band ripples of at most 3dB and stop band attenuation of at least 60dB.

Method A uses signals corresponding to Figure 4b (the signals d_1, d_2, d_3 and d_4 are filtered with the same filter) whereas method B uses the unfiltered signals corresponding to Figure 4a (corresponding to the equations above).

The simulation is performed with Octave, version 6.2.0.

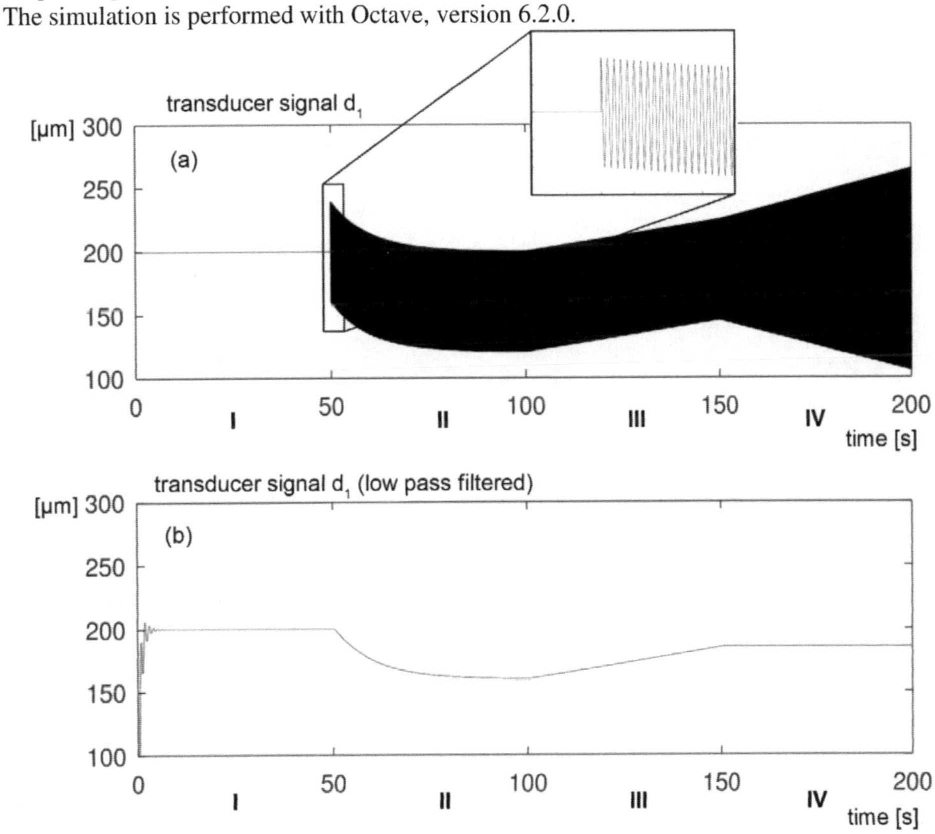

Figure 4: Simulation of transducer signal d1, **(a)** raw signal, **(b)** filtered signal); setup I – levitation of rotor; setup II – thermal expansion of rotor; setup III – transducer offset drift; setup IV – transducer amplification drift

3.2 Results

Simulation results are presented in Figure 5. Figure 5a shows simulation results for method A, that is, a function graph representing the difference $\bar{d}_3 - \bar{d}_1$. More specifically, in setup I (rotor floating without rotation, no drift) the difference $\bar{d}_3 - \bar{d}_1$ equals zero. The same applies to setup II (thermal expansion of the rotor). However, a drift of the transducer offset is indicated in setup III (transducer s_1 exhibits an offset drift). The drift of the AC amplification as simulated in setup IV, however, is not detected by method A, that is, the difference $\bar{d}_3 - \bar{d}_1$ remains at the same level as at the end of setup III.

Figure 5b shows simulation result for method B, that is, a function graph representing the difference $\Delta d_y - \Delta d_x$. For setups I to III the result of method B is similar to the result of method A. However, in contrast to method A, method B indicates an AC offset drift, that is, for setup IV the difference signal $\Delta d_y - \Delta d_x$ exhibits a signal component with an oscillation according the rotational frequency and an increasing amplitude (due to the simulation setup).

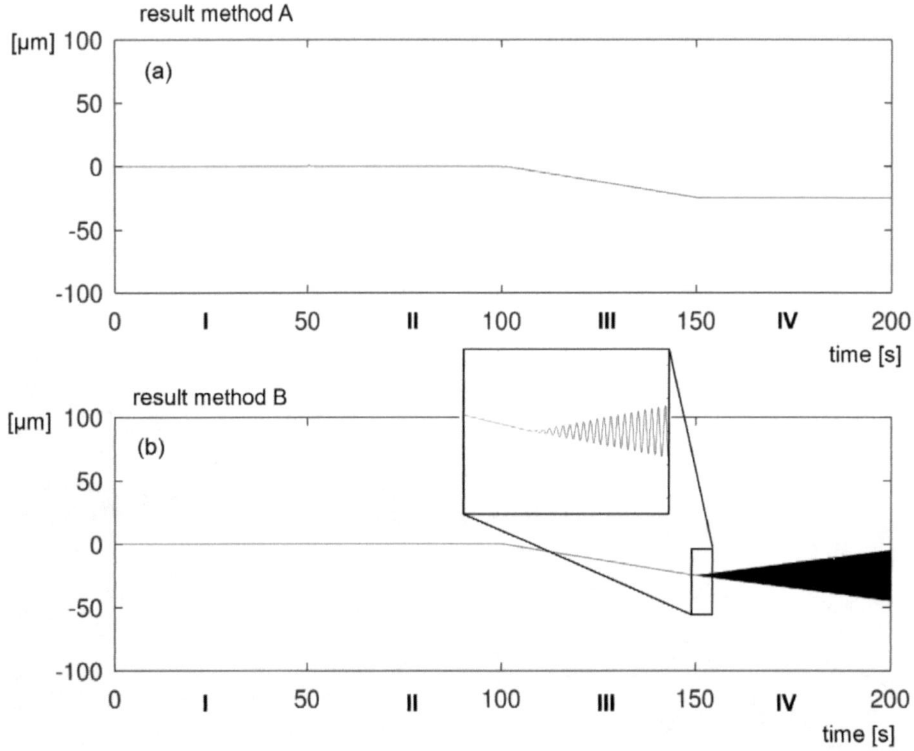

Figure 5: **(a)** Simulation result for method A, the depicted function graph is the difference $\bar{d}_3 - \bar{d}_1$. **(b)** simulation result for method B $\Delta d_y - \Delta d_x$

4 Discussion

Both, method A and method B are based on the same principle, that is, a comparison of the transducer signals of the different control axes. In general, method B seems to be more flexible than method A and provides a means to detect, in addition to a transducer offset drift, a transducer amplification drift.

The offset drift, for example as simulated in setups II and III, may comprise low frequency components in case of a slow offset drift and a broad band component including high frequencies in case of offset steps. In particular low frequency offset drifts may be detected with method A, wherein method B may, due to the absence of low pass filtering, detect both low and high frequency components of the offset drift.

Offset drift may be due to a movement of one or more of the transducers. Such a transducer movement may occur, for example, in case of a thermal expansion of a machine part to which the transducer is attached. Such a thermally induced movement of the transducer position may cause different drifts in all four transducer locations. In this sense, it is important to make the transducer fixation as stationary as possible.

In general, offset drift may also be related to a transducer amplification drift. In this paper we have neglected this effect for the sake of simplicity of the analysis by limiting the transducer amplification drift to AC components. The simulation results of setup IV show that AC amplification drift is detectable only with method B. In addition, from Figure 5b one may conclude that a detection of an oscillation or a changing oscillation in the signal Δd_x alone (i.e. without comparison to Δd_y) may indicate an amplification drift in the x axis.

With the proposed methods A and B, it is not possible to determine which of the transducers is defective. In order to obtain this kind of information, further tests would be required. However, it is at least possible to detect inconsistencies within the sensorial information. Since, in general, more than one transducer could be defective, it is not possible to estimate the extent of the defectiveness without an uncertainty. This uncertainty, however, would, due to closed loop control, also occur in a similar way if a completely independent monitoring system for monitoring the transducer integrity were in use.

It should also be noted that information about a temperature of the rotor may be obtained with method B. In particular, it is possible to estimate the thermal expansion of the rotor (relative to the stator, wherein a thermal expansion of the stator may be estimated independently) in the plane where the transducers are located. Since machine temperatures (including the rotor temperature) are usually monitored in order to detect abnormal situations, method B gives additional information that can be used in general machine operation monitoring.

5 Conclusion

The methods proposed in this paper provide simple and efficient means to protect machinery against non-obvious malfunctions of magnetic bearing position transducers such as offset drift, amplification drift and sporadic dropouts. It is possible to recognize a potential problem of the transducers and to bring, if necessary, the machinery into a safe state. Further investigations about a root cause of the detected problem may be started subsequently. For the implementation of the method there is no need for extra hardware. By solely a small software routine it is hence possible to increase the robustness, reliability and safety level of the machinery.

6 References

[1] B.-U. Koehler, J. Denk, G. van Maanen, M. Lang (2017): Applying Standard Industrial Components for Active Magnetic Bearings, Actuators, 6(1), 8; doi:10.3390/act6010008

[2] B.-U. Köhler (2018): Method for Monitoring A Magnetic Bearing Device, EP3511585B1, Siemens AG

[3] J. Denk, B.-U. Köhler, D. Stoiber, F. Viering (2018): Method for Monitoring A Magnetic Bearing Device, EP3511584B1, Siemens AG

Approaches to Efficiency and Life Cycle Assessment of Magnetic Bearing Systems

Hubert Mitterhofer[1] and Stefan Fragner[1]
[1] Linz Center of Mechatronics GmbH
hubert.mitterhofer@lcm.at

Abstract

The scientific method for assessing a product system's impact on the environment is the Life Cycle Assessment. This work points out why the magnetic bearing community will benefit from embracing this method for the design of magnetically levitated systems. A brief introduction to LCA following the ISO14040 standard and an interpretation for the application to magnetic bearings is given. Links to popular databases, literature and software used in the field of LCA are equally provided. Due to the material similarity, some data from the field of electric motor LCA are given and the special role of rare earth permanent magnets is highlighted.

1 Motivation and Introduction

Global warming due to man-made climate changes and the environmental impacts of our industrial societies are very likely the biggest challenges of our time. Their importance stems from the large number of fields which are concerned, combined with the severity of the consequences if the challenges in the respective fields are not met. Significant hopes rest on technical improvement, striving for increased efficiency, reduced material usage, better reusability or recyclability, etc. In addition to their valuable inputs to increasing system lifetime, rotor speeds, or improved data harvesting and digitization opportunities, magnetic bearings can also contribute to these environmental goals.

The method for evaluating the related environmental aspects throughout the entire life cycle including raw material extraction, manufacturing, transportation, use, and the end of life in a cradle-to-grave or cradle-to-cradle approach is the life cycle assessment (LCA) which will be shortly described in Chapter 2. LCA serves as a key method in modern product design and helps to distinguish seemingly ecologically good design options from actually good ones.

Both, the international research community and, especially, the industrial players active in the field of magnetic bearings have got used to arguing the advantages of magnetic bearings through five typical points:

- Lubrication-free operation,
- friction-free and abrasion-free operation,
- long lifetime,

- hermetic sealing capacity and thus suitability for vacuum or cryogenic conditions, and
- suitability for high rotational speeds.

However, the energy and efficiency aspect of magnetic bearings is hardly discussed – with the exception of high speed systems where the absence of friction originates an efficiency advantage. A solid comparison of the overall environmental impact in the form of an LCA can help to establish a further strong point for the beneficial use of magnetic bearings in highly performant systems. This technique is, however, new to many researchers and the literature research at the beginning of the work for this paper yielded no example. Meanwhile, one single example of such a study was published in early 2021: In [1], Byrd et. al. have compared a magnetic bearing system to an oil lubricated bearing system in the application of large water cooled chillers.

It is, therefore, not the goal of this work to perform and present an LCA of a specific magnetic bearing. It is rather to bring the issue to the magnetic bearing community - the designers of these systems - and to present the related aspects in Chapter 3.

2 Life cycle assessment – a brief introduction

In the year 1995, the authors of the Nordic Council report on Guidelines on Life-Cycle Assessment [2] described the LCA as follows: "*The LCA methodology evaluates holistically the environmental consequences of a product system or activity, by quantifying the energy and materials used, the wastes released to the environment, and assessing the environmental impacts of those energy, materials and wastes.*" While the statement was clear and comprehensible, the mentioned "LCA methodology" was diverse. An increasing number of different implementation methods led the International Organization for Standardization to publish the ISO 14040 standard [3] in 1997 in its first edition in order to unify this methodology*. This unification of the different approaches defines the principles and structure of an LCA in order to achieve comparability between individual LCA results.

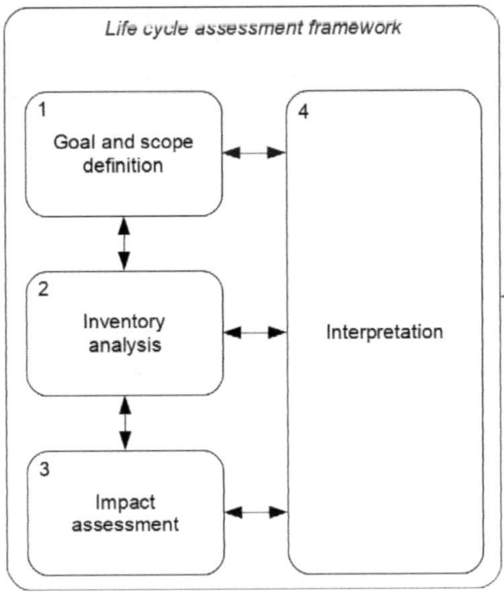

Figure 1 Stages of a life cycle assessment according to ISO 14040:2006:1

* ISO 14044 [30] comprises directives and technical specifications added later.

The ISO 14040 standard defines four stages in an LCA. As it can be seen in Figure 1, the stages are interconnected and each stage influences every other stage in the LCA framework. Below, a short description explains the content of the individual stages. Many further guidelines and explanations for the application of ISO 14040 can be found in the literature, e.g. in [4] or [5].

Goal and scope definition stage

The first task of an LCA study is to formulate its goal. Then the functional unit[†] as well as the system boundaries must be defined. Additionally, the handling of (missing) data has to be clarified and the required data quality has to be stated.

Life Cycle Inventory (LCI) analysis stage

The inventory analysis deals with the description, acquisition, modelling and calculation of material and energy flows to create an inventory of input/output data. The numerical result contains only input flows from the natural environment und output flows to the natural environment. Materials and pre-products of the technical sphere have to be retraced to get the input streams from the ecosphere. The graphical result of the inventory is the process flow chart which visualizes the system boundaries and enables traceability.

The main part of an LCI analysis is the data acquisition. Data can either be measured, calculated, or estimated. Their accuracy influences the quality of the inventory considerably. With the mass and energy balance, the inventory can be validated. Unless every single process is balanced, the data quality is not sufficient.

Life Cycle Impact Assessment (LCIA) stage

In the impact assessment stage, the data from the LCI analysis are bundled with regard to their potential environmental impacts in order to achieve a better understanding of the effects on the natural environment.

The goal is to assess all theoretical environmental impacts and to make them comparable. The combination of scientific-based assessment methods and subjectively generated compare mechanisms leads to an impact assessment consisting of mandatory and optional components.

While CO_2 emissions causing global warming are the most dominant environmental influence in today's public perception, there are many different impact categories that may be considered in an LCIA. In fact, there is a wide range of LCIA *methods*, each defining a certain set of categories and characterization factors. Amongst the most popular methods are *ReCiPe* [6], *IPCC 2013*, *ILCD 2011* [7], *CML 2012* [8], and *Cumulative energy demand* [9] according to a recent survey [10]. A comprehensive description of the most popular method sets and the included impact categories has been published by Acero et al. in [11]. Table 2 shows typical impact categories present in many of the mentioned LCIA methods.

Interpretation stage

In the fourth and last stage of the LCA study, the results of LCI analysis and the LCIA are compared with the goals defined in the first stage to generate conclusions, limitations, and recommendations. An important part is also the critical review, which is intended to ensure the validity of the study.

[†] A functional unit is a quantified description of the function of a product that serves as the reference basis for all calculations regarding impact assessment.

Impact category	Indicator unit
Climate change	kg CO2 equivalent
Stratospheric ozone depletion	kg CFC-11 equivalent
Ecotoxicity for aquatic fresh water	CTUe (comparative toxic unit for ecosystems)
Human Toxicity – cancer effects	CTUh (comparative toxic unit for humans)
Ionising radiation – human health effects	kg U^{235} equivalent (to air)
Acidification	mol H+ equivalent
Eutrophication - aquatic	Fresh water: kg P equivalent Marine: kg N equivalent
Land transformation	kg soil organic matter deficit

Table 1: Common LCIA impact categories [28]

3 Application to Magnetic Bearings

3.1 Goal and Scope definition

This initial stage of an LCA is a highly individual task, depending on why an LCA shall be conducted. Typical tasks could be the comparison of one bearing technology to another one within a certain application as presented in [1] for a chiller compressor bearing system.

Furthermore, also the scope, i.e. the system boundaries have to be chosen. For a bearing system, the *physical boundary* may be drawn around the bearings alone, but more often, an assessment of a complete drive unit or even the entire process may be more meaningful.

The *life cycle boundary* can be selected according to the previously chosen goal. Focusing on raw material extraction and end of life may be sufficient for a purely material-oriented examination while additional inclusion of manufacturing, transportation, and use, would be necessary for an energy and emission investigation.

The defined goal and scope will later on determine the impact assessment perspectives and the final interpretation.

3.2 LCI analysis

3.2.1. Materials and manufacturing

The subsequent inventory analysis stage contains the most intuitive considerations for the magnetic bearing engineer confronted with an LCA: The assessment of the involved components for tracing their energy and material flows. Of course, this highly depends on the magnetic bearing topology. Figure 2 gives an exemplary collection of typical parts of an active magnetic bearing (AMB).

These components compare well to those of a permanent magnet synchronous machines (PMSM) electric machine: solid shaft parts, electric steel lamination, insulated copper wire, power electronic components, and rare-earth permanent magnets (REPM), i.e. NdFeB or SmCo type magnets, etc. These high energy density REPMs are also the predominant material aspect for an LCI of passive magnetic bearings (PMB) and will be given special consideration in Section 3.3.

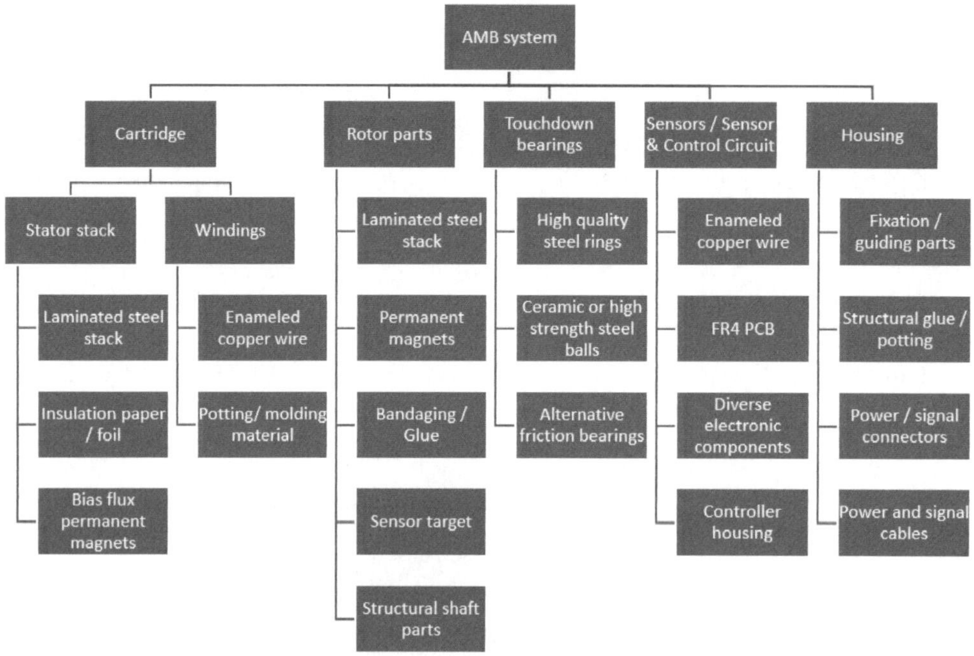

Figure 2 Exemplary component structure of AMB system.

Some recent publications have tracked the lifecycle issue for entire PMSMs [12], [13]. The authors of those two publications have presented a scalable LCI for automotive traction motors. Due to the mentioned material similarities between PMSMs and AMBs, the content gives valuable references for LCI magnetic bearing systems.

The next step after analyzing and listing the involved components is the energy and material flow, starting from raw material production to product assembly, i.e. a *cradle-to-*(factory) *gate*. **Figure 3** depicts such a flow diagram for PMSMs, demonstrating the impressive amount of aspects to consider.

3.2.2. LCI databases

Merging the component analysis and the material- and flow diagram allows to extract the masses of the required materials – not only of the finished magnetic bearing alone but of all process steps throughout its lifetime. For this step, the data on the necessary raw material and energy input per functional unit of a material are required. This information is typically found in LCI databases. A substantial number of such databases is offered online for free or under commercial license. A comprehensive list can be found at the openLCA Nexus [14], a short non-exhaustive selection according to the relevant topics may serve as orientation for magnetic bearing designers (information based on [14]):

- *ecoinvent* – Commercial database covering: agriculture, building and construction materials, chemicals, electricity, fishing, metals, refineries, textiles, tourism, transport, waste treatment and recycling, and water supply
- IDEMAT – Industrial Design & Engineering Materials database, also available as smartphone app giving low complexity information in form of eco-costs in € and carbon footprint in kg CO_2 equivalent

- Environmental Footprints (EF) – Partly free, partly commercial database offered by the European Commission covering amongst others: agriculture and foods, textile, thermal insulation, metal sheets, copper production, paints, IT equipment, photovoltaic electricity production
- NEEDS - New Energy Externalities Developments for Sustainability focusing on future electricity supply systems, material supply, transport services

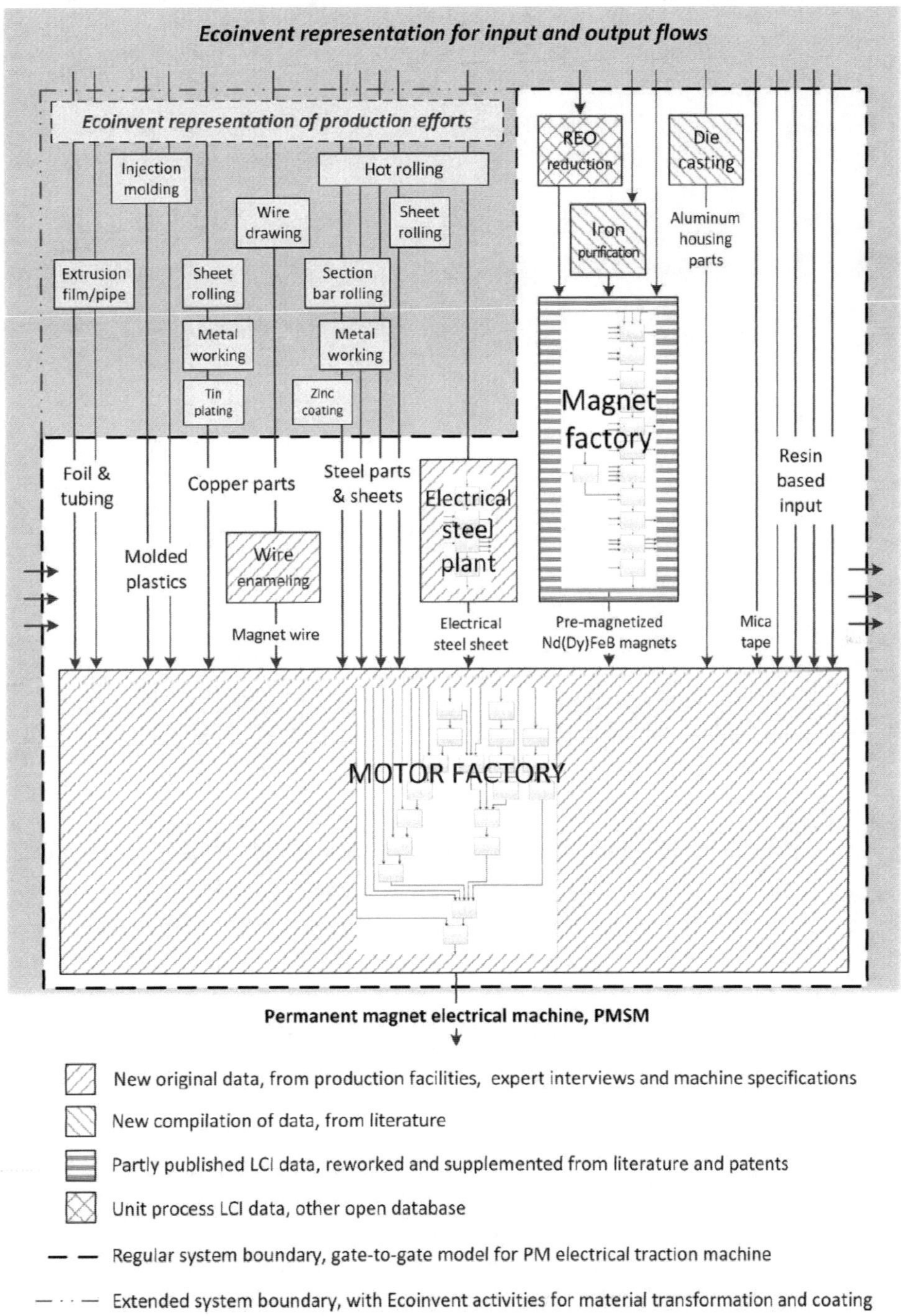

Figure 3 Flow diagram of material and energy for PMSM components presented in [13]

In order to demonstrate the granularity appropriate LCIs, Table 2, Table 3, and Table 4 present the LCI for NdFeB REPM, silicon steel sheets, and enameled copper. It has to be stressed that the mentioned electricity input and raw materials only cover the described processes – large shares are included in the raw products e.g. the unalloyed steel or the uncoated copper wire. This dependencies are resolved when evaluating the LCA in suitable software tools (such as openLCA [15], GaBi [16], or SimaPro [17], etc.).

Production of 1 kg of NdFeB nickel coated REPM		
Process Input	Amount	Unit
Unalloyed steel for electrolytic iron	913	g
Electricity, iron electrolysis	2	kWh
Boron carbide	15	g
Neodymium	310	g
Nickel	11	g
Dysprosium	91	g
Electricity, sintering	8.4	kWh
Electricity, other	5.6	kWh
Hydrogen	0.6	kg
Caustic soda	1	g
Sulfuric acid	1.4	g
Water	6	kg
Emissions to air	Amount	Unit
Nickel	4.2	mg
Emissions to water	Amount	Unit
Nickel sulfamate (23% nickel)	5.1	mg
Solid waste	Amount	Unit
NdFeB scrap	200	g
Sludge (ferrous chloride)	250	g
Sludge (97% nickel sulfamate, 3% nickel chloride, 24% nickel in total)	2.2	g

Table 2: Process inputs, emissions, and wastes of NdFeB REPM, data retrieved from [13]

Enameling of 1 kg of copper wire		
Process Input	Amount	Unit
Copper wire, uncoated	0.96	kg
Liquid enamel, polyester share	43	g
Liquid enamel, xylene solvent share	23	g
Electricity, enameling	0.5	kWh
Emissions to air	Amount	Unit
Xylene	23	g

Table 3: Process inputs, emissions, and wastes of enameling copper wire, data retrieved from [13]

Production of 1 kg of electrical silicon steel sheets		
Process Input	Amount	Unit
Unalloyed steel	1.116	kg
Ferrosilicon	31	g
Aluminum	5	g
Electricity (starting from alloyed steel)	0.63	kWh
Propane / LPG	12	g
Sulfuric acid	19	g
Rolling / lubricating oil	0.4	g
Quicklime powder	0.8	g
Phenolic resin	1	g
Emissions to air	Amount	Unit
CO_2	36	g
Nitrogen oxides	0.1	g
Sulfur oxides	0.06	mg
Solid waste	Amount	Unit
Steel scrap	114	g
Sludge (60% $CaSO_4$, 40% $Fe(OH)_2$)	3.3	g

Table 4: Process inputs, emissions, and wastes of electrical steel sheet, data retrieved from [13]

3.2.1. End of life

In the end of life consideration, the used materials will contribute either to the waste fraction or, in case of reuse or recycling, to the emission sum through negative emission factors. Table 5 gives values for end of life greenhouse gas emission values for the different scenarios.

Material	Reuse emissions	Recycle emissions	Landfill emissions	Unit
Steel	-3.58	-1.81	0.01	kg CO_2 / kg
Aluminum	-15.33	-7.19	0.72	kg CO_2 / kg
Copper	-7.48	-4.71	0.01	kg CO_2 / kg

Table 5: Emission factors of reuse, recycling, or landfill scenario of selected materials from [1].

3.2.2. Use phase energy

Magnetic bearings do not require any operation medium other than electric energy. Due to their contact-free operation, there is no need for lubrication, i.e. no need for operating fluids that have to be exchanged in certain intervals, that would have to be filtered, cooled, or in some other way conditioned. The same is true for wastes; Ideal magnetic bearings do not produce any wastes during operation – neither gaseous, fluid nor solid. As the only exclusion, abrasion from touchdown bearings may be considered, especially if low-cost friction type bearings are used.

This absence of operation media and use phase wastes reduces the material and energy flows in production and operation which is reflected in the LCI.

The actual magnetic bearing energy consumption is primarily dominated by parasitic losses such as windage losses or eddy current losses. The structural and material measures for mitigating these losses are well understood and frequently described in the literature. Also the control of AMBs has significant influence on the power consumption and a number of measures can be applied in order to reduce the mean input power, e.g. disturbance rejection through notch filters [18], zero-force control achieved by offsetting the axis of rotation against the direction of gravity or a (quasi) constant process force [19], or by applying bias magnets in the form of passive lift bearings or compensation bearings [20] [21].

Chiller compressor example

The biggest and most important efficiency aspect stems from an intrinsic capacity of magnetic bearings: the friction-less operation. In addition to the absence of mechanical bearing losses, this also permits higher speeds at which radial compressors reach higher efficiency levels and driving motors can be built more compact and thus, with less material input. Adding to the energy savings for not having to condition the lubrication oil, these advantages give the magnetically levitated chiller compressor described in [1] an astonishing 38.8% advantage in consumed use energy over the comparison compressor with oil bearings. This amounts to a mean input power reduction of 54.5kW.

Power plant turbine example

Another impressive example has been published in [22] by Grund et al. In 2014, Siemens has installed the world's first magnetically levitated power plant turbine (see Figure 4) in the lignite-fueled plant of Jänschwalde, Germany. Its 2.5 ton rotor is levitated by 2 radial AMBs and an axial AMB. The magnetic bearings have been recognized to increase the turbine efficiency by 1%. A large contribution to that increase, however, does not come from advanced bearing efficiency but stems from the fact that the lubrication oil conditioning including filtering, preheating, etc. can be completely eliminated from the system.

Ironically yet reasonably, it needs to be said that the complete lignite power plant will be shut down until 2028 in order to reduce carbon emissions.

Figure 4 10MW feedwater pump on AMB installed in a fossile fuel power plant. Image: courtesy of Siemens (Turbinenwerk Görlitz)

High-purity pump and spin processor example

The process steps in wafer fabrication involve multiple pumping of liquids into or out of the process chamber. Since any contamination must be prevented, perfect sealing of the pumps involved is required. Magnetically levitated pumps allow hermetic separation of rotor and stator. Originally used mainly in the chemical mechanical planarization (CMP) process for pumping slurry and as blood pumps in medical technology, these pumps (Figure 5, left image) are increasingly being used in the semiconductor, medical and pharmaceutical industries.

Another field of AMB application are so-called spin processors (Figure 5, right image), i.e. drives which rotate the wafer to distribute photoresist or to spin off cleaning liquids in a process chamber conditioned by vacuum or process gas filling. Due to the hermetic separation of rotor and stator by the process chamber wall, the spin process can be performed in the closed atmosphere, which reduces impurities and saves time and energy by eliminating the conditioning process.

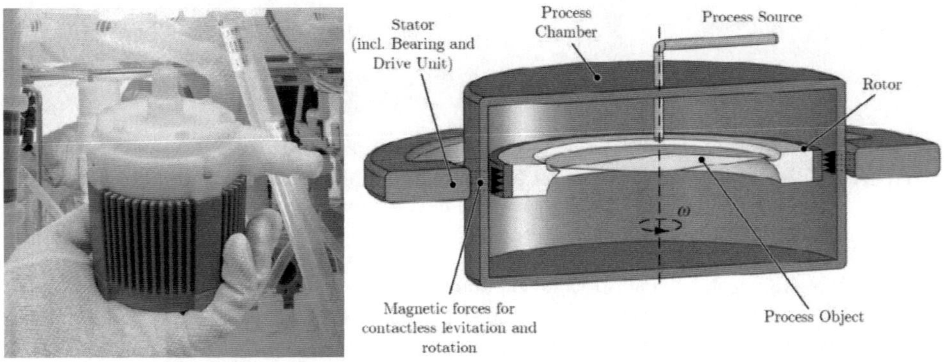

Figure 5 High-purity pump (left) and spin processor model (right) [29]

Both mentioned systems alone may be disadvantages in terms of life cycle impact when directly compared to pumps and spin processors with conventional ball bearings. However, due to their high-purity operation, an increased chip yield, reduced processing time, or significantly reduced process gas consumption may be achieved. The authors of this article do not dispose of concrete data on these aspects, it is, however, highly likely, that a significantly larger system boundary definition would be necessary when comparing such complex systems. Additional filters, external purification systems, complex sealing systems, and the longer plant operation time due to lower chip yield of the compared conventional bearing solution would also have to be considered. In that case, the life cycle impact of the magnetically levitated systems will very likely outperform the conventional solution.

3.3 LCIA

According to the selected LCIA method (cf. LCIA methods, Section 2), the LCIA will express the material and energy flow of the LCI in the respective set of impact categories. This evaluation is typically provided by an LCA software tool in combination with appropriate LCI databases.

Regardless of the precision of the data provided in a database, there are certain dependencies of the impact values, e.g. to the geographic location selected in the scope of the LCA. The biggest geographic influence on magnetic bearing systems originates from the carbon intensity of the electric energy at the place of operation. Values differ from >600 g CO2 equivalent / kWh in coal-intensive regions to <100 g CO2 equivalent / kWh in regions which exploit wind and hydro power. This of course can heavily influence the LCIA result concerning global warming potential.

Also REPM materials play a particular role for some selected impact categories. They constitute a central element of both, AMBs with PM bias flux and PMBs, and their environmental impact is widely discussed in the general public. Especially the human toxicity and ecotoxicity aspects make the use of REPM very critical.

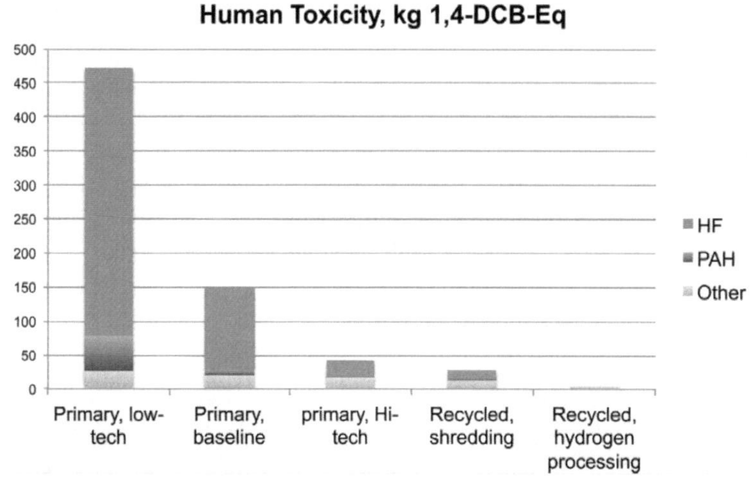

Figure 6 Comparison of human toxicity of different REPM production routes presented in [23]. (HF: hydrogen fluoride, PAH: polycyclic aromatic hydrocarbons)

Therefore, numerous recent projects and publications deal with the LCA of REPM ([23], [24]) and with the comparison between magnet production from virgin, "primary" raw material and production from recycled magnet material [25].

In Figure 6, the authors of [25] show the potential of REPM recycling and the possible hazards of low-tech REPM primary production. This is a clear reminder to use REPM materials only where they can contribute a real benefit in terms of an LCIA and to avoid distributing them for price reasons only to household or other consumer products which will end up in landfills and incinerators.

3.4 Interpretation

As with the Goal and Scope definition stage, the interpretation is a highly individual task, depending on the defined goal, the used LCIA method and the overall purpose of the conducted LCA. Therefore, this stage shall not be further considered in this work.

4 Conclusion and Outlook

LCA has become a buzzword in today's efforts to drive technical development for combatting detrimental environmental impacts. A large community of researchers has been dealing with the environmental impact assessment for the past 30 years. There are countless publications, standards, databases, software suits, methods, etc. around this topic.

Nevertheless, only during the work on this article, the first scientific publication on an LCA for a magnetic bearing system was published, showing how little this important topic has been perceived by the magnetic bearing community.

Despite the fact that many aspects of an LCA for a magnetic bearing system can be translated from LCAs for electric motors due to similarity of the used raw materials and manufacturing processes, significant effort has still to be made to establish widely-accepted, comparable, and potentially scalable models for the most dominant types such as radial and axial AMBs.

Regardless of these difficulties ahead, the authors of this article hope to have presented a first insight into LCA from and for the viewpoint of a magnetic bearing designer. The merits of embracing this methodology into the workflows of magnetic bearing design are high:

The analyses of [1] and [22] have shown, that the overwhelming advantage of the magnetically levitated compressor or turbine systems lies in the better system efficiency in combination with long on-times, i.e. in the consumed electrical use energy.
The same conclusion is drawn by Auer and Meincke in [26] who assessed induction machines with efficiency classes IE2, IE3, and IE4 and by Rothboeck in [27] when analyzing ferrite-type and NdFeB-type PMSM in different use case scenarios. It is, therefore, very clear, that in addition to the five classical arguments for magnetic bearing systems, an efficiency advantage will materialize in better LCA performance in systems with long lifetime and long on-time. At the same time, a clear LCA result will also help to identify the fields where magnetic bearings will not be beneficial.

The given examples have worked on high power systems but the future may also bring examples for much smaller systems in industrial production machines, public or building infrastructure or the HVAC sector as all of these fields dispose of machines with long on-times.

A less evident but potentially even more rewarding field lies in applications where magnetic bearings enable new solutions or processes due to their special features (e.g. lubrication-free, hermetic, cryogenic operation capability). When conducting an LCA for system comparison, the conventional reference solution may dispose of significantly more system parts to achieve the same performance as the magnetically levitated one. Depending on the application, the environmental impact can differ not only by a few percentage points, but by whole orders of magnitude.

References

[1] E. Byrd, B. Netzel, D. B. Adams and H. Zhang, "Life cycle GHG assessment of magnetic bearing and oil lubricated bearing water cooled chillers," *Journal of Industrial Ecology,* 2021.

[2] L.-G. Lindfors, K. Christiansen, L. Hoffman, Y. Virtanen, V. Juntilla, O.-J. Hanssen, A. Ronning, T. Ekvall and G. Finnveden, Nordic Guidelines on Life-Cycle Assessment, Copenhagen: Nordic Council of Ministers, 1995.

[3] ISO 14040:2006-1, Environmental management - life cycle assessment - principles and framework, International Standardization Organization.

[4] I. V. Muralikrishna and V. Manickam, "Life Cycle Assessment," *Environmental Management,* 2017.

[5] W. Klöpffer and B. Grahl, Life Cycle Assessment (LCA): A Guide to Best Practice, ISBN: 978-3-527-32986-1: Whiley, 2014.

[6] National Institute for Public Health and the Environment, Netherlands, "LCIA: the ReCiPe model," [Online]. Available: https://www.rivm.nl/en/life-cycle-assessment-lca/recipe. [Accessed June 2021].

[7] S. Sala, M.-A. Wolf and R. Pant, "Characterisation factors of the ILCD Recommended Life Cycle Impact Assessment methods. Database and supporting information," *European Commission Report EUR25167EN,* 2012.

[8] Universiteit Leiden, "CML-IA Characterisation Factors," 2016. [Online]. Available: https://www.universiteitleiden.nl/en/research/research-output/science/cml-ia-characterisation-factors. [Accessed July 2021].

[9] R. Frischknecht, F. Wyss, S. Knöpfel-Büsser, T. Lützkendorf and M. Balouktsi, "Cumulative energy demand in LCA: the energy harvested approach," *The International Journal of Life Cycle Assessment 20 (7),* 2015.

[10] iPoint GmbH, "Life Cycle Impact Assessment – which are the LCIA indicator sets most widely used by practitioners?," [Online]. Available: https://www.ipoint-systems.com/blog/lcia-indicator/. [Accessed June 2021].

[11] A. P. Acero, C. Rodriguez and A. Ciroth, "LCIA methods - Impact assessment methods in Life Cycle Assessment and their impact categories," 11 May 2016. [Online]. Available: https://www.openlca.org/wp-content/uploads/2016/08/LCIA-METHODS-v.1.5.5.pdf. [Accessed July 2021].

[12] A. Nordeloef, E. Grunditz, A.-M. Tillman, T. Thiringer and M. Alatalo, "A scalable life cycle inventory of an electrical automotive traction machine - Part I: design and composition," *International Journal of Life Cycle Assessment,* 2017.

[13] A. Nordeloef and A.-M. Tillman, "A scalable life cycle inventory of an electrical automotive traction machine - Part II: manufacturing processes," *International Journal of Life Cycle Assessment,* 2017.

[14] GreenDelta GmbH, "openLCA Nexus," [Online]. Available: https://nexus.openlca.org/databases. [Accessed June 2021].

[15] GreenDelta GmbH, "openLCA Website," [Online]. Available: http://www.openlca.org/. [Accessed June 2021].

[16] Sphera Solutions GmbH, "GaBi Website," [Online]. Available: https://gabi.sphera.com. [Accessed June 2021].

[17] PRé Sustainability, "SimaPro Website," [Online]. Available: https://simapro.com/. [Accessed June 2021].

[18] H. Rong and K. Zhou, "Disturbance Rejection for a Zero-bias Controlled Active Magnetic Bearing Based on Disturbance Observer and Notch Filter," *Journal of the Applied Computational Electromagnetic Society,* 2019.

[19] S. Zheng, H. Li, B. Han and J. Yang, "Power Consumption Reduction for Magnetic Bearing Systems During Torque Output of Control Moment Gyros," *IEEE Transactions on power electronics,* 2017.

[20] M. Hedlund, J. Abrahamsson, P.-L. J. J. Lundin and H. Bernhoff, "Passive Axial Thrust Bearing for a Flywheel Energy Storage System," *Proc. Brazilian Workshop on Magnetic Bearings,* 2013.

[21] J. Perez-Loya, C. Abrahamsson, F. Evestedt and U. Lundin, "Performance Tests of a Permanent Magnet Thrust Bearing for a Hydropower Synchronous Generator Test-Rig," *Journal of the Applied Computational Electromagnetics Society,* 2017.

[22] C. Grund and D. Haje, "Qualification of Magnetic Bearings for Industrial Turbines," *Proc. 14th International Symposium on Magnetic Bearings,* 2014.

[23] B. Sprecher, Y. Xiao, A. Walton, J. Speight, R. Harris, R. Kleijn, G. Visser and G. J. Kramer, "Life Cycle Inventory of the Production of Rare Earths and the Subsequent Production of NdFeB Rare Earth Permanent Magnets," *Environmental Science & Technology 48 (7),* pp. 3951-3958, 2014.

[24] N. Adibi, Z. Lafhaj and J. Payet, "New resource assessment characterization factors for rare earth elements: applied in NdFeB permanent magnet case study," *The International Journal of Life Cycle Assessment,* 2019.

[25] H. Jin, P. Afiuny, T. McIntyre, Y. Yih and J. W. Sutherland, "Comparative Life Cycle Assessment of NdFeB Magnets: Virgin Production versus Magnet-to-Magnet Recycling," *Proc. 23rd CIRP Conference on Life Cycle Engineering,* 2016.

[26] J. Auer and A. Meincke, "Comparative life cycle assessment of electric motors with different efficiency classes: a deep dive into the trade-offs between the life cycle stages in ecodesign context," *The International Journal of Life Cycle Assessment,* 2018.

[27] K. Rothboeck, Model-based sustainable product development applied on electric drives, Master Thesis LCM - FH Upper Austria, 2019.

[28] H. K. Stranddorf, L. Hoffmann and A. Schmidt, Impact categories, normalisation andweighting in LCA, Danish Ministry of the Environment, 2005.

[29] P. Karutz, W. Gruber, T. Nussbaumer and J. W. Kolar, "The Bearingless 2-Level Motor," *Proc. 7th International Conference on Power Electronics and Drive Systems,* 2007.

[30] ISO 14044:2006, Environmental management - Life cycle assessment - Requirements and Guidelines, International Standardization Organization.

Dynamic suspension performance of an ultra-compact 5-DOF controlled axial gap type self-bearing motor for use in pediatric ventricular assist devices

Masahiro Osa[1*], Toru Masuzawa[1], Toshihiro Yamashita[1] and Eisuke Tatsumi[2]
[1] Ibaraki University, Japan
[2] National Cerebral and Vascular Center Research Institute, Japan
Masahiro.osa.630@vc.ibaraki.ac.jp

Abstract

Research interest in miniaturization of magnetically suspended motor is increased to develop next generation ventricular assist devices (VADs) for pediatric heart disease treatment. In this study, an ultra-compact 5-degrees of freedom (DOF) controlled axial gap type self-bearing motor which employ double stator mechanism has been proposed. A six-slot and four-pole self-bearing motor which has 22 mm in diameter and 34 mm in height, and centrifugal pump was fabricated. This paper investigated pump characteristics, and then dynamic impeller suspension performance with frequency response evaluation. The developed maglev pump for pediatric VAD demonstrated non-contact pump drive producing flow rate of 0.5-3.0 L/min against head pressure of around 100 mmHg at rotating speed of 4000-5000 rpm. In addition, the maglev motor indicated sufficient bandwidth for impeller suspension control during pump drive.

1 Introduction

Mechanical circulatory support (MCS) technology is strongly required for treatment of severe heart disease pediatric patients due to the shortage of donor hearts. However, these pediatric patients have to depend on extracorporeal pulsatile blood pumps which have high risk of adverse events such as blood clotting and infection. One of the most significant reasons above is that there are no implantable compact rotary blood pumps due to technical difficulty for miniaturization of impeller suspension technique with high durability and blood compatibility. Our research group has

[*] Masterminded and created the first stable version of this document

developed an ultra-compact magnetically levitated pediatric ventricular assist device (VAD) with double stator axial gap permanent magnet synchronous motor. Maglev VADs have significant advantages such as high durability, higher speed operation, and low hemolysis as well as less thrombosis compared to conventional blood pumps which have mechanically contacting bearings for impeller suspension. In ultra-compact maglev systems, which tend to have deterioration of magnetic suspension force, a guaranty of impeller suspension stability of the blood pump over operating conditions is significantly important. Especially, evaluation of dynamic response of the magnetic suspension control system must be needed for maglev rotary VADs. In this study, magnetic suspension performance of the developed ultra-compact maglev motor for pediatric VADs was investigated.

2 Material and Methods

2.1 Overview of 5-DOF controlled axial gap type self-bearing motor and principles of magnetic suspension control and motor drive

The maglev pediatric VAD consists of two six-slot axial flux motor stators and a four-pole centrifugal impeller as shown in Fig. 1. The impeller is magnetically suspended between the stators that have an identical structure as shown in Fig. 2. Concentrated windings for both magnetic suspension and rotation are integrated and wound on each stator tooth. A double stator mechanism enhances rotating torque production and provides 5-DOF active control of impeller postures. An axial position (z) of the impeller is controlled with field strengthening and field weakening of impeller four-pole magnetic fields by d-axis current regulation as shown in Fig. 3. A rotating speed (ωz) is controlled with conventional q-axis current regulation. Inclination angles (θx, θy) and radial positions (x, y) of the levitated impeller are controlled by utilizing P±2 pole algorithm. Inclination torque and radial suspension force were numerically and experimentally found in a previous study. Both control principles for the inclination angle around y-axis and the radial position in x direction with the double stator mechanism are shown in Fig. 4. When the stators produce inclination torque which has same amplitude and direction, the inclination angle of the levitated impeller can be regulated without the radial suspension force. In contrast, the radial position of the impeller can be regulated with radial suspension force when the inclination torque produced by both stators is regulated in opposite direction. The inclination around x-axis and the radial position in y direction can be controlled in a similar manner. The inclination and radial position of the levitated impeller can be controlled independently with double stator mechanism by regulating magnitude and direction of excitation current supplied to the stators to generate the two-pole rotating magnetic field.

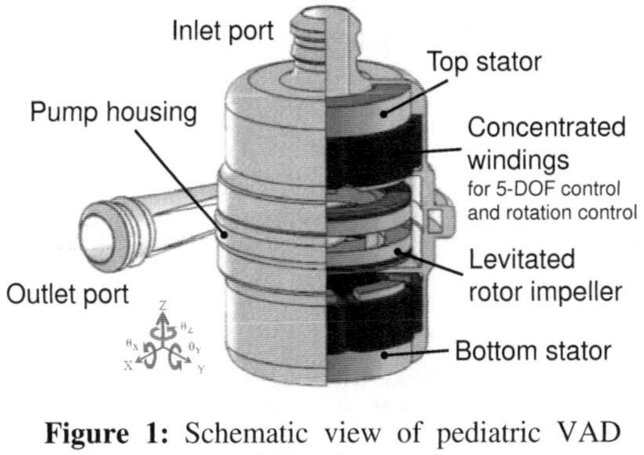

Figure 1: Schematic view of pediatric VAD with axial gap type self-bearing motor.

Figure 2: Developed ultra-compact double stator self-bearing motor.

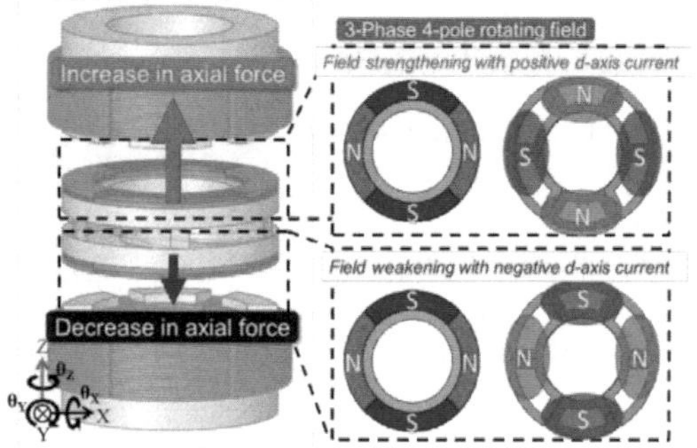

Figure 3: Axial position control with field strengthening and field weakening.

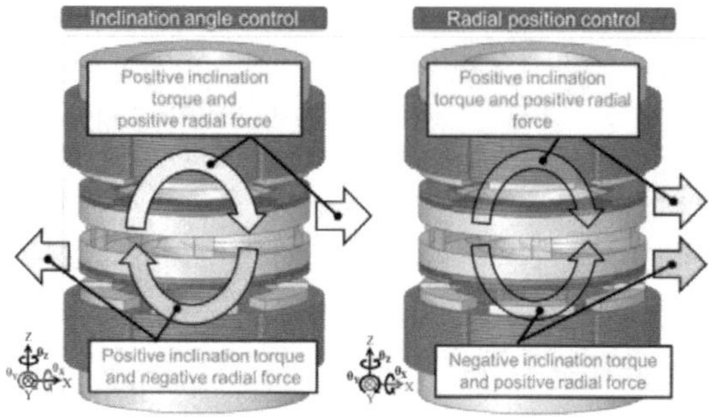

Figure 4: Inclination angle and radial position control with P+/- 2 pole algorithm.

2.2 Fabricated self-bearing motor for pediatric centrifugal VAD

Motor geometries listed in Table 1 were designed with 3D finite element method magnetic field analysis. A fabricated pediatric VAD with the 5-DOF controlled self-bearing motor are shown in Fig. 5. The motor is 22 mm in outer diameter and 34 mm in height. Sintered magnetic powder core (EU-69) is used for the motor stator core, and soft magnetic iron (SYU-1) is used for the rotor back iron. Neodymium permanent magnets have coercivity and residual flux density of 907 kA/m and 1.36 T. Concentrated copper windings of 123 turns are wound on each stator pole. The diameter of isolated copper wire is 0.3 mm. A double volute type centrifugal housing and six vanes closed impeller were designed with a computational fluid dynamics simulation. The inlet and outlet port diameter are 6 mm. The movable ranges of the impeller in axial direction, in radial direction and in inclination angle were set to +/-0.3 mm, +/-0.5 mm, and 1.4 degrees.

Table 1 Motor geometric parameters.

Inner diameter (mm)	16.8
Outer diameter (mm)	22
Pole height (mm)	9.3
Slot width (mm)	3.8
PM thickness (mm)	0.8
Rotor back iron thickness	2

Figure 5: Developed centrifugal blood pump with axial gap self-bearing motor for pump testing.

2.3 Control system for magnetic suspension and rotation of 5-DOF controlled self-bearing motor

The developed rotary blood pump with the self-bearing motor was combined with a digital feedback control system. A schematic diagram of a 5-DOF PID control system is shown in Fig. 6. Digital PID controllers are implemented on a micro-processor board MicroLabBox (dSPACE GmbH,

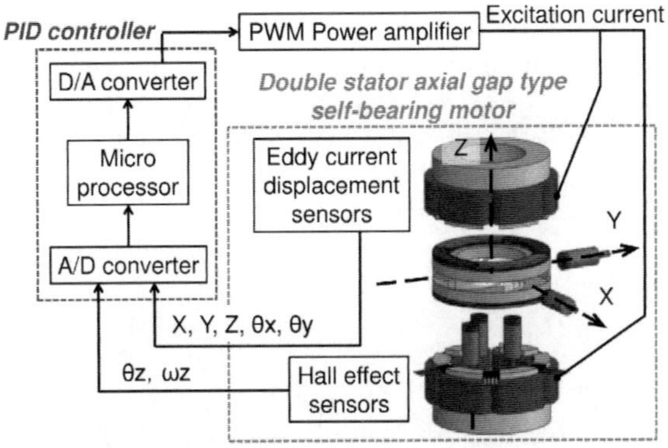

Figure 6: 5-DOF control and motor drive system with digital PID controller.

Table 2 PID control gains for rotor position regulation.

	P	I	D
Z	10 (A/mm)	0.05 (A/(mm· s))	0.002 (A s/mm)
θx, θy	1.5 (A/deg)	0 (A/(deg· s))	0.002 (A/(deg· s))
ωz	0.0005 (A/rpm)	0.0018 (A/(rpm· s))	—

Paderborn Germany) with MATLAB/Simlink for the active 5-DOF control of the impeller postures. Three eddy current displacement sensors (PU-03A, Applied Electronics Corporation) are located inner side of the stator poles to measure an axial position and inclination angles around x and y axes of the levitated impeller. Impeller radial positions in x and y direction are measured by other two eddy current displacement sensors located around outside of the levitated impeller. three Hall Effect sensors (Asahi KASEI Corporation) are placed on the stator slots to measure a rotating angle of the impeller with a sensitivity of 30 degrees electrical angle by detecting the magnetic flux produced by the rotor permanent magnets. The rotating speed is calculated by time derivative of the impeller rotating angle. Control current is calculated by PID controllers, and it is supplied by PWM power amplifier (JSP-090-10, Copley Controls) to the windings of the stators. Sampling and control frequency are 10 kHz. PID/PD gains were manually tuned to minimize the rotor vibration amplitude as listed on Table II.

2.4 Frequency response of magnetic suspension control system in self-bearing motor with 3-DOF control during pumping

5-DOF of impeller postures at different rotating speeds were evaluated to demonstrate a pump characteristic and non-contact impeller suspension during pump operation. Closed loop circulation circuit for pump drive testing is shown in Fig. 7. The levitated impeller was magnetically suspended and accelerated from 4000 rpm to 5000 rpm in water. After that, the frequency response of the axial position control system and the inclination control system was measured with a frequency sweeping

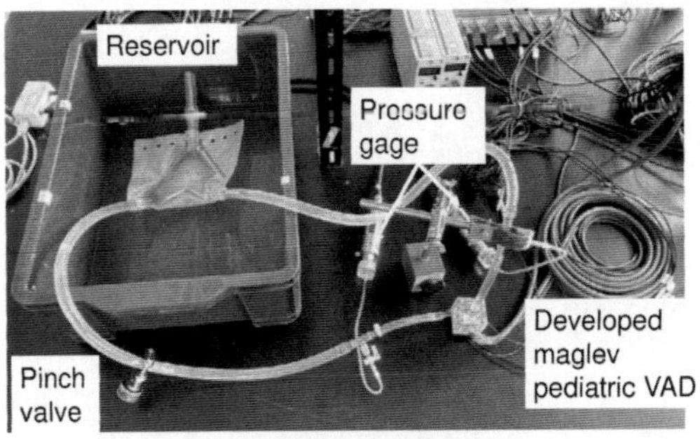

Figure 7: Experimental setup for pump testing and frequency response measurement.

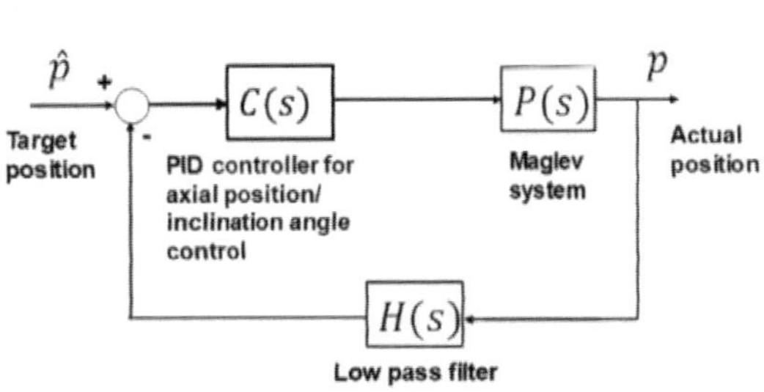

Figure 8: Block diagram for frequency response measurement.

method by using an FFT analyzer (DS-2100, DS-0264, Ono Sokki Corporation) at representative pump operating point of the speed of 4000 rpm and the flow rate of 1.5 L/min against the pressure head of around 90 mmHg for pediatric circulation. Amplitude of sinusoidal target signal which was determined as 0.1 mm in axial position and 0.1 degrees in inclination angle added to the control system as shown in Fig 8. The signal frequency was periodically varied from 10 Hz to 1 kHz. An input and an output are the target position and the actual levitated rotor displacement.

3 Results

Relationship between head pressure and flow rate produced by the developed maglev centrifugal blood pump is shown in Fig. 9. The developed pump produced the wide flow rate range of 0.5-3.0 L/min against the head pressure of 60-100 mmHg at the rotating speed of 4000-5000 rpm. The levitated impeller fully magnetically suspended at every operating condition. Mean oscillation amplitude in axial position, radial position and inclination angle were approximately 30 μm, 100 μm, 0.3 degrees, respectively. The oscillation was enough small to avoid mechanical contact between the levitated impeller and the pump housing. Bode diagram for the axial position control system and the inclination angle control system during pumping at rotating speed of 4000 rpm are shown in Fig. 10 and Fig. 11. Bandwidth in the axial position control system and the inclination control system were approximately 110 Hz and 50 Hz. In the inclination control system, the gain plots once increased around 60-150 Hz. However, this is not caused due to the inclination in response to input sinusoidal signal to the control system, and it should be excluded from the discussions.

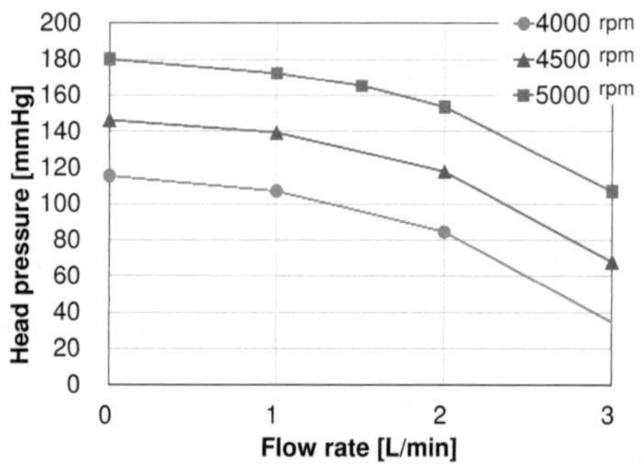

Figure 9: Pump characteristics of the developed maglev centrifugal blood pump.

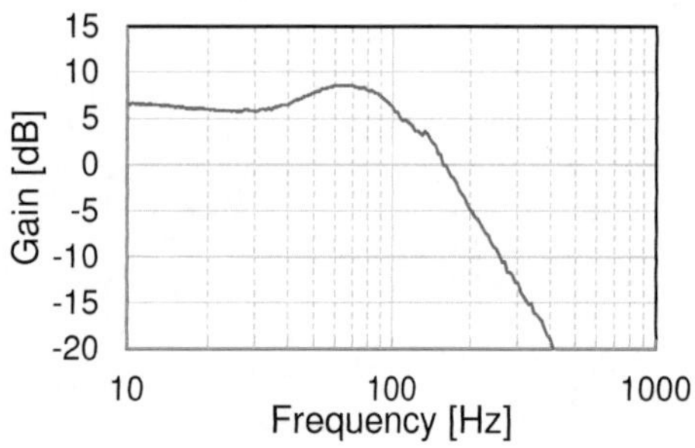

(a) Gain plot

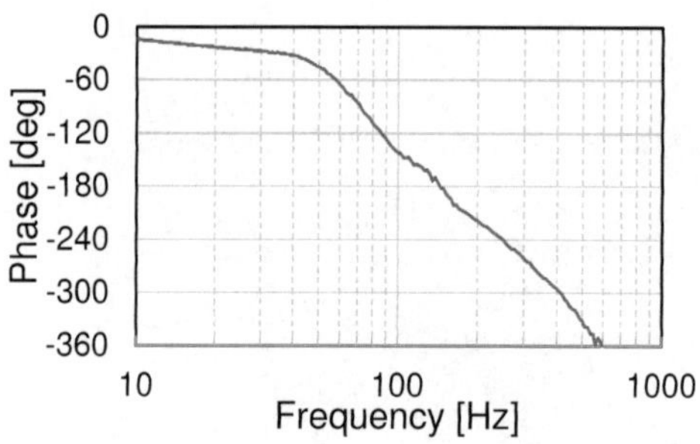

(b) Phase plot

Figure 10: Bode diagram for axial position control system.

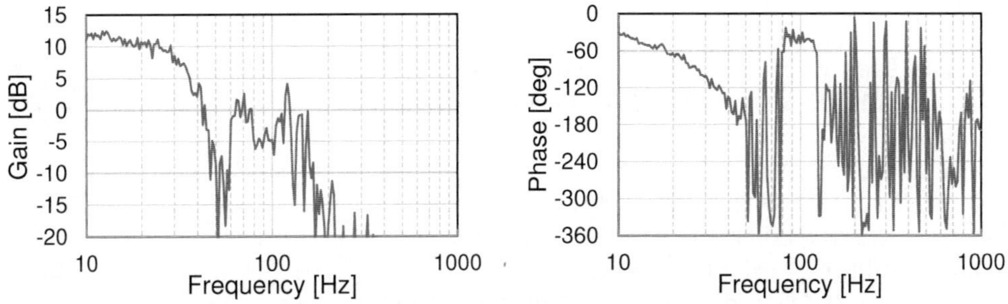

(a) Gain plot (b) Phase plot

Figure 11: Bode diagram for inclination angle control system.

4 Discussion

Magnetic suspension is one of the strongest candidates to achieve durable and biocompatible MCS devices. The developed ultra-compact self-bearing motor has a significantly contribute to progress of the VADs development for small pediatric patients. Investigation of the dynamic response in pump drive with the self-bearing motor is much important to indicate potential of the next generation maglev pediatric VADs. Sufficient pump performance with non-contact impeller suspension demonstrated that the developed pump will be available in pediatric circulation according to patients growing up. The magnetic suspension stability overall pump speeds would guarantee less hemolysis and anti-thrombogenicity. The wide bandwidth more than 100 Hz in axial position control system gives stable impeller suspension within the range of pump operating speed of 4000-5000 rpm (67-83 Hz). In contrast, although the inclination angle in pump testing was able to be sufficiently suppressed with active control, relatively small bandwidth of 50 Hz in the inclination control system was observed. This would be because that inclination stiffness produced by the active inclination control was little bit small, in addition, increase in damping effect in water. Further investigation of motor dynamic characteristics should be needed by carrying out a shaking test to completely demonstrate feasibility of the proposed self-bearing motor concept.

5 Conclusion

The ultra-compact double-stator axial gap type maglev motor with active 5-DOF control was developed. The developed pump demonstrated sufficient pump performance for the pediatric circulatory support. The feasibility of non-contact impeller suspension against the external disturbance in the maglev pediatric VAD operation was indicated by frequency response evaluation. A shaking test of the developed maglev pediatric VAD will be carried out to verify sufficient magnetic suspension stability with respect to external acceleration disturbance.

References

Baldwin, J.T.; Borovetz, H.S.; Duncan, B.W.; Gartner, M.J.; Jarvik, R.K.; Weiss, W.J.; Hoke, T.R. Weiss and Tracey R. Hoke. (2006), The National Heart, Lung, and Blood Institute Pediatric Circulatory Support. Circulation, 113, pp. 147–155.

Baldwin, J.T.; Borovetz, H.S.; Duncan, B.W.; Gartner, M.J.; Jarvik, R.K.; Weiss, W.J. (2011), The National, Heart, Lung, and Blood Instituted Pediatric Circulatolory Support Program: A Summary of the 5-year Experience. Circulation, 123, 1233–1240.

Miera, O.; Schmitt, K.R.; Delmo-Walter, E.; Ovroutski, S.; Hetzer, R.; Berger, F. (2014), Pump size of Berlin Heart EXCOR pediatric device influences clinical outcome in children. J. Heart Lung Trans., 33, pp. 816–821.

Lorts, A.; Zafar, F.; Adachi, I.; Morales, D.L. Morales. (2014), Mechanical Assist Devices in Neonates and Infants. Semin. Thoracic Cardiovasc. Surg. Pediatric Cardiac Surg. Annu. 17, pp. 91–95.

Gibber, M.; Wu, Z.J.; Chang, W.B.; Bianchi, G.; Hu, J.; Garcia, J.; Jarvik, R.; Griffith, B.P. (2010), In Vivo Experience of the Child-Size Pediatric Jarvik 2000 Heart: Update. Asaio J., 56, pp. 369–376.

Mehra, M.R.; Naka, Y.; Uriel, N.; Goldstein, D.J.; Cleveland Jr, J.C.; Colombo, P.C.; Walsh, M.N.; Milano, C.A.; Patel, C.B.; Jorde, U.P.; et al. (2017), A Fully Magnetically Levitated Circulatory Pump for Advanced Heart Failure. N. Engl. J. Med., 376, pp. 440–450.

Schoeb, R.; Barletta, N.; Fleischli, A.; Bourque, K.; Gernes, D.; Loree, H.; Richardson, J. S.; Poirier, V. (2000), Heartmate III: Bearingless Motor Design for A Malev Centrifugal LVAD. ASAIO Journal, 46, 191.

Wu, G.; Yang, L.; Li, H.; et al. (2019), Establishment of Ovine Model for CH-VAD Implantable Ventricular Assist Device. Journal of Mechanics in Medicine and Biology, 19, 1-10.

Emmanuel, S.; Watson, A.; Connellan, M.; et al. (2020), First in Man Anatomical Fitting Study of the BiVACOR Total Artificial Heart. The Journal of Heart and Lung Transplantation, 39, S189

Nicholas, A. G.; Daniel, L. T.; Nobuyuki, K.; Edward, W. P.; Toru, M. (2010), Axial Magnetic Bearing Development for the BiVACOR Rotary BiVAD/TAH. IEEE Trans. Biomed. Eng., 57, pp. 714-721.

Kurita, N.; Ishikawa, T.; Saito, N.; Masuzawa, T.; Daniel, L. T. (2018), A Double-Sided Stator Type Axial Bearingless Motor Development for Total Artificial Heart. IEEE Trans. Ind. Appl., 55, pp. 1516-1523.

Osa, M.; Masuzawa, T.; Tatsumi, E. (2012), 5-DOF Control Double Stator Motor for Paediatric Ventricular Assist Device. Proceedings of ISMB13; University of Virginia: Charlottesville, VA, USA.

Osa, M.; Masuzawa, T.; Orihara, R.; Tatsumi, E. (2019), Performance Enhancement of a Magnetic System in a Ultra Compact 5-DOF-Controlled Self-Bearing Motor for a Rotary Pediatric Ventricular-Assist Device to Diminish Energy Input. Actuators, 8, pp. 1-14.

Hybrid Active-Passive Operation of a Passively Levitated Self-Bearing Machine

Joachim Van Verdeghem[1,2], Eric L. Severson[2], and Bruno Dehez[1]

[1] Dept. of Mechatronic, Electrical Energy and Dynamic Systems,
Université catholique de Louvain, Louvain-la-Neuve, Belgium
[2] Dept. of Electrical and Computer Engineering,
University of Wisconsin-Madison, Madison (WI), USA

Abstract

Passively levitated self-bearing machines relying on a combined armature winding have been recently developed so as to allow the systems that implement them to reach higher reliability and compactness while lowering their cost and complexity. The axial guidance being achieved through electrodynamic phenomena based on induced suspension currents, the restoring force is small at low speeds. External loads may therefore prevent the rotor from being magnetically suspended below a threshold speed. In this context, this extended abstract introduces and validates an innovative hybrid active-passive operation of these self-bearing machines. It consists in regulating the rotor axial position, and thence the restoring force, through the direct-axis component of the currents flowing in the combined winding until the threshold speed beyond which passive levitation can be achieved is reached. The proposed active operation solely requires an additional axial position sensor. As a result, the benefits offered by passive suspension are not significantly impacted.

1 Introduction

Self-bearing machines gather the motor function and rotor magnetic levitation within a single structure, improving the power density of the systems that integrate them. The position sensors, controllers and power electronics linked to the actively stabilised degrees of freedom may be restrictive in applications calling for high compactness and reliability as well as low cost. This set the trend towards the adoption of alternative solutions based on passive phenomena.

Recent researches disclosed that electrodynamic thrust bearings can produce both the passive axial restoring force and the drive torque through a single armature winding thanks to an appropriate connection to the power supply [1, 2]. An electromechanical model allowing to predict the rotor axial and spin dynamics of these novel self-bearing machines was derived and experimentally validated [1]. Their combination with passive permanent magnet bearings ensuring the rotor radial and tilt stabilisations conduced to the first fully passively levitated self-bearing machines relying on a combined winding [3]. However, the axial suspension resulting from passively induced currents, the restoring force and the underlying stiffness are small at low speeds and even null at standstill. Hence, the rotor can not be levitated below a threshold speed, determined by the axial load and the destabilising stiffness exerted on the machine, which may represent an issue for applications necessitating frequent start-up and stop phases.

This extended abstract of the paper [4], accepted for publication in the IEEE-IAS Special Issue on Magnetically Levitated Motor Systems, proposes and validates a hybrid active-passive actuation approach enabling these thrust self-bearing machines to levitate in their entire speed range. Specifically, at low spin speeds, the restoring force, and thus the rotor axial position, is actively controlled through the currents flowing in the combined armature winding and supplied by the same N-phase inverter as that employed for the drive. Beyond the levitation threshold speed, the axial suspension is switched from active to passive operation.

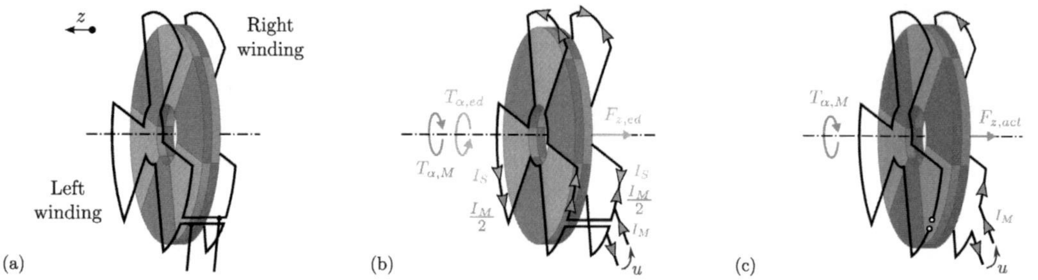

Figure 1: Self-bearing machine under study (only one phase represented and $p = 3$). (a) Structure. (b) Passive axial levitation operation. (b) Active axial levitation operation.

2 Structure & Principle

The machine under study, depicted in Fig. 1(a), ensures both the drive torque and axial restoring force production. The rotor consists of a permanent magnet arrangement creating an axial magnetic field with p pole pairs. The stator encompasses three phases constituted by the parallel connection of a left and a right winding that each includes p coils connected in series.

2.1 Suspension passive operation

In centred position, i.e. $z = 0$, the machine behaves as a conventional double-sided permanent magnet motor, thus only developing a drive torque $T_{\alpha,M}$ through the quadrature-axis current $I_{M,q}$. As soon as the rotor leaves its nominal position, i.e. $z \neq 0$, a passively induced suspension current I_S, circulating in the short-circuited path formed by the series connection of the upper and lower windings, adds to the classic motor current I_M supplied by the voltage source u, as illustrated in Fig. 1(b). The former current generates the axial restoring force $F_{z,ed}$ and a drag torque $T_{\alpha,ed}$ whereas the motor current solely produces the drive torque $T_{\alpha,M}$. At low speeds, the axial force $F_{z,ed}$ may not be sufficient to compensate for an axial load or destabilising stiffness, precluding the rotor from being magnetically suspended below a threshold speed ω^*.

2.2 Suspension active operation

The active operation addresses the levitation issue arising below the threshold speed ω^*. As represented in Fig. 1(c), it relies on the right winding while the left one is disconnected from the voltage source, reducing the machine to a single-sided structure identical to that investigated in [5]. The axial force $F_{z,act}$ and the drive torque $T_{\alpha,M}$ are regulated independently through the direct $I_{M,d}$ and quadrature $I_{M,q}$ axis components of the current supplied by the three-phase inverter. As a result, this operation mode only requires an additional position sensor, thus not substantially affecting the advantages ensuing from the passive levitation.

3 Experimental Setup

Fig. 2 depicts the self-bearing machine prototype, controlled and supplied through an AC driver, within the dedicated test rig. The latter comprises two air bearings that ensure the rotor radial and tilt guidance. A laser displacement sensor provides the measurement of the rotor axial position z. In suspension active operation, electronic switches disconnect the left windings while an axial position control loop, relying on a PID regulator that generates the force reference $F_{z,ref}$ converted into a direct-axis motor current setpoint $I_{M,d,ref}$, is activated. The prototype can be tilted by an angle σ with respect to the frame in order to exert an axial load $F_e = mg \sin \sigma$ on the machine, m being the rotor mass and g the gravitational acceleration.

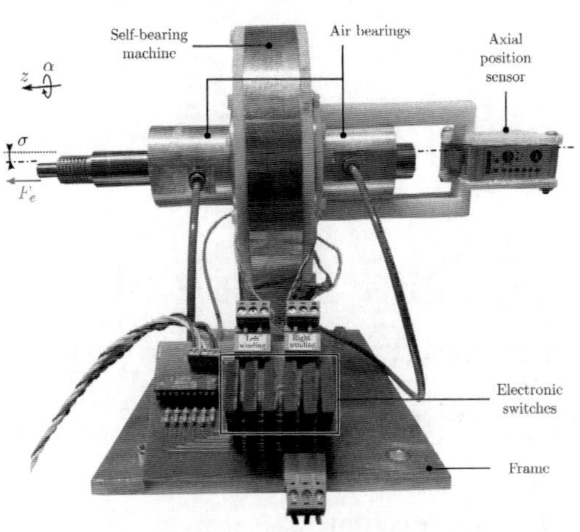

Figure 2: Self-bearing machine prototype and the dedicated test rig.

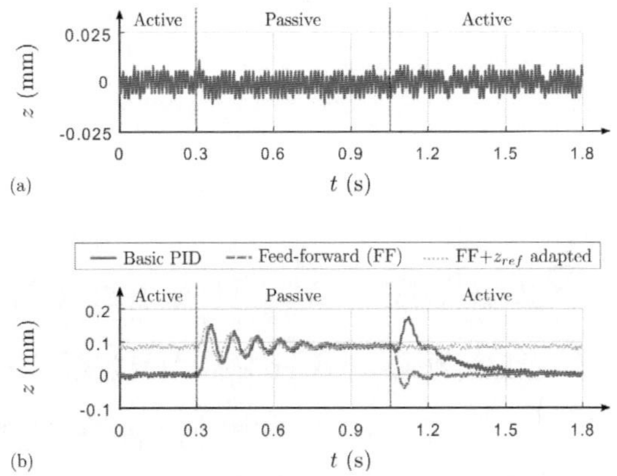

Figure 3: Time evolution of the axial position z during transitions from active to passive operation and vice versa. (a) No axial load ($F_e = 0$). (b) Axial load $F_e = 0.5$ (N).

4 Experimental Results

Fig. 3(a) represents the time evolution of the rotor axial position z during transitions from active to passive operation and vice versa when there is no axial load and the speed setpoint is fixed to 4000 (rpm). Throughout the sequence, the rotor levitates around its centred position ($z = 0$) with a constant amplitude of oscillations, demonstrating that, in these specific conditions, the passages between suspension modes do not affect the rotor axial behavior.

Fig. 3(b) illustrates the transitions when an axial load $F_e = 0.5$ (N) is applied on the rotor ($\sigma = 2.7°$). As shown in solid line, passing from active to passive operation leads to significant transient rotor oscillations. This arises from the step of axial load to which the machine is subjected at the time of the transition due to the sudden disappearance of the active force while the passive one is not yet created. Similarly, the shift from passive to active levitation results in an axial deviation before the controller brings the rotor back to the reference position $z_{ref} = 0$. The transition to active operation can be considerably enhanced by feed-forwarding the axial force exerted on the rotor in the integral term of the position regulator. In this way, as illustrated in dashed line, the axial disturbance affecting the self-bearing machine is limited to the setpoint step towards the centred position ($z = 0$). The axial behavior during the shifts can be further improved by setting the position reference z_{ref} to the axial displacement at which the rotor stabilizes in passive mode. As shown in dotted line, removing the position setpoint step allows to proceed with the shift to active suspension without creating any axial disturbance.

References

[1] J. Van Verdeghem *et al.*, "Experimental investigations on passively levitated electrodynamic thrust self-bearing motors," *IEEE Trans. Ind. App.*, vol. 55, pp. 4743–4753, Sep. 2019.

[2] G. Cavalcante Rubio and A. Chiba, "Design and analysis of a bearingless motor with passive axial suspension through null-flux coils," in *2019 IEEE IEMDC*, pp. 779–786, 2019.

[3] J. Van Verdeghem and B. Dehez, "Fully passively levitated self-bearing machines with combined windings," in *2020 IEEE Energy Conversion Congress and Exposition*, pp. 254–261, 2020.

[4] J. Van Verdeghem, E. L. Severson, and B. Dehez, "Hybrid active-passive actuation approach of passively levitated thrust self-bearing machines," *Accepted in IEEE Trans. Ind. App.*, 2021.

[5] J. Asama *et al.*, "Development of a one-axis actively regulated bearingless motor with a repulsive type passive magnetic bearing," in *2014 IEEE IPEC*, pp. 988–993, May 2014.

Random Vibration Simulation and Testing of a Compact, Magnetic Bearing Supported Blower for Space Applications

Larry Hawkins[1], Rasish Khatri[1], Alexei Filatov[1]
Chris DellaCorte[2], and S. Adam Howard[2]
[1]Calnetix Technologies, Cerritos, CA, USA
[2]NASA Glenn Research Center, Cleveland, OH, USA
lhawkins@calnetix.com

Abstract

NASA is developing a next-generation CO_2 removal system, the Four Bed Carbon Dioxide Scrubber (4BCO2), which will use the International Space Station (ISS) as a testbed. A key component of both the existing and the new system is the blower that provides the airflow through the CO_2 sorbent beds. To improve performance and reliability, magnetic bearings will be used in lieu of more conventional bearings (e.g. ball bearings or hydrodynamic bearings) to improve resistance to contaminants and enable extensibility with regards to blower speed, pressure rise and mass flow rate. The new blower features a high-efficiency permanent magnet motor, a five-axis, active magnetic bearing system, and a compact controller containing both a variable speed drive and magnetic bearing controller. The blower uses a centrifugal impeller to pull air from the inlet port and drive it through an annular space around the motor and magnetic bearing components to the exhaust port. Technical challenges of the blower and controller development include survival of the blower system under launch random vibration loads, operation in microgravity, packaging under strict size and weight requirements, and successful operation during 4BCO2 operational changeovers. Each serial build of the blower must pass a severe random vibration test to prove it will survive launch conditions. This paper discusses the vibration test requirements and provides details of a simulation performed to estimate the peak backup bearing displacements and loads. Results from the successful random vibration testing of the prototype build are discussed and compared to the simulation.

Introduction

A new magnetically-levitated blower was recently developed for use in a new 4BCO2 to be tested on the ISS. The blower, shown in Figure 1, features a high-efficiency permanent magnet motor, a five-axis, active magnetic bearing system, and a compact controller containing both a variable speed drive and magnetic bearing controller. Design considerations for the blower system active magnetic bearings (AMB) and backup bearings were previously discussed in detail [1], including magnetic finite element analysis (FEA) of the actuator/sensor system, rotordynamics/controls analysis, and backup bearing drop simulations. The design is briefly reviewed here to provide background to the reader.

All equipment that is delivered to the ISS is exposed to a significant level of launch vibration which is characteristically random in nature. NASA requires such equipment to pass a random vibration proof test to ensure it can survive the flight environment. The vibration level and frequency spectrum depend on the nature of the equipment and how it is packaged. Although the blower will be launched under soft-stow conditions, the low frequency vibration environment is still quite severe. The blower and controller will be deactivated during the launch so the magnetic bearings will be de-levitated which for standard AMB applications means the rotor is supported on the backup bearings rather than the magnetic bearings. However, due to the high imposed vibration level during the launch, consideration was given to using a locking mechanism (launch locks) to keep the rotor from bouncing in the backup bearing clearance space and potentially causing impact damage to the backup bearings. Several analysis steps were undertaken to assess the need for launch locks and to evaluate potential locking strategies. This included a transient simulation using a forcing function derived from the NASA specified Power Spectral Density (PSD) random vibration level. The simulation methodology and results are discussed here along with test results from the shaker table testing of the blower. The Magnetic Bearing Controller (MBC) was activated during the vibration testing to allow collection of rotor/housing relative displacement test data from the AMB position sensors. This data is discussed and compared to the simulation results.

Blower Design

The design of the blower, magnetic bearings and backup bearings is discussed in detail by Hawkins [1] but is described here in limited detail for easy reference. Figure 1 shows the overall layout of the blower. The centrifugal impeller at the inlet (right side of the figure) draws in cabin air and pushes it through an annular passage to the outlet. This primary flow (blue arrows) provides cooling for the motor stator as it passes through axial slots in the laminated stator. A secondary flow (orange arrows) cools the bearing elements in the central part of the machine as well as the motor end turns near the outlet. The blower is designed to be operated at any speed from rest up to 60,000 rpm.

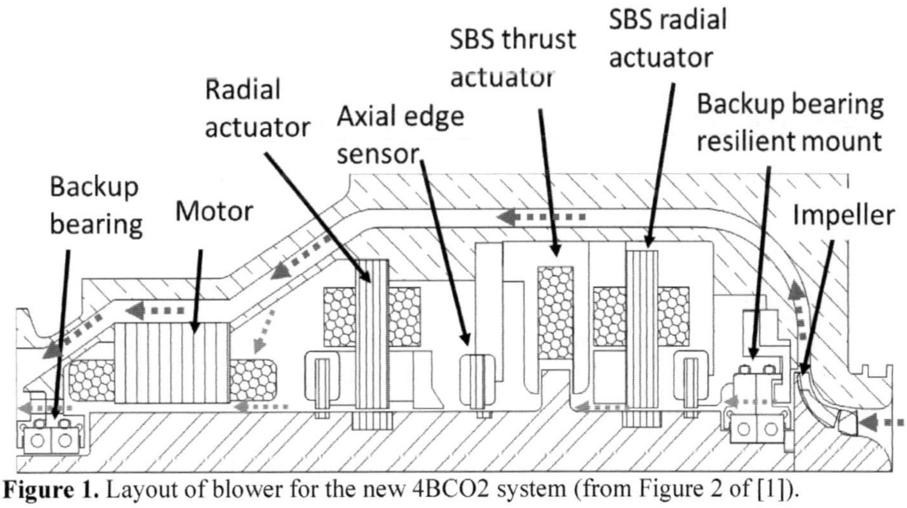

Figure 1. Layout of blower for the new 4BCO2 system (from Figure 2 of [1]).

The homopolar, permanent magnet (PM) biased magnetic bearings consist of a three-axis combination radial/thrust active magnetic bearing on the impeller side and a two-axis radial active magnetic bearing on the motor side. The basic design parameters for the magnetic bearings are summarized in Table 1. The radial position sensors are located just outboard of the actuators and the axial position sensor is between the two magnetic bearing sets. The backup bearings are comprised of resiliently mounted, angular contact ball bearings. The backup bearings serve to support the shaft when the magnetic bearings are de-activated, in the case of overload, or in the case of a critical component failure.

This new magnetic bearing blower is designed to replace a heritage foil bearing supported blower and must fit into the same length, width, and height envelope of 7.6 x 5 x 8 inches (193 x 127 x 203 mm). Additionally, the new blower was required to adapt to the inlet and outlet interface flanges of the existing system, 1.5 inch (38 mm) and 2.75 inch (70 mm) respectively. The overall design layout followed naturally from the need to meet the existing envelope and interface dimensions.

Table 1. Magnetic Bearing Design Parameters

Parameter	Radial, Brg 1	Radial, Brg 2	Axial
Minimal load capacity, N (lbf)	44 (10)	44 (10)	89 (20)
Force constant, N/A (lbf/A)	18.5 (4.17)	18.9 (4.25)	38.3 (8.61)
Negative stiffness, N/mm (lbf/in)	173 (990)	89 (510)	70 (400)
Magnetic air gap, mm (mil)	0.76 (30)	0.76 (30)	0.76 (30)
Slew corner frequency, Hz	1250	1250	282

Rotordynamic Structural Model

The rotordynamic structural model of the blower, rotor, and housing are shown in Figure 2. The rotor stiffness is modelled with Timoshenko beam elements with mass for each beam lumped at the nodes. The impeller is modelled as a lumped inertia at its center-of-gravity. The magnetic bearing and sensor laminations are assumed to contribute mass but no lateral stiffness due to the flexibility of the lamination bonding. The lateral stiffness contribution of the motor permanent magnets is derated, consistent with the authors' experience for this type of rotor construction. The springs to ground in the figure show the PM negative stiffness locations for the motor and magnetic bearings. The predicted AMB negative stiffnesses are in Table 1; the motor negative stiffness of 7.4 N/mm (42 lbf/in) is negligible by comparison but has been included for completeness.

The first forward free/free bending mode of the rotor is predicted to be at 139,200 cpm (2.3 kHz) at 0 rpm and 143,300 (2.4 kHz) at 60,000. The large critical speed margin and small gyroscopic effect allow for a relatively simple magnetic bearing control design for this machine. As rotor launch locks are to be investigated below, it is also important to consider the lowest natural frequency and corresponding mode shape of the rotor for several potential end conditions. Figure 3 shows two cases, a) rotor pinned to the housing at both ends, and b) the rotor connected to the housing with 40,000 lbf/in at the motor (left) end and the rotor pinned to the housing at the impeller (right) end. For the case of the rotor pinned on both ends, the lowest natural frequency is 37,370 cpm (620 Hz). For the second case the lowest natural frequency is 31,900 cpm (530 Hz).

Backup Bearing System

All active magnetic bearing systems include a backup (touch-down) bearing system to support the rotor during non-operation, overload of the AMBs, or component failure. A few specifications exist (or are being developed) that provide guidance on how to select and/or evaluate backup bearings (e.g. [2, 3]), but the practical process of backup bearing selection remains heavily experience and test driven within the AMB community. Schweitzer et. al [4] provides a bird's eye view of basic principles behind backup bearing modeling, selection, and analysis. Wilkes [5] provides perhaps the most comprehensive nonlinear backup bearing dynamic model to date.

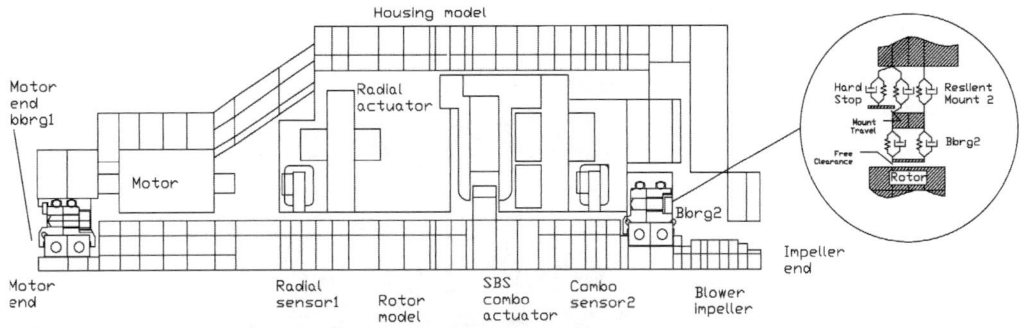

Figure 2. Geometry of the rotordynamic structural model of the blower.

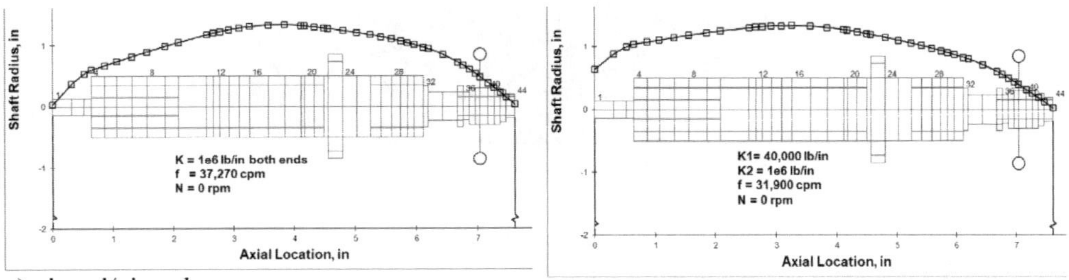

a) pinned/pinned

b) shimmed/pinned

Figure 3. Lowest 0 rpm mode shape and natural frequency with a) rotor pinned at ends, b) 40,000 klb/in at motor end (left side) and pinned at impeller end (right side).

The backup bearings chosen for this blower are cageless, angular contact, duplex pairs with steel races and ceramic balls. The motor end (radial only) backup bearing assembly is shown in Figure 4. The bearing mount is resiliently supported by elastomer O-rings that provide impact compliance, damping and reduce whirl frequency during normal backup bearing operation. The impeller end, radial/thrust (combination) backup bearing assembly uses larger bearings and includes rotor mounted thrust surfaces but is otherwise similar to the motor end. The speed factors, nd_m, of the backup bearings are 0.81E6 and 1.08E6 mm-rpm at 60,000 rpm, where n is the speed in rpm and d_m is the average of the inside and outside diameter. The speed factors are lower than usual for backup bearings, making the design very robust thermally. This compact design was possible due to the relatively low speed, 60,000 rpm, for the size of the rotating assembly. The compactness of the design was a key factor in meeting the size constraint for the new blower.

Transient, nonlinear drop simulations discussed in [1] showed that the backup bearing design should perform well in response to operating overloads and/or drop transient and spin down situation. The same simulation tool is used here to investigate the response of the system and backup bearing loads due to launch vibrations. For the launch vibration simulation discussed below, the backup bearing system is modelled as shown in the inset on the right side of Figure 2. There is a free clearance between the inner rings of the bearings and the rotor landing surfaces such that they are not normally in contact. The radial and axial free clearances are ±0.114 mm (±0.0045 in) and ±0.127 mm (±0.005 in). Each backup bearing pair is supported in a resilient mount to limit peak impact loads during the rotor drop transient. The resilient mount will also reduce impact loads in case of rotor contact due to launch vibration. Clearances and relative stiffness values for the backup bearings and resilient mount are given in Table 2. The backup bearing stiffnesses are the ball/race stiffnesses provided by the bearing manufacturer. The stiffness and damping of the resilient mounts were estimated by scaling empirical data from [6].

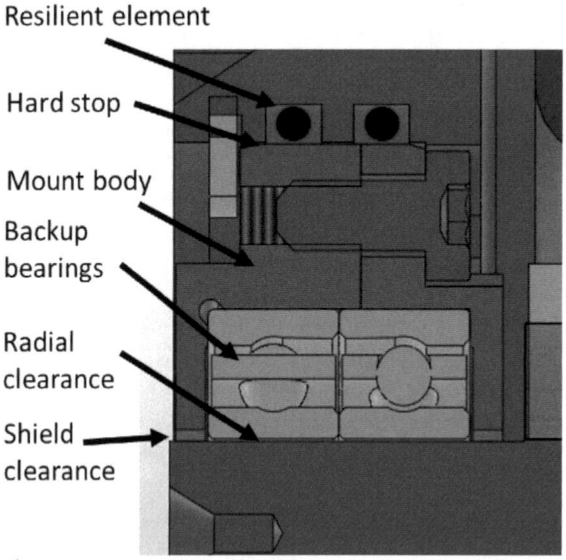

Figure 4. Radial end backup bearing assembly.

Table 2. Backup Bearing And Mount Characteristics

Parameter	Motor End (radial)	Impeller End (radial)	Impeller End (axial)
Free Clearance, mm (in)	0.114 (0.0045)	0.114 (0.0045)	0.127 (0.005)
Resilient Mount Travel, mm (in)	0.076 (0.003)	0.076 (0.003)	0.051 (0.002)
Backup Bearing Stiff (pair), N/mm (lbf/in)	22,800 (130,000)	43,800 (250,000)	9,600 (55,000)
Resilient Mount Stiffness N/mm (lbf/in)	900 (5,200)	1,100 (6,000)	--
Radial Hard Stop Stiffness, N/mm (lbf/in)	114,000 (650,000)	220,000 (1,250,000)	48,000 (275,000)

Random Vibration Requirements

NASA imposes random vibration requirements for payloads that consist of electronic equipment sent to the ISS (International Space Station) [7]. The goal is to ensure the appropriate random vibration proof test has been performed and that the item can survive the flight environment. The blower and controller will be sent as a "soft stowed" payload meaning they are not hard mounted to the launch vehicle structure. The required protoflight vibration testing requirements for this are shown in Table 3 and Figure 5. The protoflight test level is constructed from two parts: 1) the envelope of the maximum expected flight level (MEFL) + 3db margin, and 2) the minimum workmanship level (MWL). The MEFL is the maximum vibration expected during launch, including the soft-stow packaging effects. The vibration level of 0.35 G^2/Hz from 20 – 100 Hz with roll-off at 15 dB per decade at higher frequencies comes from this launch requirement. This is the portion of the curve that is of most concern for the blower and in assessing the design of the launch locks for the backup bearings. The MWL, 0.04 G^2/Hz up to 500 Hz with roll-off at 3 dB per decade at higher frequencies is an environmental stress screen and provides confidence in the assembly process and screens for latent defects.

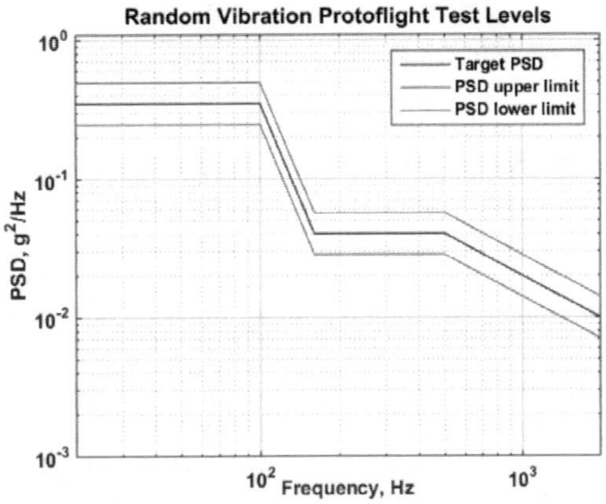

Figure 5. Random vibration protoflight test spectrum for the blower system.

Table 3. Random Vibration Protoflight Test Level

Frequency (Hz)	Protoflight Test Level (g^2/Hz)
20	0.35
100	0.35
100 - 160	-15.4 dB/octave slope
160 - 500	0.04
500-2000	-3 dB/octave slope
2000	0.01
Overall	8.8 g_{rms}
Duration	1 min/axis

Launch Locks

As the blower will be launched to the ISS with the magnetic bearings deactivated, the backup bearings could provide support of the rotor during launch. An alternative is to use launch locks to limit rotor motion and backup bearing contact during the launch. Due to the need for a compact design, design priority was given to ensuring the backup bearings were robust to typical backup bearing service such as a drop transient and spin down. It was expected at the outset that launch locks might be needed to limit the peak backup bearing loads during launch. A time simulation using the vibration level of Figure 5 was performed to evaluate the options. Several strategies were considered:

Case 1: No launch locks – rely on the magnetic bearing PM bias passive negative stiffness force of 5 lbf radially at the two backup bearing locations and 6 lbf axially to mitigate rotor motion.

Case 2: Axial/radial pinning – pin the rotor at the ends such that the rotor is held in the center of the backup bearing clearance spaces both radially and axially.

Case 3: Axial/combo radial pin, radial end shim – use a shim to take up the clearance space in the motor end backup bearing with the rotor in the center of the radial/axial clearance space on the thrust end.

Figure 6 shows a cross-section of the machine with the basic launch lock concept in place. A key requirement was that the launch locks be easy to remove, and once removed, the blower could be installed in the 4BCO2 system with no reassembly or adjustments. As such, the impeller end lock attaches with captive screws to the impeller housing. The motor end lock is attached with the same V-band clamp used to connect the blower to the 4BCO2 ducting. In the axial direction, there is an adjustment screw to clamp the rotor axially against a stiff spring. During a one-time shop set up process, the rotor radial positions can reliably be set to the backup bearing center (or to the side) using the magnetic bearing position sensors for reference. However, there is some concern that the V-band clamp is not precise enough to locate the rotor in the desired position if it is removed and reattached. For this reason, an alternate option (Case 3 above) was created to shim the motor end backup bearing to take up the free clearance space between the rotor journal and backup bearing inner ring. This is possible because the motor end backup bearing clearance space is accessible with the machine assembled. The planned vibration testing was simulated to evaluate the three cases above. A review of the simulation approach and results are described below.

An important criterion for evaluating the launch locks is the peak transient load allowed. The goal was to keep the peak impact loads below 95 lbf (420 N) for the motor end backup bearings and below 200 lbf (890 N) for the impeller end backup bearings. The peak predicted loads for the previously reported drop simulation and spin down were 18 lbf (80 N) and 39 lbf (175 N) for the motor end and impeller end backup bearings, respectively.

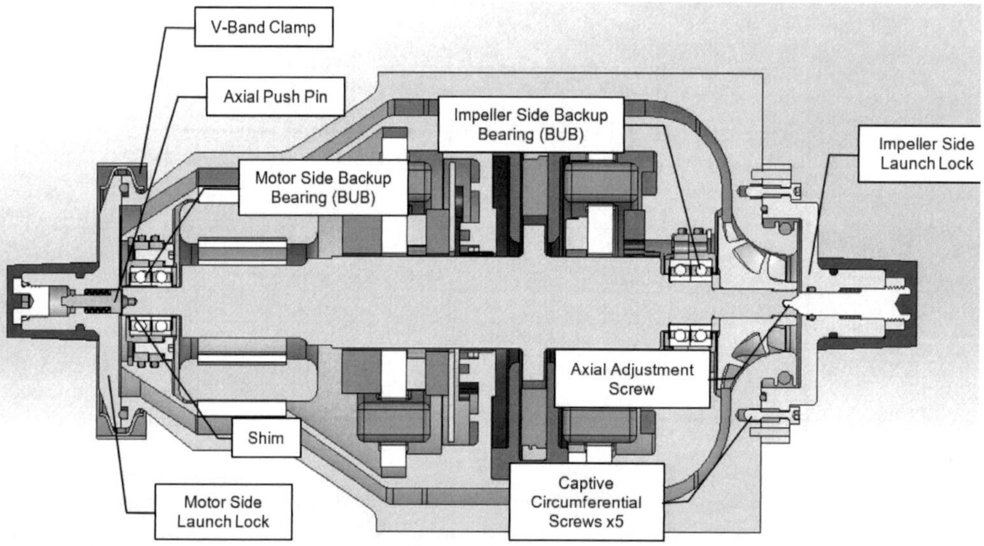

Figure 6. Blower cross-section showing launch locks.

Random Vibration Simulation

Each lock strategy was examined by using a transient, nonlinear simulation tool that is often used for rotor drop transient analysis. To drive the simulation, a forcing function time history was synthesized that would produce a base motion meeting the required PSD of Figure 5. It should be noted that there are many possible time histories that would meet the PSD specification as the relative phases are random. In general, a truly random vibration signal will contain all frequencies in the frequency band of interest and the amplitudes and phases of each frequency will vary randomly. However, the average amplitudes tend to remain relatively constant for many processes and the PSD is used to provide a statistical characterization of this.

During testing, the housing will be firmly clamped to the shaker table and is expected that the housing motion at the supports will follow the prescribed acceleration of the shaker table. It is convenient in the simulation tool to apply a driving force to the model support degrees-of-freedom that produces the prescribed acceleration. By using soft supports in the simulation model, with housing rigid body natural frequencies well below the low end of the vibration spectrum (20 Hz), the necessary driving force, f_{Sx}, can be calculated from:

$$f_{S,x} = m_{cas}\, \ddot{x}_{cas} \qquad (1)$$

where m_{cas} is the mass of the housing in the simulation model, and $\ddot{x}_{cas}$ is an acceleration time history derived from the required PSD. The simulation approach used here is very similar to the one described in detail by Hawkins [8] to model shock impact into the housing support of a magnetic bearing supported compressor.

There are several ways to synthesize a time history from a PSD. The approach used here is to follow these steps:

1) Generate a series of sine waves with random phases and with frequencies, $f_{exc,i}$, ranging from 20 Hz to 2000 Hz, separated by delta frequency, df_{exc}, of 0.5 Hz.
2) Scale the sine waves per the PSD requirement. If the required acceleration level in $Grms^2/Hz$ at frequency, $f_{exc,i}$, is ASD_i, then the amplitude of the sine wave at that frequency, X_i, is:

$$X_i = \sqrt{2\, df_{exc}\, ASD_i}$$

where the factor $\sqrt{2}$ is the conversion form RMS to peak for an individual sine.

3) Sum the sine waves point by point to get the pseudo random input time history.
4) Perform a Fourier transform and scaled to PSD to verify the synthesized time history matches the required spectrum.

The resulting time history, shown in Figure7, is not unique as phase is not part of the PSD. The time history is expected to follow a normal distribution as verified by the histogram in Figure 8. To verify that the generated time history in Figure 7 meets the requirements, its power spectrum is calculated using the built-in Matlab function *pwelch.m*. The result is shown in Figure 9 verifies good compliance with the specification except for a few excursions at higher frequencies where the PSD amplitude is lower. Another method, not used here, to generate the required time history is to start with a purely random (white noise) time history and take a Fourier transform to get spectral components, then scale these components starting at Step 2 above.

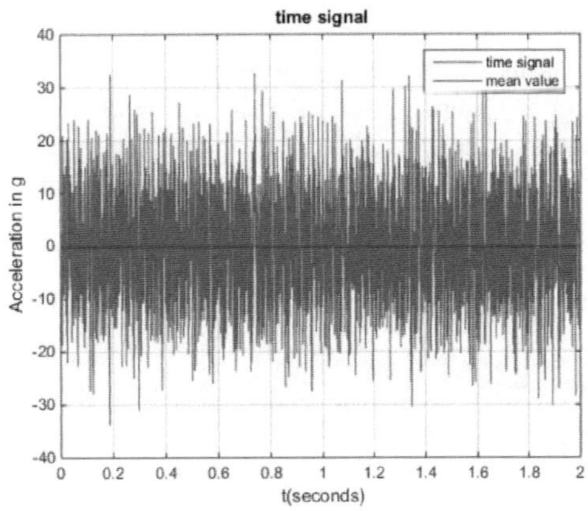

Figure 7. Accel time history synthesized to meet protoflight test level PSD.

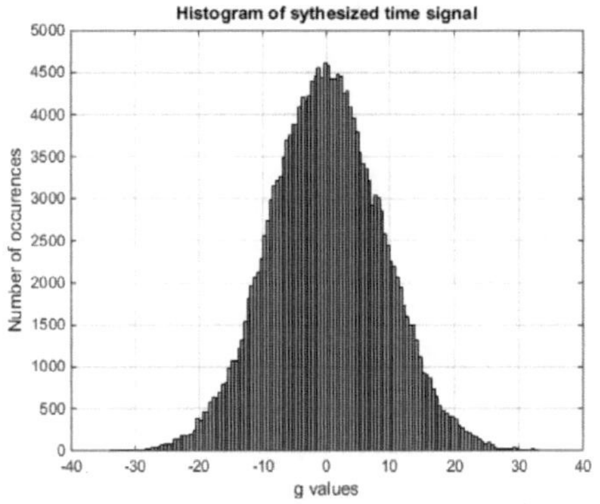

Figure 8. Histogram of synthesized random vibration time history.

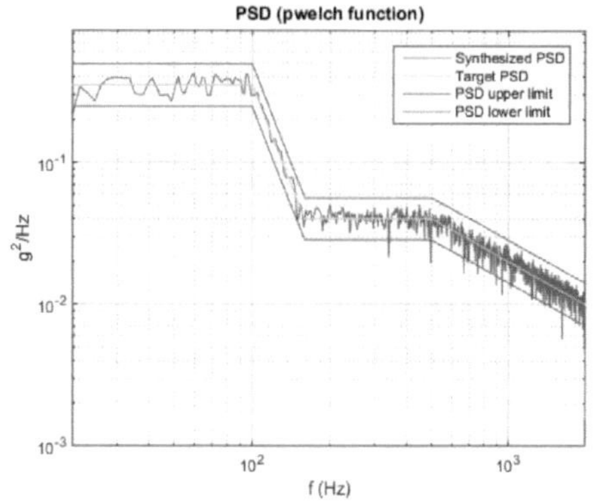

Figure 9. PSD of synthesized time history, from Matlab pwelch.m for confirmation.

Simulation Results

Figures 10-15 show simulation results for the three cases described above: Case 1: no launch locks, Case 2: axial/radial pinning, and Case 3: radial end shim, axial/combo radial pin. For each case, initial conditions were used that place the rotor in a neutral position. This was accomplished by running an initial simulation with no external excitation and using the final conditions from that run as the initial conditions for the simulation run with excitation. The rotor spin axis was horizontal for all tests (discussed below), so the same was assumed for the simulations. Simulation results are shown for vertical excitation for brevity. The horizontal excitation results differ only by the direction of the gravity vector relative to the excitation. Case 1 displacement trajectories with horizontal excitation differed from the vertical excitation but the number of excursions past 50% of design limit load were similar.

Figure 10-11 show predicted behavior for Case 1, no launch locks. Figure 10 shows relative rotor/backup bearing displacement. At start the rotor rests on the backup bearings, 0.0045 in from center as that minimum clearance was used for the simulations. Figure 10 shows many excursions where the rotor pulls out of contact with the backup bearing, crossing into the clearance space in response to the imposed motion. Figure 11 shows backup bearing loading during the simulation. The initial load of ~5 lbf at start is due to gravity and the PM negative stiffness offset force pulling the rotor against the backup bearings. For the smaller, motor end backup bearing there are ten load excursions above 47.5 lbf (half of the 95 lbf design limit). For the larger, impeller end backup bearing there are five load excursions above 100 lbf (half of the 200 lbf design limit).

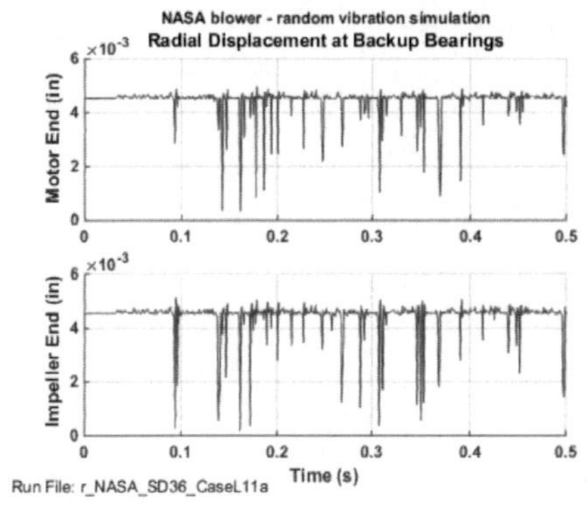

Figure 10. Predicted relative rotor/backup bearing displacement magnitudes. Case1: no launch locks.

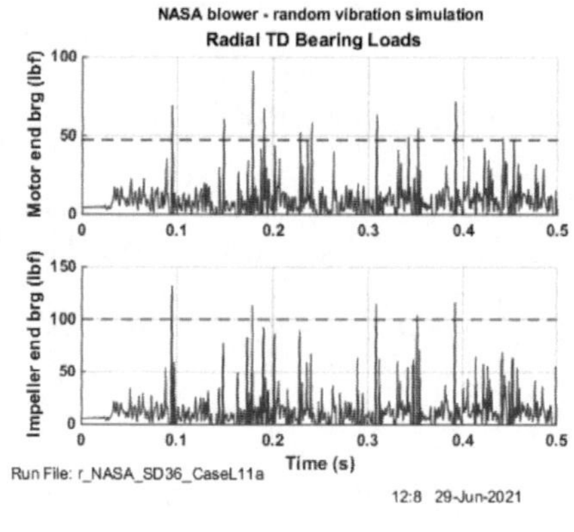

Figure 11. Predicted backup bearing loads. Case1: no launch locks.

Figure 12 shows predicted behavior for Case 2, rotor pinned at each end to the housing. This was accomplished by connecting a discrete 1e6 lbf/in stiffness between rotor end degrees-of-freedom and the corresponding housing degrees-of-freedom. Figure 10 shows relative rotor/backup bearing displacement. The peak excursion is 0.0001 in at the motor end bearing and 0.0035 in at the impeller end bearing. The impeller end bearing displacement is further from the shaft end where the rotor is pinned so the excursion should be higher. The dominant response frequency is at 690 Hz which can be compared to the prediction of 620 Hz for first pinned/pinned mode. The rotor does not contact the backup bearing so there is no load plot to show. This is the desired result, but as mentioned above, there is concern about reliably fixing the rotor at the backup bearing center on the motor end.

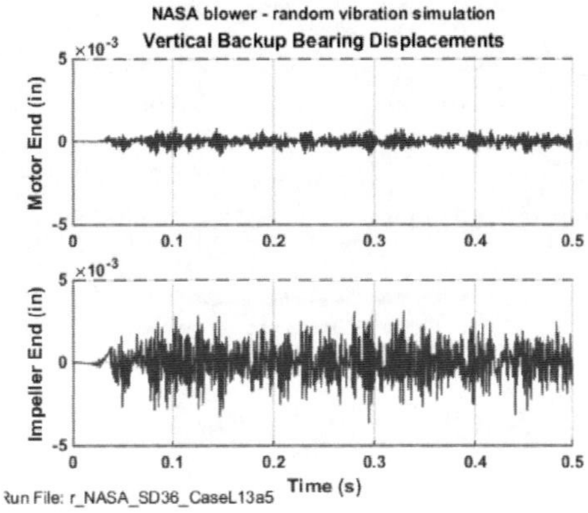

Figure 12. Predicted relative rotor/backup bearing displacements; top: motor end bearing, bottom: impeller end bearing.
Case2: motor end and impeller ends pinned

Figures 13-15 show predicted behavior for Case 3, rotor shimmed to eliminate clearance in the motor end backup bearing and rotor pinned to the housing on the impeller end. For the simulation, the shimming was modelled by removing the free clearance at the motor end bearing. Figure 13 shows relative rotor/housing displacement at two locations near the motor end. The upper plot is the displacement at the plane of the motor end (brg 1) position sensor. This is shown for comparison to the test data presented in the next section. The peak excursion is close to 0.004 in. The lower plot shows the resilient mount radial displacement which has a peak of about 0.002 in, less than the travel to the hard stop. The relative rotor to backup bearing inner ring travel is insignificant here because of the gap taken up by a shim. Figure 14 shows relative rotor/housing displacement at two locations near the impeller end. The upper plot is the displacement at the plane of the impeller end (brg 2) position sensor. The peak is excursion is close to 0.003 in. The lower plot shows radial displacement of the rotor relative to the impeller end backup bearing. which has a peak of about 0.0025 in so there is no contact with the backup bearing and the margin to contact is better than for Case 2. Figure 15 shows the loads carried by the two backup bearings. The upper plot shows the motor end backup bearing load is well controlled and peaks at about 13 lbf. Because of the shim this bearing will react whatever load is needed to move the rotor along with the housing vibration, but there is no bouncing impact loading. The lower plot confirms no load reacted by the impeller end backup bearing.

The Case 3 locking approach was chosen because it produced good results and appears to be easier to realize reliably in hardware.

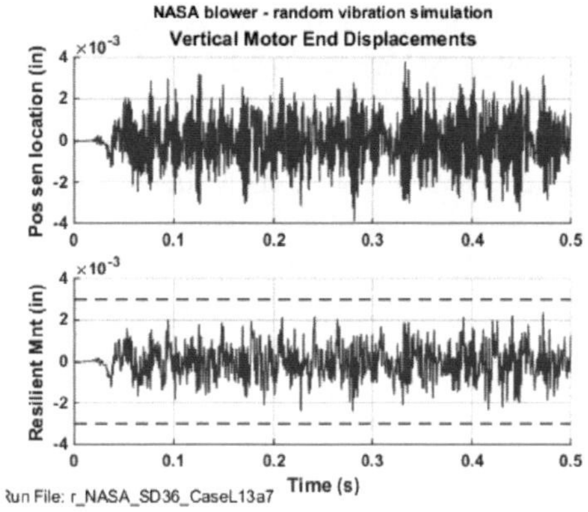

Figure 13. Predicted motor end displacements; top: position sensor location, bottom: across backup bearing resilient mount.
Case3: with motor end shim, impeller end pinned

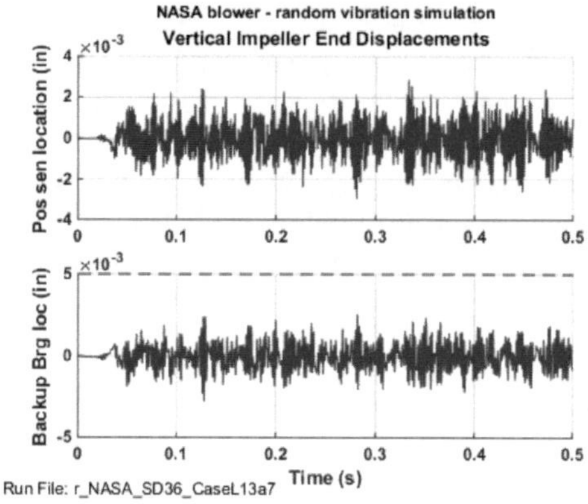

Figure 14. Predicted impeller end displacements; top: position sensor location, bottom: rotor to backup bearing relative displacements.
Case3: with motor end shim, impeller end pinned

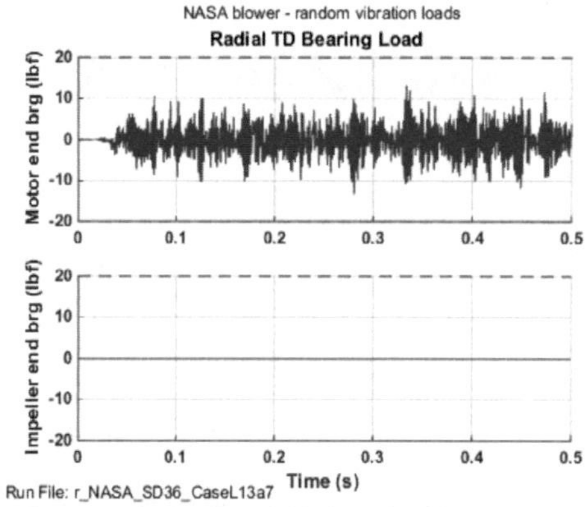

Figure 15. Predicted backup bearing loads: top) motor end; bottom) impeller end.
Case3: with motor end shim, impeller end pinned

Random Vibration Testing

Random vibration testing was performed at a commercial test lab using an electromagnetic shaker. The blower and blower controller were mounted to the shaker which was driven with a time signal that met the PSD requirements in Figure 5. Two control accelerometers mounted on the base plate are part of the shaker feedback loop used to ensure the base motion meets the required profile. The arrangement is shown in the photograph of Figure 16 with the equipment configured for shaking the blower in the horizontal direction. The blower is in the foreground left and the blower controller is in the background right. The rotor locking configuration used is Case 3, motor end backup bearing shimmed, and impeller shaft end pinned, as selected from the analysis. The combined PSD from the two control accelerometers from the vertical testing is shown in Figure 17. The measured PSD was required to be within ±1.5 dB of the specified level from 20 – 500 Hz and within ±3 dB from 500 – 2000 Hz as indicated by the tolerance bands in Figure 17.

Figures 18-21 show relative rotor/housing displacement data taken from the AMB position sensors during the vertical excitation test. Figures 18 (sensor 1) and 19 (sensor 2) show orbit plots for a 10 second period, when the measured peak amplitudes appeared to be at a maximum. The sample frequency of this measurement is 5 kHz. As can be seen in the figures, the direction of response was not purely vertical, as the peak displacements appear to be at a slight angle (approximately 9-15°) relative to the *y*-axis. The maximum peak-peak displacement is about 0.006

in at sensor 1 (motor end) and 0.005 at sensor 2 (impeller end). This can be compared to the simulation results from Figure 14 and 15 of 0.008 in at sensor 1 and 0.006 in at sensor 2. Due to the nature of random vibration, the most definitive outcome of this comparison is that the relationship between the amplitudes from sensors 1 and 2 is similar for the test and prediction. The relatively close correlation in peak amplitudes between this single simulation and the measurement run is encouraging, but repeatability of this correlation was not verified.

Figures 20 and 21 show the corresponding Fast Fourier transforms (FFTs) of the rotor-bearing relative displacement for sensor 1 and sensor 2, respectively. The FFTs are performed over a 2-second period. The FFTs shows the highest response centered around 410 Hz, which should be the lowest resonance of the system. Assuming a rigid connection at the impeller end launch lock and, the model predicted a peak response frequency of approximately 530 Hz for case 3, as noted above. This result seems to indicate that the stiffness contribution from the impeller end launch lock is somewhat lower than modelled. There is also a significant response observed in the primary, vertical axis data in the 20-100 Hz frequency range where the PSD level is 0.35 G2/Hz. This result can also be found in the simulation as well and is due mostly to response at the motor end bearing support. Overall, the test results give confidence in using the model to evaluate the launch locks and in the selection of the locking strategy.

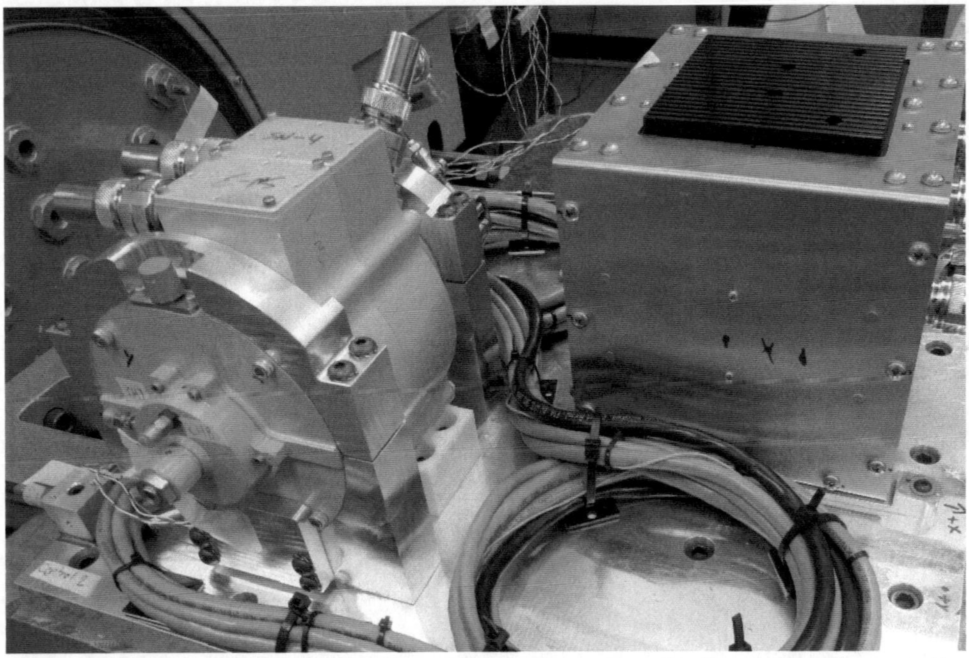

Figure 16. Blower mounted on vibration table prior to random vibration testing. The controller (background) was tested simultaneously on the same base plate.

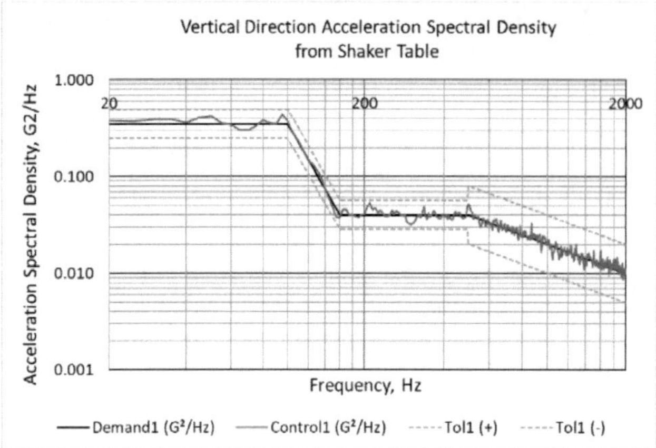

Figure 17. Acceleration spectral density (ASD or PSD) from shaker table control accelerometers, vertical excitation.

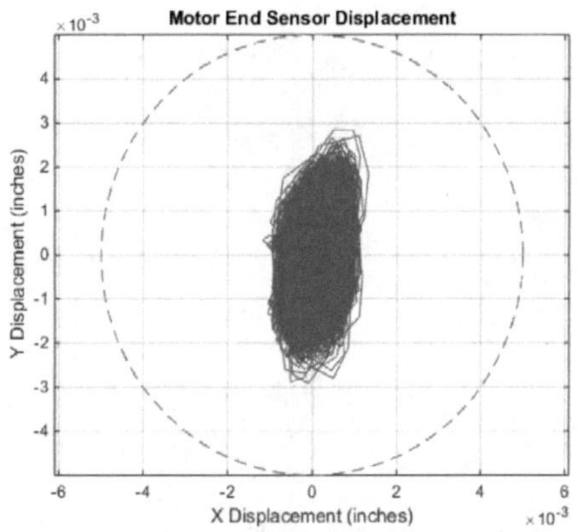

Figure 18. Measured position orbit from motor end AMB sensors (sensor 1) during vertical test.

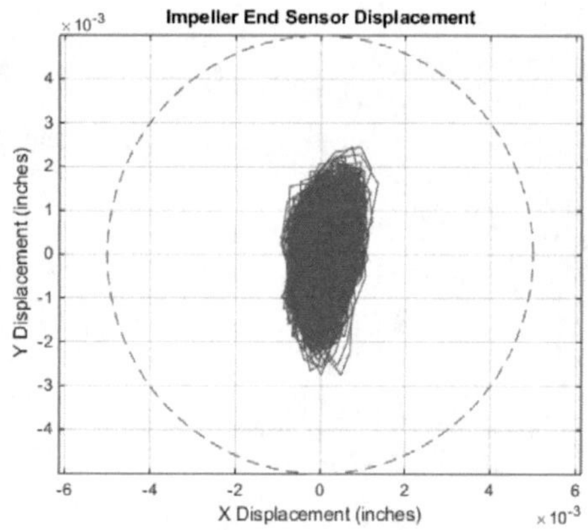

Figure 19. Measured position orbit from impeller end AMB sensors (sensor 2) during vertical test.

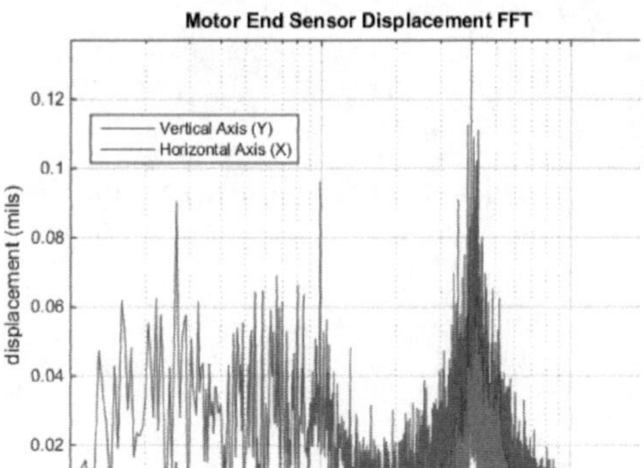

Figure 20. Fourier transform from motor end AMB sensors (sensor 1) during vertical test.

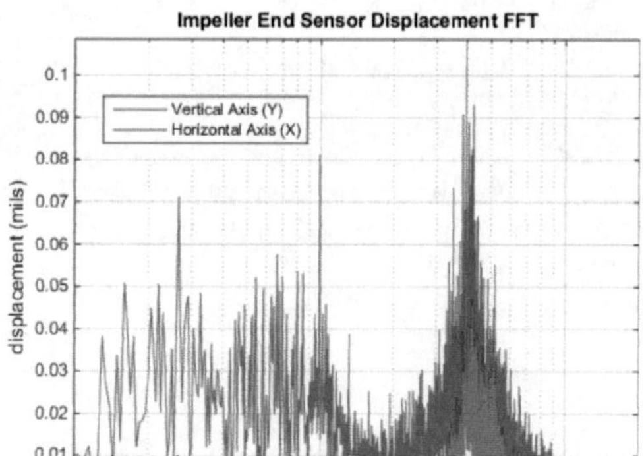

Figure 21. Fourier transform from impeller end AMB sensors (sensor 2) during vertical test.

Conclusion

Random vibration testing has been completed for a magnetically-levitated blower that will be a key component of a new 4BCO2. NASA plans to use the ISS as a test bed to evaluate the scrubber for future space missions. The vibration testing is a standard proof test that NASA imposes to validate that equipment sent to the ISS can survive the launch environment. The vibration testing is considered conservative relative to the launch environment partly because the vibration level imposed during the vibration test is intended to exceed the Maximum Expected Flight Level by 3 dB.

A transient, nonlinear rotordynamic model was developed to simulate the effects of the vibration testing (and subsequently the launch environment) on the blower backup bearings. As the blower will be launched in a de-levitated state, two different methods for locking the rotor were analyzed and compared to a "no launch lock" control case. It was determined that the rotor should be locked to limit rotor motion and backup bearing contact, with an additional shim used to fill the motor-end backup bearing clearance. For the vibration testing, this same locking configuration was used. The results of the vertical shaker test showed that there was 0.006 in. and 0.005 in. peak-peak displacement observed at the motor end sensor and impeller end sensor, respectively. While the simulation predicted somewhat larger amplitudes (0.008 in and 0.006 in), the relationship between the two sensors was the same. Measured results showed a peak displacement at approximately 410 Hz, while the simulation showed a peak response at 530 Hz, likely meaning that the launch lock mechanism was less stiff than modelled.

Overall, the results of the vibration test provide good confidence that the blower can be launched without failure due to random vibration experienced during launch. The simulation results provide confidence that the model can be used for future de-risking of the effect of launch loads on AMB-supported turbomachinery.

References

[1] Hawkins, L., Filatov, A., Khatri, R., DellaCorte, C., Howard, A., 2020, "Design of a Compact Magnetically Levitated Blower for Space Applications", GTP-21-1084, *J. of Eng. For Gas Turbines and Power*, Vol. 143, pp. 091012, September.

[2] API 617 American Petroleum Institute, 2014, *API Standard 617, Eighth Edition, Axial and Centrifugal Compressors and Expander-Compressors*, Washington, D.C., USA, API Publishing Services

[3] Keogh, P., Aeschlimann, B., Hawkins, L., Jayawant, R., Köhler, B., Maslen, E., 2018, "Activities of ISO/TC 108/SC 2/WG 7 in the Development of Standards for AMB Systems", *Proceeding of 16th International Symposium on Magnetic Bearings*, Paper ID-146, Beijing, China, August.

[4] Schweitzer, G., Maslen, E.H, 2009, *Magnetic Bearings: Theory, Design, and Application to Rotating Machinery*, Springer Verlag, Berlin, Germany, pp. 288-294.

[5] Wilkes J., Moore, J., Ransom, D. and Vannini, G., 2013, "An Improved Catcher Bearing Model and Explanation of the Forward Whirl/Whip Phenomenon Observed in Active Magnetic Bearing Transient Drop Experiments," *Proc.of ASME TurboExpo*, GT2013-94594, San Antonio, TX, June.

[6] Smalley, A.J., Darlow, M.S., Mehta, R.K., 1978, "The Dynamic Characteristics of O-Rings", *Trans. Of ASME,* Vol. 100, pp.132-138, January.

[7] NASA, 2015, *Pressurized Payloads Interface Requirements Document International Space Station Program*, SSP-57000 Rev R, October.

[8] Hawkins, L., Wang, Z., Nambiar, K., 2018, "Floating Shock Platform Testing of a Magnetic Bearing Supported Chiller Compressor – Measurements and Simulation Results", *Proc. of ASME Turbo Expo*, GT2018-77031, Oslo, Norway, June.

Bearingless Induction Motor Design with the DPNV Winding and Reduced Axial Length Pole-Specific Rotor

Jiahao Chen and Eric L Severson

Department of Electrical and Computer Engineering, University of Wisconsin-Madison, WI, USA
horychen@zju.edu.cn, eric.severson@wisc.edu

Abstract

This paper investigates the design of bearingless induction motor that uses a dual purpose no voltage (DPNV) winding and pole-specific rotor. The investigation is motivated by a previous study that showed that by integrating these two technologies into a bearingless induction motor, a high performance machine can be created for industrial compressor applications (over 95% efficiency, over 15 kNm/m^3 torque density, and less than 2.5% current for supporting the rotor weight). The use of parallel pole-specific rotor significantly reduces the rotor axial length, improving rotor dynamics. However, implementing the DPNV winding is difficult because it results in an asymmetric winding for two preferred pole combinations of the induction machine. This paper investigates these asymmetries and the challenges that they pose for the bearingless machine performance.

Complex winding factors are introduced as a tool to explore the physics behind each design in terms of sub-harmonics and winding asymmetry. Transient FEA is used to obtain steady state suspension force vectors as a function of suspension current phase angle for five selected optimal designs. The results are visualized as polar plot, where relevant performance measures are proposed. It is found that a symmetric winding (with or without sub-harmonics) has a circular force trajectory, while asymmetric winding causes an undesired elliptical force trajectory that degrades the machines force rating. Moreover, it is found that asymmetric sub-harmonics alone can also lead to this elliptical force trajectory. It is revealed by further studying the force error angle trajectory that the force error angle is generally a function of suspension current phase angle, which has implications for the optimization of these machines. In conclusion, this paper shows that the asymmetric DPNV winding can result in a high performance design, which enables higher performance pole combinations for bearingless induction motors.

Keywords: dual purpose no voltage winding, magnetic levitation, asymmetry, sub-harmonics.

Nomenclature

p, p_s	the pole pair number of torque field and suspension field, respectively.
Q_r, Q'_r	the total rotor slot number, and the cage slot number of parallel pole-specific rotor.
h, v	absolute harmonic order, and relative harmonic order.

1 Introduction

Induction motors are widely used in both commercial and industrial applications, including compressors, blowers, spindles, speed servo motors, and traction motors. The bearingless version of the induction motor technologies have been developed since the early 1990s [1, ch. 13]. Innovation in the design of this machine can be divided into the following three paths: geometry optimization, stator winding, and rotor winding. These three innovation pathways are now briefly reviewed, with emphasis placed on the stator and rotor winding approach. This paper's goal is to explore how promising innovations in the stator and rotor winding can be combined to form a practical bearingless machine and identify areas where further research is needed.

The geometry optimization of a bearingless induction motor is more complicated than synchronous machines due to rotor currents. A study presented in [2] develops techniques to optimize the machine design, including computationally-efficient modeling techniques to rapidly reach steady state and an algorithm to estimate the machine's breakdown slip.

The stator winding for a bearingless motor is responsible for creating two magnetic fields, one field for motor operation (p pole-pairs) and a second field for controlling suspension forces (p_s pole-pairs, where $p_s = p \pm 1$). Winding approaches can be broadly classified as either "separated" or "combined." *Separated windings* are actually two independent stator windings that share the same stator slots; one winding creates p pole-pairs and the second winding creates p_s pole-pairs [1]. In magnetic levitation systems, the suspension winding is typically over-dimensioned (more ampere-turn capacity is allocated than required) to build in a safety factor for the suspending the motor shaft. Typical bearingless motors with separated windings allocate 30–50% of the slot space to the levitation winding, which derates the machine's torque density by 30–50% [3]. In previous work, the authors have found that most machines actually require only $< 2.5\%$ of slot current to support the rotor's weight [3]. This means that during normal operation, the slot space that has been allocated to the suspension winding is underutilized.

As an alternate to separated windings, *combined windings* are winding approaches that use the same stator coils for both torque and force creation. This allows the stator current capacity to be used more effectively because the slot current can be dynamically allocated between creating torque and force. In a motor system where the nominal load is gravity and other force disturbance events are rare (i.e., compressor surge or passing through a critical speed), a combined winding approach can enable a significant improvement in machine performance and power density. There are several combined winding approaches that have been proposed in the literature. These include multi-phase [4, 5], dual purpose no voltage ("DPNV") [6, 7], multi-sector [8, 9], and middle-point current injection windings [10]. Of these, multi-phase and DPNV combined windings offer the highest potential for performance, because they allow independent motor d and q current control.

It was shown in [11] that the DPNV winding is a circuit-based implementation of the multi-phase winding. From a machine perspective, the two windings are identical. However, the DPNV winding has the advantage of enabling the motor operation to be controlled with a conventional three phase motor drive that has no knowledge of the machine being bearingless. For significant power and critical industrial applications, using a conventional motor drive (with no knowledge of the suspension operation) is very attractive. A machine owner can purchase a well-established and trusted commercial product for the critical power path and a specialized drive for the low power suspension operation. For this reason, the authors of this paper believe that the DPNV winding is a promising technology for the commercialization of industrial-scale bearingless motors.

Early bearingless induction motors utilized squirrel cage rotors [12, 13]. Under nominal loading, the bearingless machine's suspension field rotates at a different speed than the rotor. This induces current in squirrel cage rotor circuit which has been found to be problematic [1, p.255]. The squirrel cage leads to inevitable delay in regulation of suspension force [14], and results in inefficient magnetic levitation, because excessive stator current is needed to create the suspension magnetic field, as the suspension slip frequency gets large at high speeds [15].

Alternative rotor winding approaches have since been developed that are designed to prevent current from being induced in the rotor. For example, the highest power bearingless induction motor reported in literature is tested at 30 kW using a wound-rotor [16]. The wound-rotor induction motor has been historically proposed for soft starting [17]. If the wound-rotor winding is designed such that it has zero winding factor for the p_s pole-pair suspension field, it will not link the suspension field and therefore no current will be induced [18]. This makes the rotor structure an effective alternative to the squirrel cage rotor.

This same principle (of not linking the suspension flux) has been applied to the design of cage rotors. In [19], a "pole-specific" (PS) cage rotor is proposed that consists of multiple isolated cages, each with four bars connected in series in a manner that prevents them from linking a two pole suspension field (for a bearingless machine with $p=2, p_s=p-1$). This structure was later optimized in [20] to reduce rotor resistance. A similar structure has been developed for machines with four pole suspension fields (bearingless machine with $p=1, p_s=p+1$), where each isolated rotor cage consists of two bars placed on opposite sides of the rotor [1, 21]. These two cage rotor designs will be referred to as the "isolated PS rotor" in this paper. While the isolated PS rotor offers a high performance solution from an electric perspective, it poses mechanical challenges. Isolated end-plates must be located at each end of the rotor which adds significant axial length. Recently, a new cage rotor structure has been developed by *Chen et. al.* in [22] to reduce the number of end-plates (and therefore the axial length). This structure is referred to as a "parallel PS rotor" in this paper because it uses a common end ring to connect all rotor cages in parallel on one end of the rotor. *Chen et. al.* have generalized the concept and shown how the rotor can be designed for all pole-configurations with $p_s \geq 2$.

The authors of this paper propose developing a bearingless induction motor that utilizes both the parallel PS rotor (to improve electric performance and simplify the mechanical design) and the DPNV stator winding (to increase torque density and efficiency). Unfortunately, two potentially critical problems arise when trying to design a machine of this configuration for industrial applications. First, many DPNV windings that match the preferred slot-pole combinations of induction machines will create subharmonic airgap magnetic fields when excited with suspension currents. These fields will link the pole-specific rotor, inducing currents and potentially degrading performance. Second, these industrial-scale machines often prefer pole-combinations with either $p = 3$ or $p_s = 3$. Such machines will violate a symmetry condition of three phase DPNV windings, resulting in an asymmetric stator winding that may cause unacceptable force performance. This paper investigates DPNV windings with subharmonic fields and MMF asymmetry. The paper analyzes whether example practical machines which use these windings can still function as high performance bearingless motors and identifies areas where future research is needed to make these machines practical.

This paper will first summarize the parallel PS rotor design of [22] and explain the primary differences between isolated and parallel PS rotors in Section 2. Next, Section 3 summarizes the DPNV winding and presents the design of DPNV windings for induction machines that violate the symmetry condition. The asymmetries that result in these machine windings are explained and analyzed in Section 4 using complex winding factors. Inherent suspension sub-harmonics of the DPNV winding are investigated in Section 5. Finally, the influence of asymmetry and harmonics on the machine's suspension force performance is analyzed in Section 6.

2 Pole Specific Rotor with Reduced Axial Length

This section summarizes the parallel PS rotor design utilized in this paper (recently developed in [22]) and explains how it differs from the more established isolated PS rotor design of [19, 21].

The parallel PS rotor circuit consists of a common end-ring on one axial end and multiple end-rings on the other end. Each cage has Q_r' rotor bars that are connected in parallel so that the spacing between the bars is a suspension pole-pair. Compared with the isolated PS rotor, this type of PS rotor has significantly less axial length because all end-connections at one end of the rotor are removed (apart from the common end-ring). Example isolated PS rotor

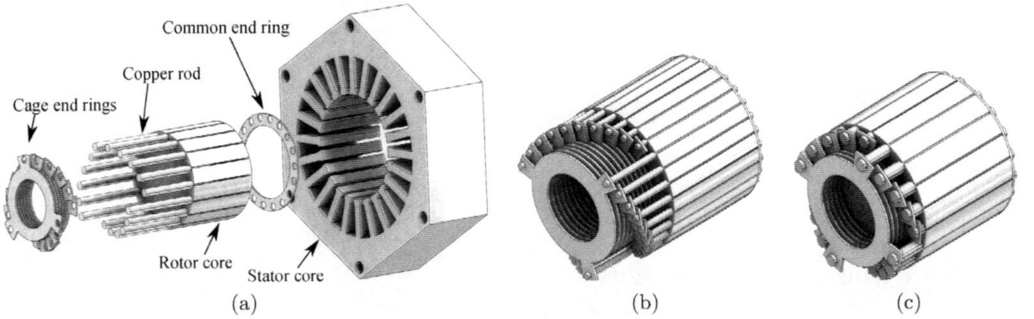

Figure 1: Parallel PS rotors with (a),(b) $Q'_r = p_s = 3$ and $Q_r = 18$, 30; (c) $Q'_r = p_s = 4$, $Q_r = 20$.

prototypes for $p = 1, p_s = 2$ can be found in [23, 24] and a prototype of the new parallel PS rotor ($Q'_r = p_s = 2$) is documented in [2, 22].

For $p = 1$ bearingless induction machines, the parallel PS rotor is recognizable as an isolated PS rotor where the axial end rings on one side have been replaced by a common end-ring. However, notable differences arise for $p \geq 2$. In $p = 2$ bearingless induction motors, if a $p_s = p - 1$ design is desired, the isolated PS rotor utilizes four series-connected rotor bars per cage [19] while the parallel PS rotor does not exist because only $Q'_r = p_s = 1$ rotor bar could be used in a cage; this design would be invalid as at least two bars are needed for rotor current to flow. Alternatively, a $p = 2$, $p_s = p + 1$ design can utilize a parallel PS rotor, but no isolated PS rotors have been documented with $p_s > 2$. In other words, if a $p = 2$ motor is desired, one can choose between a $p_s = 1$ isolated PS rotor and a $p_s = 3$ parallel PS rotor, and the latter usually has a shorter end connection region. Example CAD drawings of $Q'_r = p_s = 3$ parallel PS rotors are shown in Figure 1a with an exploded view and Figure 1b. Furthermore, the parallel PS rotor concept can be generalized for all $p \geq 3$ motors, and an example $Q'_r = p_s = 4$ parallel PS rotor for $p = 3, 5$ motor is sketched in Figure 1c.

This paper is particularly interested in developing high p_s design that can potentially reduce the rotor axial length of a bearingless induction motor. The rest of this section will summarize the key properties of the parallel PS rotor from [22].

2.1 Pole-Specific Rotor with Parallel Connected Bars

The fundamental premise of the parallel PS rotor is that a single cage is composed of bars which span a suspension pole-pair. Because the cage bars span a suspension pole-pair, they do not link the suspension field. Since a cage must consist of more than one bar, this rotor concept only works if the motor has at least 2 suspension pole-pairs, i.e. $p_s \geq 2$. The complete rotor consists of multiple cages. The cages share a common end-ring on one side of the rotor, but on the other side they have separate end-rings, as shown in the exploded view in Fig 1a.

2.2 Common End-ring

Using a common end-ring to replace the end-connections at one side of the rotor, results in 50% less end-connection space. This saves remarkable axial length for high speed applications where critical speeds limit the rotor axial length. *Chen et. al.* show in [22, Sec.II-B] that this design preserves the pole-specific property and has no performance penalty.

While the currently described parallel PS rotor does not work for $p = 2$, $p_s = 1$ machines, one can infer from this discussion that this can actually be accomplished by placing end-connections on both rotor ends as exemplified in Figure 2.

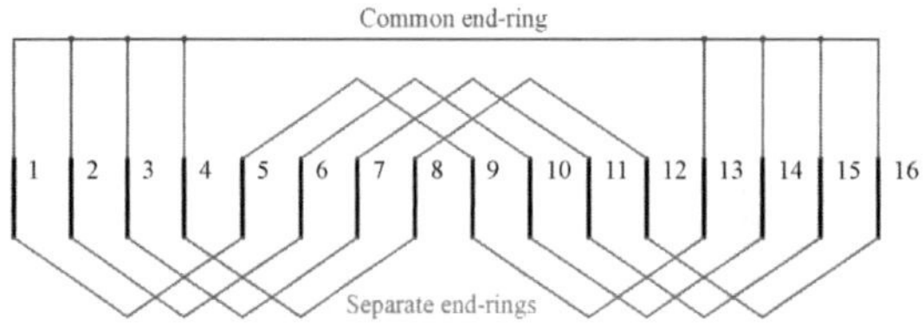

Figure 2: Common end-ring PS rotor design proposed for $Q_r = 16, p = 2, p_s = 1$.

2.3 Multiple Separate Cages

The idea of a parallel PS rotor is to not induce rotor current from the p_s pole-pair suspension field. To this end, the number of bars in a cage, denoted as Q'_r, depends on p_s. Specifically, Q'_r is determined by the fact that a cage with Q'_r bars does will not have rotor current induced by a magnetic field whose pole pair number p_s is multiple of Q'_r.

Using multiple cages, each with Q'_r bars results in a different bar spacing parameter $\alpha_c = \frac{2\pi}{Q'_r}$ from a traditional squirrel cage with Q_r bars. By analyzing the T-equivalent circuit of a parallel PS rotor, the rotor bar impedance is found to be a function of the number of cages N_c when Q_r is fixed, and the conclusion of [22] is that converting a squirrel cage rotor to a parallel PS rotor would result in comparable or even improved motor performance.

It is recommended to let one cage hold as many rotor bars as possible, as an effort to reduce rotor axial length. For instance, a parallel PS rotor with $p = 3$, $p_s = 4$ can reduce the number of end-rings by 75% as compared to a $p = 1$, $p_s = 2$ isolated PS rotor that has the same number of rotor slots, because the parallel PS rotor has one common end-ring and it groups $Q'_r = p_s = 4$ bars in a single cage.

2.4 Design rules

For a p pole-pair bearingless induction motor to be able to use the parallel PS rotor winding, it must satisfy the design rules listed in (1) according to [22]

$$p_s = p \pm 1, \quad p_s \geq 2, \tag{1a}$$

$$Q'_r = p_s/k \tag{1b}$$

where k can be any integer to make $Q'_r \geq 2$ an integer. Then, by deciding the number of cages, N_c, the total slot number is determined as $Q_r = N_c Q'_r$.

For most applications, the choice of k is recommended to be 1, such that Q'_r is maximized. For a certain Q_r value, this gives a minimal N_c as an effort to reduce rotor axial length. However, the choice of $k > 1$ allows for more values of Q_r to be usable, and such demand could emerge when a high p induction motor is desired.

3 Stator Winding Design for Induction Motors

This paper studies the DPNV winding as the combined winding approach for bearingless induction motors. The DPNV winding has been shown to have equivalent performance to multi-phase winding [11]. One major reason to use DPNV winding is that the DPNV winding allows to use any commercially available motor drive product for the maglev system, and the engineer would only need to build the suspension inverter part. However, the literature has been avoiding designing a $p = 3$ or $p_s = 3$ bearingless motors using DPNV winding (see, e.g., [25]). According to [26, Table III], the symmetry conditions state that for an m-phase motor, m and p must be co-prime and m and p_s must be co-prime. In other words, symmetric three phase DPNV winding cannot have $p = 3$ or $p_s = 3$.

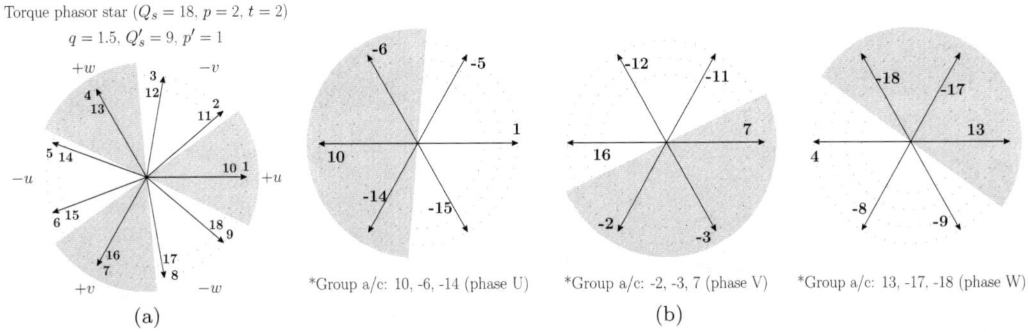

Figure 3: Star of slots plot for a winding with $Q_s = 18$, $p = 2$ $p_s = 3$, a) at torque frequency, b) at suspension frequency.

The most popular pole pair numbers for industrial scale induction motors are $p = 1, 2, 3$ [27]. Therefore, it seems that the DPNV winding approach is a good match when bearingless motors with $p = 1$ or $p = 2$ are desired. However, from last section, in order to use the parallel PS rotor (that has shorter axial length), the induction motor should have $p_s \geq 2$. In order to use DPNV winding and parallel PS rotor for a 3 phase bearingless induction machine, the only working option therefore becomes $p = 1$, $p_s = 2$, if a symmetric stator winding is highly desired. According to machine design textbook [27], a $p = 1$ induction motor design has large stator and rotor yoke depth, leading to designs with lower torque densities compared to other $p \neq 1$ motors. Torque density matters, for example, when developing bearingless induction machines for mobile applications (i.e. aviation).

This paper is interested in developing $p = 2, 3$ motors using both the DPNV winding and the parallel PS rotor. In [2], it has been reported that the DPNV winding can use less than 2.5% of rated current to support the rotor weight and that a $p = 1, 2$ bearingless induction motor using an isolated PS rotor and DPNV winding has potential to meet the requirements for industrial applications. This paper shows that the DPNV winding can also be used to improve the performance of $p \geq 2$, $p_s \geq 2$ bearingless induction motors with parallel PS rotors.

This section will now present two example asymmetric DPNV winding designs with $p = 2$, $p_s = 3$ that violate the symmetry condition of DPNV windings for symmetric suspension MMF [26, Table III], because p_s is not co-prime with $m = 3$ phases.

3.1 Example Design of an Asymmetric DPNV Winding

The DPNV design process explained in [26, 28] is used to design a stator winding with $Q_s = 18$ slots, $p = 2$ pole-pairs. The star of slots at torque frequency is shown in Figure 3a with a slot angle $\alpha_p = 40^e$. From Figure 3a, phasors belonging to zone +u and zone −u are assigned to phase U, and the phasors in zone -u suggest the corresponding conductors (i.e., $5, 6, 14, 15$) should be connected in reverse to the conductors corresponding to zone +u (i.e., $1, 10$).

For the stator winding in Figure 3a, it is feasible to choose $p_s = 1$ or $p_s = 3$. Since we are interested in using parallel PS rotor, we must select $p_s = 3$. Recall that $p_s = 3 = m$ violates a symmetry condition in [26, Table III], indicating that the suspension MMF will be assymetric. In order to allow the same winding in Figure 3a to produce a suspension field of pole pair number $p_s = 3$ that is different from p, we draw another star of slots at the suspension frequency with $\alpha_{ps} = p_s 360°/Q_s = 60^e$ for each phase, as shown in Figure 3b. The negative signs before phasor numbers account for the connection directions imposed in Figure 3a. To allocate suspension groups, the 360^e space is divided into two 180^e bands, and phase U phasors in the gray-colored 180^e band (i.e., $-6, 10, -14$) will be assigned to group a/c, while the rest (i.e., $1, -5, -15$) are

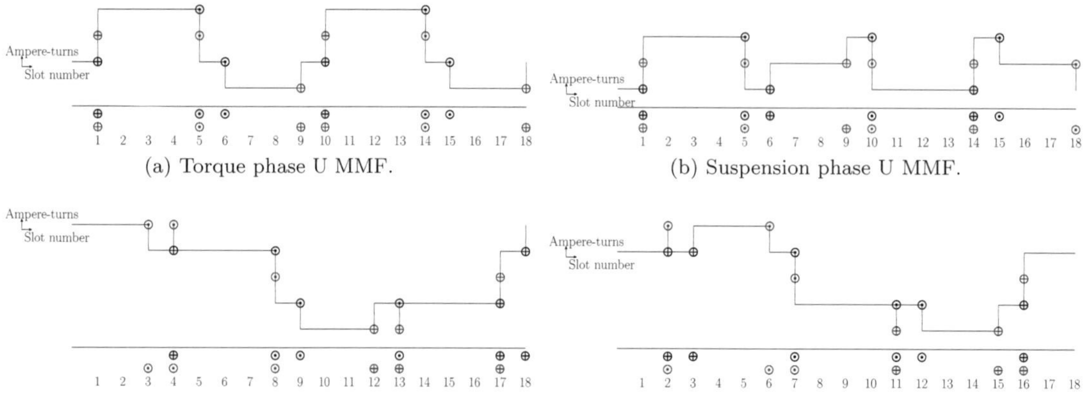

(a) Torque phase U MMF. (b) Suspension phase U MMF.

(c) Suspension phase V MMF. (d) Suspension phase W MMF.

Figure 4: MMF waveform of a DPNV winding with $Q_s = 18,\ p = 2,\ p_s = 3,\ y = 4$.

assigned to group b/d. The group name definition is consistent with [26, 28].

3.1.1 Torque MMF Waveform

The example design's torque magnetomotive force (MMF) waveform of phase U is shown in Figure 4a. The torque winding factors, $k_{Tw}(h)$, when the coil pitch y is selected to be 4 slots, are shown in Table 1. The torque winding factor is $k_{Tw}(p) = 0.945$.

3.1.2 Suspension MMF Waveform

The resulting suspension MMF waveforms of phase U, V, and W are shown in Figure 4b-4d and are clearly asymmetric between phases. The MMF of phase U is observed to have a shape that is different from phases V and W; further, the MMF waveform of phase W can be obtained by mirroring about a vertical axis and shifting the MMF waveform of phase V. The suspension winding factors, $k_{Sw}(h)$, are shown in Table 1. Interestingly, it is found that the three phases have the same $k_{Sw}(p_s)$ values of 0.577 for the working harmonic of the suspension field ($h = p_s$). However, the shape asymmetry observed in the MMF waveforms does cause unequal $k_{Sw}(h)$ among 3 phases for the sub-harmonics ($h = 1$) and higher-order harmonics ($h > 3$). The suspension sub-harmonic field has a pole-pair number that differs from p by ± 1, and by interacting with the torque field, it will produce a suspension force.

A second example design for this pole combination is studied in this paper ($p = 2, p_s = 3, Q_s = 24, y = 5$); its winding factors are listed in Table 2. Again, identical $k_{Sw}(p_s)$ is observed for all three phases. One might be tempted to conclude that, for these two particular $p = 2, p_s = 3$ winding designs, the shape asymmetry in the MMF waveform does not incur any problem if we neglect harmonics. Unfortunately, this is not true, as we have only studied the *amplitude asymmetry*, and have not studied the *phase-shift asymmetry*. Different asymmetries are discussed later, in Section 4.

4 Asymmetries in Suspension MMF Waveform

This section investigates the types of asymmetry that arise from violating the DPNV winding symmetry requirement that both p and p_s are co-prime with m listed in [26, Table III]. We care about this because in a previous optimization study [22], the DPNV winding based optimal motor designs with either $p = 3$ or $p_s = 3$ have resulted in significantly shorter rotor axial length, showing the potential to further improve the rotor dynamics for high speed applications.

Table 1: Winding Factors for DPNV winding with $p=2, p_s=3, Q_s=18, y=4$.

Order h	1	p=2	p_s=3‡	4	5	6	7
$k_{Tw}(h)^*$	-	0.945	-	0.061	-	-0.577	-
$k_{Sw,U}(h)^\dagger$	0.140	-	0.577	-	-0.061	-	-0.945
$k_{Sw,V}(h)$	0.473	-	0.577	-	-0.186	-	-0.398
$k_{Sw,W}(h)$	0.473	-	0.577	-	-0.186	-	-0.398

*$k_{Tw}(h)$ is the torque winding factor for all phases for harmonic order h.
†$k_{Sw,U}(h)$ is the suspension winding factor for phase U for harmonic order h.
‡The phase angles of $\bar{\boldsymbol{k}}_{Sw}(3)$ of the three phases are 0, 120, 240 [elec. deg].

This section is going to use complex winding factors to study asymmetry. The calculated complex number winding factor will allow us to gain insight into two types of asymmetries observed in the winding—whether it is a *phase shift asymmetry*, where the angle between phases are not 120 elec. deg, or whether it is an *amplitude asymmetry*, where not all phases produce the same magnitude of MMF.

4.1 Complex Winding Factor Calculation

This section now introduces a complex winding factor calculation that will be used in later parts of the paper for analysis of winding asymmetry.

The pitch factor for harmonic order $h = vp$ is traditionally computed as

$$k_p(vp) = \sin\left(vp\frac{\gamma}{2}\right) \tag{2}$$

where γ is the coil span in [mech. rad], and $\gamma \in (0, 2\pi)$. Only certain values of γ are allowable, based on the number of stator slots, to make the coil pitch $y = p\gamma/\alpha_u$ an integer, where $\alpha_u = p\frac{2\pi}{Q_s}$ [elec. rad] is the slot angle in star of slots.

This paper only considers the case of a double layer winding where each phasor in the star of slots corresponds to a coil. The location of the coil can then be represented by the phasor number; we can calculate a complex pitch factor for the coil of phasor number $i \in [1, Q_s]$ as

$$\bar{\boldsymbol{k}}_{p,i}(vp) = \sin\left(vp\frac{\alpha_u}{2p}y\right) e^{j\left[\frac{\pi}{2} - vp\left(\frac{\alpha_u}{2p}(2i+y)\right)\right]} \tag{3}$$

which is a function of the harmonic order $h = vp$ and the winding parameters Q_s, y, p. This definition of pitch factor has the same magnitude as (2), but has an angle indicating the location of the coil. The coil index i for the double layer winding is exactly the phasor number i in the star of slots (see e.g., Figure 3).

The final winding factor takes the winding distribution into account by calculating the "vector average" of the pitch factors of all coils in one phase. For phase U,

$$\bar{\boldsymbol{k}}_{w,U}(vp) \triangleq k_{w,U}(vp)\angle\alpha_{w,U}(vp) = \frac{1}{N_{\mathcal{U}}}\sum_{i\in\mathcal{U}} \bar{\boldsymbol{k}}_{p,i}(vp) \tag{4}$$

where $\mathcal{U}$ is the set of coil indices of phase U, and $N_{\mathcal{U}}$ is the number of coils of phase U. For double layer winding, since each phasor in star of slots corresponds to a coil, we have $\mathcal{U} = \{i \in \mathbb{N} \,|\, \text{star of slots phasor numbers of phase U}\}$, and the set $\mathcal{U}$ will have $N_{\mathcal{U}} = \frac{Q_s}{m}$ elements. For example, Figure 3 gives $\mathcal{U} = \{i \in \mathbb{N} \,|\, 1, 5, 6, 10, 14, 15\}$.

The phase winding MMF can be determined from the winding factor. Here, the vp-th harmonic is expressed for phase U, $\Theta_{U,v}(\alpha)$, as a function of the air gap location α:

$$\Theta_{U,v}(\alpha) = N_{\mathcal{U}} k_{w,U}(vp)\left[\frac{4}{\pi}\frac{\Theta_\pi}{vp}\cos\left(vp\alpha + \alpha_{w,U}(vp) - \frac{\pi}{2}\right)\right] \tag{5}$$

Table 2: Winding Factors for DPNV Winding with $p = 2, p_s = 3, Q_s = 24, y = 5$.

Order h	1	$p = 2$	$p_s = 3^*$	4	5	6	7
$k_{Tw}(n)$	-	0.933	-	-	-	-0.500	-
$k_{Sw,U}(n)$	0.427	-	0.604	-	-0.073	-	-0.427
$k_{Sw,V}(n)$	0.427	-	0.604	-	-0.073	-	-0.427
$k_{Sw,W}(n)$	0.262	-	0.604	-	-0.012	-	-0.695

*The phase angles of $\bar{\boldsymbol{k}}_{Sw}(3)$ of the three phases are 0, 90, 225 [elec. deg].

where $\Theta_\pi = \frac{1}{2} z_Q I_U$ with z_Q the conductor number in a slot and I_U the conductor current. If we calculate the winding factor of each phase, we know the amplitude of the MMF of each phase and that phase's separation from the other phases. In other words, to determine amplitude/phase asymmetry, all we need to do is to calculate the winding factors for all phases.

4.2 Phase-shift Asymmetry

The complex winding factor (4) is useful to determine if there exists *phase-shift asymmetry* in the suspension winding factors among the three phases. For a symmetric three phase winding, the working harmonic's winding factor angle referred to phase U should be a multiple of 120^e,

$$\begin{aligned} \alpha_{w,U} - \alpha_{w,U} &= 0^e \\ \alpha_{w,V} - \alpha_{w,U} &= 120^e \\ \alpha_{w,W} - \alpha_{w,U} &= 240^e \end{aligned} \tag{6}$$

where $\alpha_{w,X}$, $X = U, V, W$ is the winding factor angle defined in (4).

In Table 2, when $h = p_s = 3$, the winding factor angles are 0^e, 90^e, 225^e, which means there exists phase-shift asymmetry, though there is no amplitude asymmetry.

In Table 1, when $h = p_s = 3$, the winding factor angles are 0^e, 120^e, 240^e, which means there is no phase-shift asymmetry. It is surprising to find that there is neither amplitude asymmetry nor phase-shift asymmetry at the working harmonic $h = p_s = 3$ for a symmetry condition violating DPNV winding (recall that there is amplitude asymmetry in the subharmonics).

4.3 Amplitude Asymmetry

Symmetric three phase winding will have the same winding factor amplitude for each phase. For example, the two windings shown in Table 2 and Table 1 do not experience amplitude asymmetry for the working harmonic $h = p_s$, though there are amplitude asymmetry in other harmonics $h \neq p_s$.

We now introduce a third example winding with $p = 3, p_s = 4, Q_s = 36, y = 5$ in Table 3. This winding does exhibit amplitude asymmetry for harmonic $h = p_s$[1].

4.4 Discussion

If a symmetric suspension MMF profile for $p=3$ or $p_s=3$ motors is desired, there are at least two alternatives, 1) use a separated winding approach; 2) use a non-three-phase DPNV winding. However, as previously noted, the separated winding makes inefficient use of the slot area, significantly decreasing the machine's torque density, and the non-three-phase DPNV winding loses the merit/convenience of compatibility with existing commercial three phase motor drives.

Instead of pursuing these alternatives, this paper studies the performance impact of designing with an assymetric DPNV winding. Hints that this may be feasible can be found in Table 1, where the DPNV winding ($p = 2$, $p_s = 3$, $Q_s = 18$) has no asymmetry if harmonics are neglected.

[1] With less stator slots, the winding of $p = 3, p_s = 4, Q_s = 18, y = 3$ has no amplitude asymmetry for having the same $k_{Sw}(p_s) = 0.577$ for three phases, but phase-shift asymmetry still exists.

Table 3: Winding Factors for Design No. 7 ($p = 3, p_s = 4, Q_s = 36, y = 5$).

Order h	1	2	$p = 3$	$p_s = 4^*$	5	6
$k_{Tw}(h)$	-	-	0.933	-	-	-
$k_{Sw,U}(h)$	-	0.391	-	0.647	-	0.083
$k_{Sw,V}(h)$	-	0.503	-	0.617	-	0.144
$k_{Sw,W}(h)$	-	0.391	-	0.647	-	0.083

*The phase angles of $\bar{\boldsymbol{k}}_{Sw}(4)$ of the three phases are 0, 110, -140 [elec. deg].

5 Suspension Harmonics of DPNV Winding

Sub-harmonics are often seen in a fractional slot winding. Even if the induction motor uses integral slot torque winding, the suspension slot per pole per phase $q_s = \frac{Q_s}{2mp_s}$ can still be a fractional number. In this paper, the sub-harmonics are examined by calculating the winding factor for arbitrary harmonic order $h = vp$ or $h = vp_s$ using (4). Sub-harmonics are not caused by violation of symmetry condition, and are common phenomena observed in DPNV winding suspension winding factors when $p_s > 2$.

5.1 Attenuation of Unmodeled Suspension Force

The sub-harmonics of the suspension field can interact with the torque field to generate unmodeled suspension force as long as the rotor uses soft magnetic materials (steels). Note sometimes higher harmonics could also contribute to the unmodeled suspension force, e.g, the harmonic of order $h = 4$ of the $p = 3, p_s = 2$ winding. By "unmodeled", we mean this force contribution is not intentionally utilized during bearingless motor design phase, so it is desired to reduce this unmodeled suspension force.

Recall that with a conventional squirrel cage, the suspension force is attenuated as the stator current frequency increases [15]. If a parallel PS rotor is used, the sub-harmonics will also induce rotor currents, since the rotor was not designed to reject the number of poles of the sub-harmonic field. The induced sub-harmonic rotor currents can interact with the sub-harmonic suspension field to create asynchronous torque, which is undesired but is frequently insignificant. These rotor currents may actually partially cancel out the sub-harmonic suspension field and thereby attenuate the unmodeled suspension force.

It is not straightforward to determine whether the sub-harmonics are detrimental or beneficial to the suspension force performance. To answer this question, it is necessary to learn the rotating direction of the harmonic field. If the sub-harmonic field rotates in the same direction as the torque field, it will produce constant force; while if it is reversed, it causes a force disturbance [29, Sec. II-B2].

5.2 Asymmetric Sub-harmonics

For all windings investigated in this paper, when p or p_s is not co-prime with the phase number $m = 3$, if the sub-harmonics exist, the suspension winding factors of the sub-harmonics are asymmetric in both amplitude and phase-shift. Since the sub-harmonics of order $h = p \pm 1$ will contribute to the radial force, we expect this sub-harmonic asymmetry to cause problems. In other words, we expect that the "promising" DPNV winding from Table 1, will still show some undesired behavior because of asymmetric sub-harmonics. This undesired behavior will be revealed later in Section 6.3.

6 FEA Results in Polar Plots

The purpose of this section is to determine whether MMF asymmetries and sub-harmonics create force performance issues for bearingless induction motors.

Table 4: List of Selected Optimal Designs from [22] for 50 kW and 30,000 r/min Application.

Design No. & Name	q	q_s	Q_r'	k_{Tw}	$\bar{k}_{Sw,U}$, $\bar{k}_{Sw,V}$, $\bar{k}_{Sw,W}$	L_{stk},	L_{ew},	$L_{stk}+L_{ew}$ [mm]
1. p1p$_s$2Q$_r$14Q$_s$24y9	4	2	2	0.885	$0.592\angle 0^e$, $0.592\angle 120^e$, $0.592\angle 240^e$	**91**	30	**121**
2. p2p$_s$3Q$_r$30Q$_s$18y4	$\frac{3}{2}$	1	3	0.945	$0.577\angle 0^e$, $0.577\angle 120^e$, $0.577\angle 240^e$	110	32	142
3. p2p$_s$3Q$_r$18Q$_s$24y5	2	$\frac{4}{3}$	3	0.933	$0.604\angle 0^e$, $0.604\angle 90^e$, $0.604\angle 225^e$	105	21	126
4. p3p$_s$2Q$_r$20Q$_s$36y5	2	3	2	0.933	$0.503\angle 0^e$, $0.480\angle 130^e$, $0.503\angle 260^e$	98	33	131
5. p3p$_s$4Q$_r$20Q$_s$36y5	2	$\frac{3}{2}$	4	0.933	$0.647\angle 0^e$, $0.617\angle 110^e$, $0.647\angle 220^e$	110	**16**	126
6. p4p$_s$5Q$_r$20Q$_s$24y3	1	$\frac{4}{5}$	5	1.000	$0.604\angle 0^e$, $0.604\angle 120^e$, $0.604\angle 240^e$	-	-	-

Note: L_{stk}, L_{ew} is the axial length of the rotor lamination stack, end-ring region.

Note: q is slot per phase per torque pole, and q_s is slot per phase per suspension pole.

6.1 Optimal Designs Adapted from a Previous Research

Table 4 lists the six designs to be studied in this paper. The first 5 designs in Table 4 are the selected optimal designs from [22], meaning that this paper is an extension of the previous research in studying optimized bearingless induction motor designs for the rating of 50 kW and 30,000 r/min. As an example of a winding that has sub-harmonics but no asymmetry, a sixth design of $p = 4, p_s = 5$ is added as design 6. Note that design 6 is not optimized.

In Table 4, the slot per phase per torque pole q, the slot per phase per suspension pole q_s, the cage number N_c, torque winding factor, complex number suspension winding factors and rotor axial length are summarized. As per [22], it should be emphasized that the high p_s design, i.e., design 5, with $Q_r' = 4, Q_r = 20$ has significantly shorter eng-ring region length L_{ew}, as compared with design 1 having even less rotor slots of $Q_r = 14$.

As a companion table of Table 4, Table 5 summarizes the asymmetries and harmonics properties of each design, followed by performance results that are explained in the next subsection.

6.2 Simulation Approach and Analyzing Tool

2D transient finite element analysis (FEA) using JMAG Designer is conducted for evaluating the steady state force vectors at rated operating conditions of 50 kW and 30,000 r/min. In FEA, the motor and suspension terminals of the stator DPNV winding are both excited with sinusoidal currents at the same frequency, and there is a *phase delay*, ϕ_{sus}, between the motor currents and suspension currents. Rated coil group current is flowing with 2.5% allocated for suspension and 97.5% allocated for torque.

We use polar plots to visualize how the force vectors vary as a function of $\phi_{sus} \in [0, 360]$ [elec. deg]. The steady state force vectors will rotate along the air gap when ϕ_{sus} changes. In a first series of plots (Figure 5), we depict the average force vector amplitude against the average force vector angle as a "force polar plot" (FPP). In a second series of plots (Figure 6), we depict the force error angle against the average force vector angle as a "E_a polar plot" (E_a-PP). Here the force error angle E_a is defined as the maximum angle deviation of the force vector from the average force vector at steady state over one fundamental period. It is worth mentioning that the steady state of a design with sub-harmonics needs to be evaluated for longer time, which is related to the period of the sub-harmonics, as reported in [28].

6.3 Force Polar Plot (FPP)

A circular FPP is generally desired, but elliptical FPP can result for an asymmetric DPNV winding. For elliptic FPP, the force rating of the machine must be de-rated to correspond to the minimum radius of the ellipse.

The force polar plots of the six designs listed in Table 4 are shown in Figure 5. The phase angle of the suspension currents, ϕ_{sus}, that corresponds to the minimum or maximum of the elliptical FPP is indicated using a red or blue arrow.

Table 5: Asymmetries, Harmonics and Performance Measure of the Designs from Table 4.

Design No.	1	2	3	4	5	6
Harmonics at $h = p \pm 1$	-*	✓†	✓	✓	✓	✓
Amplitude asymmetry in working harmonics $h = p_s$	-	-	-	✓	✓	-
Phase-shift asymmetry in working harmonics $h = p_s$	-	-	✓	✓	✓	-
Asymmetry in harmonics $h = p \pm 1$	-	✓	✓	✓	✓	-
Elliptical force polar plot?	No	Yes	Yes	Yes	Yes	No
Maximal force variation in FPP, ΔF [N]	0.06	5.6	9.4	27.5	5.3	0.27
Maximal E_a variation in E_a-PP, ΔE_a [mech. deg]	**0.8**	**2.1**	5.2	9.0	**2.3**	7.4
Acceptable Design?	**Yes**	**Yes**	No	No	**Yes**	No

*Check mark "✓" means the description applies.
†Dash "-" means the description does not apply.

Based on Figure 5, the following observations can be made.

1. *Windings that have no asymmetry and no sub-harmonics have circular FPP.* E.g., the $p = 1$, $p_s = 2$ winding in Figure 5a.
2. *Symmetric windings that have sub-harmonics have circular FPP.* E.g., the $p = 4, p_s = 5$ winding in Figure 5f.
3. *Windings that have asymmetry in sub-harmonics and have no asymmetry in working-harmonics have elliptical FPP.* E.g., the $p = 2, p_s = 3, Q_s = 18$ winding in Figure 5b.
4. *Windings that have asymmetry in working-harmonics have elliptical FPP.* E.g., Figure 5c–5e.

This shows that even asymmetry in the sub-harmonics (when the working harmonics are symmetric) can cause elliptical FPP. This indirectly verifies the fact that sub-harmonics do contribute to bearingless motor's suspension force capability. In conclusion, the elliptical FPP is due to i) the (amplitude and phase-shift) asymmetries in suspension winding factor at $h = p_s = p \pm 1$, and ii) the asymmetric harmonics that satisfy $h = p \pm 1$. The above observations are compactly summarized as Table 5.

All windings that violate the DPNV winding symmetry condition by having either $p = 3$ or $p_s = 3$ show elliptical FPP. As a performance measure, let ΔF designate the difference between maximal force and minimal force of an FPP, i.e., the length difference between the blue and red arrows in Figure 5. The ΔF values for the six designs are listed in Table 5. It is desired to have zero or low ΔF. Symmetric windings have close to zero ΔF values, e.g., design 1 and 6. For asymmetric windings, design 3 and 4 show large ΔF values, but we do observe low ΔF values in design 2 that has no asymmetry in working-harmonics and design 5 that has asymmetry in both working and sub-harmonics. This provides evidence that a DPNV winding with asymmetry in both the working and subharmonics can yield acceptable force performance.

6.4 Force Error Angle Polar Plot

Force error angle E_a as a function of suspension current angle ϕ_{sus} is sketched as polar plot in Figure 6 for all six designs. Again, only the first five designs are optimal designs selected from the Pareto front in [22]. From Figure 6, it is observed that E_a is generally a function of ϕ_{sus}, and the E_a polar plot (E_a-PP) exhibits different trajectory patterns for different motor designs, and the trajectory patten is mirrored about the origin.

All E_a-PPs are drawn at the same scale to ease visual comparison of the variation in E_a values when ϕ_{sus} changes. As a performance measure, the maximal variation in E_a values in

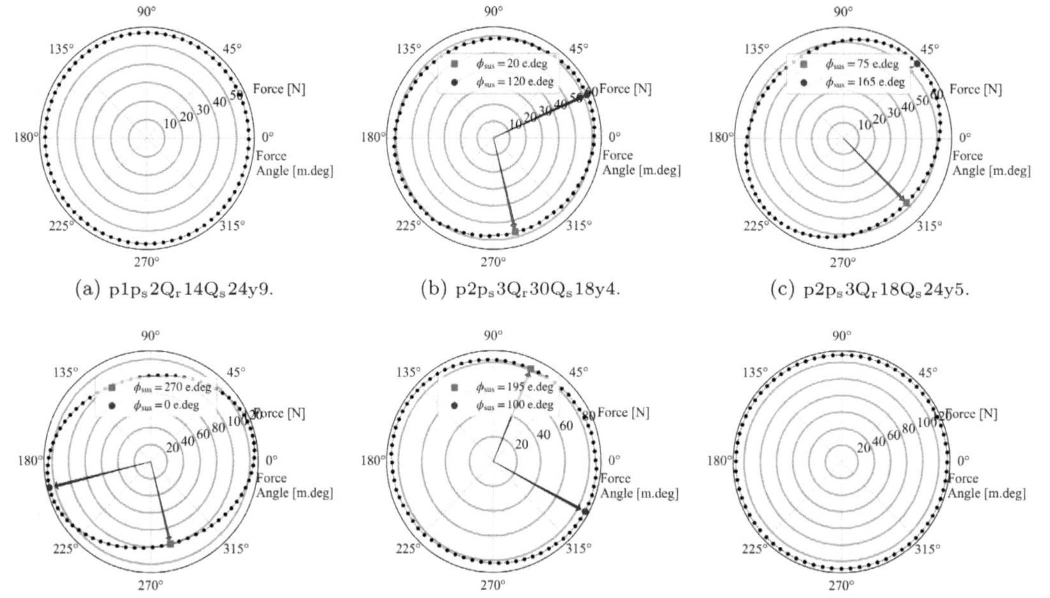

(a) $p1p_s2Q_r14Q_s24y9$. (b) $p2p_s3Q_r30Q_s18y4$. (c) $p2p_s3Q_r18Q_s24y5$.

(d) $p3p_s2Q_r20Q_s36y5$. (e) $p3p_s4Q_r20Q_s36y5$. (f) $p4p_s5Q_r20Q_s24y3$.

Figure 5: Force polar plots of the designs listed in Table 4.

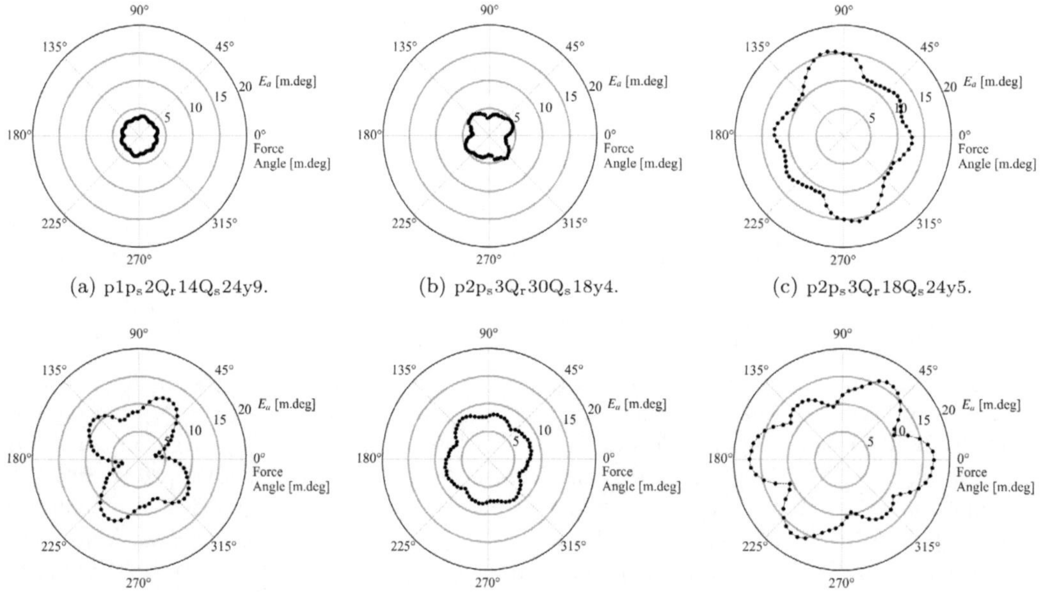

(a) $p1p_s2Q_r14Q_s24y9$. (b) $p2p_s3Q_r30Q_s18y4$. (c) $p2p_s3Q_r18Q_s24y5$.

(d) $p3p_s2Q_r20Q_s36y5$. (e) $p3p_s4Q_r20Q_s36y5$. (f) $p4p_s5Q_r20Q_s24y3$.

Figure 6: Force error angle E_a polar plots of the designs listed in Table 4.

an E_a-PP is denoted as ΔE_a and is summarized in Table 5 for all six designs. We suggest to one use a design that has low ΔE_a, e.g., design 1, 2, and 5.

In conclusion, there is no clear correlation between the shape of FPP and the shape of E_a-PP, and the asymmetric DPNV winding can result in acceptable bearingless induction motor design, in a sense that its ΔF and ΔE_a are both reasonably small, e.g., design 2 and design 5.

7 Conclusion

The DPNV winding and parallel PS rotor are two recent technologies that can be used to create industrial scale, power dense bearingless induction motors. Preliminary optimization results indicate that by using these two technologies, bearingless induction machines can meet the requirements of industrial compressor systems. However, two desired machine pole combinations result in asymmetric stator windings and require further investigation to determine if acceptable machine performance can be achieved. This paper finds that the asymmetry of the DPNV winding can be studied by using complex winding factors of each winding phase. Based on transient FEA results, this paper uses a force polar plot to visually study the steady state force vector behavior of five selected optimal designs. Observations are drawn on the influence of asymmetries combined with the DPNV winding's inherent sub-harmonic fields. By using FPP, it is found that MMF asymmetry causes an elliptical force capability that reduces the machine's force rating. The level of force reduction varied between machines from being negligible to being a significant concern. From an E_a-PP, it is revealed that force error angle of a bearingless induction motor is dependent on the suspension current phase angle. This latter finding imposes a challenge to accurately evaluate the force vector error of a bearingless motor design. Future research is needed to take the suspension current phase angle into account in order to accurately model bearingless motors and avoid under-calculating the force error angle performance metric.

Together, the findings from this paper provide evidence that asymmetric DPNV windings can be used with the parallel PS rotor to create a high performance bearingless induction machine. The nuances in performance between designs imply that careful attention must be paid in the design and optimization process of both asymmetric and symmetric designs to ensure accurate performance calculations are made.

References

[1] A. Chiba, T. Fukao, O. Ichikawa, M. Oshima, M. Takemoto, and D. G. Dorrell, *Magnetic bearings and bearingless drives*. Elsevier, 2005.

[2] J. Chen, Y. Fujii, M. W. Johnson, A. Farhan, and E. L. Severson, "Optimal design of the bearingless induction motor," *IEEE Transactions on Industry Applications*, vol. 57, no. 2, pp. 1375–1388, 2021.

[3] J. Chen, J. Zhu, and E. L. Severson, "Review of bearingless motor technology for significant power applications," *IEEE Transactions on Industry Applications*, vol. 56, no. 2, pp. 1377–1388, 2020.

[4] M. Kang, J. Huang, J.-q. Yang, and H.-b. Jiang, "Analysis and experiment of a 6-phase bearingless induction motor," in *2008 International Conference on Electrical Machines and Systems*. IEEE, 2008, pp. 990–994.

[5] M. T. Bartholet, T. Nussbaumer, S. Silber, and J. W. Kolar, "Comparative evaluation of polyphase bearingless slice motors for fluid-handling applications," *IEEE Transactions on Industry Applications*, vol. 45, no. 5, pp. 1821–1830, 2009.

[6] W. Khoo, "Bridge configured winding for polyphase self-bearing machines," *IEEE Trans on Magnetics*, vol. 41, no. 4, pp. 1289–1295, 2005.

[7] R. Oishi, S. Horima, H. Sugimoto, and A. Chiba, "A novel parallel motor winding structure for bearingless motors," *IEEE Transactions on Magnetics*, vol. 49, no. 5, pp. 2287–2290, May 2013.

[8] G. Valente, L. Papini, A. Formentini, C. Gerada, and P. Zanchetta, "Radial force control of multisector permanent-magnet machines for vibration suppression," *IEEE Transactions on Industrial Electronics*, vol. 65, no. 7, pp. 5395–5405, 2017.

[9] S. Kobayashi, M. Ooshima, and M. N. Uddin, "A radial position control method of bearingless motor based on d-q-axis current control," *IEEE Transactions on Industry Applications*, vol. 49, no. 4, pp. 1827–1835, 2013.

[10] A. Chiba, K. Sotome, Y. Iiyama, and M. A. Rahman, "A novel middle-point-current-injection-type bearingless pm synchronous motor for vibration suppression," *IEEE Transactions on Industry Applications*, vol. 47, no. 4, pp. 1700–1706, July 2011.

[11] A. Khamitov, W. Gruber, G. Bramerdorfer, and E. L. Severson, "Comparison of combined winding strategies for radial non-salient bearingless machines," *IEEE Transactions on Industry Applications*, 2021.

[12] A. Chiba, D. Power, and M. Rahman, "Characteristics of a bearingless induction motor," *IEEE Transactions on Magnetics*, vol. 27, no. 6, pp. 5199–5201, 1991.

[13] A. Salazar and R. Stephan, "A bearingless method for induction machines," *IEEE Trans on Magnetics*, vol. 29, no. 6, pp. 2965–2967, 1993.

[14] T. Katou, A. Chiba, and T. Fukao, "Magnetic suspension force in an induction bearingless motor with a squirrel cage rotor," *Electrical Engineering in Japan*, vol. 159, no. 3, pp. 77–87, 2007.

[15] J. Chen and E. L. Severson, "Design and modeling of the bearingless induction motor," in *2019 IEEE International Electric Machines Drives Conference (IEMDC)*, May 2019, pp. 343–350.

[16] C. Redemann, P. Meuter, A. Ramella, and T. Gempp, "30kw bearingless canned motor pump on the test bed," in *7th International Symp. on Magnetic Bearings, August 2000, ETH Zurich*, 2000.

[17] H. J. Herbein, *Rotating Machinery*. Rinehart Press, 1971.

[18] J. M. S. Ferreira, M. Zucca, A. O. Salazar, and L. Donadio, "Analysis of a bearingless machine with divided windings," *IEEE Transactions on Magnetics*, vol. 41, no. 10, pp. 3931–3933, Oct 2005.

[19] A. Chiba, "Transfer characteristics of radial force of induction-type bearingless motors with four-pole rotor circuits," in *The 5th International Symposium on Magnetic Bearings (ISMB5)*, Aug 1996, pp. 319–325.

[20] A. Chiba and T. Fukao, "Optimal design of rotor circuits in induction type bearingless motors," in *1998 IEEE International Magnetics Conference (INTERMAG)*, Jan 1998, pp. 225–225.

[21] A. Chiba and J. Asama, "Influence of rotor skew in induction type bearingless motor," *IEEE Transactions on Magnetics*, vol. 48, no. 11, pp. 4646–4649, 2012.

[22] J. Chen, M. Johnson, A. Farhan, Y. Fujii, and E. L. Severson, "Reduced axial length pole-specific rotor forbearingless induction machines," *IEEE Transactions on Energy Conversion*, 2021.

[23] A. Chiba and J. Asama, "Influence of rotor skew in induction type bearingless motor," *IEEE Transactions on Magnetics*, vol. 48, no. 11, pp. 4646–4649, 2012.

[24] A. Chiba and J. A. Santisteban, "A pwm harmonics elimination method in simultaneous estimation of magnetic field and displacements in bearingless induction motors," *IEEE Transactions on Industry Applications*, vol. 48, no. 1, pp. 124–131, Jan 2012.

[25] D. Dietz and A. Binder, "Design guidelines and scaling effects for bearingless pm synchronous machines accounting for eddy current reaction field," *IEEE Transactions on Industry Applications*, 2021.

[26] E. L. Severson, R. Nilssen, T. Undeland, and N. Mohan, "Design of dual purpose no-voltage combined windings for bearingless motors," *IEEE Transactions on Industry Applications*, vol. 53, no. 5, pp. 4368–4379, 2017.

[27] J. Pyrhonen, T. Jokinen, and V. Hrabovcová, *Design of rotating electrical machines*. John Wiley & Sons, 2013.

[28] J. Chen, A. Farhan, M. Johnson, and E. L. Severson, "Design of bearingless permanent magnet motors using no voltage combined windings," *IET Conference Proceedings*, pp. 803–808(5), 2021.

[29] B. Ramadas and E. L. Severson, "Suspension force design guidelines for bearingless permanent magnet machines," in *IEEE Energy Conversion Congress and Exposition (ECCE)*, 2021.

A Hall Sensor-Based Position Measurement Hybrid Magnetic Levitation System

Xing-Dong Fu[1], Guang-Zhong Cao[1*], Li-Jia Wang[1], Su-Dan Huang[1], Jian Zhou[2], and Xiao-Sheng Yang[1]
[1] Shenzhen University, Shenzhen, China.
[2] Xi'an Jiaotong University, Xi'an, China.
xingdongf@gmail.com, gzcao@szu.edu.cn, 1910294025@email.szu.edu.cn, hsdsudan@gmail.com, jzhou@mail.xjtu.edu.cn, yangxiaosheng27@gmail.com

Abstract

This paper presents a hall sensor-based position measurement of hybrid magnetic levitation system (MLS) developed in the laboratory for applications of contactless operations at a microscale. Cylindrical permanent magnets and iron-core electromagnetics are used to build a desire hybrid magnetic field of the MLS. The nonlinear magnetic flux density model of the MLS is first deduced. Based on the deduced model, the mathematical relation of the current supplied to the electromagnet and the position of the levitation object was obtained and analyzed. And then, the installation of the linear Hall-effect sensors was determined theoretically. Furthermore, the solution model for the position of the levitation unit was given. Finally, the effectiveness of the solution model of the MLS is validated by simulation analysis.

1 Introduction

As a typical application of magnetic actuators, magnetic levitation system (MLS) has been a popular research area in nowadays. MLS can be used in special environments and realize contactless operations due to the reason that there is no mechanical contact [1]. These environments which require contactless operations at a microscale include such as operating inside the human body [2]-[4], micro-manipulation robot in clean room [5] and bio-manipulation [6]. To achieve precise motion control of the MLS, the optical-based devices, such as laser sensors or camera are generally used to measure the position of the levitated object [11],[12], however, there is an obvious limitation for the operation in a nontransparent environment. In [13], a methodology to determine the position of a magnetically guided robot in horizontal planes using Hall-effect sensors is proposed. The Hall-effect sensors solve the above limitation and it is inspirational to this paper. From [13], using hall sensors to obtain the position information of MLS is an effective and significant method.

The MLS presented in this paper has the features of simple structure, low power consumption, low cost and complex magnetic fields. To give a more complete account of the research work, this paper is organized as follows. Section II describes the structure and the magnetic mode of the MLS. Section III discuss the laying position of linear Hall-effect sensors and the solution model for the position of the levitation unit was proposed. Base on the finite element simulation data, section IV shows the results of the proposed model. Conclusion of this paper is given in Section V.

2 System Description

2.1 The Structure of the Magnetic Levitation System

The MLS presented in this paper is shown in Fig. 1. The system consists of a magnetic drive unit (MDU), a levitated object with a solid cylindrical permanent magnetic, and a DSP control system. The MDU has twelve hollow cylindrical permanent magnetics and four iron-core electromagnets. The hollow permanent magnetic array of the MDU can form a permanent magnetic field (see Fig. 2) which provides part of the levitation force applied on the levitated object. Therefor the permanent magnetic array plays a key role of reducing the energy consumption of the MDU. The bottom plate which made of printed circuit board (PCB) in Fig. 2 is used to mount the sensors and the electromagnets. The four electromagnets are used to form a desired magnetic field to regulate the position of the levitated object. The linear Hall-effect sensors are installed in the center of the bottom plate to measure the two-dimensional horizontal motion of the levitated object. The DSP controller acts as a central processing unit to first calculate the position of the levitated object based on position feedback from the linear Hall-effect sensors, and second to output current to drive the electromagnets. Finally, a personal computer is used to communicate with the DSP controller base on UART and monitor the levitation system.

2.2 The Magnetic Model of the Magnetic Levitation System

In this paper, the magnetic field of the MDL is produced by four iron-core electromagnets, twelve hollow cylindrical permanent magnets and one solid cylindrical permanent magnet. On account of the magnetic field superposition principle, the analysis of the magnetic field can be simplified to a problem that to analyze the magnet field of a solid cylindrical permanent magnet and an iron-core electromagnet. Using the current model, the magnetic field produced by a solid cylindrical permanent magnet at a random point *P (x, y, z)* in free space is:

$$\begin{cases} B_x = M\dfrac{\mu_0}{4\pi}\displaystyle\int_0^h\int_0^{2\pi}\dfrac{R\cos\theta_0(z-z_0)}{((x-R\cos\theta_0)^2+(y-R\sin\theta_0)^2+(z-z_0)^2)^{3/2}}d\theta_0 dz_0 \\ B_y = M\dfrac{\mu_0}{4\pi}\displaystyle\int_0^h\int_0^{2\pi}\dfrac{R\sin\theta_0(z-z_0)}{((x-R\cos\theta_0)^2+(y-R\sin\theta_0)^2+(z-z_0)^2)^{3/2}}d\theta_0 dz_0 \\ B_z = M\dfrac{\mu_0}{4\pi}\displaystyle\int_0^h\int_0^{2\pi}\dfrac{-R(\sin\theta_0(y-R\sin\theta_0)+\cos\theta_0(x-R\cos\theta_0))}{((x-R\cos\theta_0)^2+(y-R\sin\theta_0)^2+(z-z_0)^2)^{3/2}}d\theta_0 dz_0 \end{cases} \quad (1)$$

where B_x, B_y, and B_z are the x-, y- and z-component magnetic flux density of the solid cylindrical permanent magnet at point *P (x, y, z)*, *R*, *M* and *h* are the radius, the magnetization and the height of the permanent magnet, μ_0 is the permeability of vacuum. In order to simplify the following expression, as the laying position of the cylindrical permanent magnet is given, (1) can be rewritten as a function of R:

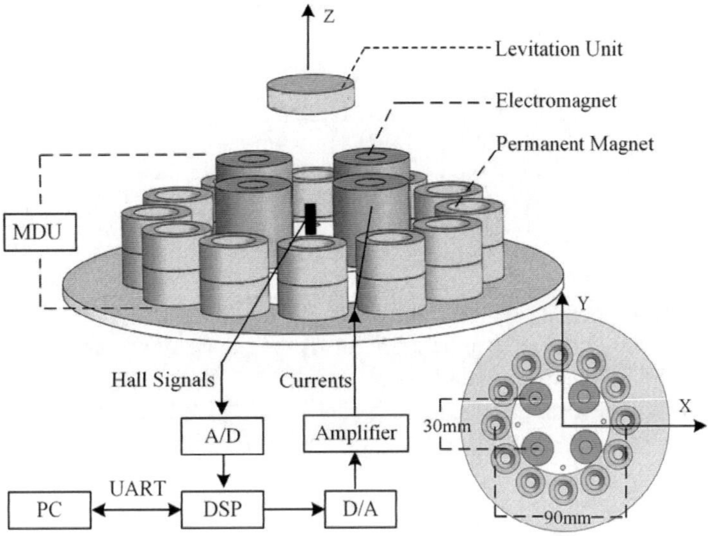

Figure 1: Structure of the magnetic levitation system

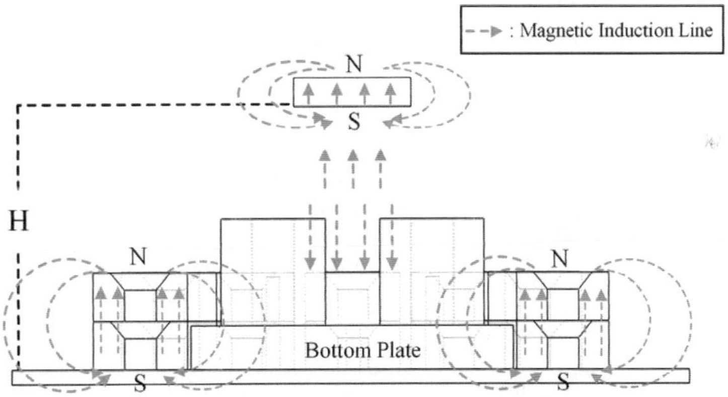

Figure 2: The magnetic field diagram of the MLS with currents applied to the electromagnets were all zero

$$\begin{cases} B_x = f_1 x_{pm}(R,x,y,z) \\ B_y = f_1 y_{pm}(R,x,y,z) \\ B_z = f_1 z_{pm}(R,x,y,z) \end{cases} \tag{2}$$

Base on (2), the magnetic field produced by a hollow cylindrical permanent magnet use in the MDU is:

$$\begin{cases} B_x = f_1 x_{pm}(R_{out} - R_{in}, x, y, z) \\ B_y = f_1 y_{pm}(R_{out} - R_{in}, x, y, z) \\ B_z = f_1 z_{pm}(R_{out} - R_{in}, x, y, z) \end{cases} \tag{3}$$

where R_{out} and R_{in} are the outside radius and inside radius of the hollow cylindrical permanent magnet. Here, the magnetic model of the permanent magnetics in the MLS is defined. And then the magnetic field produced by a single iron-core electromagnet at point *P (x, y, z)* is [11]:

$$\begin{cases} B_x = \beta \dfrac{\mu_0 \sigma I}{4\pi} \displaystyle\int_{r_{in}}^{r_{out}} \int_0^L \int_0^{2\pi} \dfrac{R\cos\varphi(z-h)}{((x-R\cos\varphi)^2+(y-R\sin\varphi)^2+(z-h)^2)^{3/2}} d\varphi dh dR \\ B_y = \beta \dfrac{\mu_0 \sigma I}{4\pi} \displaystyle\int_{r_{in}}^{r_{out}} \int_0^L \int_0^{2\pi} \dfrac{R\sin\varphi(z-h)}{((x-R\cos\varphi)^2+(y-R\sin\varphi)^2+(z-h)^2)^{3/2}} d\varphi dh dR \\ B_z = -\beta \dfrac{\mu_0 \sigma I}{4\pi} \displaystyle\int_{r_{in}}^{r_{out}} \int_0^L \int_0^{2\pi} \dfrac{\cos\varphi(x-R\cos\varphi)+\sin\varphi(y-R\sin\varphi)}{((x-R\cos\varphi)^2+(y-R\sin\varphi)^2+(z-h)^2)^{3/2}} d\varphi dh dR \end{cases} \tag{4}$$

where β is the magnetic flux density factor of the iron-core effect in enhancement of the magnetic field, and σ is the winding density of the electromagnet. Its unit is turns per unit area, *I* is the current supplied to the electromagnet, r_{in} and r_{out} are the inner and outer radius of the electromagnet, respectively, and is the length of each coil.

Base on the above discussions, the magnetic model of the MLS is given by (2), (3) and (4). How to use the model and the hall sensors to determine the position of the levitation unit will be described in the next.

3 Hall Sensor-Based Position Measurement

3.1 Installation of the Linear Hall-effect Sensors

There are three magnetic field sources in the MLS, a solid cylindrical permanent is the levitated unit, the permanent magnets and the electromagnets are used in the MDU. In fact, when the Hall-effect sensors were installed in the magnetic drive unit, only the position of the levitated object change and current supply to the electromagnets change cause the magnetic field measured by the linear Hall-effect sensors change. Therefore, the key to the installation of the linear Hall-effect sensors is to find a position where will not be affected by the electromagnet's magnetic field, so that the position of the levitated object can be obtained from the change of the magnetic field measured by the linear Hall-effect sensors.

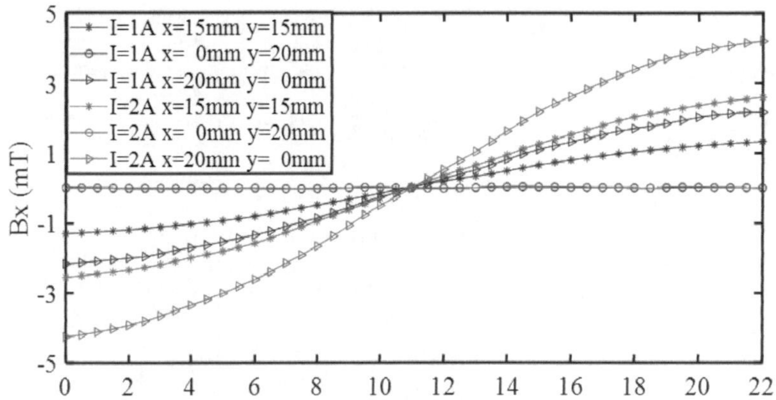

(a)

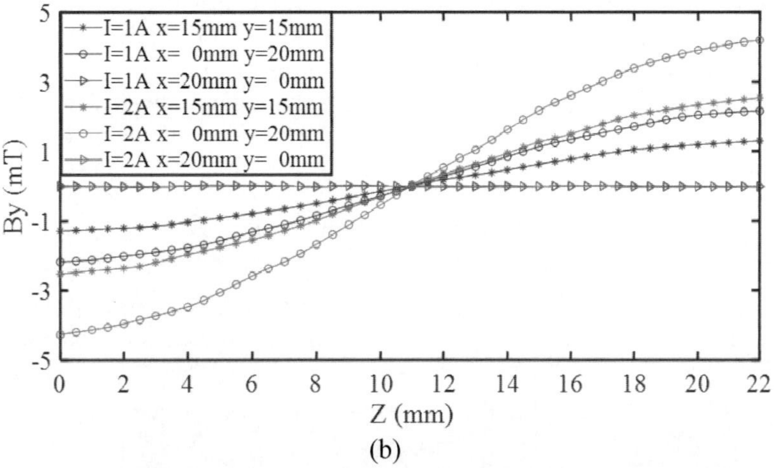

(b)

Figure 3: The horizontal component of the electromagnet's magnetic flux density under different current and different horizontal position versus z-axis. (a) x-component. (b) y-component

Figure 4: The laying position of the Hall sensors in the experimental device

To find the laying position of the linear Hall-effect sensors, using (4) to explore the magnetic field distribution of an electromagnetic. The electromagnet used in the calculation model has the following geometric parameters: $N = 500$ turns, $r_{out} = 10.5$ mm, $r_{in} = 4.5$ mm, and $L = 22$ mm. Fig.3 shows that no matter how the current applied to the electromagnet change, the horizontal components of the magnetic field at a height of L/2 are always zero. The laying position of the linear Hall-effect sensors is determined (see Fig.4) base on this characteristic of the electromagnet. Hall sensor H1 and H2 are used to measure the magnetic flux density in horizontal direction, H3 is used to the measure the magnetic flux density in vertical direction, by the way.

3.2 Solution Model for the Position of Levitation unit

According to the installation of the hall sensors, the interference of electromagnets to the sensors can be ignored. On the other hand, due to the fact that the magnetic field generated by the permanent magnet in MDU is a static magnetic field, the measured values of hall sensors only change with the position of the levitation unit. In fact, the nonlinear relationship between the position of the levitation unit and the measured value of the hall sensors satisfies formula (1). However, due to the complexity of formula (1), it is difficult to obtain the position of the levitation unit directly from the magnetic flux density information. Therefore, a solution model for the position of levitation unit based on B-P(Back-Propagation) neural network is designed and it process structure is shown in Fig.5.

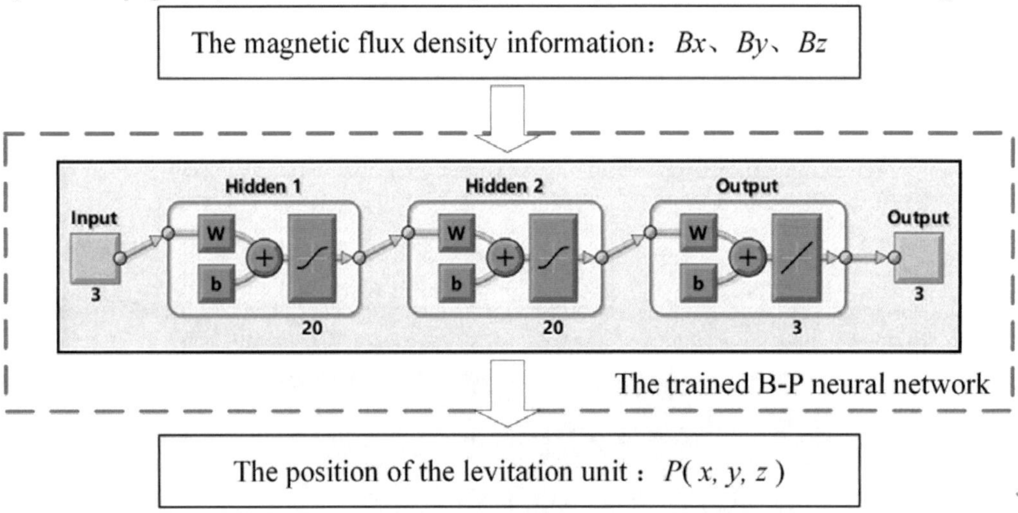

Figure 5: The process structure of the proposed model

The training data (B_x, B_y, B_z, x, y, z) of the network come from the finite element model (FEM) via ANSYS Maxwell software. The value range of FEM is -60mm< x <60mm, -60mm< y <60mm, 37mm< z <47mm. Fig.6(a), Fig.2(b) and Fig.2(c) show the magnetic field of the HMLS at z=37mm, I=0A.

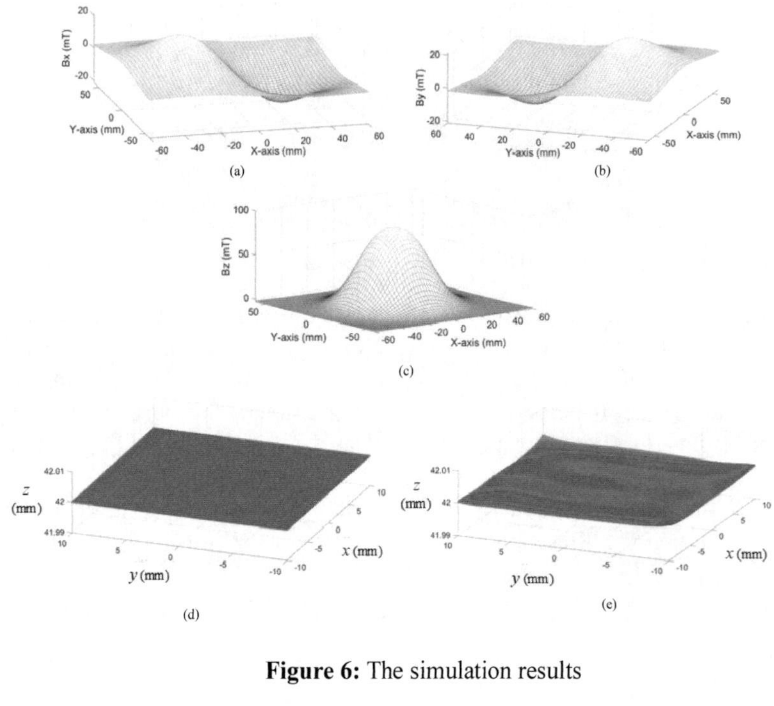

Figure 6: The simulation results

Significance	Symbol	Value	Unit
Intensity of magnetization	M	8900	kA/m
The inner diameter of a cylindrical hollow permanent magnet	R_{in}	2	mm
The outer diameter of a cylindrical hollow permanent magnet	R_{out}	5	mm
The height of a cylindrical hollow permanent magnet	h	20	mm
The inner diameter of the electromagnetic coil	r_{in}	4.5	mm
The outer diameter of the electromagnetic coil coi	r_{out}	10.5	mm
The current in the electromagnetic coil	I	1	
The number of turns of the electromagnetic coil	N	500	Am
The length of the electromagnetic coil	L	22	mm

Table 1: Parameters of hybrid repulsive magnetic levitation system

4 Simulation Results and Analysis

The proposed position measurement of the HMLS is validated by FEM analysis and the error caused by the noise in sampling circuit is ignored in the simulation. Parameters of the HLMS is listed in Table 1.

Fig.6 (d) shows the position of the sampling point for the magnetic flux density of the levitation unit. By analyzing the sampled data, Fig.6 (e) shows the results of the proposed solution model. According to the simulation results in Fig.6 (d) and Fig.6 (e), the solution position approaches to the desired position, where the maximum absolute errors of x-axis, y-axis and z-axis are 1.3×10-3mm, 1.2×10^{-3}mm and 1.3×10^{-3}mm, respectively. The maximum relative errors are 0.0065%, 0.0065% and 0.0180%, respectively. The effectiveness of the proposed method is verified.

5 Conclusion

In this paper, the analytical nonlinear modelling and the installation of the linear Hall-effect sensors were presented for the MLS developed in the laboratory, which is to show the potential application in the contactless operation for the magnetic levitation system. The results of the finite elements analysis verified the effectiveness of the nonlinear modelling of the MLS. Experimental device is developed based on the analytical nonlinear modelling. The future work will be focused on the effective control method of the developed MLS by employing the linear Hall-effect sensors to detect the position based on the analytical nonlinear model

References

[1] S. Yim, K. Goyal, and M. Sitti, "Magnetically actuated soft capsule with the multimodal drug release function," IEEE/ASME Trans. Mechatronics, vol. 18, no. 4, pp. 1413–1418, Aug. 2013

[2] F. Munoz, G. Alici, W. Li and M. Sitti, "Size Optimization of a Magnetic System for Drug Delivery With Capsule Robots," in IEEE Transactions on Magnetics, vol. 52, no. 5, pp. 1-11, May 2016.

[3] Y. Zhang, M. Chi and Z. Su, "Critical Coupling Magnetic Moment of a Petal-Shaped Capsule Robot," in IEEE Transactions on Magnetics, vol. 52, no. 1, pp. 1-9, Jan. 2016.

[4] Y. Zhang et al., "Design, analysis and experiments of a spatial universal rotating magnetic field system for capsule robot," 2012 IEEE International Conference on Mechatronics and Automation, Chengdu, 2012, pp. 998-1003.

[5] K. Park, K. Ahn, S. Kim, and Y. Kwak, "Wafer distribution system for a clean room using a novel magnetic suspension technique," IEEE/ASME Trans. Mechatronics, vol. 3, no. 1, pp. 73–78, Mar. 1998.

[6] S. Martel, J.-B. Mathieu, O. Felfoul, A. Chanu, E. Aboussouan, S. Tamaz, P. Pouponneau, L. Yahia, G. Beaudoin, G. Soulez, and M. Mankiewicz, "Automatic navigation of an untethered device in the artery of a living animal using a conventional clinical magnetic resonance imaging system," Appl. Phys. Lett., vol. 90, art. no. 114105, 2007.

[7] A. M. Benomair, A. R. Firdaus and M. O. Tokhi, "Fuzzy sliding control with non-linear observer for magnetic levitation systems," 2016 24th Mediterranean Conference on Control and Automation (MED), Athens, 2016, pp. 256-261.

[8] H. R. Karampoorian and R. Mohseni, "Control of a nonlinear magnetic levitation system by using constraint generalized model predictive control," ICCAS 2010, Gyeonggi-do, 2010, pp. 48-51.

[9] H. Chiang, C. Fang, W. Lin and G. Chen, "Second-order sliding mode control for a magnetic levitation system," 2011 8th Asian Control Conference (ASCC), Kaohsiung, 2011, pp. 602-607.

[10] J. Yang, Y. Lee and O. Kwon, "Development of magnetic force modeling equipment for magnetic levitation system," ICCAS 2010, Gyeonggi-do, 2010, pp. 29-33.

[11] E. Shameli, "Design, implementation and control of a magnetic levitation device," Ph.D. dissertation, Dept. Mech. Eng., University of Waterloo, Waterloo, ON, Canada, 2008.

[12] P. Mane, P. Dugwekar and A. Ingole, "Position control of magnetic levitation system," 2015 International Conference on Industrial Instrumentation and Control (ICIC), Pune, 2015, pp. 1626-1629.

[13] X. Zhang, M. Mehrtash and M. B. Khamesee, "Dual-Axial Motion Control of a Magnetic Levitation System Using Hall-Effect Sensors," in IEEE/ASME Transactions on Mechatronics, vol. 21, no. 2, pp. 1129-1139, April 2016.

Identification of cracks in shafts using magnetic bearings

Cesar Augusto Fonseca[1], Michael Kreutz[2], and Thomas Thümmel[2]

[1] Post-Graduation Center of the Brazilian Navy, CIAW,
Rio de Janeiro, Brazil
cesar.linhares@marinha.mil.br

[2] Dept. of Mechanical Eng., Technische Universität München (TUM),
Garching bei München, Germany
m.kreutz@tum.de, th.th.thuemmel@gmail.com

Abstract

This paper shows an experimental approach to the identification of cracks on magnetically levitated rotors. The crack is artificially done to shaft in such a way that the breathing crack model is no longer representative and the responses are compared against the normal shaft operating in the same conditions. The experiments have been conducted the shaft rotates at 400 RPM, away from any natural frequency, and the influence of this severe crack was assessed using white noise excitation added to the control signal of one of the magnetic bearings. It has been observed shifts on the higher resonances when rotor is either rotating or not of around 10 Hz lower for the damaged rotor. A Finite Element Model is proposed and, using the same identified parameters of the test rig, it reproduced the same changes in the frequency domain.

Keywords: Crack detection, Magnetic Bearing, Modal Analysis

1 Introduction

Magnetically levitated rotors have become more common in recent years, due to its frictionless features and the capacity to achieve high angular velocities, causing cost reduction in maintenance and efficiency. They are found in centrifuges, vacuum pumps, flywheels, and compressors.

Normally, the behavior a rotor on an ordinary bearing is well known, but the demand for higher angular velocities and lower to none friction power loss makes the magnetic bearing a powerful solution [1]. The downside is that the low damping may not be able to contain occasional instabilities and the rotor is constantly dependent on the power supply for the active control system [2]. In the event of power loss or a complete failure of the monitoring system, a magnetic bearing requires mechanical protection such as safety bearings to support a possible rotor drop [3].

The most common type of failure found in rotatory systems is the unbalance, which increases the lateral vibration causing the excessive forces in the bearings [4]. Other eventualities can happen at electrical and control systems, with problems regarding the sensors and/or actuators [5, 6]. Therefore, magnetically levitated rotors are endowed with a monitoring system that is constantly checking potential signs of damage or any kind of problems.

The monitoring system is active while the rotor is running and the faster the rotor spins, the more precise it has to be to perform a diagnosis of the machine. Each type of damage can create a clue or a signal that, compared to a model-based program, is easily identified. In the work of Roßner [7] many types of failures on a rotor are addressed such as unbalance, axle curvature and roundness failure of the shaft. Guangming and Jin [8] approach the same problem on a still standing rotor using a finite element model.

The detection of failures of rotors through the use magnetic bearings has been addressed by [9, 10, 11] and other published papers studied the effects of crack on the dynamics of a rotor for identification and diagnosis [12, 13, 14, 15]. The cracks are usually represented by the "breathing crack" model, see ref. [16], as it opens and closes following the bending of the shaft as it is turning. These models are extensively studied over the last years.

The present work describes a study conducted experimentally on a fully monitored rotor with two magnetic bearings, whose shaft has a crack close to its middle point. The rotor was excited by white noise on one of the magnetic bearings and displacement and force signals are captured on other sections of the shaft. The response is compared to an undamaged rotor upon the same conditions. Then, a finite element model is created to reproduce the overall behavior of the test rig. Unlike previous works, the crack is too wide to closes itself while the rotor is turning, so the effects of the crack on the shaft are investigated through a shift of higher natural frequencies on the frequency response functions and time evolution of orbit plots, either experimentally or numerically.

2 The test bench

The test bench is shown in Figure 2a and its components in Figure 2b. The machine is constituted of a rotor, which is levitated by two magnetic bearings close to its extremities, an 8mm diameter/700mm long shaft. It is powered by a servo motor on a Cardan joint and on the other end is free. The servo motor is controlled by a real time system dSPACE board which also serves to gather data from the set of sensors assembled to the test rig and emits to and received data from a Matlab/Simulink program. Inside the magnetic bearing there is a mechanical gap between the rotor and the backup bearing which is comprised of 1.4mm and 2.5mm at the disk. The whole system is levitated by active magnetic bearings and the lateral displacement is captured at the bearings and at the disk by eddy current sensors. The sensors are mounted in a way that the coordinates x and y are orthogonally measured and their middle values are given as set point of the PID-Controller for the magnetic bearings to adjust the electrical current of the magnetic coils. The position at the disk is measured by laser sensors and the angular velocity by an encoder close the joint. Additionally to the controller forces, an excitation signal is provided by the magnetic bearings to serve as input force for the conditioning monitoring system see Figure 1. It is a burst random noise applied by the magnetic bearing on the right near the extremity. Besides, the force components applied by the magnets are also measured since the bearing is mounted on a force transducers. For further details of the test rig construction are available in [19].

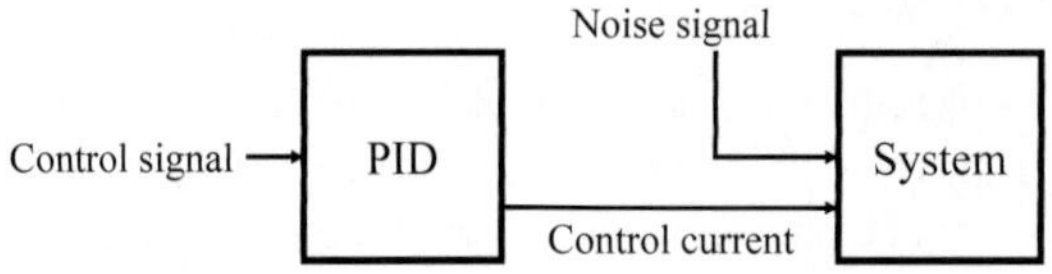

Figure 1: Added noise to the magnetic bearing current control.

Since the displacements and the input forces are assessed, then it is possible to perform an Experimental Modal Analysis (EMA) of the rotor, either if it is rotating or at standstill. Previous analysis of the rotor had been performed by [18, 19] comparing the experimental results with a Finite Element Model (FEM). The Frequency Response Function of the rotor is determined in two conditions; when it is rotating at 400 RPM and at standstill, 0 RPM. Now, another shaft with the same characteristic is deliberately sawn to create a defect, see Figure 3.

A 3mm deep and 1 mm wide crack is created on the shaft close to the middle point of the shaft near the disk. The damaged rotor replaces the healthier one and the experiment is run with the conditions and the added white noise.

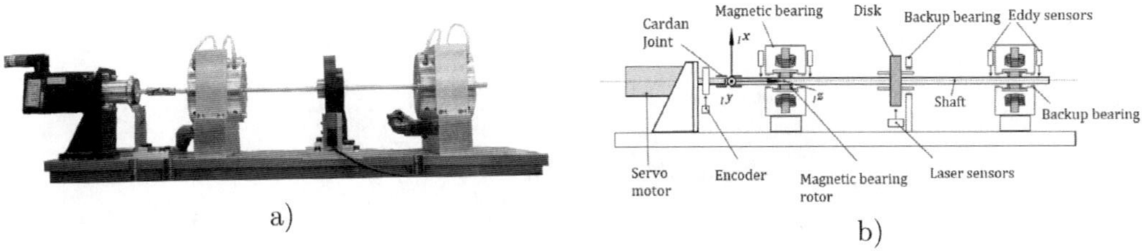

Figure 2: a) Photo of the test bench. b) Rotor scheme. Adapted from [18] and [19].

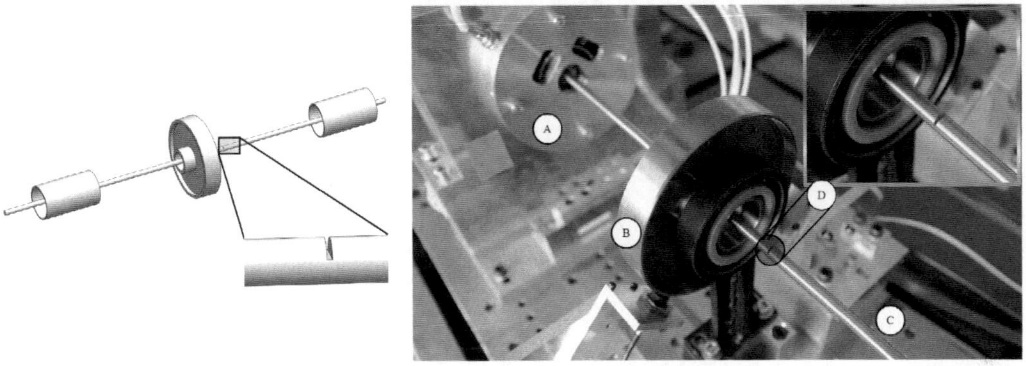

Figure 3: Left: CAD drawing with detail of the crack on the shaft. Right: Detail of the test rig. A - Magnetic Bearing; B - Disk; C - Shaft; D - Crack on the shaft.

3 Experimental run-up and results

The rotor is set to spin at an angular velocity of 400 RPM $\approx$ 6.66 Hz for 600 seconds and the signals from the displacements at the bearings and the disk and the forces are acquired. The orbit plots at position 67.5 mm from the coupling is shown in Figure 4 after the transient attenuates. A smoothed line is drawn to mitigate the noisy raw data. Additionally, one may notice that inner loops orbits related to the breathing crack cases do not appear.

The Frequency Response Functions (FRF) of the lateral displacement at position 630 mm is calculated and plotted to investigate variations of the natural frequencies on the presence of the crack. In Figure 5a and b one can see the FRFs of the normal rotor and the damaged rotor respectively. In both figures, the red plot is the standstill case and the blue the rotating case. The FRF spans up to 250 Hz, because above this threshold the coherence falls sharply to zero, due to the limited bandwidth of the excitation signal till this frequency.

The experiments show that the vibration of the rotor increased considerably due to the existing serious damage. The FRFs of the rotor of the lateral displacement show that there is a shift of some resonances around 110Hz, 160 Hz, and 210 Hz, but some remain practically the same.

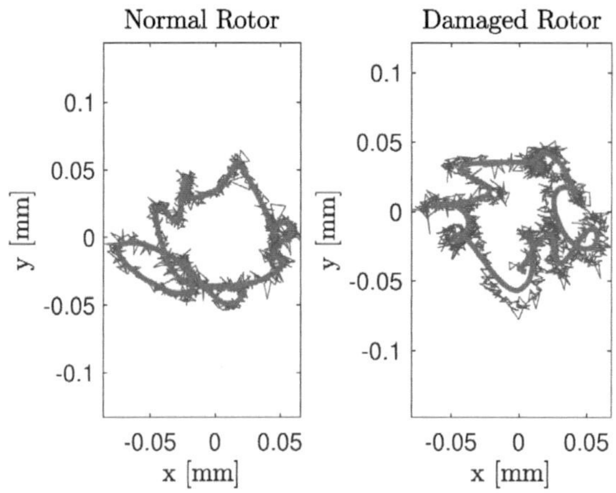

Figure 4: Experimental orbit plots.

At standstill, the angular position of the crack showed almost no influence compared to the normal shaft, but as the shaft rotates, then it is possible to see changes on the spectra. The main difference between the normal shaft and the damaged shaft is observed at the resonance at 210 Hz. The peak decreases considerably and it moves clearly to the left.

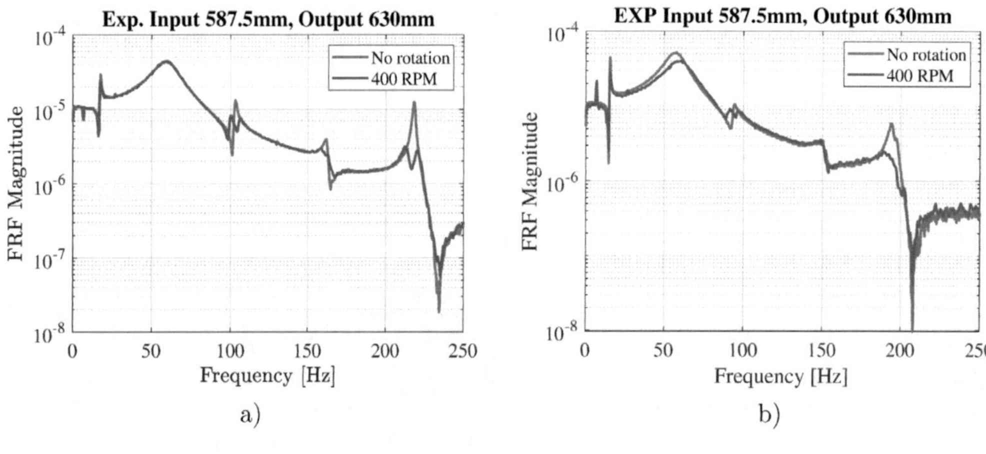

Figure 5: FRF of both rotors. a) Normal shaft. b) Damaged shaft.

4 Mathematical Model

The Finite Element Method (FEM) is implemented in order to achieve a representative numerical model for the rotor-bearing lateral vibration. The equations of motion are defined as:

$$M\ddot{X} + (D - G(\Omega))\dot{X} + KX = f(t), \tag{1}$$

where M is the global mass matrix, D is the damping matrix and K is the stiffness matrix. Each element has two nodes with four degrees of freedom, two linear and two angular displacements, and all of them are represented by the vector X. The mass matrices of shaft elements due to linear and angular movements are given as

$$M_e^T = \frac{\rho A l}{420}\begin{bmatrix} 156 & 0 & 0 & 22l & 54 & 0 & 0 & -13l \\ 0 & 156 & -22l & 0 & 0 & 54 & 13l & 0 \\ 0 & -22l & 4l^2 & 0 & 0 & -13l & -3l^2 & 0 \\ 22l & 0 & 0 & 4l^2 & 13l & 0 & 0 & -3l^2 \\ 54 & 0 & 0 & 13l & 156 & 0 & 0 & -22l \\ 0 & 54 & -13l & 0 & 0 & 156 & 22l & 0 \\ 0 & 13l & -3l^2 & 0 & 0 & 22l & 4l^2 & 0 \\ -13l & 0 & 0 & -3l^2 & -22l & 0 & 0 & 4l^2 \end{bmatrix}, \tag{2}$$

$$M_e^R = \frac{\rho A r^2}{120 l}\begin{bmatrix} 36 & 0 & 0 & 3l & -36 & 0 & 0 & 3l \\ 0 & 36 & -3l & 0 & 0 & -36 & -3l & 0 \\ 0 & -3l & 4l^2 & 0 & 0 & 3l & -l^2 & 0 \\ 3l & 0 & 0 & 4l^2 & -3l & 0 & 0 & -l^2 \\ -36 & 0 & 0 & -3l & 36 & 0 & 0 & -3l \\ 0 & -36 & 3l & 0 & 0 & 36 & 3l & 0 \\ 0 & -3l & -l^2 & 0 & 0 & 3l & 4l^2 & 0 \\ 3l & 0 & 0 & -l^2 & -3l & 0 & 0 & 4l^2 \end{bmatrix}, \tag{3}$$

where the global mass matrix $M = M_e^T + M_e^R$. The gyroscopic and stiffness matrices of the shaft elements due to bending are presented respectively as follows:

$$G_e = 2\Omega\frac{\rho A r^2}{120 l}\begin{bmatrix} 0 & -36 & 3l & 0 & 0 & 36 & 3l & 0 \\ 36 & 0 & 0 & 3l & -36 & 0 & 0 & 3l \\ -3l & 0 & 0 & -4l^2 & 3l & 0 & 0 & l^2 \\ 0 & -3l & 4l^2 & 0 & 0 & 3l & -l^2 & 0 \\ 0 & 36 & -3l & 0 & 0 & -36 & -3l & 0 \\ -36 & 0 & 0 & -3l & 36 & 0 & 0 & -3l \\ -3l & 0 & 0 & l^2 & 3l & 0 & 0 & -4l^2 \\ 0 & -3l & -l^2 & 0 & 0 & 3l & 4l^2 & 0 \end{bmatrix}, \tag{4}$$

$$K_e = \frac{EI}{l^3}\begin{bmatrix} 12 & 0 & 0 & 6l & -12 & 0 & 0 & 6l \\ 0 & 12 & -6l & 0 & 0 & -12 & -6l & 0 \\ 0 & -6l & 4l^2 & 0 & 0 & 6l & 2l^2 & 0 \\ 6l & 0 & 0 & 4l^2 & -6l & 0 & 0 & 2l^2 \\ -12 & 0 & 0 & -6l & 12 & 0 & 0 & -6l \\ 0 & -12 & 6l & 0 & 0 & 12 & 6l & 0 \\ 0 & -6l & 2l^2 & 0 & 0 & 6l & 4l^2 & 0 \\ 6l & 0 & 0 & 2l^2 & -6l & 0 & 0 & 4l^2 \end{bmatrix}. \tag{5}$$

The parameters ρ, E, I, A and r stand for the linear density, the Young modulus for steel, the area moment of inertia, the cross section area, and the radius of the shaft, respectively. The term l is the length of the element. The shaft is divided into elements of 1cm each, except for the region near the crack where the length reduces to 1mm. Smaller sizes of the elements did not increased the accuracy analytical results, see ref. [18]. Since the magnetic bearing aligns the center of the shaft and it develops small displacements, the magnetic force is considered linear with a constant spring coefficient and a damping term:

$$F_{\text{mag}} = k_{\text{mag}} X|_{\text{bearing}} + d_{\text{mag}} \dot{X}|_{\text{bearing}}. \tag{6}$$

These coefficients are inserted into the specific node of the magnetic bearing. The properties of the disk and the bearings are considered punctual and are added on their corresponding nodes. The elements affected by these components have their radii increased by 0.8mm to reproduce the stiffening caused by the connection between the components and the shaft.

The global damping matrix follows the Rayleigh model $D = \alpha M + \beta K$. The determination of α and β comes from the experimental data, which resulted in $\alpha = 0$ and $\beta = 10^{-5}$.

To determine the rotor's frequency response function analytically, one has to analyze the transfer function of equation (1) in the frequency domain as:

$$H(\omega) = \left| \left[(2\pi i\omega)^2 M + (D - G(\Omega))\, 2\pi i\omega + K \right]^{-1} \right|. \tag{7}$$

4.1 The crack in the Finite Element Model

The usual representation of cracks in rotors is the breathing model in which the crack exhibits a dynamics of its own, as the rotor executes a full rotation. The crack is completely open when it is positioned downwards and closes when upwards. This is usually modeled by time-varying stiffness terms. In the present work, the crack does not close itself, due to its width and depth. Thus, the damage is described by removing the mass and gyroscopic terms on the cracked node from the mass and gyroscopic matrices. The element's stiffness is affected by taking the moment of inertia of the remaining element piece. Therefore, the effect is no longer represented as time varying.

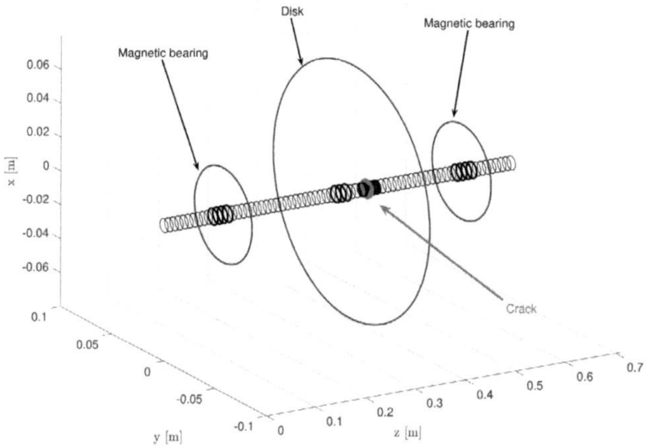

Figure 6: FEM mesh of the rotor.

The proposal of the model is backed up by the experimental data. In Figure 7, one can see the FFT of the lateral vibration of the normal and damaged rotors. The breathing crack case would exhibit peaks at $\Omega/2$ and $\Omega/3$, which do not appear in the frequency domain.

4.2 Numerical analysis

The shaft's parameters shown in Table 1 are used to assemble the matrices necessary for equation (1). Then, Table 2 summarizes the features of the finite element mesh, in which the nodal positions of the crack, disk and bearings are presented.

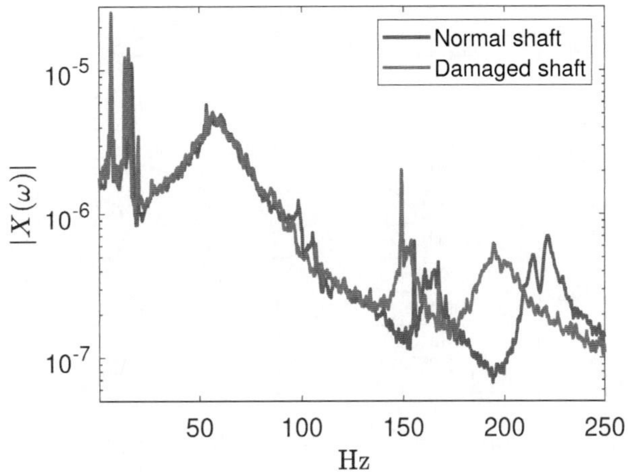

Figure 7: Experimental FFT of the lateral displacement of the rotor. $\Omega = 400$ RPM.

Table 1: Physical parameters of the rotor.

Parameter	Value
ρ	7860 kg/m^3
E	2.11×10^{11} Pa
A	5.0265×10^{-5} m^2
I	2.01×10^{-10} m^4
r	4×10^{-3} m
k_{mag}	1.1×10^{5} N/m
d_{mag}	20 Ns/m
Ω	400 RPM
I_{crack}	1×10^{-11} m^4
A_{crack}	33×10^{-6} m^2
M_{disk}	2276 g
$I_{\text{disk},x,y}$	1.5×10^{-3} kg$\cdot$ m^2
$I_{\text{disk},r}$	2.8×10^{-3} kg$\cdot$ m^2
M_{bearing}	851 g
$I_{\text{bearing},x,y}$	4.84×10^{-4} kg$\cdot$ m^2
$I_{\text{bearing},r}$	4.03×10^{-4} kg$\cdot$ m^2

Figures 8 illustrates the modal analysis of the rotor for two scenarios, one whose angular velocity is zero and the second one in which it is 400 RPM. The FEM shows a shift of the higher natural frequencies, which follows the same trend as the one captured experimentally as may be seen in Figure 5. For the standstill case, the natural frequency around the 210 Hz reduces to 200 Hz. Likewise, the lower natural frequency of 110 Hz decreases slightly, but the one in the middle does not shift as it does for the experimental set up. Similarly, it can be seen in Figure 8, *i.e.* $\Omega = 400$ RPM, that the peaks are lower in the case of the damaged shaft.

A time integration of equation (1) is performed using a Newmark algorithm, [20]. The force vector $f(t)$ includes an unbalance force located at the position of the disk and whose magnitude

Table 2: Mesh features of both normal and damaged shaft.

		Normal	Damaged
# Elements		70	97
DoF		284	397
crack	node	-	51
	position	42 cm	
disk	node	39	39
	position	39 cm	
bearing	node	13 and 63	13 and 89
	position	12 cm and 62 cm	
maximum element length		1 cm	1 cm
minimum element length		1 cm	1 mm

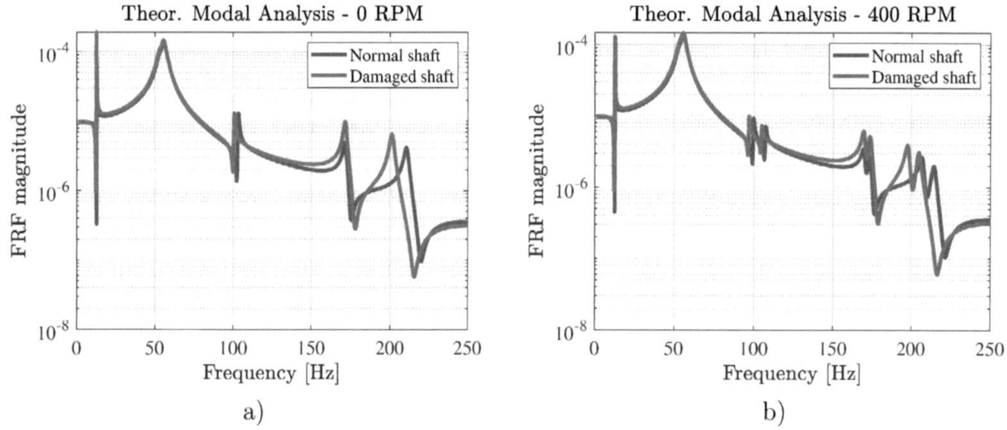

Figure 8: FEM modal analysis. a) Zero angular velocity. b) 400 RPM angular velocity.

is $U = 2.805 \times 10^{-4}$ kg·m. The time integration iterates every 10^{-3} s and spans up to 60 s. The initial conditions for all nodal degrees of freedom are set to zero except for the last linear velocity at the free tip of the shaft, which begins with 1.0 m/s. Besides, the rotor is has the same angular velocity of 400 RPM.

The orbit plots from the normal shaft and the damaged one are shown in Figures 9a and 9b. The transient is clearly seen as the rotor slowly settles into a circular trajectory, Figure 9c. As expected, the damaged rotor executes a larger orbit.

5 Conclusion

The present work investigated the effects caused by a deep and wide crack upon the dynamics of a rotor supported by magnetic bearings. These cracks do not follow the conventional breathing crack model as it stays open during the whole rotation of the shaft. The experiments showed that overall the natural frequencies were affected by the damage, which were observed to be found at least 10 Hz lower than the healthier shaft. A finite element model was employed to reproduce the phenomena and could approximate the FRF numerically with the one calculate with experimental data.

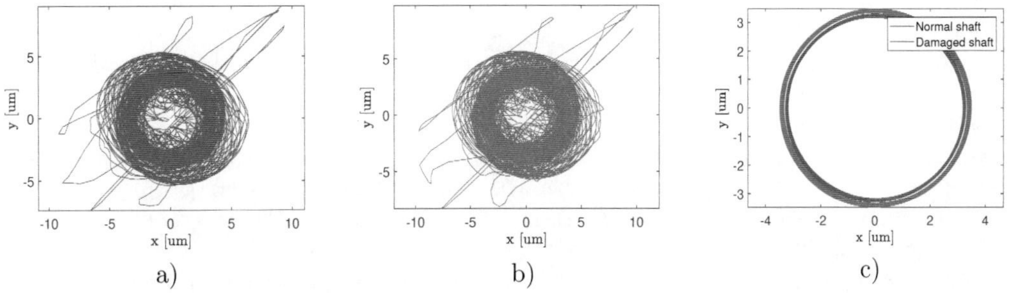

Figure 9: Orbit plots of the two scenarios at 400 RPM. a) the normal shaft. b) the damaged shaft. c) Orbits after the transient is damped.

The model is by and large a linear approach to the problem in question, but it eventually manages to reproduce the phenomena. Also, this set up indicates the possibility of using magnetic bearings to identify a problem without stopping the machine by adding a noise as an additional excitation to the control signal. Further improvement can be done to the model, by taking into account that the location of the crack creates an asymmetric rotor and nonlinearities of the magnetic forces. Besides, the coupling between the driving motor and the shaft is not considered, which indeed does influence the dynamics of the system.

References

[1] Maslen E.H., Schweitzer G., (2009) Magnetic bearings. Springer, Heidelberg. https://doi.org/10.1007/978-3-642-00497-1

[2] Schweitzer, G. *Applications and Research Topics for Active Magnetic Bearings*, IUTAM Symposium on Emerging Trends in Rotor Dynamics pp 263-273

[3] Fonseca, C. A., Santos, I. F., Weber, H. I. *An experimental and theoretical approach of a pinned and a conventional ball bearing for active magnetic bearings*, December 2020, Mechanical Systems and Signal Processing, 138, page 106541, DOI: 10.1016/j.ymssp.2019.106541

[4] Mogal, S.P., Lalwani, D.I.: *A brief review on fault diagnosis of rotating machineries.* Appl. Mech. Mater. 541–542, 635–640 (2014)

[5] Zhang, Y.Z., Zhang, L.Z., Luo, X.Y., Guan, X.P.: *Sensor/actuator faults detection for networked control systems via predictive control.* Int. J. Autom. Comput. 10(3), 173–180 (2013)

[6] Tsai, C., King, Y.H., Lee, R.M.: *Fault diagnosis for magnetic bearing.* Mech. Syst. Signal Process. 23(4), 1339–1351 (2009)

[7] Roßner, M. *Modellbasiertes Monitoring von Rotoren mit mehreren gleichzeitigen Fehlern.* PhD Thesis. Technische Universität München, 2015, 166 pages. ISBN: 9783843923965

[8] Guangming Dong, Jin Chen, *Vibration analysis and crack identification of a rotor with open cracks*, Japan J. Indust. Appl. Math. (2011) 28:171–182 DOI: 10.1007/s13160-011-0031-3

[9] Mani, G., Quinn, D.D., Kasarda, M.: *Active health monitoring in a rotating cracked shaft using active magnetic bearings as force actuators.* J. Sound Vib. 294, 454–465 (2006)

[10] Silva, G.M., Pederiva, R.: *Diagnosis in a rotor supported by active magnetic bearings.* In: Proceedings of the 7th IFToMM International Conference on Rotor Dynamics, International Federation for the Promotion of Mechanism and Machine Science, 2006 Vienna, Austria

[11] Thümmel, T., Rossner, M., Wagner, C., Maierhofer, J. and Rixen, D., *Rotor Orbits at Operation Speed and Model-Based Diagnosis of Multiple Errors.* In: Cavalca K., Weber H. (eds) Proceedings of

the 10th International Conference on Rotor Dynamics – IFToMM. Vol. 2 (2018). DOI: 10.1007/978-3-319-99268-6_16.

[12] Al-Shudeifat, M.A., Butcher, E.A.: *New breathing functions for the transverse breathing crack of the cracked rotor system: approach for critical and subcritical harmonic analysis.* J. Sound Vib. 330, 526–544 (2011)

[13] Cavalini Jr., A.A., Sanches, L., Bachschmid, N., Steffen Jr., V.: *Crack identification for rotating machines based on a nonlinear approach.* Mech. Syst. Signal Process. 79, 72–85 (2016)

[14] Guo, C., Yan, J., Yang, W.: *Crack detection for a Jeffcott rotor with a transverse crack: an experimental investigation.* Mech. Syst. Signal Process. 83, 260–271 (2017)

[15] Bachschmid, N., Pennacchi, P., Tanzi, E.: *A sensitivity analysis of vibrations in cracked turbogenerator units versus crack position and depth.* Mech. Syst. Signal Process. 24, 844–859 (2010)

[16] Gasch, R.: *A survey of the dynamic behaviour of a simple rotating shaft with a transverse crack.* J. Sound Vib. 160, 313–332 (1993)

[17] Silva, G.M., Pederiva, R. *Unbalance Identification in a Rotor Supported by Active Magnetic Bearing* IFToMM 2018, pp. 82–96, (2019) DOI: 10.1007/978-3-319-99262-4_7

[18] Kreutz, M., Maierhofer, J., Thümmel, T., Rixen, D., 2020. *Modaler Modellabgleich eines Rotors in Magnetlagern.* Sechste IFToMM D-A-CH Konferenz 2020. DOI: 10.17185/duepublico/71192

[19] Maierhofer, J., Thümmel, T., Rixen, D., *Modellbasiertes Monitoring an magnetgelagerten Rotoren.* In: SIRM 2017 (2017), S. 1–10

[20] Cunha, A., Soize, C. & Sampaio, R. Computational modeling of the nonlinear stochastic dynamics of horizontal drillstrings. Comput Mech 56, 849–878 (2015). DOI: 10.1007/s00466-015-1206-6

Model and Control of a Six-Pole Magnetic Bearing

Laura Julia Martins Mothé[1], Vinícius Ramos Vasco[2], Yago Pessanha Corrêa[3], Domingos de Farias Brito David[4], and Afonso Celso Del Nero Gomes[5]

[1] Universidade Federal do Rio de Janeiro (UFRJ)
lauramothe@gmail.com
[2] Marinha do Brasil (*Brazilian Navy*)
viniciusramosvasco@gmail.com
[3] Instituto Federal Fluminense (IFFluminense)
yago.correa@iff.edu.br
[4] Universidade Federal Fluminense (UFF)
domingos@vm.uff.br
[5] Universidade Federal do Rio de Janeiro (UFRJ)
nero@coep.ufrj.br

Abstract

Mechanical Bearings are machine elements that support the rotating axes and promote smooth movement between contact surfaces. In order to eliminate friction between the parts and, consequently, the need for lubrication, Active Magnetic Bearings (AMB) were developed based on the generation of reluctance electromagnetic forces. Due to the unstable nature of this equipment, it is necessary to implement a control strategy to avoid displacements and keep the axis aligned. There are studies about different geometries of magnetic bearings regarding the number of poles they present. The eight and four-pole configurations are the most usual, although other alternatives are being studied, like the three and six-pole geometry. The main objective of this paper is to present a mathematical model for the six-pole magnetic bearing applied to a vertical rotor system, propose control strategies to stabilize the system and compare its performance to the four and eight-pole bearings. In addition, comparisons between rotational speeds are carried out to define low, medium and high rotational ranges, and a study on the system's response to control designed for a certain rotational speed in the system operating at different speeds.

1 Background

Bearings are mechanical transmission elements designed to support loads applied to a rotating shaft, promoting movement between the contact surfaces by reducing friction between them. They keep the axis aligned, preventing perpendicular displacements. Since the bearing is a fixed component in contact with another that moves, the friction generated results in a loss of efficiency in the transference of mechanical energy. In order to reduce this loss and increase the operating life of the bearings, lubrication is necessary periodically. However, the use of lubricating oils presents burdens such as time demands for maintenance, high costs, risks of environmental contamination and restrictions on use in some industrial sectors [1].

Alternatively, with the objective of mitigating the disadvantages mentioned with the application of conventional mechanical bearings, a new field of research has emerged and has become increasingly prominent: Magnetic Bearings. Based on the generation of restoring forces capable of preventing radial and axial movements of the rotating axis, its main characteristic is the absence of mechanical contact between the surfaces and, consequently, the elimination of friction [2]. The restorative forces that govern the operation of Magnetic Bearings are obtained through the injection of voltage or current in electromagnets strategically located radially to the axis.

To determine these input variables and keep the axis aligned, it is necessary to apply a control strategy, which leads to calling this devices Active Magnetic Bearings (AMBs).

There are some possible geometries for Magnetic Bearings in relation to the stator pole number. Currently, the three best known geometries are the eight, four and three-pole. These three geometries differ in heat dissipation, magnetic flux loss, interconnection of fluxes, and the reluctance force equations, which leads to different models for each one [3].

The main objective of this work, following the research area of Magnetic Bearings, is to develop and present a mathematical model for the six-pole magnetic bearing applied to a system with a vertical rotor and propose control strategies to stabilize the system. Furthermore, a comparative study between the six, four and eight-pole bearings performance is presented from simulations. Two other studies are also carried out: comparisons between rotational speeds to define low, medium and high rotational ranges, and the response to control laws designed for a specific rotational speed in the system operating at different speeds.

2 Discussion

At present, there are two AMBs configurations that are better known and applied: the most common and traditional, eight-pole configuration [4][5][6], and also the four-pole configuration [7][8]. It can be said that these two cases have symmetry of 4, since the reluctance forces applied to the rotor come from 4 points located on the canonical x (horizontal) and y (vertical) axes.

There are other possible topologies for AMBs, such as three [1][9], six and twelve-pole, however, they are found less frequently. These geometries have symmetry of 3, where the magnetic forces applied to the rotor come from 3 coplanar points, none of them in the canonical x, y directions, in general; one of these points, at most, can lie in either the x or the y directions.

This paper adresses the development of the reluctance forces and modelling the dynamic system in state space for the six-pole AMB which will be simulated in order to compare its performance with the eight and four-pole configuration.

In the six-pole AMBs, there are three channels with two poles each, spatially separated by $(2\pi/3)$rad angles. A basic illustration of this scheme is presented in Figure 1. The rotor shaft is to be positioned at the origin of an x, y canonical coordinate system. This position is controlled using the magnetic forces $\mathbf{F}_k(t)$ for $k = 1, 2, 3$, generated by electromagnetic principle from the injection of currents in each pair of poles of the stator.

The main influence on the magnetic forces is the square of the magnetic flux $\phi_k(t)$ [7], as shown in Equation 1:

$$F_k = \frac{\phi_k^2(t)}{2\mu_0 A} \quad \text{for} \quad k = 1, 2, 3, \tag{1}$$

where μ_0 is the magnetic permeability and A is the winding cross area through which the flux passes.

The three fluxes ϕ_k are said to be decoupled, since each one of them does not intercept the other two. All the flux injected in the rotor by one of the poles in channel k is absorbed by the other pole in the same channel.

Considering $d_k(t)$ for $k = 1, 2, 3$ the gap width at pole k, Equation 1 can be rewritten so that the magnitude of the reluctance force generated is [10]:

$$F_k = K\left(\frac{i_k(t)}{d_k(t)}\right)^2 \quad \text{with} \quad K = \frac{\mu_0 A n^2}{4} \quad \text{for} \quad k = 1, 2, 3, \tag{2}$$

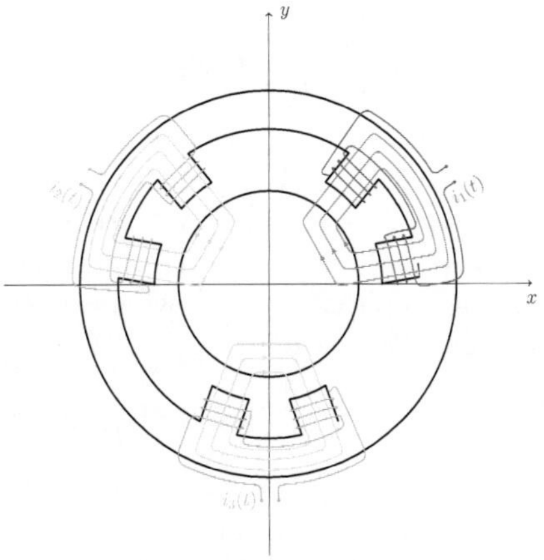

Figure 1: Six-pole configuration for Active Magnetic Bearings. Channel 1 is composed of the pair of poles above and to the right, making an angle of $(\pi/3)$rad with the x axis.

where i_k is the injected current and n is the number of coil windings. Therefore, the flux ϕ_k depends only on the current i_k and the distance d_k.

The spatial directions of the three magnetic forces generated in the six-pole AMB, projected in the x and y directions, lead to

$$F^x(t) = F_1(t)\cos\frac{\pi}{6} - F_2(t)\cos\frac{\pi}{6} + F_3(t)\cos\frac{3\pi}{2} \tag{3}$$

$$F^y(t) = F_1(t)\,\mathrm{sen}\,\frac{\pi}{6} + F_2(t)\,\mathrm{sen}\,\frac{\pi}{6} + F_3(t)\,\mathrm{sen}\,\frac{3\pi}{2}. \tag{4}$$

Applying Equation 2 into Equations 3 e 4, we obtain the resultant magnetic forces on the rotor. Omitting (t) in the time-varying signals, the resultants are:

$$F^x = \frac{K\sqrt{3}}{2}\left[\left(\frac{i_1}{d_1}\right)^2 - \left(\frac{i_2}{d_2}\right)^2\right] \tag{5}$$

$$F^y = \frac{K}{2}\left[\left(\frac{i_1}{d_1}\right)^2 + \left(\frac{i_2}{d_2}\right)^2 - 2\left(\frac{i_3}{d_3}\right)^2\right]. \tag{6}$$

The rotor's distances to the poles $d_k(t)$ can be denoted by Equation 7, where h is the nominal gap width and e_k measure the rotor's displacement from the origin in the direction of pole k.

$$d_k(t) = h - e_k(t) \quad \text{for} \quad k = 1, 2, 3. \tag{7}$$

It can be seen that the e_k displacements are redundant to determine the rotor's position in the plane and only two values are sufficient for this purpose. Projecting the displacements on the canonical directions x and y leads to:

$$e_1(t) = x(t)\cos\frac{\pi}{6} + y(t)\,\mathrm{sen}\,\frac{\pi}{6} \tag{8}$$

$$e_2(t) = -x(t)\cos\frac{\pi}{6} + y(t)\,\mathrm{sen}\,\frac{\pi}{6} \tag{9}$$

$$e_3(t) = x(t)\cos\frac{3\pi}{2} + y(t)\,\mathrm{sen}\,\frac{3\pi}{2} = -y(t). \tag{10}$$

Applying equations 7, 8, 9 e 10 to the reluctance forces equations 5 and 6 and linearizing around the operating point $OP = (i_1^0, i_2^0, i_3^0, x^0, y^0) = (0,0,0,0,0)$ leads to an uncontrollable model. To solve this problem, it is possible to apply the base and differential currents concept: $i_k(t) = i_0 + v_k(t)$ for $k = 1,2,3$, where i_0 is a fixed base current and v_k are differential currents.

At first, it is reasonable to consider a particular case where the 3 currents v_k can be expressed in terms of only 2 differential currents i_x and i_y:

$$i_1(t) = i_0 + i_x(t) + i_y(t) \tag{11}$$

$$i_2(t) = i_0 - i_x(t) + i_y(t) \tag{12}$$

$$i_3(t) = i_0 - i_y(t). \tag{13}$$

Combining Equations 5 to 13, we can obtain a complete mathematical model for the reluctance forces generated in the 6-pole AMB. The linearization of this model around the operating point $O = (i_x^0, i_y^0, x^0, y^0) = (0,0,0,0)$ results in:

$$F^x(t) = k_p x(t) + 2\sqrt{3}k_v i_x(t) \tag{14}$$

$$F^y(t) = k_p y(t) + 4k_v i_y(t) \tag{15}$$

$$k_p = 3\frac{\mu_0 A n^2 i_0^2}{4h^3} \quad \text{and} \quad k_v = \frac{\mu_0 A n^2 i_0}{4h^2}. \tag{16}$$

The model presents decoupled structure, which is a desirable fact and will make the control of 6 poles AMBs easier.To further understand how these magnetic forces interact with the x and y displacements of the rotor, a dynamic mathematical model is necessary.

In the Magnetic Bearing, the stator is responsible for generating the magnetic forces while all dynamic aspects depend on the shaft. For this work, it will be considered a rigid and homogeneous rotor in the vertical position. Center of mass displacements can be determined by the angles α and β. The dynamics of the shaft is described by

$$J\ddot{\boldsymbol{p}}(t) + G\dot{\boldsymbol{p}}(t) = \boldsymbol{E}(t), \tag{17}$$

where J is the inertia coefficient (or inertia matrix JI_2), $\boldsymbol{p} = \begin{bmatrix}\beta & -\alpha\end{bmatrix}$ is the position vector based on the angular movements of the shaft, G is the gyroscopic matrix:

$$G = \begin{bmatrix} 0 & \omega I_z \\ -\omega I_z & 0 \end{bmatrix} = \omega I_z \begin{bmatrix} 0 & 1 \\ -1 & 0 \end{bmatrix}, \tag{18}$$

which depends on the rotation speed ω, and $\boldsymbol{E}$ is the external excitation vector. Three external excitations are considered: magnetic, gravitational and supporting bearing [7].

From Equation 17, as can be seen in [7], it is possible to obtain the state space system model, shown in equation 19 allowing the application of modern control techniques:

$$\dot{\boldsymbol{x}}(t) = A\boldsymbol{x}(t) + B\boldsymbol{v}(t), \tag{19}$$

where A and B are, respectively, 4×4 and 4×3 matrices structured as follows:

$$A = \begin{bmatrix} 0_2 & I_2 \\ A_{21} & A_{22} \end{bmatrix} \qquad B = \begin{bmatrix} B_1 \\ B_2 \end{bmatrix}, \tag{20}$$

in which

$$A_{21} = J^{-1} b^2 k_p I_2 = A_{21}(k_p) \qquad A_{22} = -J^{-1}(G + C_a) = A_{22}(\omega) \tag{21}$$

$$B_1 = 0_{2\times3} \qquad B_2 = J^{-1} b d k_v V = B_2(k_v), \tag{22}$$

where I is an identity matrix, the other variables are constructive parameters of the prototype and the state variables vector $\boldsymbol{x}$ is chosen to be

$$\boldsymbol{x} = \begin{bmatrix} x_s \\ y_s \\ \dot{x}_s \\ \dot{y}_s \end{bmatrix}. \tag{23}$$

Since the injected currents i_k are the control variables of the system, the control vector $\boldsymbol{v}$ is defined considering equations 11, 12 and 13

$$\boldsymbol{v}(t) = \begin{bmatrix} v_1(t) \\ v_2(t) \\ v_3(t) \end{bmatrix} = \begin{bmatrix} i_x(t) + i_y(t) \\ -i_x(t) + i_y(t) \\ -i_y(t) \end{bmatrix} = \begin{bmatrix} 1 & 1 \\ -1 & 1 \\ 0 & -1 \end{bmatrix} \begin{bmatrix} i_x(t) \\ i_y(t) \end{bmatrix}. \tag{24}$$

Making the vector $[i_x(t) \;\; i_y(t)]^T = \boldsymbol{u}(t)$, equation 19 can be rewritten into a structure with only two control variables, the differential currents $i_x(t)$ and $i_y(t)$

$$\dot{\boldsymbol{x}}(t) = A\boldsymbol{x}(t) + B^d \boldsymbol{u}(t), \qquad \text{where } B^d = \frac{bdk_v}{J} \begin{bmatrix} 0 & 0 \\ 0 & 0 \\ 2\sqrt{3} & 0 \\ 0 & 4 \end{bmatrix}. \tag{25}$$

The output vector is formed by the signals measured by sensors $x_s(t)$ and $y_s(t)$.

$$\boldsymbol{z}(t) = \begin{bmatrix} x_s(t) \\ y_s(t) \end{bmatrix} = \begin{bmatrix} 1 & 0 & 0 & 0 \\ 0 & 1 & 0 & 0 \end{bmatrix} \dot{\boldsymbol{x}}(t) = C\dot{\boldsymbol{x}}(t) \tag{26}$$

In order to keep the shaft centralized, an active control strategy is needed to find an input capable of stabilizing the system. This paper addresses the optimal control based on the Linear Quadratic Regulator (LQR) theory, considering two types of schemes: centralized and decentralized.

The block diagram in Figure 2 represents the closed-loop system achieved for the six-pole AMB. In this case, the objective of the system is to reach the zero state to stabilize the rotor at the center of the coordinate system.

As shown in the block diagram, the control signal is defined by $\boldsymbol{u} = F\boldsymbol{x}$, since the reference value $\boldsymbol{r} = 0$. Therefore, the state equation can be rewritten as:

$$\dot{\boldsymbol{x}} = A\boldsymbol{x} + B^d F \boldsymbol{x} \tag{27}$$

where F is the feedback gain matrix, responsible for changing the system's closed-loop poles, once the system is controllable.

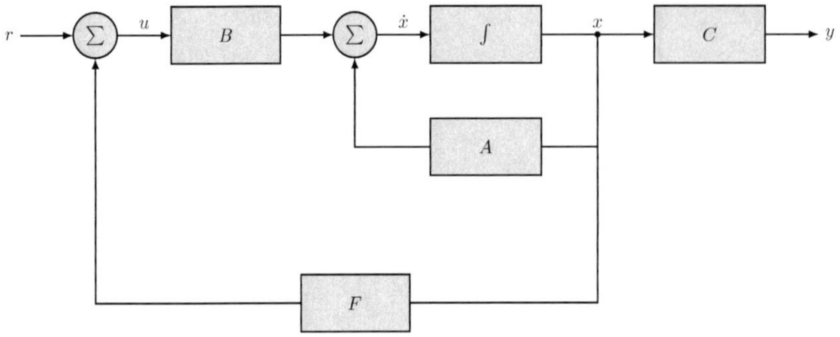

Figure 2: Full state feedback block diagram.

The LQR method has the advantage of providing a systematic model for calculating the gain matrix F [11]. The quadratic functional J expressed in equation 28 describes the index to be minimized:

$$J = \int_0^{\infty} (\boldsymbol{x}^T(t)Q\boldsymbol{x}(t) + \boldsymbol{u}^T(t)R\boldsymbol{u}(t))\, dt \tag{28}$$

where Q is a symmetric positive semidefinite matrix and R is a symmetric positive definite matrix, which detemine the state error and the energy consumption, respectively.

Considering that this is a controllable system with all states observable and available, the control law that minimizes J can be written:

$$\boldsymbol{u}^* = F^*\boldsymbol{x}. \tag{29}$$

Matrix F^* is given by:

$$F^* = -R^{-1}B_d^T P, \tag{30}$$

where P is the solution of an algebraic Riccati equation:

$$A^T P + PA - PBR^{-1}B_d^T P + Q = 0. \tag{31}$$

The optimal performance index of J is given by:

$$J = \boldsymbol{x_0}^T P \boldsymbol{x_0}, \tag{32}$$

where $\boldsymbol{x_0}$ represents the initial state.

The LQR control technique just introduced is sometimes called centralized LQR (LQRc). In this case, when performing state feedback, each component of the control vector $\boldsymbol{u}$ depends on all state variables [10], as shown in Equation 33:

$$\boldsymbol{u} = F_c^*\boldsymbol{x} \implies \begin{bmatrix} u_1 \\ u_2 \end{bmatrix} = \begin{bmatrix} f_{11} & f_{12} & f_{13} & f_{14} \\ f_{21} & f_{22} & f_{23} & f_{24} \end{bmatrix} \begin{bmatrix} x_s \\ y_s \\ \dot{x}_s \\ \dot{y}_s \end{bmatrix}. \tag{33}$$

Therefore, it is expected that the feedback matrix F_c^* to have all nonzero elements, but this fact requires greater computational effort.

To overcome this inconvenience, decentralized LQR (LQRd) control can be implemented. The presence of null elements in the state feedback matrix prevents interactions between different states in the control application. These elements are located strategically so that the system has a decoupled behavior, that is, each of the two control variables is independent of the other, as shown in 34:

$$u_1 = f_{11}x_s + f_{13}\dot{x}_s \qquad \text{and} \qquad u_2 = f_{22}y_s + f_{24}\dot{y}_s. \tag{34}$$

Thus, the decentralized feedback gain matrix F_d^* assumes the following structure:

$$\begin{bmatrix} u_1 \\ u_2 \end{bmatrix} = \begin{bmatrix} f_{11} & 0 & f_{13} & 0 \\ 0 & f_{22} & 0 & f_{24} \end{bmatrix} \begin{bmatrix} x_s \\ y_s \\ \dot{x}_s \\ \dot{y}_s \end{bmatrix} = F_d^* \boldsymbol{x}. \tag{35}$$

Since each control variable is only related to one state variable, the state space vector $\boldsymbol{x}$ can be subdivided in two. Further details can be found in [10]. Thus, equation 35 can be rewritten so that the structure of the decentralized proposal is:

$$\boldsymbol{u} = \begin{bmatrix} p_1 & 0 & d_1 & 0 \\ 0 & p_2 & 0 & d_2 \end{bmatrix} \boldsymbol{x} = F_d^* \boldsymbol{x}, \tag{36}$$

where the notation used in the gain matrix refers to the PD (proportional-derivative) control laws, since each control variable relates only to one output and its derivative.

Equations 37 to 39 present the necessary conditions for F to satisfy 36 and minimize the performance index $J = \boldsymbol{x_0}^T P_d \boldsymbol{x_0}$, where $X_0 = \boldsymbol{x_0}\boldsymbol{x_0}^T$ is the matrix associated with the initial conditions.

$$r_i F_i C_i X C_i^T + B_i^T P_d X C_i^T = 0 \qquad \forall i = 1, 2 \tag{37}$$

$$A_0 X + X A_0^T + X_0 = 0 \tag{38}$$

$$A_0^T P_d + P_d A_0 + Q + \sum_{i=0}^{2} C_i^T F_i^T r_i F_i C_i = 0 \tag{39}$$

In [12] it was shown that such equations are necessary for the solution of the decentralized problem and that they are interdependent conditions. Then, to determine the decentralized feedback gain matrix F_d^*, it is necessary to use an algorithm. It can be said that the LQRd case is simpler to implement than the LQRc due to the amount of null elements, however, the characterization of the gain matrix is more laborious.

3 Results

To compare the performances of the 8, 4 and 6 pole Magnetic Bearing models, simulations were performed submitting the systems to the control laws presented: centralized and decentralized LQR, varying the rotation speed ω and the base current i_0 to portray different operating scenarios.

To find the matrices of the system's dynamic model and the constants k_p and k_v, physical parameters of the Magnetic Bearing prototypes are used [8], together with the base current i_0.

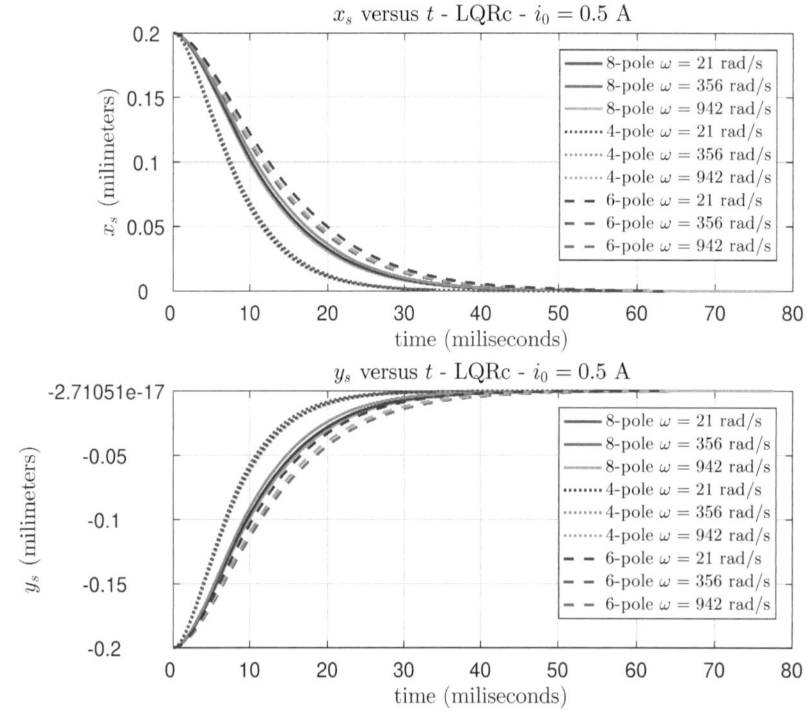

Figure 3: Centralized LQR response.

For the first set of simulations, the base current is considered $i_0 = 0.5$ A. The operating speeds ω used are 21 rad/s (approximately 200 rpm), 356 rad/s (approximately 3400 rpm) and 942 rad/s (approximately 9000 rpm). Figure 3 presents the LQRc response curves (x_s and y_s displacements) for each speed and geometry:

It is observed that the 4-pole bearing is the fastest to return to the equilibrium point, followed by the 8-pole and, lastly, the 6-pole bearing. The existing differences are small. It can also be noticed that the settling time increases as the operating speed increases. This fact suggests that lower operating speeds are better to stabilize the system. However, the variation in settling times for the 3 speeds used is much smaller for the 4-pole geometry than for the others.

Table 1: Closed loop eigenvalues - LQRc - $\omega = 356$ rad/s - $i_0 = 0.5$ A

Closed loop eigenvalues		
8 polos	**4 polos**	**6 polos**
$-191.9 \pm 5.74i$	$-285.7 \pm 6i$	$-167.7 \pm 5.58i$
$-149.4 \pm 4.47i$	$-200.7 \pm 4.22i$	$-128.1 \pm 4.26i$

In Table 1, the closed loop eigenvalues of the 3 geometries submitted to the operating speed

$\omega = 356$ rad/s can be observed. The real part of the eigenvalues in the 4-pole bearing is greater (in module) than the others, which confirms the shorter stabilization time indicated in the graph.

A simulation with the same scenario is carried out for the decentralized LQR control strategy, as shown in Figure 4:

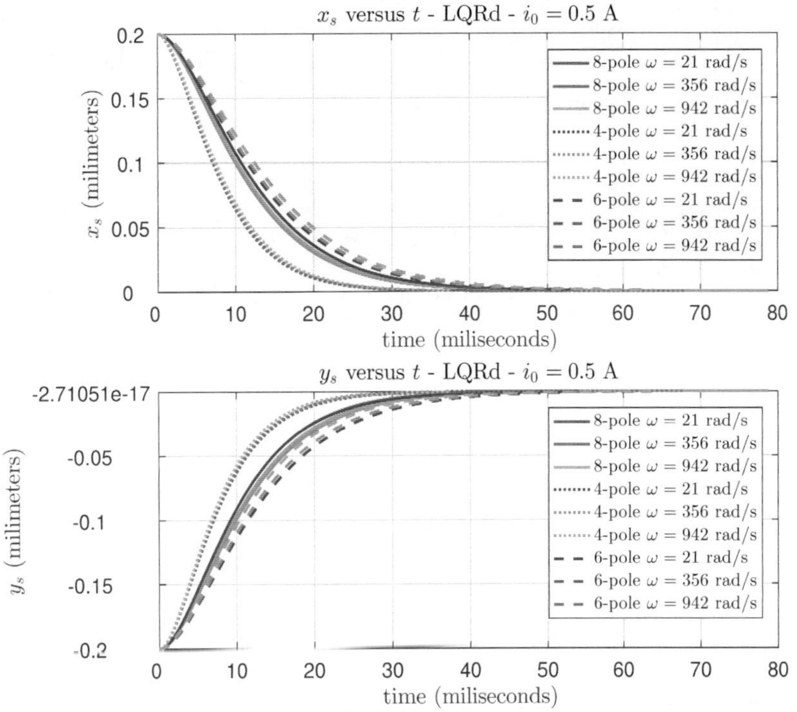

Figure 4: Decentralized LQR response.

The same characteristics found in the simulation of the LQRc control are observed, however, when comparing the two control strategies, the LQRd presents a minimal decrease in the real part of the eigenvalues, which generates a small increase in the stabilization time, as expected in this technique. Table 2 presents the closed loop eigenvalues:

Table 2: Closed loop eigenvalues - LQRd - $\omega = 356$ rad/s - $i_0 = 0.5$A

Closed loop eigenvalues		
8 polos	**4 polos**	**6 polos**
$-190.2 \pm 25.2i$	$-283.5 \pm 20.2i$	$-168 \pm 25.4i$
-133.9	-187.2	-115.1
-165	-216.4	-140

It can be stated that among the 3 compared geometries, the 4-pole one presents the best performance, followed by the 8-pole and finally the 6-pole geometry. But still, the performances are quite similar.

Comparisons on the behavior of the system for the three speeds used (21 rad/s, 356 rad/s and 942 rad/s) were also carried out with the aim of analyzing whether such values can be defined as low, medium or high speed, as these values were determined based on previous works.

The procedure adopted was to compare the feedback matrices F and the closed-loop eigenvalues for the three rotation speeds ω. The centralized LQR feedback matrices for the 6-pole bearing and base current $i_0 = 0.5$ A are presented, respectively, in the equations 40 to 42:

$$F_c^{*6}\ (21\ \text{rad/s}) = \begin{bmatrix} -1168.9 & -2.1 & -8 & 0 \\ 2.4 & -1012.3 & 0 & -7 \end{bmatrix} \tag{40}$$

$$F_c^{*6}\ (356\ \text{rad/s}) = \begin{bmatrix} -1167.5 & -35.3 & -8 & 0 \\ 40.8 & -1011.1 & 0 & -7 \end{bmatrix} \tag{41}$$

$$F_c^{*6}\ (942\ \text{rad/s}) = \begin{bmatrix} -1159 & -93 & -8 & 0.1 \\ 107.4 & -1003.7 & 0.1 & -6.9 \end{bmatrix}. \tag{42}$$

The matrices have a similar structure. However, the elements f_{12} and f_{21} present a considerable difference, which suggests an impact promoted by the change in operating speed values. The eigenvalues for this situation are presented in table 3:

Table 3: Closed loop eigenvalues - LQRc - Six-pole MB - $i_0 = 0.5$ A

Closed loop eigenvalues		
21 rad/s	**356 rad/s**	**942 rad/s**
-166.27	$-167.76 \pm 5.58i$	$-167.26 \pm 15,29i$
-169.42	$-128.10 \pm 4.26i$	$-127.56 \pm 11.66i$
-129.39		
-126.98		

The eigenvalues present real part with very similar values, however, as the operating speed increases, the imaginary part also increases (in module), which generates a considerable impact on the oscillation frequency of the system response.

Therefore, with the significant changes observed in some elements of the feedback matrices and in the closed-loop eigenvalues of the system with the increase in the operating speed, we can consider the values used in this work - 21 rad/s (approximately 200 rpm), 356 rad/s (approximately 3400 rpm) and 942 rad/s (approximately 9000 rpm) - as low, medium and high speeds, respectively.

Another interesting study to be carried out is to understand how an optimal control law designed for a 6-pole magnetic bearing operating in a given angular speed range ω behaves when applied to other values of rotational speed.

The comparison was made considering the base current $i_0 = 1.5$ A and the centralized LQR control strategy. The optimal control law designed for the six-pole bearing operating at low speed, $\omega = 21$ rad/s, was applied in the mathematical models of low, medium ($\omega = 356$ rad/s) and high ($\omega = 942$ rad/s) speed. Figure 5 shows the response curves x_s and y_s for the three mathematical models:

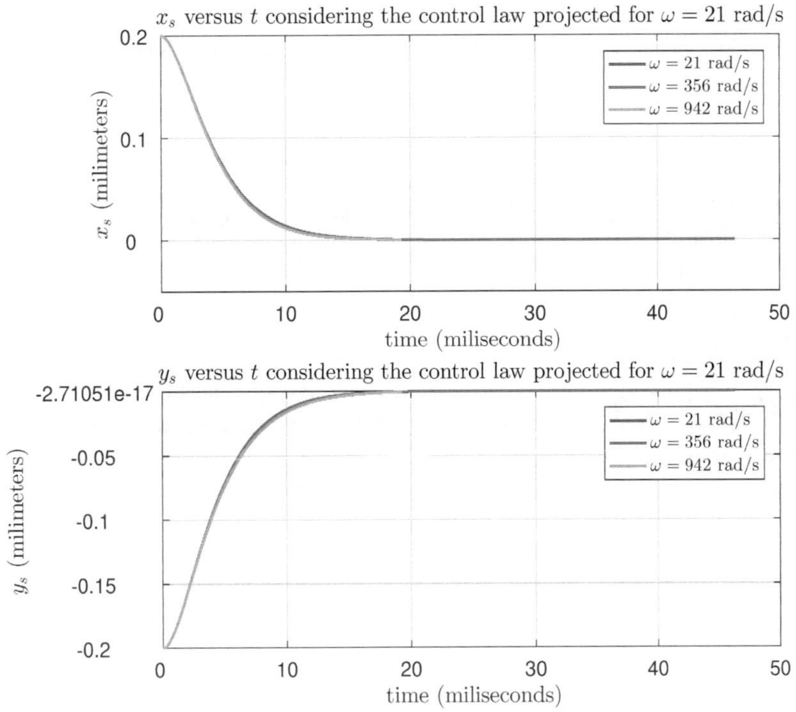

Figure 5: Comparison on the control law for low rotation speed $\omega = 21$ rad/s.

As can be seen, the three models behave well when subjected to a low speed control law, with no large discrepancy between the response curves. It can be said then that the system responds satisfactorily and very similarly when it operates at a rotation speed different from which the control law was designed.

4 Conclusions

The linear dynamic model of the 6-pole magnetic bearing was developed and shown in state space structure, presenting the decoupling characteristic between the control variables. Simulations to compare the three geometries were performed for two control strategies (centralized LQR and decentralized LQR) varying the values of the rotation speed ω.

It is concluded that the increase in rotation speed implies a small increase in the stabilization time necessary for the bearing to return to the central position. As for the results of the six-pole geometry presented, it was found that it had a slightly slower response time than the eight- and four-pole bearings.

In addition to comparing the existing geometries, a comparison was made between the rotation speeds used in this work and it was possible to define three operating speed ranges: low, medium and high. It was also studied the behavior of the system when applying a control law designed for a given rotation speed in the system operating at different speed, from which it was concluded that the system responds satisfactorily.

References

[1] V. R. Vasco, "Modelagem e controle ótimo de um mancal magnético de três polos," Master's thesis, UFRJ, 2019.

[2] H. BLEULER, *Decentralized control of magnetic rotor bearing systems.* PhD thesis, Swiss Federal Institute of Technology, 1984.

[3] W. Zhang and H. Zhu, "Radial magnetic bearings: An overview," *Results in Physics*, 2017.

[4] A. CHIBA, T. FUKAO, and e. ICHIKAWA, O., *Magnetic Bearings and Bearingless Drives.* Newnes-Elsevier, 2005.

[5] G. SCHWEITZER, H. BLEULER, and e. TRAXLER, A., *Magnetic Bearings.* vdf Hochschulverlag, 1994.

[6] G. Schweitzer, "Active magnetic bearings - chances and limitations," *6th International Conference on Rotor Dynamics*, 2002.

[7] D. F. B. DAVID, J. A. SANTISTEBAN, and A. C. D. N. GOMES, "Modeling and testing strategies for an interconnected four-pole magnetic bearing," *Actuators*, vol. 6, no. 3, 2017.

[8] P. H. S. PINTO, "Comparações teóricas e práticas sobre geometrias de mancais magnéticos," Master's thesis, COPPE, Universidade Federal do Rio de Janeiro, 2018.

[9] S. CHEN and C. HSU, "Optimal design of a three-pole active magnetic bearing," *IEEE Transactions on Magnetics*, vol. 38, no. 5, p. 3, 2002.

[10] R. M. Stephan, F. C. Pinto, A. C. D. Gomes, J. A. Santisteban, and A. O. Salazar, *Mancais Magnéticos - Mecatrônica sem atrito.* Editora Ciência Moderna Ltda., 1 edição ed., 2013.

[11] K. Ogata, *Engenharia de Controle Moderno.* Pearson, 2011.

[12] L. S. RODRIGUES, "Controle ótimo descentralizado a dois darâmetros para mancais-motores magnéticos," Master's thesis, COPPE, Universidade Federal do Rio de Janeiro, 2005.

Criticality of rare earths for application in electric vehicles and wind energy in Brazil

Sérgio Eduardo Meirelles de Paula Junior[1], Mariana Figueredo Jacques de Souza[1], Cristiano Nunes da Silva[1], Dejair de Pontes Souza[1] and Virgílio José Martins Ferreira Filho[1*]

[1]Laboratory of Systems Advanced of Management of Production (SAGE-COPPE-UFRJ) Federal University of Rio de Janeiro, Av. Moniz Aragão, 360 - Technology Center - Block 2 (CTII) University City - Rio de Janeiro, RJ - Zip code: 21941-901
virgilio@ufrj.br

Abstract

Rare earth elements (REEs) are important components in technologies aimed at decarbonising the economy, such as electric cars and wind turbines. In 2010, China, the world's largest producer of these elements. The dependent countries on Chinese supply have suffered an increase of the prices and were worried about the provision. Thus, several criticality studies were carried out to evaluate whether these elements could be a bottleneck for the consolidation of the fleet expansion plans for electric vehicles and wind energy. Brazil has the second largest reserve of these elements and intends to create a rare earth productive chain with products with high added value. Therefore, the present study has the objective of evaluating if the REEs could become a bottleneck for the Brazilian plans to expand the fleet of electric vehicles and wind energy.

1 Introduction

The rare earth elements (REEs) are technically defined as the fifteen elements of the lanthanide family - elements with the atomic number (Z) between 57 and 71 - plus two associated, yttrium (Y) and scandium (Sc). The term "earth" is due to the fact that throughout the 18th and 19th centuries, when such elements were isolated, in the form of oxides, at that time, "earth" designated metallic oxides (eg alkaline-earth metals). The term "rare", on the other hand, was due to the late discovery of the first elements and their complex separation (GUPTA e KRISHNAMURTHY, 2005; SOUSA FILHO e SERRA, 2014).

However, contrary to what one imagines when reading "rare earth", these elements are abundant in the crust, cerium (Ce) and lanthanum (La), for example, are more abundant than nickel (Ni) and lead

* Created the first draft of this document

(Pb), respectively . While neodymium (Nd) has a higher concentration in the crust than copper (Cu). In the less abundant group, such as dysprosium (Dy) and erbium (Er) have a higher occurrence than gold (Au) (MCLENNAN, 2001; RUDNICK e GAO, 2003).

The REEs are critical components for the operation of many devices that are part of our daily lives such as: smartphones, cars, computers, televisions and light bulbs. The unique properties present in these elements are essential for the expansion of so-called green technologies, which aim to reduce dependence on non-renewable fuels and improve energy efficiency. With the constant increase in energy demand and the search for cleaner sources, REEs have gained economic, political and academic interest in recent years. Mainly after China, the world's largest exporter of REEs since the 1990s, announced a reduction in export quotas in 2010.

The importance of REEs in the industry lies in the fact that many of these elements are present in wind power systems, electric motors and high-efficiency lighting systems, hybrid and electric vehicles, as well as in the aerospace and defense industries. REEs resources are distributed all over the planet, including Brazil, where the second largest reserve in the world is present. Although they are abundant in the earth's crust, they are found in low concentrations, making their extraction costly and technically difficult. In fact, the term “rare” was not given by the degree of scarcity, which ends up generating some confusion. He was assigned to this group because of the late discovery of these elements.

One of the main problems, or possibly the cause of the others, is the concentration of production in a few countries, where the largest producer is responsible for more than 85% of world production. So, to supply the demand for REEs, the nations with the biggest economies - USA, Japan and EU countries depend on imports.

The forecast of increased demand for green technologies in the energy and transport sectors is noticeable. Expansion plans for the implantation of wind turbines for the generation of electric energy and for the hybrid and electric car fleet are outlined by several countries. Many of the technologies applied nowadays are dependent on permanent magnets that use REEs for their manufacture.

One example is the Maglev Cobra project, a vehicle based on magnetic levitation designed for urban transport, developed by the Superconductor Applications Laboratory (LASUP), at the Federal University of Rio de Janeiro (UFRJ). The vehicle, which has advantages when compared to the wheel-rail and VLT (light rail vehicle) systems (SOUSA, 2016), uses REEs magnets along its track. In this case, the magnet used is produced with neodymium (neodymium-iron-boron – NdFeB).

After the announcement of the reduction of export quotas by the Chinese government, prices, which were previously modest, reached peaks with an appreciation of more than 600% for some elements such as dysprosium, europium and terbium, between 2010 and mid-2011 (MASSARI e RUBERTI, 2013). Part of the extreme rise in the price of REEs happened for reasons of speculation, as the expert explains LIFTON (2012).

There is a lot of uncertainty about the issues of REEs resource, exploration and market, since most of it is controlled by China, which was sometimes justified by environmental issues or by a possible shortage of internal supply, which may still have been motivated by an oversupply ZEPF (2013). The author also mentions that the information available in the reports and studies is based on incomplete or unreliable data, despite the large amount of published reports. The REEs offer data is basically provided by a source, which under scientific aspects is not reliable and cannot be proven by validation. This creates a space for discussing how reliable the supply-demand relationship is.

Regarding production, information on reserves and market values is almost not publicly available. United States Geological Survey data (USGS), they are the most used, but they do not clearly present their sources. Several consulting firms offer reports that analyze the REEs market, but access to these documents is paid, making it difficult to obtain the information.

With the rise in prices triggered by the 2010 crisis, many REEs exploration projects in new mines were launched in different parts of the world. Speculators with little experience in the REEs market bought small mining companies, which caught the attention of investors, but the projects did not go ahead. Lack of knowledge about the market led investors to analyze the REEs as a whole, not considering that each element has its own market and its own application, as he explains LIFTON (2012).

For many REEs production exceeds demand, only a few are critical or rare indeed (LIFTON, 2012). What happens is a problem of balance in the supply, since these elements are not found in isolation (BINNEMANS et al., 2013).

Brazil, which was a relevant producer until the mid-twentieth century and currently does not produce, has plans to create a value chain for the production of REEs with high added value products, such as neodymium (Nd) magnets and, thus, position as a worldwide supplier of the material. In this way, there is the opportunity to create a sustainable value chain, which can be a competitive differential in relation to other countries.

The REE sector is at risk of short-term shortages of some elements due to growing demand, particularly in clean energy technology applications such as wind. With supply constraints due to the complex and unpredictable market, many elements have been identified as critical. Recycling and reuse, which currently have almost no investment in research and development, emerge as promising solutions to face scarcity.

The difficulties of inserting itself as a relevant actor in the REEs value chain stem mainly from the control of the market that China manages to exercise. In fact, the creation of a national value chain aims to guarantee the supply of REEs for application in green technologies and other technologies without depending on the external market. To meet Brazil's expectations, a multidisciplinary research on the subject is necessary, since little is said about the subject in national research centers. In addition, the country has plans to expand the fleet of hybrid and electric vehicles and wind energy, increasing the importance of this study in academia. Thus, the demand for REEs, which are present in NdFeB magnets, tends to increase. Thus, the problematization of this study arises: would the REE be a bottleneck for the execution of national plans?

The objective of this work is to verify whether the supply of REEs present in NdFeB magnets would be an obstacle to the execution of plans to expand the fleet of hybrid and electric vehicles and wind energy in Brazil by 2050. For this, establish the sectorial panorama of the REEs, as well as verifying the existing technological applications help to fulfill this proposal.

2 Methods

The methodology of the present study was based on MOSS et al. who used the Critical Metals in the Path towards the Decarbonization of the EU Energy Sector study. It consisted of a bibliographic search of papers using Portal Periódicos Capes as database, in addition to books, reports produced by national and international government agencies, business reports, law and regulations. In the first moment, the main articles about REEs present in journals and in several government agencies such as USGS and British Geological Survey (BGS) were identified. Reports and studies from the National Department of Mineral Production (DNPM), United States Department of Energy (US DoE), United States Environmental Protection Agency (USEPA), Joint Research Center (JRC), publications and presentations from national conferences and international, studies carried out by consultancies, conversations with specialists, among others, were also sources of research.

Thus, the analysis of the REEs market was performed based on the data and information collected. The main studies on criticality, scarcity, supply risk were surveyed, in addition to the data on the worldwide supply of REEs. The state of the art of the hybrid technologies, electric vehicles and wind farm was also identified. Thus, a scenario was defined based on the study “Greenhouse Gas Emissions-2050: Economic and Social Implications of the Government Plan Scenario” developed by CENTRO CLIMA/COPPE/UFRJ and in the National Energy Plan 2050 (NEP 2050) to project the expansion of the fleet of hybrid and electric vehicles and the generation of wind farm in Brazil. However, the number of REEs present in the neodymium magnet required for each of these applications was calculated. The REEs present on the magnet in addition to the neodymium are praseodymium (Pr), dysprosium (Dy) and terbium (Tb).

First, the main studies on criticality, scarcity, supply risk, in addition to data on the world supply of REEs were raised. The state of the art of technologies used in hybrid and electric vehicles and in wind energy generation was also identified. Finally, the role of creating the national REE production chain to guarantee the country's supply and be able to export value-added products to countries seeking supply alternatives was discussed.

To determine the number of hybrid and electric vehicles to be sold in Brazil by 2050, the considerations of the GHG emission scenario for the transport sector present in the study were adopted as a study IES-50. For the projection of sales of light vehicles of the automobile and light commercial type, the average growth rate of 3.43% p.a., from 2018 to 2050, of the IES-50 was adopted. For 2015 and 2016, data from the National Association of Automotive Vehicle Manufacturers (ANFAVEA, 2017). In view of the 2016 economic crisis, there was a 19.3% drop in the number of licenses for these vehicles. On the other hand, the sector forecasts a 2.4% growth (FENABRAVE, National Federation of Automotive Vehicle Distribution) at 4% (ANFAVEA) for this segment in 2017, so the same growth rate of 3.43% p.a. for the year 2017 (EXAME, 2017).

For hybrid cars, the IES-50 adopted a 0.05% share of sales in 2015. Then, linear growth was adopted until 2045, when this technology will reach a 90% share of sales. In 2050, the share of sales will be 82%, due to the greater share of electric vehicles. As for electric cars, the IES-50 adopted a share of 0.001% of sales in 2015. Then, a linear growth was considered until 2045, when it will reach a share of 10% of sales. From 2045 to 2050, an exponential growth in the share of sales was considered, in which this technology will reach 18% of the share in 2050.

According to Global EV Outlook 2016, BEV participated with 59% of the Electric Vehicle (EV) market in the countries analyzed in 2015 (IEA, 2016). On the other hand, in Europe, in 2015, the participation of PHEVs of 60% of the total EV (JRC, 2016). Therefore, the share of BEV (Batery Electric Vehicle) and PHEV (Plug-in Hybrid Electric Vehicle) in the EV market, for the Brazilian reality, was considered 50% for each of them.

To project data referring to hybrid light commercials, the start of commercialization in 2020 was considered, with a share of 1.99% of sales, as in the IES-50. However, the document indicates an average growth of 119% in sales until 2045, when such technology would reach 100% of the share. However, such growth margin proved absurd when calculated, extrapolating, in 2026, the total number of light commercials sold. Therefore, for this projection, an average growth of approximately 21% p.a. in sales was considered in order to reach 100% of the share of sales in 2045.

With the results of the projection of sales of hybrid and electric vehicles, the necessary quantities of NdFeB magnets to meet the demand in the 2050 scenario were calculated. Demands for each of the REEs of the magnet were also calculated.

For the projection of the amount of REEs to be used in wind turbines for 2050 in Brazil, considerations were adopted from the reference scenario of the IES-Brazil 2050 Project for the energy supply sector. According to this, the share of wind energy could reach 11% of the total installed capacity in 2050, making it the second main source of electricity generation.

In view of the iNCD measures, forecasts for the evolution of installed capacity by energy source were presented, where data were generated year by year through the average growth rate calculated every five years. The forecast is that wind energy will grow at an average of approximately 10% p.a. in the first five years. Between 2020 and 2050 this rate does not reach 4% p.a.

For this study, the difference between the current year's installed capacity and the previous year's installed capacity (= increased capacity in that year) was used as a reference, between 2025 and 2035, for example, the highest installed capacity will occur in the last year of the 5-year series, that is, in the years ending with "0" or "5". Thus, the greatest expansion of capacity remains for the year under review.

This was a choice to bring the prediction closer to reality. If linear growth were adopted, in the 5 years of each interval the expansion would be the same in each one of them.

3 Results

3.1 - Electric vehicles

The results of the projection for electric vehicles in Brazil are shown in **Figure 1**. With the results of the projection of sales of hybrid and electric vehicles, the necessary quantities of NdFeB magnets to meet the demand in the 2050 scenario were calculated. Demands for each of the magnet's REE were also calculated. The results are shown in **Table 1.**

For comparison, the ratio between the total demand each year and the supply for each REE in 2015 was established. In 2040, for example, the demand for Nd for vehicle engines will be 528.8t, which represents 2.84% of world Nd production in 2015.

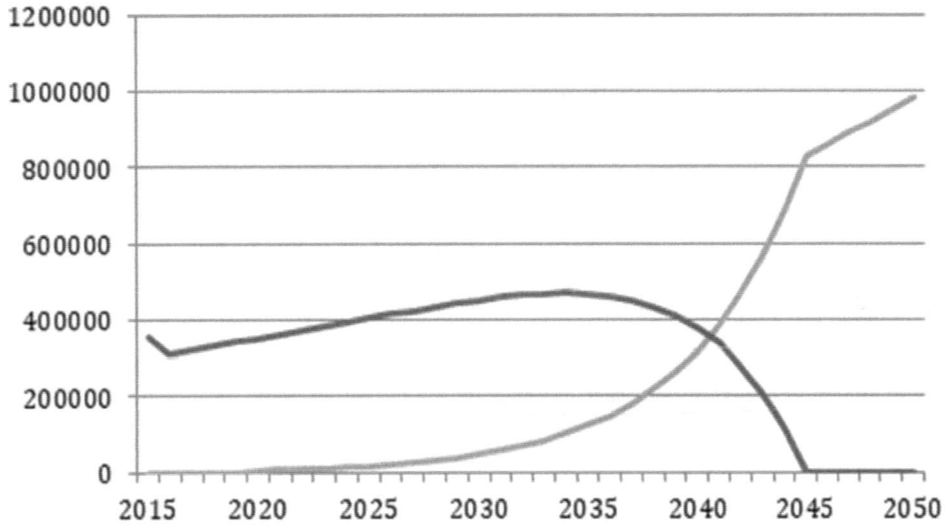

Figure 1: Graph of total sales of commercial vehicles (-Hybrids and -Others) until 2050

According to **Table 1**, the element that will present the worst scenario in 2050 will be dysprosium. For the transport sector in Brazil alone, demand would be 43% of all Dy produced in 2015. Of course, this estimate does not consider the increase in the supply of REEs. Furthermore, efforts to reduce the dependence of REEs on engines will bring positive results. This comparison serves to alert decision makers about a possible shortage in supply and, thus, seek alternatives, whether they are: reducing the use of these elements, replacing or even recycling.

	Nd total	% Nd*	Pr total	% Pr*	Dy total	% Dy*
2015	0.16	0.00%	0.05	0.00%	0.05	0.01%
2016	8.72	0.05%	2.91	0.05%	2.89	0.37%
2017	17.91	0.10%	5.97	0.10%	5.93	0.76%
2018	27.73	0.15%	9.24	0.16%	9.18	1.18%
2019	38.19	0.21%	12.73	0.22%	12.65	1.63%
2020	50.35	0.27%	16.78	0.29%	16.67	2.14%
2021	62.44	0.34%	20.81	0.36%	20.67	2.66%
2022	75.32	0.40%	25.11	0.44%	24.94	3.20%
2023	89.05	0.48%	29.68	0.52%	29.49	3.79%
2024	103.68	0.56%	34.56	0.60%	34.33	4.41%
2025	119.26	0.64%	39.75	0.70%	39.49	5.07%
2026	135.86	0.73%	45.29	0.79%	44.99	5.78%
2027	153.54	0.83%	51.18	0.90%	50.84	6.53%
2028	172.36	0.93%	57.45	1.01%	57.07	7.33%
2029	192.42	1.03%	64.14	1.12%	63.72	8.19%
2030	213.79	1.15%	71.26	1.25%	70.79	9.10%
2031	236.58	1.27%	78.86	1.38%	78.34	10.07%
2032	260.89	1.40%	86.96	1.52%	86.39	11.10%
2033	286.84	1.54%	95.61	1.67%	94.98	12.21%
2034	314.56	1.69%	104.85	1.84%	104.16	13.39%
2035	344.23	1.85%	114.74	2.01%	113.98	14.65%
2036	376.01	2.02%	125.34	2.19%	124.51	16.00%
2037	410.10	2.20%	136.70	2.39%	135.80	17.45%
2038	446.75	2.40%	148.92	2.61%	147.93	19.01%
2039	486.22	2.61%	162.07	2.84%	161.00	20.69%
2040	528.84	2.84%	176.28	3.09%	175.11	22.50%
2041	574.97	3.09%	191.66	3.35%	190.39	24.47%
2042	625.05	3.36%	208.35	3.65%	206.97	26.60%
2043	679.59	3.65%	226.53	3.96%	225.03	28.92%
2044	739.19	3.97%	246.40	4.31%	244.77	31.45%
2045	804.57	4.33%	268.19	4.69%	266.41	34.24%
2046	840.75	4.52%	280.25	4.90%	278.39	35.78%
2047	878.30	4.72%	292.77	5.12%	290.83	37.37%
2048	917.28	4.93%	305.76	5.35%	303.73	39.03%
2049	957.72	5.15%	319.24	5.59%	317.13	40.75%
2050	999.69	5.37%	333.23	5.83%	331.02	42.54%

*Comparison with estimated production in 2015 (considering only legal production)

Table 1: Future demand for REEs for application in hybrid and electric vehicle engines in Brazil

3.2 - Energy sector

The results of the projection for wind energy in Brazil are shown in **Figure 2**, graph of the evolution of the installed capacity of wind energy in Brazil. To calculate the total amount of NdFeB magnet required, the data in **Table 2** were used. The magnets used in wind turbines have a different composition from those used in electric vehicle engines, since the application conditions are not the same. **Table 3** presents data from the literature and those considered in this study.

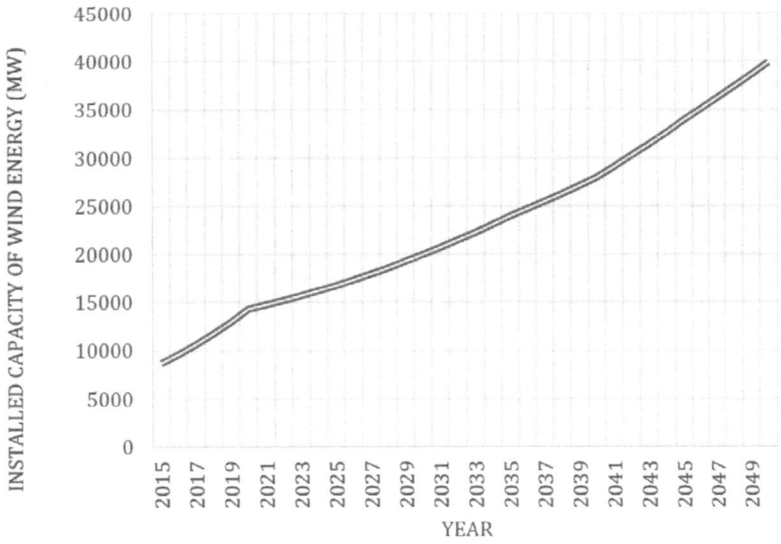

Figure 2: Projected growth of installed wind energy capacity by 2050

	DD-PMG	MS-PMG	HS-PMG
ZEPF (2013)	675	0.12	0,12
PAVEL et al. (2016)	0.65	0.16	0,08
Considered in the study	0.65	0.12	0.12

Table 2: NdFeB magnet mass per MW (t/ NdFeB/MW)

REE	ZEPF (2013)	LACAL-ARÁNTEGUI (2015)	Considered in the study
Nd	22.5	20.3-22.4	22
Pr	7.5	8.7-9.32	8
Dy	4.5	3.0-6.0	4.5

Table 3: Composition of NdFeB magnets used in wind turbines (% mass of REE in the NdFeB)

The demand for total REEs (**Table 4**), as well as in the analysis of hybrid and electric vehicles, was compared with the world production of REE in 2015. It is noted that, even using a relatively high amount of NdFeB magnet, its application in wind energy for Brazil's plans it does not bring great concerns.

	Nd total	% Nd*	Pr total	% Pr*	Dy total	% Dy*
2015	9.74	0.05%	3.54	0.06%	1.99	0.26%
2016	14.95	0.08%	5.44	0.10%	3.06	0.39%
2017	22.15	0.12%	8.06	0.14%	4.53	0.58%
2018	30.70	0.17%	11.16	0.20%	6.28	0.81%
2019	40.79	0.22%	14.83	0.26%	8.34	1.07%
2020	52.65	0.28%	19.15	0.34%	10.77	1.38%
2021	18.91	0.10%	6.88	0.12%	3.87	0.50%
2022	20.96	0.11%	7.62	0.13%	4.29	0.55%
2023	23.12	0.12%	8.41	0.15%	4.73	0.61%
2024	25.40	0.14%	9.24	0.16%	5.19	0.67%
2025	27.79	0.15%	10.11	0.18%	5.68	0.73%
2026	35.01	0.19%	12.73	0.22%	7.16	0.92%
2027	38.25	0.21%	13.91	0.24%	7.82	1.01%
2028	41.68	0.22%	15.16	0.27%	8.53	1.10%
2029	45.32	0.24%	16.48	0.29%	9.27	1.19%
2030	49.16	0.26%	17.87	0.31%	10.05	1.29%
2031	52.77	0.28%	19.19	0.34%	10.79	1.39%
2032	55.43	0.30%	20.16	0.35%	11.34	1.46%
2033	58.21	0.31%	21.17	0.37%	11.91	1.53%
2034	61.13	0.33%	22.23	0.39%	12.50	1.61%
2035	64.18	0.35%	23.34	0.41%	13.13	1.69%
2036	56.80	0.31%	20.65	0.36%	11.62	1.49%
2037	59.28	0.32%	21.56	0.38%	12.13	1.56%
2038	61.86	0.33%	22.50	0.39%	12.65	1.63%
2039	64.55	0.35%	23.47	0.41%	13.20	1.70%
2040	67.34	0.36%	24.49	0.43%	13.77	1.77%
2041	87.84	0.47%	31.94	0.56%	17.97	2.31%
2042	91.32	0.49%	33.21	0.58%	18.68	2.40%
2043	94.94	0.51%	34.52	0.60%	19.42	2.50%
2044	98.70	0.53%	35.89	0.63%	20.19	2.59%
2045	102.61	0.55%	37.31	0.65%	20.99	2.70%
2046	88.99	0.48%	32.36	0.57%	18.20	2.34%
2047	91.93	0.49%	33.43	0.59%	18.80	2.42%
2048	94.97	0.51%	34.53	0.60%	19.42	2.50%
2049	98.10	0.53%	35.67	0.62%	20.07	2.58%
2050	101.34	0.54%	36.85	0.64%	20.73	2.66%

*Comparison with the estimated production of 2015 (considering only the legal production)

Table 4: Total REEs demand

4 Conclusions

With the result of the vehicle fleet projections, it became clear that, in 2050, the Dy is the most critical, produced through the criticality matrix developed by the National Research Council (NRC). In that year, approximately 42.54% of the world production of the element in 2015 would be necessary to supply only the national needs in this application. As for the expansion of wind energy, the results were positive. Although turbines require a large amount of REEs per unit, for national plans this demand does not represent a significant step for world production.

Recycling can be an alternative to guarantee the supply of REEs. However, the technological difficulties in the processes, which are often inefficient, and the lack of incentives are the main problems. In addition, the amount required for these types of applications is high, so the waste to be recycled must contain a large volume of REEs to become viable, making it an advantageous alternative only if the cost of the material already recycled is lower or equivalent to that of original.

References

ANFAVEA. Anuário da Indústria Automobilística Brasileira, p. 62, 2017. Disponível em: http://www.virapagina.com.br/anfavea2017/#62/z Acesso em mai. de 2017.

BEN - BALANÇO ENERGÉTICO NACIONAL 2017: Ano base 2016 / Empresa de Pesquisa Energética. Rio de Janeiro: EPE, 2017

Binnemans, Koen et al. Recycling of rare earths: a critical review. *Journal of Cleaner Production*, [s.l.], v. 51, p.1-22, jul. 2013.

Binnemans, K.; Jones, P. T. Rare Earths and the Balance Problem. *Journal of Sustainable Metallurgy*, v. 1, n. 1, p. 29–38, 2015.

BGS. Risk list 2015. Disponível em: https://www.bgs.ac.uk/mineralsuk/statistics/riskList.html Acesso em mar. de 2017.

DNPM – Sumário Mineral, Edição 2015. Ministério de Minas e Energia - Secretaria de Geologia, Mineração e Transformação Mineral - Departamento Nacional De Produção Mineral. Março de 2016

Exame, Revista. Alfavea prevê crescimento de 4% na venda de veículos em 2017. Exame.com, 5 de jan. de 2017. Disponível em: http://exame.abril.com.br/economia/anfavea-preve-crescimento-de-4-na-venda-de-veiculos-em-2017/ Acesso em mai. de 2017.

Gupta, C. K.; Krishnamurthy, N.. Extractive metallurgy of rare earths. *International Materials Reviews*, [s.l.], v. 37, n. 1, p.197-248, jan. 2005.

IEA - International Energy Agency, Clean energy ministerial and Electric vehicles initiative. Global EV outlook 2016. Beyond one million electric cars; 2016. Disponível em: https://www.iea.org/publications/freepublications/publication/Global_EV_Outlook_2016.pdf

Joint Research Centre (JRC). 2017. JRC Wind Energy Status Report 2016 Edition - Market, technology and regulatory aspects of wind energy. União Europeia, 2017. Disponível em: https://setis.ec.europa.eu/sites/default/files/reports/wind_energy_status_report_2016.pdf

Lifton, Jack. The only five rare earth elements that matter: depoiment [19 de junho, 2012]. Streetwise reports:The Gold Report. Entrevista concedida a Alec Gimurtu. Disponível em: http://www.theaureport.com/pub/na/the-only-five-rare-earth-elements-that-matter-jack-lifton Acesso em jan. de 2017.

Massari, S.; Ruberti, M. Rare earth elements as critical raw materials: Focus on international markets and future strategies. *Resources Policy*, v. 38, n. 1, p. 36–43, 2013.

Mclennan, S. M., 2001. Relationships between the trace element composition of sedimentary rocks and upper continental crust. *Geochem. Geophys. Geosyst.* 2 (4), 2017.

Moss R, Tzimas E, Willis P, Arendorf J, Tercero Espinoza L. Critical Metals in the Path towards the Decarbonisation of the EU Energy Sector: Assessing Rare Metals as Supply-Chain Bottlenecks in Low-Carbon Energy Technologies. EUR 25994. Luxembourg (Luxembourg): Publications Office of the European Union; 2013. JRC82322

Rudnick, R.l.; Gao, S.. Composition of the Continental Crust. Treatise On Geochemistry, [s.l.], p.1-64, 2003.

Sousa Filho, Paulo C. de; SERRA, Osvaldo A.. Terras raras no Brasil: histórico, produção e perspectivas. *Química Nova*, São Paulo , v. 37, n. 4, p. 753-760, 2014 .

Sousa, Wescley T. B. de et al . Projeto MagLev Cobra - Levitação Supercondutora para Transporte Urbano. *Rev. Bras. Ensino Física*., São Paulo , v. 38, n. 4, e4308, 2016

U.S. Geological Survey, Mineral Commodity Summaries. Rare Earths. P. 134 -135, jan. 2017. Disponível em: https://minerals.usgs.gov/minerals/pubs/commodity/rare_earths/mcs-2017-raree.pdf Acesso em mar. 2018.

Zeff, V. A new approach to the Nexus of supply, demand and use: Exemplified along the use of neodymium in permanent magnets. Berlim, Alemanha: Springer-Verlag. 2013.

Summary of Machine Design and Current Regulation for the Parallel DPNV Bearingless Motor Winding

Nathan Petersen, Anvar Khamitov, Timothy Slininger, and Eric L. Severson

Department of Electrical and Computer Engineering, University of Wisconsin-Madison, WI, USA
eric.severson@wisc.edu

Abstract

This paper provides a summary of investigations into the design of combined windings and precision current regulation techniques for bearingless motors. The parallel dual-purpose, no-voltage (DPNV) winding topology is used to realize both torque and force creation. For significant power applications, it is often advantageous to implement suspension operation independent of the main motor drive. This paper shows that reasonable decoupled operation (force and torque) of combined windings using independent drives is possible through two means: 1) careful design of winding parameters such as number of slots, poles, winding factors, and geometric parameters, 2) careful current regulator design to reject disturbance voltages between the suspension and torque systems. For drives that are not independent (i.e., can transmit data between suspension and torque controllers in real-time), this paper shows that a controls-based direct voltage decoupling approach can be used to ideally decouple force and torque creation. To achieve the highest power density of the bearingless motor, the paper shows that the decoupling approach should be used. Analytical, simulation, and experimental results are presented to verify the results.

Keywords: combined winding, winding design, current regulation, bearingless motor.

This paper summarizes the key contributions from [1].

1 Introduction

Bearingless (self-bearing) motors are machines with the ability to generate both forces and torque. In this paper, radial force bearingless motors are considered for the application of levitating the rotor. Historically, these machines used two separate windings: a "torque" winding and a "suspension" winding. More recently, clever machine designers make use of combined windings where one winding creates both torque and force. The combined winding allows the drive to dynamically allocate slot current to either force or torque, depending on the system needs. This improves slot space utilization, leading to more power dense bearingless motors.

This paper studies the parallel dual-purpose no-voltage (DPNV) bearingless motor winding, depicted in Figure 1a. There are two terminal sets: three-phase torque terminals and three-phase suspension terminals. Reference [1] derives the electrical model of the parallel DPNV winding; the result is depicted in Figure 1b. In the circuit, the torque inverter $\vec{v}_t$ creates torque currents $\vec{i}_t$ which are responsible for standard motoring operation. The suspension inverter $\vec{v}_s$ injects currents into the "negative" side of coil group a which act to create forces for levitation.

Unfortunately, due to the winding topology, there is inherent cross-coupling between force and torque generation – by flowing currents which create force, inadvertently, torque will be created. Reference [1] derives this in detail from an electrical prospective. Intuitively, this can be observed by noticing that half of the suspension voltage $\vec{v}_s$ appears in the torque system circuit. The suspension operation directly creates a voltage disturbance in the torque system which, if not fully rejected, will create extraneous torque currents which can cause torque ripple.

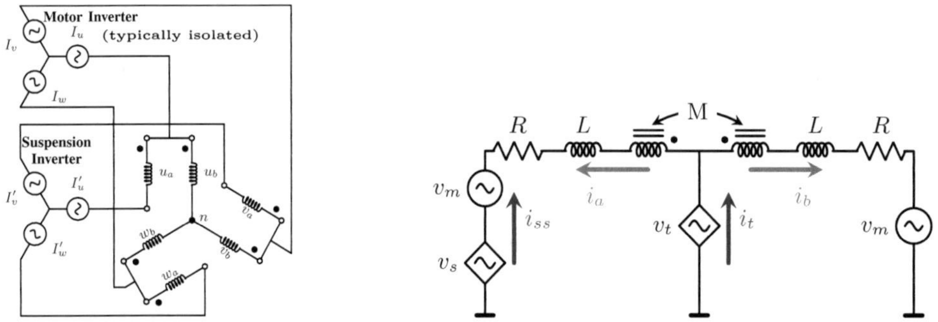

Figure 1: Parallel DPNV winding (a) drive connections (b) space vector equivalent circuit.

The amount of voltage disturbance acting from the suspension to torque system is defined as $\vec{v}_{cs}$. The $\vec{i}_t$ current regulator must reject the disturbance voltage $\vec{v}_{cs}$ in order to eliminate electrical cross-coupling. Reference [1] first derives analytical expressions for $\vec{v}_{cs}$ in terms of the machine design parameters. It is shown that, indeed, all designs which can create force inherently also create $\vec{v}_{cs}$. However, some designs are preferable due to lower cross-coupling. Overall, the magnitude of $\vec{v}_{cs}$ is entirely determined by the machine design. By carefully designing the machine, $\vec{v}_{cs}$ can be minimized.

While all machine designs will exhibit some electrical cross-coupling $\vec{v}_{cs}$ due to force generation, the overall impact of the cross-coupling is determined by the controls implementation. If no real-time communication between the the suspension and torque current regulators is possible, the overall response of the $\vec{v}_{cs}$ cross-coupling is determined by the torque current regulator's disturbance rejection capabilities. However, when real-time communication is possible (e.g., both current regulators are implemented in the same processor), direct electrical decoupling is possible to entirely eliminate the $\vec{v}_{cs}$ effects.

2 Machine Design

The voltage cross-coupling term $\vec{v}_{cs}$ is proportional to the product of the suspension currents $\vec{i}_s$ and the winding impedance $\bar{z}_s$. The required suspension current $\vec{i}_s$ is determined by the force requirement through the force proportionality constant $\bar{k}_i$ since $\vec{F} = \bar{k}_i\vec{i}_s$. Therefore, by understanding how the machine design impacts both $\bar{z}_s$ and $\bar{k}_i$, insight can be given towards minimizing $\vec{v}_{cs}$. Reference [1] provides a detailed derivation of normalized versions of both the winding impedance z_{sn} and force constant k_{in}. Normalization allows two candidate machine designs to be compared directly in terms of suspension/torque pole-pairs and winding factors. To minimize the voltage cross-coupling $\vec{v}_{cs}$, the winding impedance z_{sn} should be minimized while the force constant k_{sn} should be maximized. Unfortunately, it is shown that the two performance metrics are in conflict: $z_{sn} \propto k_{in}^2$.

To further study the performance metrics z_{sn} and k_{in}, 17 unique bearingless motors are designed and analyzed. Each machine is designed with a different number of slots, number of slots/pole/phase, and coil span. The rated specification for all machines is: 50 kW power, 30 kRPM speed, 480 Vrms line-to-line back-EMF, 0.7 T magnetic loading, and 75 kArms/m electric loading. Figure 2a shows the performance metrics of the designs, relative to Design 1. Notice that the magnitude of the voltage cross-coupling $|\vec{v}_{cs}|_n$ varies widely between designs. For details about each design variant, refer to [1].

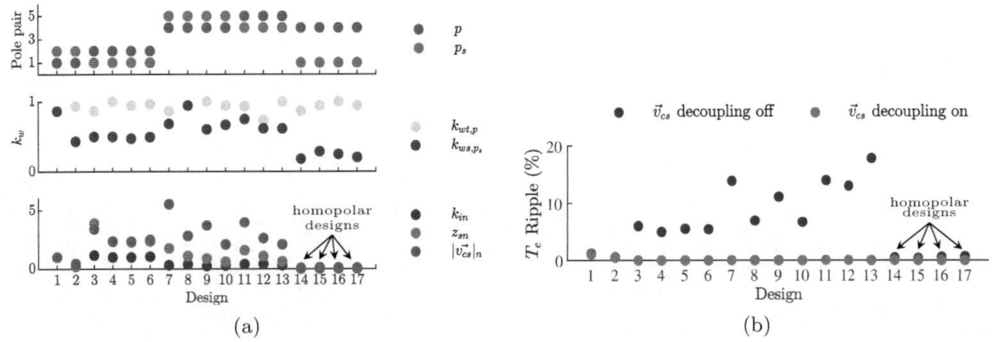

Figure 2: Various bearingless motor design performance metrics: (a) metrics based on machine design (b) resulting torque ripple due to current regulation capabilities.

3 Current Regulation

The machine design directly determines the voltage cross-coupling between the force and torque systems via $\vec{v}_{cs}$. However, the true impact of $\vec{v}_{cs}$ is entirely determined by the controls implementation. Precision current regulation is required to achieve ideal decoupling between force and torque creation. Even if real-time decoupling is not possible due to independence between motoring and suspension operation, the force-induced torque ripple can still be minimized by improving the disturbance rejection capabilities of the torque current regulator.

Figure 2b shows the simulated torque ripple of the 17 designs from Figure 2a when operating at rated speed, rated torque, and supporting the rotor weight. Standard synchronous frame PI current regulation is implemented in the discrete-time controller (referred to as SFPI). When direct electrical decoupling is used, the force-induced torque ripple vanishes. However, without decoupling, the torque ripple rises to nearly 20% for some designs.

Bearingless motors have the greatest advantage over conventional motors when operating at ultra high speeds. For precise actuation of both force and torque at these high speeds, direct digital current regulation must be deployed. Reference [1] derives the direct digital complex vector current regulator (referred to as CVCR). Theoretically, the CVCR completely decouples the rotational speed from the current regulation capability, thus enabling ultra high speeds. Figure 3 shows simulation results which compares SFPI vs. CVCR for high speed operation. Notice the deviations in the suspension currents $i_{\mathrm{s,xy}}$ at increasing speeds.

4 Experimental Results

The proposed precision current regulation techniques are implemented on an experimental bench for validation. Two prototype machines are used: 1) RL load configured as a parallel DPNV winding, 2) bearingless ac homopolar motor. First, the RL load is used. Figure 4 shows experimental results of the CVCR controlling currents at 1 kHz fundamental frequency with 1 kHz bandwidth and the controller digital sample rate of 10 kHz. With $\vec{v}_{cs}$ decoupling, the torque currents $i_{\mathrm{t,dq}}$ are not affected by the step changes in the suspension currents $i_{\mathrm{s,xy}}$. However, without $\vec{v}_{cs}$ decoupling, excessive torque current ripple occurs.

Finally, the proposed current regulation techniques are implemented to levitate and rotate the ac homopolar bearingless motor. Figure 5 shows the torque and suspension currents in their respective synchronous reference frame. Notice that, with no $\vec{v}_{cs}$ decoupling, the torque currents $i_{\mathrm{t,dq}}$ show noticeable ripple which vanishes when decoupling is enabled.

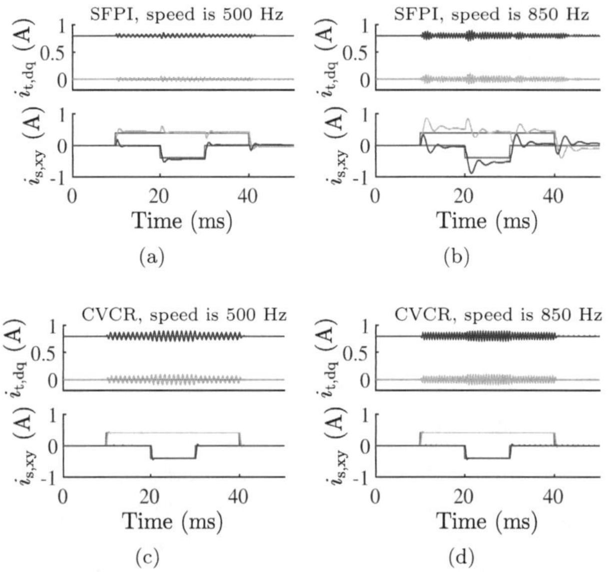

Figure 3: Simulated current regulation (torque $i_{\mathrm{t,dq}}$ and suspension $i_{\mathrm{s,xy}}$) comparing the Synchronous Frame PI (SFPI) controller with the Complex Vector Current Regulator (CVCR).

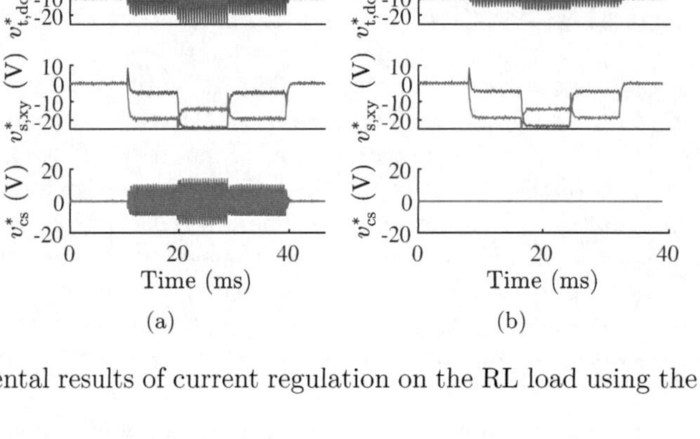

Figure 4: Experimental results of current regulation on the RL load using the proposed CVCR.

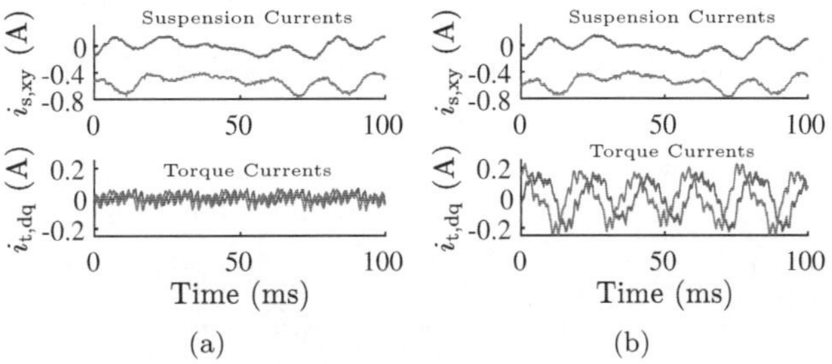

Figure 5: AC homopolar motor operating with CVCR and $\vec{v}_{cs}$ decoupling: (a) on (b) off.

5 Conclusion

This paper summarizes the contributions of Reference [1]. The parallel dual-purpose no-voltage (DPNV) combined winding bearing motor topology is investigated in detail. The electrical cross-coupling between the force and torque systems is the key topic; the paper proposes solutions to reduce or eliminate the cross-coupling. It is shown that careful machine design can result in reduced cross-coupling, however, the controls implementation ultimately determines the overall impact of the force-induced cross-coupling. The motoring current regulator's disturbance rejection properties are directly responsible for reducing the force-induced torque ripple. If real-time communication between the suspension and torque current controllers is possible, direct electrical decoupling is possible. Overall, the paper highlights the need for codesign of both the machine and controls to achieve the best bearingless motor solution.

References

[1] Nathan Petersen, Anvar Khamitov, Timothy Slininger, and Eric Loren Severson. Machine design and precision current regulation for the parallel dpnv bearingless motor winding. *IEEE Transactions on Industry Applications*, pages 1–1, 2021.

A Comparative Study of an Optimal Control Scheme in Different Magnetic Bearings' Geometries

Yago Pessanha Corrêa[1], Laura Julia Martins Mothé[2], Vinícius Ramos Vasco[3], Domingos de Farias Brito David[4], and Afonso Celso Del Nero Gomes[5]

[1] Instituto Federal Fluminense (IFFluminense)
yago.correa@iff.edu.br
[2] Universidade Federal do Rio de Janeiro (UFRJ)
lauramothe@gmail.com
[3] Marinha do Brasil (*Brazilian Navy*)
viniciusramosvasco@gmail.com
[4] Universidade Federal Fluminense (UFF)
domingos@vm.uff.br
[5] Universidade Federal do Rio de Janeiro (UFRJ)
nero@coep.ufrj.br

Abstract

Active Magnetic Bearings (AMB) are devices that support a shaft, showing the advantage of having no contact with any rotating parts, thus eliminating wear and the need of lubrication. Intrinsically, these equipments are unstable and, therefore, a control strategy is necessary in order to stabilize the system. One of the most used state-space feedback method is the Linear Quadratic Regulator (LQR). This paper aims to simulate and compare the eight, four and three-pole AMB with the centralized, decentralized and two-parameter decentralized LQR control strategies. After deducting the reluctance force equations and the system dynamics equations for each of the geometries, several simulations were performed. The results were conclusive and indicated the preference to use the two-parameter decentralized controller, due to the shorter stabilization time and the control variables decoupling, among other factors.

1 Background

Bearings are fixed devices used as a support for rotating axes. Besides that, they allow the relative movement between two surfaces. In a mechanical bearing, due to the contact between the shaft surface and the bearing surface, frictional forces show up. This is a considerable problem as it wears out the machine elements. A possible solution to mitigate this issue is the use of lubricating oils and bushings [1]. In some applications, such as pharmaceutical and food industries, employing general-purpose lubricants is not allowed due to possible contamination [2].

Active Magnetic Bearings (AMB), commonly known as Magnetic Bearings, come as a reasonable solution to overcome the problems presented above. The component's overheating caused by friction is eliminated in this scenario since there is no direct contact between the rotating parts. This type of non-contact bearing is based on the generation of restorative forces caused by electromagnets placed in strategic points. An active control scheme is necessary to stabilize the system considering its inherently unstable feature [3].

Currently, three widespread Magnetic Bearings configurations are addressed in the literature: eight-pole, four-pole and three-pole geometries [4]. In some specific applications, six-pole and twelve-pole bearings can also be used. The differences refer to heat dissipation, energy consumption, manufacturing cost and symmetry [5]. Each of them presents a different modelling as a result of its own reluctance forces and (un)coupled magnetic fluxes, as will be shown later in this paper.

This work aims to present different comparisons between the eight-pole, four-pole and three-pole geometries of AMBs. In order to carry out the comparative study, three optimal control schemes (all of them based on the Linear Quadratic Regulator approach) will be used – centralized, decentralized and two-parameter decentralized. Simulations are done, and compared in each of the proposed geometries, to visualize if the rotor is kept in the central position when the active control, based on state space analysis, is used.

2 Discussion

The first step in order to compare the geometries is to find the reluctance forces in each possible configuration – i.e., it is necessary to model the system. Consider an 8-pole Magnetic Bearing shown in Figure 1. This configuration is the most used in industrial applications [5].

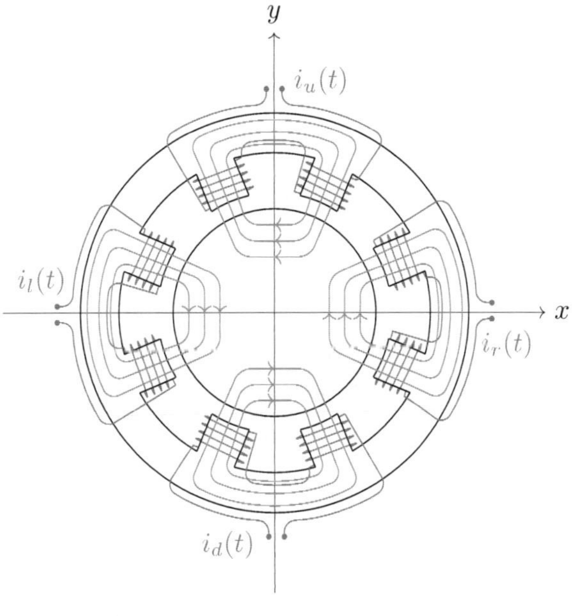

Figure 1: 8-pole Magnetic Bearing

As one can see in Figure 1, this configuration has uncoupled magnetic fluxes (represented in green). The following development is based on references such as [6, 2]. A position control scheme can be implemented independently – one for the x direction and one for the y – due to electromagnets' symmetry.

Reluctance forces equations rely on two control variables – $i_l(t)$ and $i_r(t)$, for the x direction; $i_u(t)$ and $i_d(t)$, for the y. This situation can be problematic in a control point of view. Differential currents approach may be applied so that $i_x(t)$, or $i_y(t)$, will be the only control variables [7].

The resulting equations are nonlinear, but considering that the bearing operates near the operating point ($i_x(t) \approx x(t) \approx 0$) under normal conditions, Taylor series can be applied, which allows a linearization of the proposed model for the x direction. Likewise, the same procedure is done for the y.

The linearized reluctance forces for the 8-pole AMB are presented in (1):

$$F_x(t) = g_p x(t) + g_i i_x(t) \qquad \text{and} \qquad F_y(t) = g_p y(t) + g_i i_y(t) \tag{1}$$

where

$$g_p = \frac{\mu_0 \, N_a^2 \, A_a \, i_b^2}{h^3} \qquad \text{and} \qquad g_i = \frac{\mu_0 \, N_a^2 \, A_a \, i_b}{h^2} \tag{2}$$

are the magnetic constants.

A more compact geometry, allowing a cheaper structure and more space for heat dissipation, called 4-pole Magnetic Bearing, is being used in many applications [8, 9]. Consider a 4-pole Magnetic Bearing shown in Figure 2. The following concepts are based on [6].

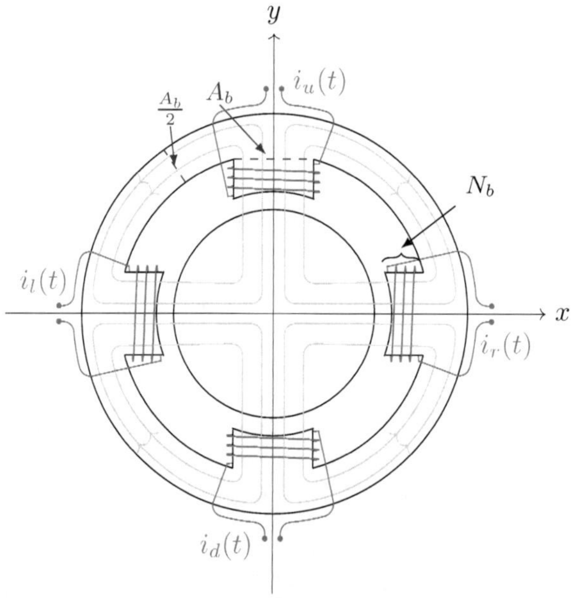

Figure 2: 4-pole Magnetic Bearing

The main difference between the 8-pole and 4-pole geometries is that the latter has an interconnection of magnetic fluxes [10]. It is important to mention that the symmetry detected in an 8-pole Magnetic Bearing is kept here. The same mathematical procedure to obtain the reluctance forces may be developed, but now taking into account the fluxes interconnection.

One may notice in Figure 3 the interconnected magnetic flux distribution caused by current $i_r(t)$. The notation ϕ_{mn} refers to the magnetic flux that crosses the m pole due to the current that circulates at the n pole. The reluctance forces in this geometry rely on the total magnetic flux ϕ_m (for each pole m) which is a function of the partial fluxes ϕ_{mn}, as can be seen in (3):

$$\begin{aligned} \phi_r &= \phi_{rr} + \phi_{ru} - \phi_{rl} + \phi_{rd} \qquad & \phi_u &= -\phi_{ur} - \phi_{uu} - \phi_{ul} + \phi_{ud} \\ \phi_l &= -\phi_{lr} + \phi_{lu} + \phi_{ll} + \phi_{ld} \qquad & \phi_d &= -\phi_{dr} + \phi_{du} - \phi_{dl} - \phi_{dd} \end{aligned} \tag{3}$$

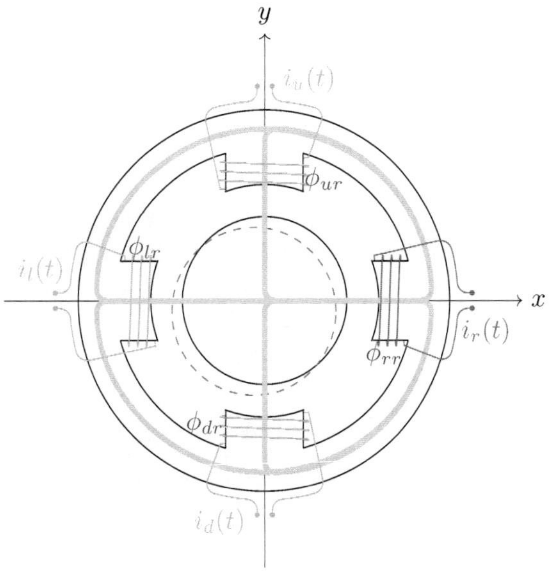

Figure 3: 4-pole interconnected magnetic flux distribution associated only with the current $i_r(t)$

A step by step mathematical development can be seen in [6]. The linearized reluctance forces for the 4-pole AMB are presented in (4):

$$F_x(t) = g_p x(t) + g_i i_x(t) \qquad \text{and} \qquad F_y(t) = g_p y(t) + g_i i_y(t) \tag{4}$$

where

$$g_p = \frac{2\,\mu_0\,N_b^2\,A_b\,i_b^2}{h^3} \qquad \text{and} \qquad g_i = \frac{2\,\mu_0\,N_b^2\,A_b\,i_b}{h^2} \tag{5}$$

are the magnetic constants. These values have an extra factor of 2 when compared to the first topology, shown in (2); this means that the magnetic constants of a 4-pole AMB are, at least, 2 times higher than in the 8-pole case.

Although this geometry uses interconnected fluxes, the reluctance forces are decoupled from each other, as demonstrated in (4).

An even more compact structure, the 3-pole AMB, has been studied over the past few years [2, 5]. Consider a 3-pole AMB in an optimal configuration shown in Figure 4. In this design, we shall consider a horizontal rotor being supported by the device. This scheme has the minimum number of poles able to balance the rotor, allowing a better heat dissipation and lower iron losses [11, 5].

In this asymmetric configuration (due to the odd number of poles), it is possible to use one less actuator as two poles share the same coil wiring [11, 12]. The modelling process is very similar compared to the other geometries but interesting since an odd number of poles causes a disbalancing in the reluctance forces. One may notice in Figure 5 the interconnected magnetic flux distribution caused by current $i_1(t)$. The total magnetic flux ϕ_m (for each pole m) is a function of the partial fluxes ϕ_{mn}, as can be seen in (6):

$$\phi_1 = \phi_{11} + \phi_{12} + \phi_{13} \qquad \begin{aligned} \phi_2 &= -\phi_{21} - \phi_{22} + \phi_{23} \\ \phi_3 &= -\phi_{31} + \phi_{32} - \phi_{33} \end{aligned} \tag{6}$$

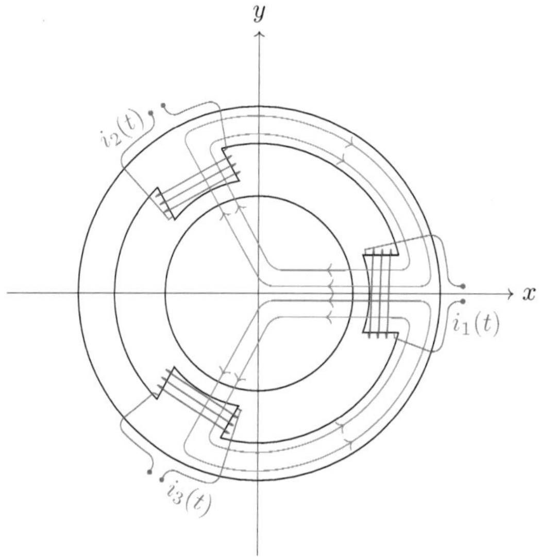

Figure 4: Optimal configuration of a 3-pole Magnetic Bearing

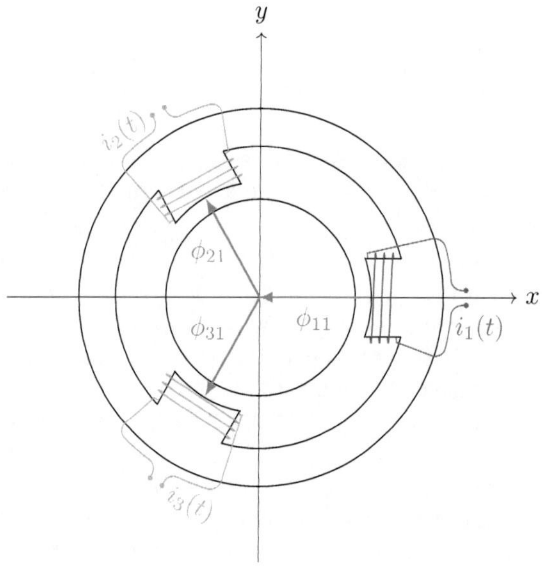

Figure 5: 3-pole interconnected magnetic flux distribution associated only with the current $i_1(t)$

A step by step mathematical development can be seen in [13, 11, 2]. The linearized reluctance forces for the 3-pole AMB are presented in (7):

$$F_x(t) = \frac{1}{2} g_p x(t) + \frac{\sqrt{3}}{3} g_i i_{d3} \qquad \text{and} \qquad F_y(t) = \frac{1}{2} g_p y(t) + g_i i_{d1} \tag{7}$$

where

$$g_p = \frac{\mu_0 \, N_c^2 \, A_c \, i_{B1}^2}{h^3} \qquad \text{and} \qquad g_i = \frac{\mu_0 \, N_c^2 \, A_c \, i_{B1}}{h^2} \tag{8}$$

are the magnetic constants. It is not straightforward to compare the 3-pole AMB reluctance forces to the other results, since the differential currents are not restricted to just one axis. However, keeping the currents in the same magnitude order, the reluctance forces in the 3-pole AMB are lower than those of 8-pole and 4-pole AMBs.

State space models are an important tool, considering a later application of a linear control technique. The dynamic analysis considers a rigid and homogeneous shaft. Further details are presented in [1, 2, 10].

The first two geometries (8-pole and 4-pole AMBs) share the same model pattern. State space representation is shown in (9):

$$\dot{\boldsymbol{x}}(t) = A\boldsymbol{x}(t) + B\boldsymbol{u}(t) + D\boldsymbol{v}(t) \tag{9}$$

where the matrices A, B and D have structures defined in (10):

$$A = \begin{bmatrix} 0 & I \\ A_{21} & A_{22} \end{bmatrix} \qquad B = \begin{bmatrix} 0 \\ B_2 \end{bmatrix} \qquad D = \begin{bmatrix} 0 \\ D_2 \end{bmatrix} \tag{10}$$

where the blocks A_{21}, A_{22}, B_2 and D_2 are $2{\times}2$ matrices that depend on the constructive parameters of the existing prototypes [1, 2, 10]; block A_{22} depends on the rotor speed. The prototypes are equipped with sensors that measure the rotor position in two orthogonal directions: x_s and y_s. These signals, together with the corresponding velocities, are the state vector components, as shown in (11):

$$\boldsymbol{x} = \begin{bmatrix} x_s \\ y_s \\ \dot{x}_s \\ \dot{y}_s \end{bmatrix} \qquad \boldsymbol{u} = \begin{bmatrix} i_x \\ i_y \end{bmatrix} \qquad \boldsymbol{v} = \begin{bmatrix} \cos(\ t) \\ -\sin(\ t) \end{bmatrix} \tag{11}$$

The control vector $\boldsymbol{u}$ components are the differential currents injected in the AMB coils. An extra (small) mass can be attached to the rotor of the examined prototypes; this allows the study of unbalanced masses in the rotational dynamics. The net effect of this lack of symmetry is the appearance of the harmonic disturbances in vector $\boldsymbol{v}$ that enter the model through D.

A different procedure must be done for the 3-pole AMB as the reluctance forces are different and the torque from gravitational force should be taken into account since the rotor is in the horizontal position. The standard state space representation shown in (9) and matrices presented in (10) are valid here as well.

The state space vector $\boldsymbol{x}$, control inputs $\boldsymbol{u}$ and disturbance vector $\boldsymbol{v}$ are chosen to be the ones shown in (12):

$$\boldsymbol{x} = \begin{bmatrix} x_s \\ y_s \\ \dot{x}_s \\ \dot{y}_s \end{bmatrix} \qquad \boldsymbol{u} = \begin{bmatrix} \frac{\sqrt{3}}{3} i_{d3} \\ i_{d1} \end{bmatrix} \qquad \boldsymbol{v} = \begin{bmatrix} \cos(\ t) \\ -\sin(\ t) \end{bmatrix} \tag{12}$$

As we are working with an approximate linear model, these equations are accurate only around the operation point. Besides that, one can represent this as a linear and time-invariant system if and only if a constant value is fixed for .

Some control strategies can be used in an AMB system, such as PID [14], pole placement technique [15] and $\mathcal{H}_\infty$ regulator [16]. This paper addresses the use of an optimal control strategy based on Linear Quadratic Regulator (LQR) theory [1, 17, 18].

Consider a block diagram, in Figure 6, that illustrates a full state feedback. The controller's goal is to drive the closed-loop poles to a desired position, stabilizing the rotor.

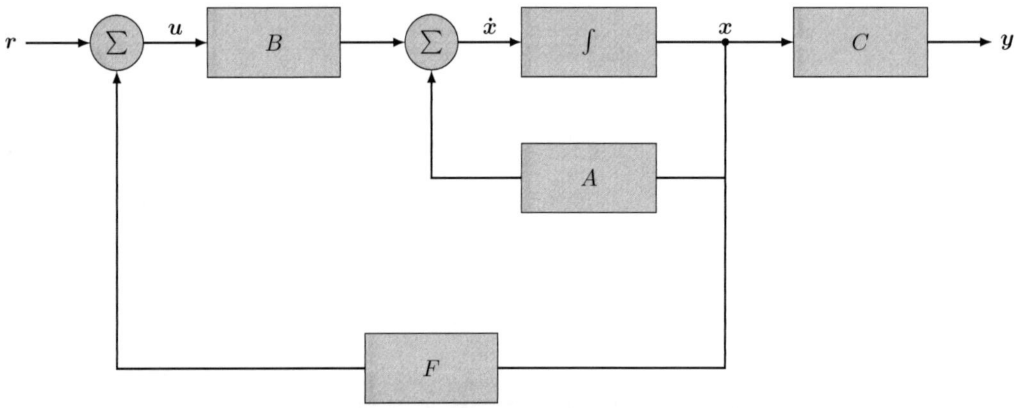

Figure 6: Full state feedback block diagram

LQR control has an advantage, compared to pole placement technique, of providing a systemic and automatic procedure to calculate the feedback gain matrix [17]. LQR perfomance index J is defined by two quadratic functions and can be expressed by (13):

$$J = \int_0^\infty (\boldsymbol{x}^T(t)Q\boldsymbol{x}(t) + \boldsymbol{u}^T(t)R\boldsymbol{u}(t))\ \mathrm{d}t \tag{13}$$

where the weighting matrix Q (state error) is symmetric positive semidefinite and the weighting matrix R (energy consumption) is symmetric positive definite.

Considering a controllable system and that all states are observable and available (here x and y are measured and $\dot{x}$ and $\dot{y}$ are estimated), we can write (14):

$$\boldsymbol{u}^* = F^*\boldsymbol{x} \tag{14}$$

where $\boldsymbol{u}^*$ is the optimal control law that minimizes J, through the feedback gain matrix F^*, given by (15):

$$F^* = -R^{-1}B^TP \tag{15}$$

in which P is the solution of an algebraic matrix Riccati equation, as can be seen in (16):

$$A^T P + PA - PBR^{-1}B^T P + Q = 0 \tag{16}$$

In this work, three LQR control techniques are designed and simulated. The first one, named here as centralized LQR (LQRc), refers to the "traditional" LQR, where all states are fedback and interact with each other, which can be seen in (17):

$$\boldsymbol{u} = F_c^* \boldsymbol{x} \implies \begin{bmatrix} u_1 \\ u_2 \end{bmatrix} = \begin{bmatrix} f_{11} & f_{12} & f_{13} & f_{14} \\ f_{21} & f_{22} & f_{23} & f_{24} \end{bmatrix} \boldsymbol{x} \tag{17}$$

In most cases, all F_c^* (centralized) matrix elements are nonzero, since there is a well-defined connection between every single state. This leads to a greater computational effort.

An alternative and interesting solution would be the presence of some zero elements in the feedback gain matrix, as this would alleviate the control implementation. This is possible using the decentralized Linear Quadratic Regulator (LQRd). In a gain matrix F_d^* (decentralized), each one of the two input variables is decoupled and independent of the other. That is to say, each input is related only to an output and its derivative, as shown in (18):

$$u_1 = f_{11}x_s + f_{13}\dot{x}_s \qquad \text{and} \qquad u_2 = f_{22}y_s + f_{24}\dot{y}_s \tag{18}$$

Further details are presented in [7, 19]. The decentralized structure can be expressed by (19):

$$\boldsymbol{u} = F_d^* \boldsymbol{x} \implies \begin{bmatrix} u_1 \\ u_2 \end{bmatrix} = \begin{bmatrix} f_{11} & 0 & f_{13} & 0 \\ 0 & f_{22} & 0 & f_{24} \end{bmatrix} \boldsymbol{x} \tag{19}$$

The necessary conditions for F to satisfy (19) and minimize J in (13) are in [19, 20]; they lead to an algorithm that will be used in the sequel.

The work developed in [19] comes up with an even more particular structure. As it relies on two parameters, this control law is called two-parameter decentralized Linear Quadratic Regulator ($\text{LQRPD}_{2\text{p}}$). Similar to the previous case, input variables are decoupled, taking into account only their position and speed on the interest direction. The difference here is that an additional restriction is imposed on the control law, making it simpler. The $\text{LQRPD}_{2\text{p}}$ structure is as follows:

$$\boldsymbol{u} = F_{2p}^* \boldsymbol{x} \implies \begin{bmatrix} u_1 \\ u_2 \end{bmatrix} = \begin{bmatrix} p & 0 & d & 0 \\ 0 & p & 0 & d \end{bmatrix} \boldsymbol{x} \tag{20}$$

Only two parameters appear in the feedback matrix: p for the x_s and y_s and d for the derivatives. All the existence conditions for this problem, as well as an algorithm leading to the solution can be found in [19]. These results will be used in the next section, but will be not presented here.

3 Results

The magnetic constants g_p and g_i are determined using the prototype' geometric parameters [6, 10], in addition to the base current i_b (8-pole and 4-pole) or i_{B1} (3-pole).

Considering $i_b = i_{B1} = 0.52$ A [2], the first simulation set is performed, comparing all three geometries in different control schemes. The operating speed used is 21 rad/s (approximately 200 rpm), 356 rad/s (about 3400 rpm) and 942 rad/s (roughly 9000 rpm). Figure 7 shows the centralized LQR response for each speed and geometry.

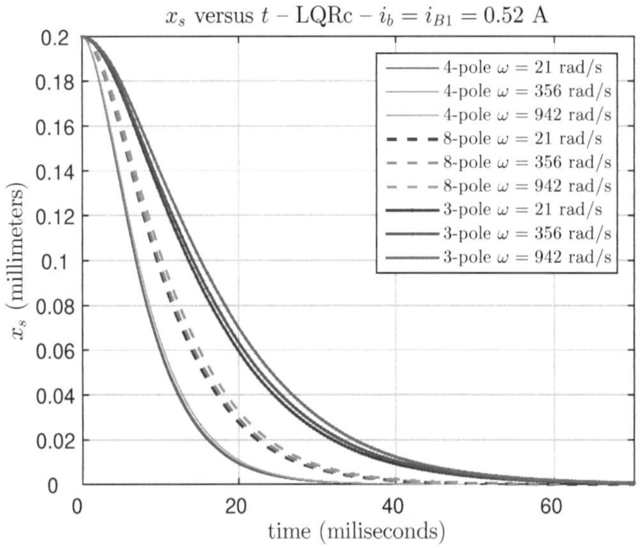

Figure 7: Centralized LQR response

As can be seen, the 4-pole AMB is the fastest to return to equilibrium. In second place, comes out the 8-pole Magnetic Bearing, and finally the 3-pole. However, when the rotor operating speed is increased from 200 rpm to 3400 rpm (and after from 3400 rpm to 9000 rpm), we have a slightly longer settling time – this is noticed in all geometries. Four-pole geometry settling time variations (after increasing the operating speed) are much smaller than the others, as can be seen in Figure 7 and in Table 1.

Table 1: Comparison between LQRc, LQRd and LQRPD$_{2p}$: settling time – $i_b = i_{B1} = 0\ 52$ A

Settling time (in ms) – $i_b = i_{B1} = 0.52$ A			
	$\omega = 21$ **rad/s**	$\omega = 356$ **rad/s**	$\omega = 942$ **rad/s**
LQRc	8-pole: 42.9 4-pole: 30.8 3-pole: 61.7	8-pole: 43.9 4-pole: 31.3 3-pole: 63.6	8-pole: 45.2 4-pole: 32.0 3-pole: 65.3
LQRd	8-pole: 43.0 4-pole: 30.9 3-pole: 61.8	8-pole: 44.1 4-pole: 31.6 3-pole: 64.6	8-pole: 44.7 4-pole: 32.9 3-pole: 70.3
LQRPD$_{2p}$	8-pole: 42.7 4-pole: 30.6 3-pole: 61.3	8-pole: 40.0 4-pole: 29.4 3-pole: 56.0	8-pole: 36.2 4-pole: 27.3 3-pole: 50.7

For the decentralized LQR control strategy, a similar simulation is done, as can be noticed in Figure 8.

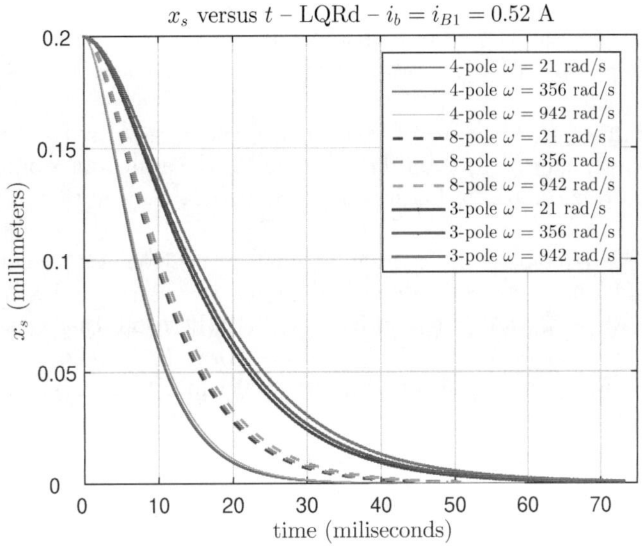

Figure 8: Decentralized LQR response

When one compares LQRc to LQRd response, the latter presents a small decrease in the eigenvalues' real part (the only exception is the 8-pole AMB with = 942 rad/s) – therefore, a minimal settling time increase is expected in this control strategy, as exposed in Table 1.

The last control method to be simulated is the two-parameter decentralized LQR ($\text{LQRPD}_{2\text{p}}$). Figure 9 exhibits its response.

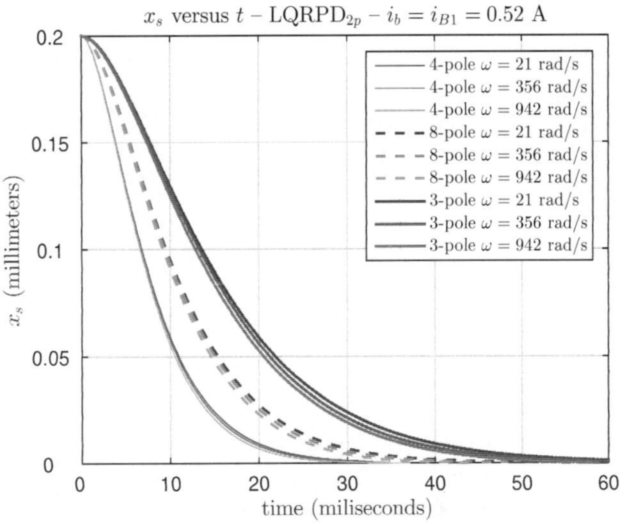

Figure 9: Two-parameter decentralized LQR response

Similarly to LQRc and LQRd strategies, the 4-pole AMB is the one that returns faster to the center position (regardless of), followed by the 8-pole, and finally the 3-pole geometry. In this strategy, one can see, in Table 1, a shorter settling time, for all geometries and . In fact, LQRPD_{2p} control strategy provides an optimal gain matrix, and the closed-loop system response is fast enough. Notice that as the rotor speed increases, the centralized and decentralized strategies present a longer settling time, whereas in the LQRPD_{2p} case this time decreases.

In order to compare the prototype's performance against disturbances, a 1 gram unbalance mass is placed on the rotor, causing an orbital motion due to the appearance of harmonic forces. Hence, x_s would not go to zero, unlike Figure 7 (where disturbances were not considered), for example.

Each control strategy is compared, in order to analyze a given AMB geometry. For the centralized LQR, considering a low rotation speed, Figure 10a shows x_s by y_s curves for the eight, four and three-pole geometries. It can be seen that the orbital motions for the 4-pole geometry are smaller than the 8-pole, which are smaller than the 3-pole. The lower are the orbital movements, the more rigid is the suspension, featuring a better performance against disturbances.

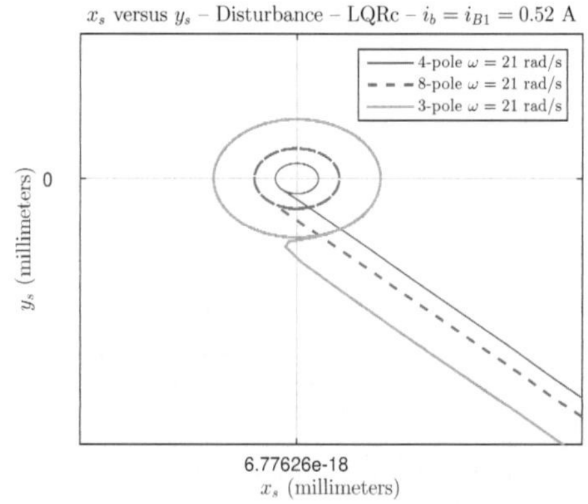

(a) Centralized LQR

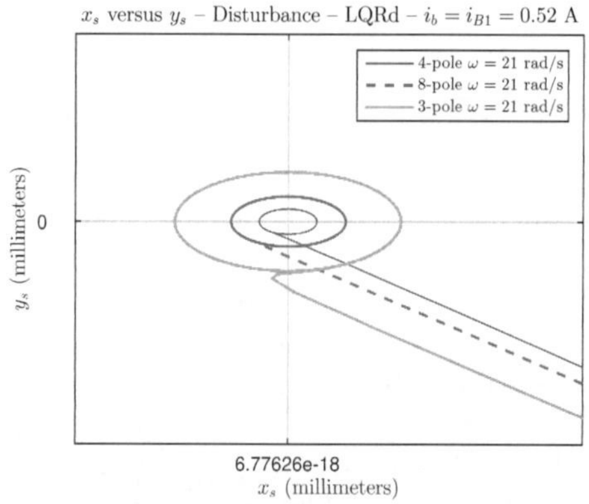

(b) Decentralized LQR

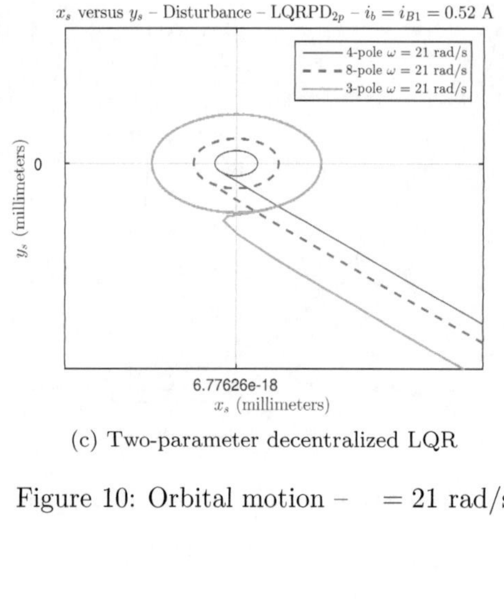

(c) Two-parameter decentralized LQR

Figure 10: Orbital motion – = 21 rad/s

For the decentralized LQR (Figure 10b) and the two-parameter decentralized LQR (Figure 10c) cases, it is clear that the response curves, in the three control strategies, are nearly the same, given the low rotation speed.

With the increase of the rotation speed to 356 rad/s, it is apparent the higher influence of harmonic forces on the x_s and y_s response curves. This happens because the torque caused by the unbalance mass depends on the square of the rotor speed. Nevertheless, the 4-pole AMB is the one with the best performance, as can be seen in Figure 11a.

The same comments made, when considering the speed of 21 rad/s, are valid here. When looking at Figure 11b, the decentralized control strategy does not show any difference in the orbital motion, when compared to the centralized control. In a like manner, the two-parameter decentralized control (Figure 11c) displays a very similar response due to disturbances.

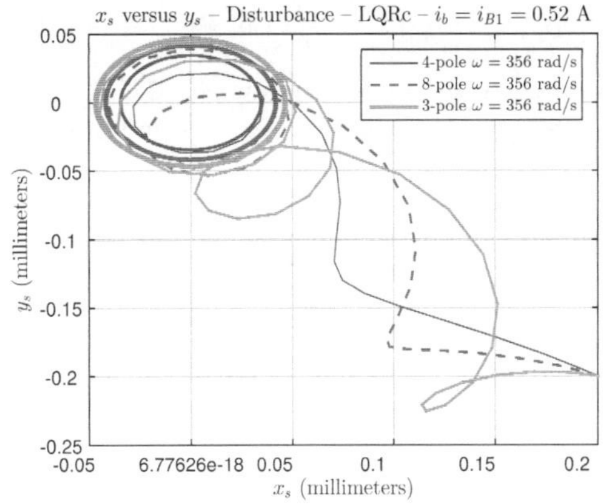

(a) Centralized LQR

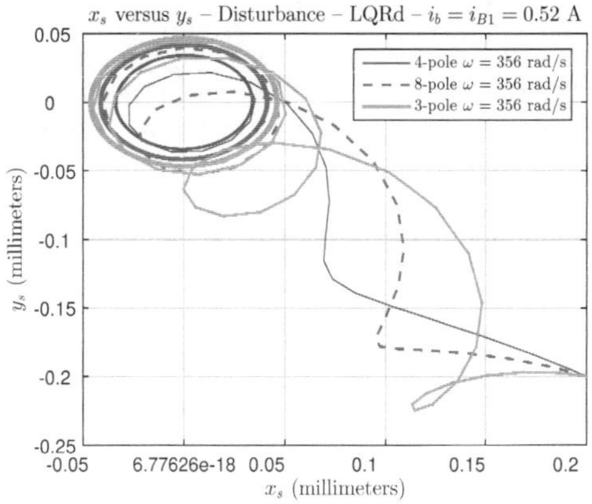

(b) Decentralized LQR

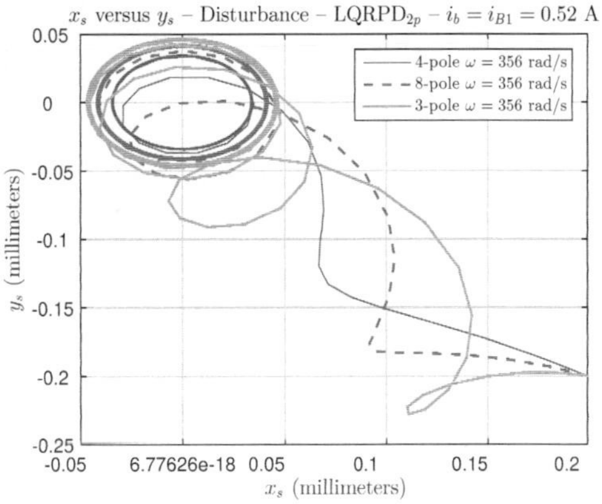

(c) Two-parameter decentralized LQR

Figure 11: Orbital motion – = 356 rad/s

Another important fact concerns the number of iterations necessary to stabilize the system: the two-parameter decentralized strategy proved to be equally or more effective in almost all simulations and geometries (7 out of 9 cases). Table 2 shows this fact.

Table 2: Comparison between LQRd and $LQRPD_{2p}$: number of iterations – $i_b = i_{B1} = 0\ 52$ A

Number of iterations – $i_b = i_{B1} = 0.52$ A			
	$\omega = 21$ **rad/s**	$\omega = 356$ **rad/s**	$\omega = 942$ **rad/s**
LQRd	8-pole: 2 4-pole: 8 3-pole: 2	8-pole: 6 4-pole: 11 3-pole: 8	8-pole: 15 4-pole: 15 3-pole: 28
$LQRPD_{2p}$	8-pole: 3 4-pole: 2 3-pole: 3	8-pole: 5 4-pole: 4 3-pole: 5	8-pole: 8 4-pole: 6 3-pole: 10

With what has been shown so far in this paper, it seems that the two-parameter decentralized control strategy is the best of all simulated. In order to analyze the control signal of the three strategies, another simulation is performed, using as reference the 4-pole geometry and a rotor speed of 356 rad/s. This choice is justified by the fact that the 4-pole AMB is the one with the shortest settling time and the most rigid suspension among all studied. Thus, it is interesting to analyze how the control signal varies over time for this geometry. Moreover, an average operating speed (= 356 rad/s) is adopted, since, as presented in the simulations, changing   does not significantly change the settling time.

As the $LQRPD_{2p}$ was the one that presented the shortest settling time of all strategies, it is interesting to note whether this good transient response requires a greater control effort when compared to other LQR techniques, as shown in Figure 12.

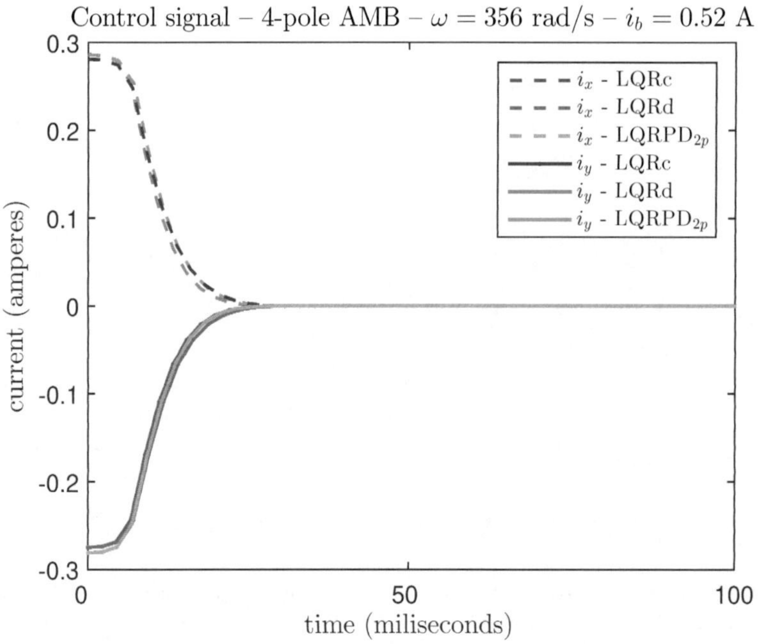

Figure 12: 4-pole Magnetic Bearing control signal

Immediately, one notices that the electric currents i_x and i_y (that is, the control signal) necessary to stabilize the 4-pole AMB are approximately equal to 0.28 A. This represents about 54% of the base current used (0.52 A). Another fact to be highlighted is that the three curves (for each axis) are basically superimposed. Although the two-parameter decentralized strategy has the shortest settling time, no additional control effort is required.

4 Conclusions

We have shown that the 4-pole Magnetic Bearing was the fastest (or equally fast) geometry, followed by the eight, and finally the three-pole AMB. The two-parameter decentralized Linear Quadratic Regulator ($LQRPD_{2p}$) is a viable alternative, as it presents a structure with only two different parameters, with total decoupling between the variables. Besides that, two-parameter decentralized LQR presented the shortest settling time among all strategies and the lowest number of iterations in 77.8% of the cases, with the same control effort compared to the other strategies. An increase in leads to a longer stabilization time (the exception is the two-parameter decentralized strategy, in which there is a reduction in this settling time). Furthermore, simulations with harmonic disturbances were made, and revealed that the change in the control strategy does not impact these results – in all simulations, the 4-pole AMB was the one with the best results. It should be mentioned that in these comparisons the 3-pole AMB had an horizontal rotor while the other geometries used a vertical one. Further research is needed for the case of all structures holding rotors in the same position, be it vertical and/or horizontal.

References

[1] R. M. Stephan, F. C. Pinto, A. C. D. Gomes, J. A. Santisteban, and A. O. Salazar, *Mancais Magnéticos - Mecatrônica sem atrito*. Editora Ciência Moderna Ltda., 1 edição ed., 2013.

[2] V. R. Vasco, "Modelagem e controle ótimo de um mancal magnético de três polos," Master's thesis, COPPE/UFRJ, 2019.

[3] W. Jones, "Earnshaw's theorem and the stability of matter," *Eur. J. Phys 1*, pp. 85–88, 1980.

[4] A. CHIBA, T. FUKAO, and e. ICHIKAWA, O., *Magnetic Bearings and Bearingless Drives*. Newnes-Elsevier, 2005.

[5] W. Zhang and H. Zhu, "Radial magnetic bearings: An overview," *Results in Physics*, 2017.

[6] P. H. S. Pinto, "Comparações teóricas e práticas sobre geometrias de mancais magnéticos," Master's thesis, COPPE/UFRJ, 2018.

[7] G. Schweitzer, H. Bleuler, and A. Traxler, *Active Magnetic Bearings*. vdf Hochschulverlag, 1994.

[8] J. M. Krodkiewski and R. B. Zmood, "Use of programmed magnetic bearing stifness and damping to minimize rotor vibrations," *ISMB*, 1992.

[9] K. Kjolhede and I. F. Santos, "Experimental contribution to high-precision characterization of magnetic forces in active magnetic bearings," *Gas Turbines: Structures and Dynamics*, 2006.

[10] D. F. B. David, J. S. Santisteban, and A. C. D. Gomes, "Modeling and testing strategies for an interconnected four-pole magnetic bearings," *Actuators*, 2017.

[11] S. Chen and C. Hsu, "Optimal design of a three-pole active magnetic bearing," *IEEE Transactions on Magnetics*, 2002.

[12] S. Chen, S. Chen, and S. Yan, "Experimental validation of a current-controlled three-pole magnetic rotor-bearing system," *IEEE Transactions on Magnetics*, 2005.

[13] D. F. B. David, J. S. Santisteban, A. C. D. Gomes, and V. R. Vasco, "Considerations on the mechanical dynamics of interconnected flux three-pole magnetic bearings," *ISMB*, 2018.

[14] R. R. Gomes, "Motor mancal com controle implementado em um dsp," Master's thesis, COPPE/UFRJ, 2007.

[15] G. Schweitzer and E. H. Maslen, *Magnetic Bearings.* Springer-Verlag Berlin Heidelberg, 2009.

[16] A. R. FRANCO, "Posicionamento de eixo com motor mancal magnético utilizando controle h-infinito," Master's thesis, COPPE/UFRJ, 2013.

[17] K. Ogata, *Engenharia de Controle Moderno.* Pearson, 2011.

[18] N. S. Nise, *Engenharia de Sistemas de Controle.* LTC, 2017.

[19] L. S. Rodrigues, "Controle ótimo descentralizado a dois parâmetros para mancais-motores magnéticos," Master's thesis, COPPE/UFRJ, 2005.

[20] H. Bleuler, *Decentralized control of magnetic rotor bearing systems.* Doctor of technical sciences dissertation, Swiss Federal Institute of Technology, Zurich, 1984.

Development of Magnetically Levitated Motor Driven at Extremely Low Temperature

Mochimitsu Komori[1], Hirohisa Kato[1], Ken-ichi Asami[1], Nobuo Sakai[1]
[1] Kyushu Institute of technolgy
komori_mk@yahoo.co.jp

Abstract

A cryogenic motor with active magnetic bearings (AMBs) which works in liquid nitrogen is discussed in this paper. The cryogenic motor is composed of two AMBs, four displacement sensors, a permanent magnet (PM) motor, etc. In order to check the stability for the rotor, FFT (fast Fourier transform) analysis for the levitated rotor in the radial direction is performed. Displacement of rotor in the axial direction is studied during rotation.

1 Introduction

A cryogenic motor with active magnetic bearings (AMBs) which works in liquid nitrogen (-196 degrees centigrade) is discussed in this paper. The cryogenic motor is composed of a rotor, 6-pole motor, AMBs, passive magnetic bearing (PMB), two displacement sensors, etc. Some experiments for the AMB are performed in liquid nitrogen. This paper discusses static and dynamic characteristics of the cryogenic motor.

The cryogenic motor working in liquid nitrogen is very interesting and special, because conditions in liquid nitrogen are very different from those at room temperature. Thus, there are few papers concerning cryogenic motors with AMBs working in liquid nitrogen (Kajikawa, Teion Kogaku, 2012), (Komori, IEEE Trans. on Applied Superconductivity, 2004), (Komori, Mechanical Engineering Journal, 2017).

2 Cryogenic Motor Structure

Figure . 1 shows a photo of the cryogenic motor with two radial AMBs working in liquid nitrogen (-196 degrees centigrade). The cryogenic motor is composed of a rotor, two AMBs, four displacement sensors, a permanent magnet (PM) motor, etc. The upper and lower black parts are the same AMBs. The PM motor is set in the center of the cryogenic motor. The stator of the AMB has eight electromagnets with 250 turns. The displacement sensors are inductance-type sensors based on

1-1 Sensui, Tobata, Fukuoka 804-8550, Japan

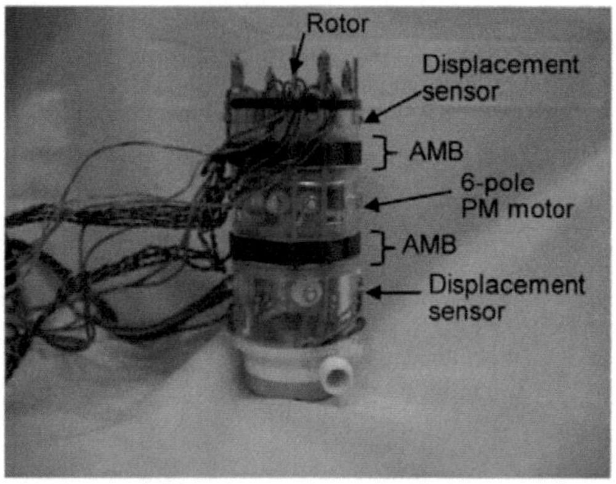

Figure 1: Photo of a cryogenic motor which works in liquid nitrogen.

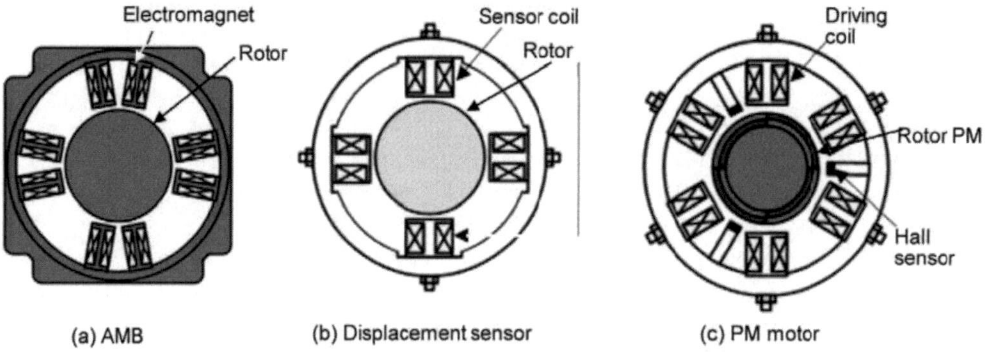

Figure 2: Cross sections of the cryogenic motor

detecting the inductance difference between two sensing coils. The detail of the inductance-type sensor is reported in Refs. (Hirose, IEEE Trans. on Magnetics, 2015), (Wang, Sensors and Actuators, 2010). A PMB composed of a couple of PMs facing each other is set at the bottom of the rotor. Thus, the rotor is levitated completely without any mechanical contacts.

Figure 2 shows the cross sections of the cryogenic motor showing (a) the cross section of AMBs, (b) the cross section of displacement sensors and (c) the cross section of PM motor. The AMB shown in Figure 2(a) has the same structure as other commercial magnetic bearings. The AMB has eight electromagnets with 250 turns. The cross section of displacement sensor is shown in Figure 2(b). A couple of sensor coils in the vertical direction are set for detecting the displacement in the vertical direction, and a couple of coils in the horizontal direction are set for detecting the displacement in the horizontal direction. The cross section of PM motor is shown in Figure 2(c). Six electromagnets for stator and four arc-shaped PM for rotor are set as shown in Figure 2(c). Hall sensors are used for detecting the rotor angle to drive the electromagnets.

Detecting the rotor displacements accurately is very important in the temperature range from room temperature to liquid nitrogen temperature. Thus, the relationship between sensor output and displacement are studied at room temperature and liquid nitrogen temperature. In the experiment, a laser displacement sensor is used to measure the correct displacement. The experimental result is shown in Figure 3 representing the relationship between sensor output voltage and displacement. In

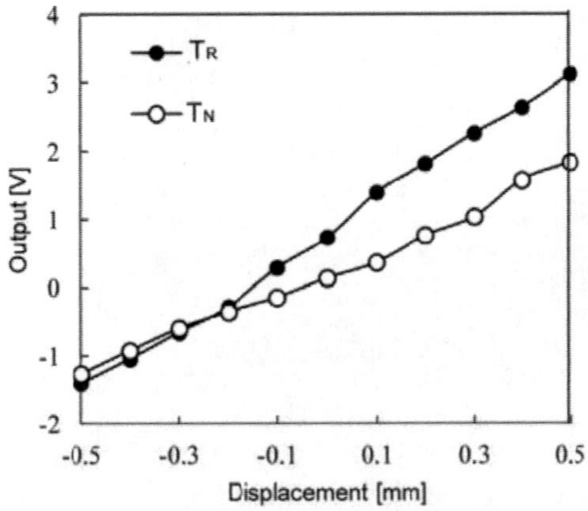

Figure 3: Relationship between output voltage from the displacement sensor and rotor displacement.

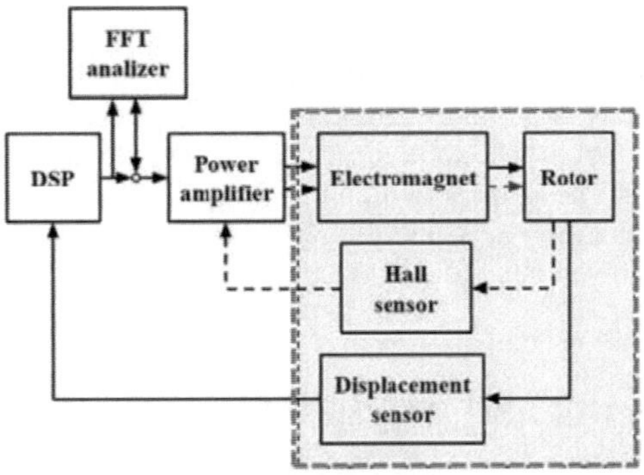

Figure 4: Block diagram of the cryogenic motor with power amplifier, DSP controller and FFT analyzer.

the Figure ure, TR and TN mean room temperature and liquid nitrogen temperature, respectively. From the result in Figure 3, both relationships at TR and TN are linear in the displacement range from -0.5 mm to 0.5 mm. However the difference between the output at TR and that at TN is observed. Thus, the displacement sensor in liquid nitrogen has to be adjusted to the same gain as that at room temperature.

Figure 4 shows the block diagram of cryogenic motor with power amplifier, DSP (Digital Signal Processor) controller and FFT (fast Fourier transform) analyzer. The cryogenic motor is shown inside the dotted line which is composed of electromagnets, rotor, Hall sensor and displacement sensor. The driving current is applied to the electromagnet using the power amplifier. The DSP applies the control signal to the power amplifier according to rotor displacement.

3 Characteristics

Figure 5 shows a photo of cryogenic motor in liquid nitrogen. The rotor is set in the center of the motor. Four sensor coils are set around the rotor. The motor is immersed in liquid nitrogen. That is,

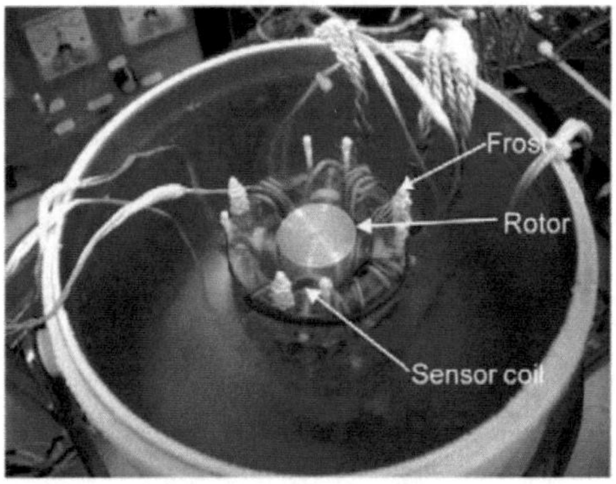

Figure 5: Photo of cryogenic motor in liquid nitrogen. The rotor is set in the center of the motor.

the level of liquid nitrogen is almost equal to the top of the cryogenic motor. Dynamic experiments are performed in the same condition as Figure 5.

Figure 6 shows the relationship between rotor displacement and rotation speed for the cryogenic motor working in liquid nitrogen. x1 and y1 represent the displacements in the x and y directions at the upper part of rotor, respectively. x2 and y2 represent the displacements in the x and y directions at the lower part of rotor, respectively. The rotor displacements are measured in the rotation speed range from 1,000 rpm to 4,500 rpm. The result shows that each displacement is smaller than 0.2 mm, which is very small. There is no vibration peak in the speed range from 1,000 rpm to 4,500 rpm, which means that the rotor spins stably in this speed range.

4 Some Experiments and evaluation

Figure 7 shows the experimental setup for studying dynamic performances of the cryogenic motor in liquid nitrogen. The experimental setup of cryogenic motor is the same as that shown in Figure 5. In the Figure, the cryogenic motor is connected to turbine blades at the bottom for pump performance.

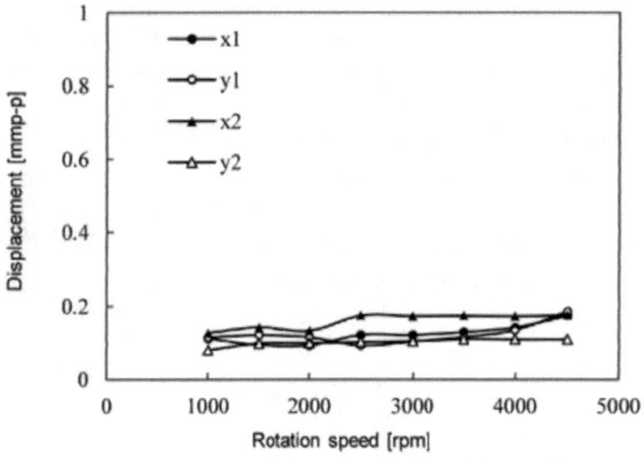

Figure 6: Relationship between rotor displacement and rotation speed for the cryogenic motor working in liquid nitrogen.

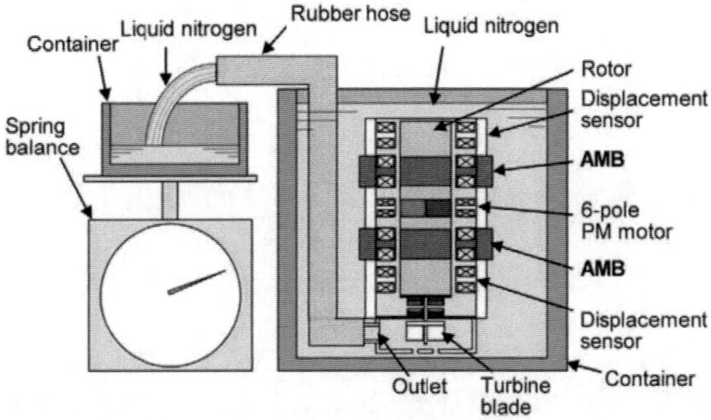

Figure 7: Experimental setup for studying dynamic performances of cryogenic motor in liquid nitrogen.

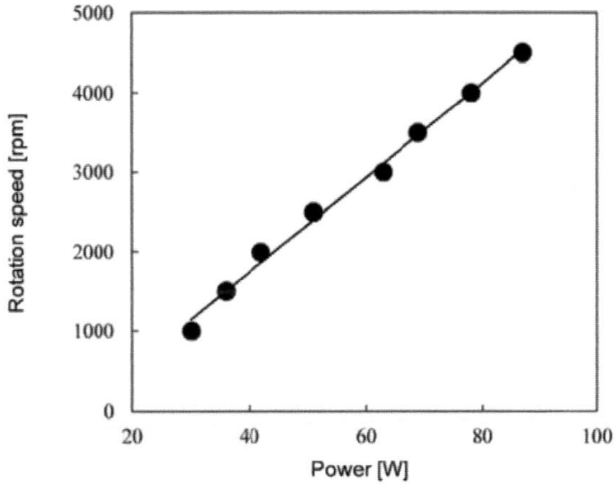

Figure 8: Relationship between pump flow and rotation speed for the cryogenic pump working in liquid nitrogen.

The outlet of the cryogenic motor is connected to a rubber hose. Thus, liquid nitrogen flows through the rubber hose into the small container which is set on a spring balance. The weight of liquid nitrogen in the container for a minute is measured in order to estimate the pump flow of liquid nitrogen.

During the pump performance, the applied power to the cryogenic motor is measured in the rotation speed range from 1,000 to 4,500 rpm. The applied power is estimated by measuring the voltage and the current of electromagnet. Figure 8 shows the relationship between rotation speed and applied power. The result shows the linear relationship in the speed range from 1,000 and 4,500 rpm. The cryogenic motor does not spin in the speed range more than 4,500 rpm, because the rotation torque is limited in the speed range higher than 4,500 rpm.

Figure 9 shows the relationship between pump flow and rotation speed. The pump flow is measured in the speed range from 2,000 rpm and 4,500 rpm. The result shows the good linear relationship between pump flow and rotation speed. The pump flow is not measured in the speed range more than 4,500 rpm, because the rotation torque is limited at speeds higher than 4,500 rpm.

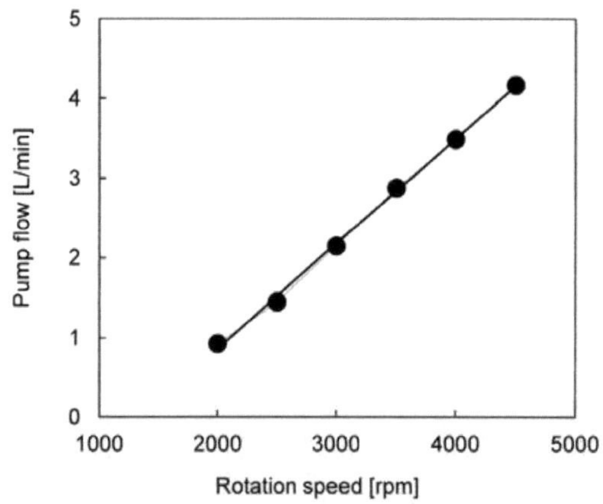

Figure 9: Relationship between pump flow and rotation speed for the cryogenic pump working in liquid nitrogen.

5 Summary

This paper discusses the cryogenic motor which works in liquid nitrogen (-200 C). The cryogenic motor is composed of two AMBs, four displacement sensors, a PM motor, etc. Some static and dynamic characteristics are studied in liquid nitrogen.

Some experiments are performed at room temperature and in liquid nitrogen. From the result of FFT analysis, it is found that the rotor is stabilized in the radial direction and that the rotor spins stably in the speed range 1,000 rpm to 4,500 rpm. From the result of spin test, it is found that the displacements in the x and y directions are smaller than 0.2 mm and that there is no vibration peak in the speed range from 1,000 rpm to 4,500 rpm. From the result of pump performance, it is found that the relationship between rotation speed and applied power is linear and the relationship between pump flow and rotation speed is also linear.

References

Kajikawa, K. (2012), Development of a liquid hydrogen transfer pump system with MgB2 wire, TEION KOGAKU (Vol.47, pp.674-679).

Komori, M. (2004), Development of a liquid nitrogen pump using superconducting bulk motor, IEEE Trans. on Applied Superconductivity (Vol. 14, pp. 1659-1662).

Komori, M. (2017), Basic study on active magnetic bearing working in liquid nitrogen and its application to cryogenic pump, Mechanical Engineering Journal (Vol. 4, pp.16-24)

K. Hirose, K. (2015), Application of one-axis-controlled magnetic bearing with a hollow shaft to noncontact rotation drive, IEEE Trans. on Magnetics (Vol. 51, pp. 1-4).

Wang, P. (2010), A frameless eddy current sensor for cryogenic displacement measurement, Sensors and Actuators (Vol. A159, pp. 7-11)

Higuchi, T. (1982), Control system design for totally active DC-type magnetic bearings, Transactions of the Society of Instrument and Control Engineers (Vol. 18, pp. 507-513)

Linear Observers Design for a Three-Pole Magnetic Bearing

Vinícius Ramos Vasco[1], Yago Pessanha Corrêa[2], Laura Julia Martins Mothé[3], Domingos de Farias Brito David[4], and Afonso Celso Del Nero Gomes[5]

[1] Marinha do Brasil (*Brazilian Navy*)
viniciusramosvasco@gmail.com
[2] Instituto Federal Fluminense (IFFluminense)
yago.correa@iff.edu.br
[3] Universidade Federal do Rio de Janeiro (UFRJ)
lauramothe@gmail.com
[4] Universidade Federal Fluminense (UFF)
domingos@vm.uff.br
[5] Universidade Federal do Rio de Janeiro (UFRJ)
nero@coep.ufrj.br

Abstract

In this work, linear observers for a three-pole active magnetic bearing (AMB) were designed and simulated, considering the AMB modelled as a linear dynamical system. Although the system being nonlinear, under the assumption of lower speeds, the linearized model can behave similarly to the nonlinear model. The linearized model was made from the reluctance forces equations and, as the system is naturally unstable, Linear Quadratic Regulator (LQR) was used to stabilize the system. The state vector contains both the horizontal and vertical sensor measurements, and also their derivatives, that are not measured. For the state feedback with LQR, it is necessary to have all the state vector data, and linear observers are a way to obtain the non-measured states. Observation errors for the full and reduced order linear observers are then compared.

1 Background

Active magnetic bearings are mechatronic devices that allow spinning rotors to operate contactless, providing the possibility to achieve high speeds with low bearing losses, with a controllable dynamic that can change the suspension's stiffness to different applications or states of operation [1]. Magnetic bearings are suitable for situations that demand the conditions mentioned above, as well as high efficiency, low maintenance and green operation, some examples are in turbomachinery, flywheel energy storage applications, machines in aerospace, and industries that require clean environments, like food and pharmaceutical industry [2].

The majority of magnetic bearings installed in industries uses the eight poles geometry, that produces four uncoupled magnetic fluxes to balance the rotor [3]. However, further research suggested other configurations, such as 4-pole and 3-pole, both with coupled magnetic fluxes. One way magnetic bearings can be modelled is through reluctance forces, and analytically is possible to verify that they have very strong nonlinearities in relation to the base current, the control current, the number of turns, the air gap and the rotor's positions. In [4], the reluctance forces from the 8-poles and 4-poles magnetic bearings were analytically obtained, linearized and compared.

The main advantages of the 3-pole AMB in comparison to other geometries is that it requires fewer actuators, has smaller remagnetization frequency resulting in lower iron losses and allows more space to place sensors and to dissipate heat, being very convenient for low-cost applications [5]. The main disadvantage is the higher complexity in the design and implementation of nonlinear controllers. One of the first studies proposed to overcome this problem was

[6], providing an exact linearization via feedback. Even though the techniques and the implementation of nonlinear controllers have been more feasible over the time, the study of linear control implementation are valid in some conditions, like constant and low rotor speed, and smaller rotor displacements.

Through the equations of motion and the relations between the linearized reluctance forces and control currents it is possible to model the system with rigid rotor and active magnetic bearings in the state-space form [4, 7]. The proper modelling allows to directly apply some techniques like state-space analysis with linear state-feedback, as well as optimal control and state observers, subjects discussed in detail in the control literature [8, 9, 10].

2 Discussion

This paper addresses the observation errors comparison of a three-pole magnetic bearing, modelled as a linear dynamical system with linear observers to estimate the states not measured. The responses using the full and reduced order observer will be compared. The 3-pole magnetic bearing considered is in the optimal configuration, as proposed in [5]. This optimal design considers the bearing supporting a rotor in a horizontal position and $\pi/6$ orientation, ensuring stability and base currents that minimizes the iron loss. Besides that, it also allows to reduce one actuator from the scheme by using only one coil wiring to the two upper poles, as shown in Figure 1.

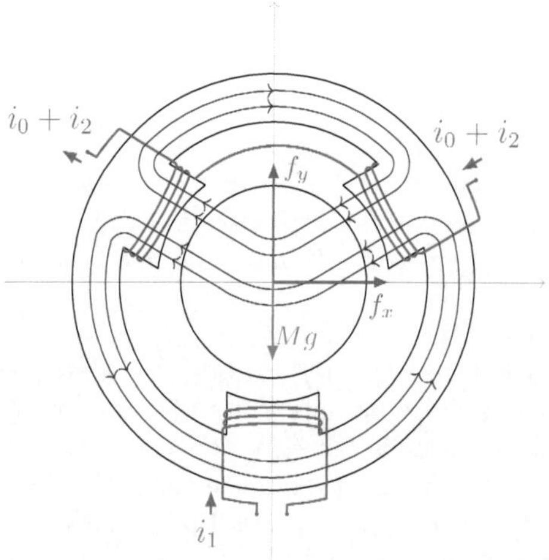

Figure 1: 3-Pole Magnetic Bearing in optimal design.

The stabilization problem in this work refers to control the differential currents in order to view how the rotor's center position changes over the time. As the region of study is close to the desired center point, linearization and linear control appears as valid tools for the analysis. Some examples of studies that have been made with the nonlinear control approach are in [11], [12] and [13]. For the linear approach, the detailed procedure on how to obtain the horizontal and vertical reluctance forces (f_x and f_y), here presented in (1), and how to model the system using these equations are in [7][1]. The linearized reluctance forces for the 3-pole AMB are:

$$f_x = \frac{1}{2} g_p x + \frac{\sqrt{3}}{3} g_i i_1, \quad f_y = \frac{1}{2} g_p y + g_i i_2 \tag{1}$$

where

$$g_p = \frac{\mu_0 \; A \; n^2 \; i_0^2}{h^3}, \quad g_i = \frac{\mu_0 \; A \; n^2 \; i_0}{h^2}, \tag{2}$$

μ_0 is the permeability of free space, A is the pole's surface area, n is the coils turns number, h is the gap's size, i_0 is the coil base current, i_1 and i_2 are the coils differential currents, used as inputs to the rotor's position control.

The mechanical system model in state space for the 3-pole magnetic bearing considered the operation in a condition with disturbances, very close to what should be in a real condition. The external actions considered were the ones due to the torques from: the magnetic force, the gravitational force and the supporting bearing. Using the equations from the classic rotational dynamic and doing some algebrism, all detailed in [7] and [4], it is possible to set the model in state space form:

$$\dot{\boldsymbol{x}}(t) = A(\omega)\boldsymbol{x}(t) + B\boldsymbol{u}(t), \quad \boldsymbol{y} = C\boldsymbol{x} \tag{3}$$

in which matrices $A(\omega)$, B and C, the state vector $\boldsymbol{x}$ and the control inputs $\boldsymbol{u}$ are shown in (4), (5) and (6) :

$$A(\omega) = \begin{bmatrix} 0_2 & I_2 \\ K_1 I_2 & -G_e(\omega) \end{bmatrix}, \quad B = \begin{bmatrix} 0_2 \\ K_2 I_2 \end{bmatrix}, \tag{4}$$

$$C = \begin{bmatrix} I_2 & 0_2 \end{bmatrix}, \quad \boldsymbol{x} = \begin{bmatrix} x_s & y_s & \dot{x}_s & \dot{y}_s \end{bmatrix}^T, \tag{5}$$

$$\boldsymbol{u} = \begin{bmatrix} \frac{\sqrt{3}}{3} i_1 & i_2 \end{bmatrix}^T \tag{6}$$

where the constants K_1, K_2 depends on design parameters, the 2x2 matrix G_e depends on design parameters and also rotor speed, as shown in [7], the 0_2 is the 2x2 zero matrix, the I_2 is the 2x2 identity matrix, x_s and y_s are the sensors' displacement measurements.

This open loop model is unstable, as the state matrix has poles with positive real part, so a closed loop control is necessary to stabilize the system.

With the linearized model in state space form, it is very straightforward to verify that the system is observable, controllable, and the LQR is an option to stabilize the system, as this state feedback control can relocate the closed loop eigenvalues to have negative real parts. The LQR control in magnetic bearing has potential problems with robustness due to parameters uncertainties and instability with different speeds [1]. As in this work we consider the system with negligible modelling uncertainties and with constant rotational speed, LQR should be viable for analysis.

The performance index V for the infinite horizon LQR is the integral of two quadratic forms sum:

$$V(\boldsymbol{x}, \boldsymbol{u}, t_0) = \int_{t_0}^{\infty} (\boldsymbol{x}^T Q \boldsymbol{x} + \boldsymbol{u}^T R \boldsymbol{u}) dt, \tag{7}$$

where Q is a symmetric semi-positive definite matrix and R is symmetric positive definite. As the origin is considered the reference position ($\boldsymbol{r} = \boldsymbol{0}$), for this problem the LQR control input $\boldsymbol{u}$ is given by

$$\boldsymbol{u} = F\boldsymbol{x}, \tag{8}$$

[1] In this reference there are typos in equations (41), (51) and (67), the negative signs should be positive

where

$$F = -R^{-1}B^T P, \tag{9}$$

with P being the solution to the Algebric Ricatti Equation (ARE). The diagram for the LQR state feedback is shown in Figure 2.

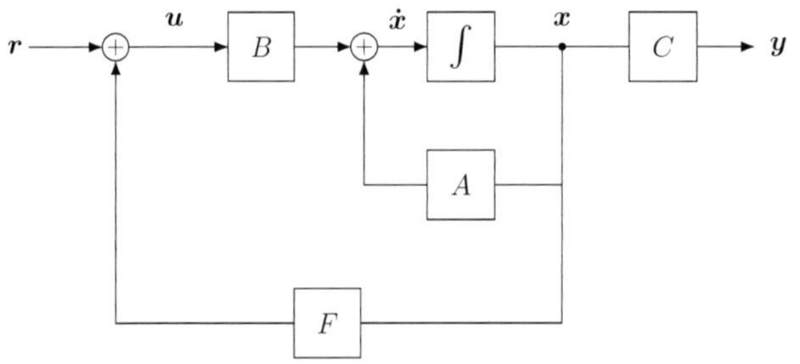

Figure 2: State feedback diagram.

But the implementation of the LQR control assumes that the states $\dot{x}_s$ and $\dot{y}_s$ are available, which usually is not the case. To obtain them, one option is to use sensors for direct measure, but it adds more cost to the set. The other option is to differentiate the signals of x_s and y_s, but with the inconvenience of amplifying the high frequency noise. In practical applications, digital computers are used, so the differentiation is obtained by subtracting the value of one sample to the value of one prior and dividing the result to the sample time. It is also common the use of filters to minimize the effect of noise. In this work we will design the full-order observer and the reduced-order observer theoretically and compare the simulated responses.

Both observers use the system's input and output to estimate the states [9], with the full order observer estimating all the states, regardless if they are already available or not, and the reduced-order observer estimating only the non-measured states. Then the estimated state vector $\hat{\boldsymbol{x}}$ is used to feedback to the system's input. For the full-order observer, the diagram is in Figure 3.

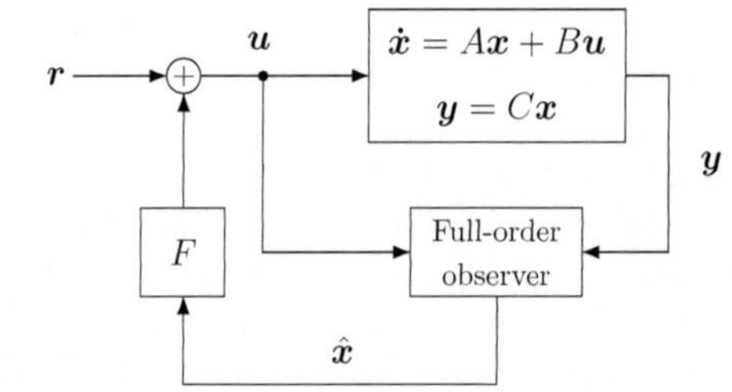

Figure 3: Estimated state feedback with full-order observer.

Detailing the Observer block in Figure 4, the full-order observer will contain a replica of the system matrices A, B and C, with the estimation error $\boldsymbol{e}$ being used by the replica system. The observer gain L must be chosen to allow $\boldsymbol{e}$ to exponentially decay to 0 for any error initial condition [14], and the observer dynamic also needs to be faster than the system dynamic. In this work we choose the eigenvalues of $A - LC$ to be 5 times faster than the closed-loop system eigenvalues.

As the system in study is fully controllable and observable, the separation principle allows to independently do the pole placement design and the observer design [9], as the extended

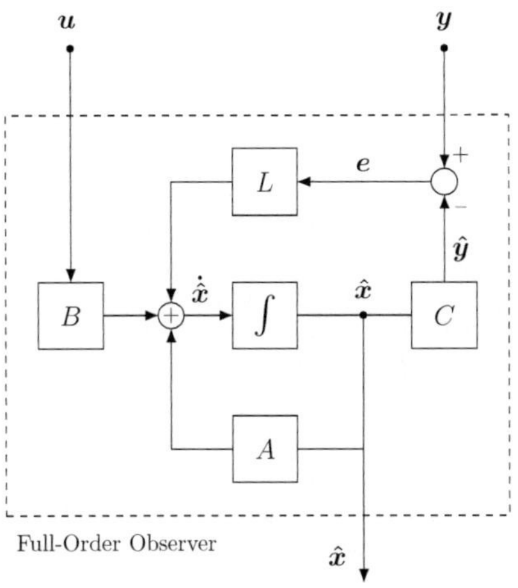

Figure 4: Detailed full-order observer block.

system equation is given by equation (10):

$$\begin{bmatrix} \dot{\boldsymbol{x}} \\ \dot{\boldsymbol{e}} \end{bmatrix} = \begin{bmatrix} A+BF & -BF \\ 0 & A-LC \end{bmatrix} \begin{bmatrix} \boldsymbol{x} \\ \boldsymbol{e} \end{bmatrix}, \tag{10}$$

where the estimation error $\boldsymbol{e}$ is the difference between the system output $\boldsymbol{y}$ and the estimation output $\hat{\boldsymbol{y}}$,

$$\boldsymbol{e} = \boldsymbol{y} - \hat{\boldsymbol{y}} = C(\boldsymbol{x} - \hat{\boldsymbol{x}}). \tag{11}$$

The reduced order observer allows estimating only the non-available states, by separating the state $\boldsymbol{x}$ and matrices A, B and C in two parts: the available part and the unavailable part, here identified with subindices a and u, respectively. The diagram for the state feedback with reduced order observer is presented in Figure 5.

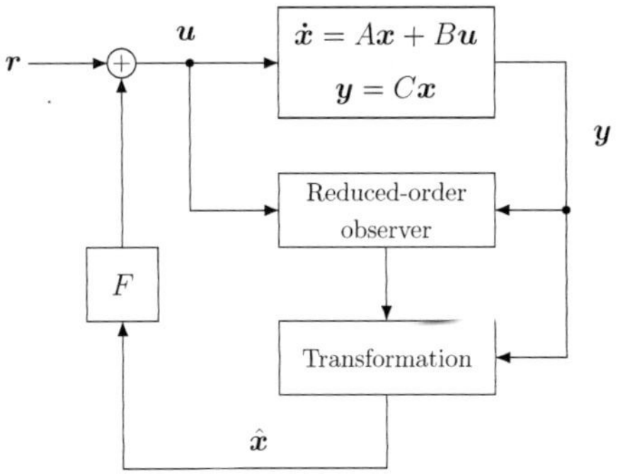

Figure 5: Estimated state feedback with reduced-order Observer.

Detailing the reduced-order observer and the transformation block in Figure 6, the matrices $\hat{A}, \hat{B}, \hat{C}, \hat{D}$ and $\hat{G}$ are expressions related to the available and unavailable parts of the system matrices as well as the observer gain for the reduced system, here given by L_r.

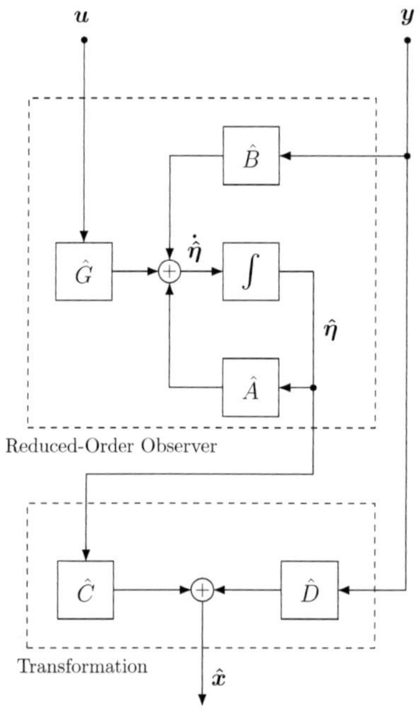

Figure 6: Detailed reduced-order Observer block and Transformation block.

Following [9], the dynamical system can be rewritten in the following form:

$$\dot{\boldsymbol{x}} = \begin{bmatrix} \dot{\boldsymbol{x}}_{\boldsymbol{a}} \\ \dot{\boldsymbol{x}}_{\boldsymbol{u}} \end{bmatrix} = \begin{bmatrix} A_{aa} & A_{au} \\ A_{ua} & A_{uu} \end{bmatrix} \begin{bmatrix} \boldsymbol{x}_{\boldsymbol{a}} \\ \boldsymbol{x}_{\boldsymbol{u}} \end{bmatrix} + \begin{bmatrix} B_a \\ B_u \end{bmatrix} \boldsymbol{u}, \tag{12}$$

$$\boldsymbol{y} = C \begin{bmatrix} \boldsymbol{x}_{\boldsymbol{a}} \\ \boldsymbol{x}_{\boldsymbol{u}} \end{bmatrix} = \begin{bmatrix} I_2 & 0_2 \end{bmatrix} \begin{bmatrix} \boldsymbol{x}_{\boldsymbol{a}} \\ \boldsymbol{x}_{\boldsymbol{u}} \end{bmatrix} = \boldsymbol{x}_{\boldsymbol{a}}. \tag{13}$$

The matrices $\hat{A}, \hat{B}, \hat{C}, \hat{D}$ and $\hat{G}$, indicated in Figure 6, are given by:

$$\hat{A} = A_{uu} - L_r A_{au}, \tag{14}$$

$$\hat{B} = \hat{A} L_r + A_{ua} - L_r A_{aa} \tag{15}$$

$$\hat{C} = \begin{bmatrix} 0_2 & I_2 \end{bmatrix}^T, \quad \hat{D} = \begin{bmatrix} I_2 & L_r \end{bmatrix}^T \tag{16}$$

$$\hat{G} = B_u - L_r B_a \tag{17}$$

The matrices $\hat{C}$ and $\hat{D}$ are necessary to reconstruct the full state with the available part coming from the output and the unavailable part coming from the reduced-order observer.

3 Results

The rotor speed initially considered for the simulation was $\omega = 356$ rad/s, the base current $i_0 = 0.23$ A, and the design parameters K_1, K_2 and G_e from equation (4) are given by:

$$K_1 = 3101.24, \quad K_2 = 19.76, \tag{18}$$

$$G_e = \begin{bmatrix} 0.51 & 10.22 \\ -10.22 & 0.51 \end{bmatrix}. \tag{19}$$

For $Q = I_4$ and $R = I_2$, the feedback gain matrix F is given by:

$$F = \begin{bmatrix} 311.27 & 28.24 & 5.68 & 0 \\ -28.24 & 311.27 & 0 & 5.68 \end{bmatrix}, \tag{20}$$

and the system's closed-loop eigenvalues are $-46.45 \pm j4.21$ and $-66.2 \pm j6.01$.

A simulation was made to compare the responses for the nonlinear model when subjected to the same linear control used by the linear model, to evaluate if the responses were similar. The schemes used are shown in Figure 7.

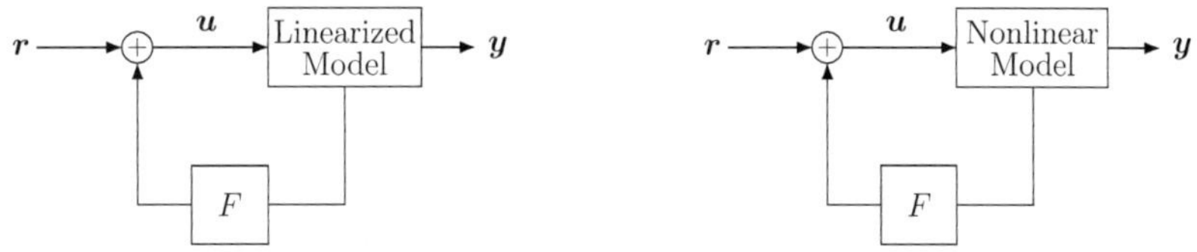

Figure 7: LQR control applied to the linearized and nonlinear model

The differences between the linearized and the nonlinear control responses depends on the rotor speed. They are negligible when $\omega = 356$ rad/s, as can be seen in Figure 8a, but for $\omega = 3204$ rad/s, for example, the differences become significant, as can be seen in Figure 8b.

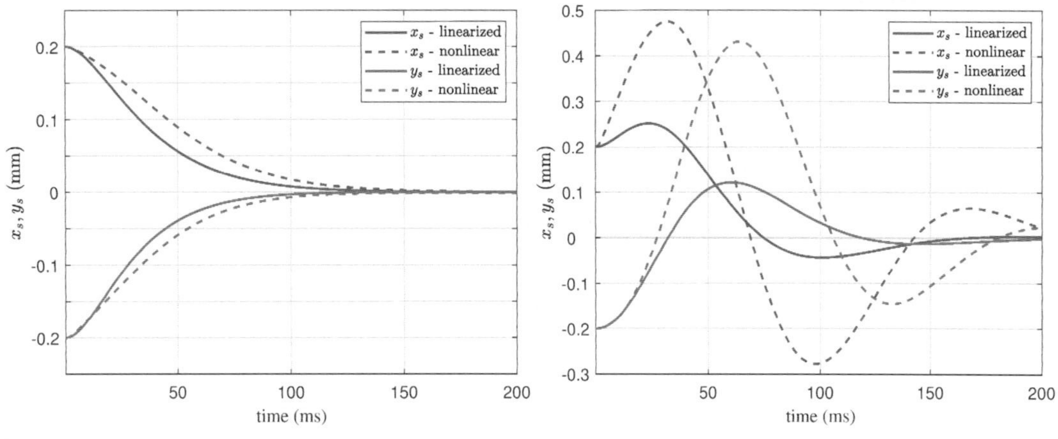

(a) When $\omega = 356$ rad/s (b) When $\omega = 3204$ rad/s

Figure 8: LQR Outputs to the linearized and nonlinear models

For the linearized system with rotor speed $\omega = 356$ rad/s, the full-order observer gain L was designed considering the observer eigenvalues 5 times faster than the closed-loop system eigenvalues, resulting in

$$L = \begin{bmatrix} 580.20 & -21.77 \\ -8.33 & 545.49 \\ 8.59 \times 10^4 & -6.14 \times 10^3 \\ 2.65 \times 10^3 & 7.49 \times 10^4 \end{bmatrix}. \tag{21}$$

For an initial condition $x_0 = \begin{bmatrix} 0.2 & -0.2 & 0 & 0 \end{bmatrix}^T$ and an observer initial condition $x_{0-obs} = \begin{bmatrix} 0 & 0 & 0 & 0 \end{bmatrix}^T$, the observation error for each state of the full order observer is presented in Figure 9a.

As for the reduced-order observer design, the observer gain L_r was designed considering the eigenvalues of the unobservable part being 5 times faster than the eigenvalues of the observable part, resulting in:

$$L_r = \begin{bmatrix} 231.75 & -10.22 \\ 10.22 & 231.75 \end{bmatrix}. \tag{22}$$

From equations (14), (15), (16) and (17), the reduced-order observer matrices $\hat{A}, \hat{B}, \hat{C}, \hat{D}$ and $\hat{G}$ are:

$$\hat{A} = \begin{bmatrix} -232.26 & 0 \\ 0 & -232.26 \end{bmatrix}, \tag{23}$$

$$\hat{B} = 10^4 \times \begin{bmatrix} -5.07 & 0.24 \\ -0.24 & -5.07 \end{bmatrix} \tag{24}$$

$$\hat{C} = \begin{bmatrix} 0_2 & I_2 \end{bmatrix}^T, \quad \hat{D} = \begin{bmatrix} I_2 & L_r \end{bmatrix}^T, \tag{25}$$

$$\hat{G} = \begin{bmatrix} 19.76 & 0 \\ 0 & 19.76 \end{bmatrix}, \tag{26}$$

and the corresponding observation errors for each state, for the same initial conditions mentioned above are shown in Figure 9b.

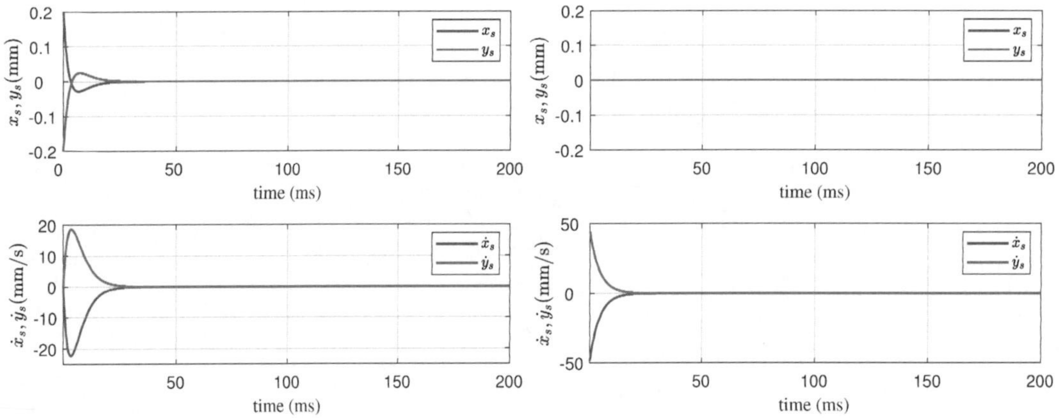

(a) full-order (b) reduced-order

Figure 9: Observation errors when $\omega = 356$ rad/s

In the noiseless conditions simulated, the reduced-order observer presented slightly better transient response performance over the full-order observer, with observation errors for the non-measured states about 5 ms faster than the full-order observation errors. In noisy environment, it is expected to see fewer advantages from the reduced-order observer due to more non-limited bandwidth noise in x_u [10], partly because of the memoryless transformation of y. It was not seen differences in estimation errors from figures 9a and 9b when simulating with speeds lower $\omega = 1780$ rad/s, approximately. For larger speeds, the difference between the nonlinear and linear outputs become significant, and the linear model isn't a good approximation.

4 Conclusions

We have shown how to design linear Luenberger observers for the 3-pole active magnetic bearing. The use of observers in magnetic bearings applications was considered because they can be modelled as a linear dynamical system via linearization of the reluctance forces. Both the full-order and the reduced-order observers could estimate the non-available states in close times, with the full-order observer in about 20 ms and the reduced-order observer in 15 ms, although it had larger initial error than the full-order. The shaft's rotation speed affected the linearized model performance, with it being very close to the nonlinear model for lower speeds, until about $\omega = 1780$ rad/s. For larger speeds, the linearized model responses have larger differences to the nonlinear model responses. This analysis shows that the use of the observers with the linearized model is suitable for problems where the application doesn't require very high speeds. More studies involving observers and digital differentiation in magnetic bearings will be done in future work.

References

[1] G. Schweitzer, H. Bleuler, E. H. Maslen, M. Cole, P. Keogh, R. Larsonneur, E. Maslen, R. Nordmann, and Y. Okada, *Magnetic Bearings: Theory, Design, and Application to Rotating Machinery.* : Springer Berlin Heidelberg, 2009.

[2] A. Chiba, T. Fukao, O. Ichikawa, M. Oshima, M. Takemoto, and D. G. Dorrell, *Magnetic Bearings and Bearingless Drives.* : Elsevier, 1 ed., 2005.

[3] W. Zhang and H. Zhu, "Radial magnetic bearings: An overview," *Results in Physics*, vol. 7, pp. 3756–3766, 2017.

[4] D. F. B. David, J. A. Santisteban, and A. C. D. N. Gomes, "Modeling and Testing Strategies for an Interconnected Four-Pole Magnetic Bearing," *Actuators*, vol. 6, SEP 2017.

[5] S. Chen and C. Hsu, "Optimal design of a three-pole active magnetic bearing," *IEEE Transactions on Magnetics*, vol. 38, pp. 3458–3466, SEP 2002.

[6] C. T. Hsu and S. L. Chen, "Exact linearization of a voltage-controlled 3-pole active magnetic bearing system," *IEEE Transactions on Control Systems Technology*, vol. 10, pp. 618–625, JUL 2002.

[7] V. R. Vasco, A. C. D. N. Gomes, D. d. F. B. David, and J. A. Santisteban, "Considerations on the mechanical dynamics of interconnected flux three-pole magnetic bearings," in *Proceedings of ISMB16*, p. 8, 2018.

[8] T. Kailath, *Linear Systems.* Pearson Education (US), 1979.

[9] K. Ogata, *Modern Control Engineering.* : Prentice Hall, fifth edition ed., 2010.

[10] B. D. O. Anderson and J. B. Moore, *Optimal Control: Linear Quadratic Methods (Dover Books on Engineering).* : Dover Publications, Inc., 1990.

[11] C.-T. Hsu and S.-L. Chen, "Nonlinear control of a 3-pole active magnetic bearing system," *Automatica*, vol. 39, no. 2, pp. 291 – 298, 2003.

[12] S.-L. Chen, "Nonlinear Smooth Feedback Control of a Three-Pole Active Magnetic Bearing System," *IEEE Transactions on Control Systems Technology*, vol. 19, pp. 615–621, MAY 2011.

[13] S. M. Darbandi, M. Behzad, H. Salarieh, and H. Mehdigholi, "Linear Output Feedback Control of a Three-Pole Magnetic Bearing," *IEEE-ASME Transactions on Mechatronics*, vol. 19, pp. 1323–1330, AUG 2014.

[14] V. Radisavljevic-Gajic, "Linear observers design and implementation," in *Proceedings of the 2014 Zone 1 Conference of the American Society for Engineering Education*, IEEE, apr 2014.

REVIEW ABOUT CONTROL AND ESTIMATION STRATEGIES APPLIED TO BEARINGLESS MACHINES

José Á. De Paiva[1], Werbet L. A. Da Silva[2], Jossana M. D. S. Ferreira[2], André L. Maitelli[2], Andrés O. Salazar[2], and Elmer R. L. Villarreal[3]

[1] Federal Education, Science and Technology of Rio Grande do Norte (IFRN), Natal-RN, Brazil
alvaro.paiva@ifrn.edu.br
[2] Federal University of Rio Grande do Norte (UFRN), Natal-RN, Brazil
werbethluizz@hotmail.com
jossana@ect.ufrn.br
maitelli@dca.ufrn.br
andres@dca.ufrn.br
[3] Federal Rural of Semiarid (UFERSA), Mossoró-RN, Brazil
elmerllanos@ufersa.edu.br

Abstract

Motors with magnetic bearings and bearingless machines have as a basic principle magnetic levitation to reduce wear on mechanical parts to increase their useful life and reduce the need for preventive and corrective maintenance, especially in equipment that operates in difficult-to-access environments. However, in order for them to achieve the desired performance, their respective control systems must operate efficiently and synchronously on: the radial rotor position, the torque, speed, and supply currents of these machines. In this context, the first systems developed used in a broad and unrestricted way the classic controllers PI (Proportional-Integrative), PD (Proportional-Derivative), and PID (Proportional-Integrative-Derivative), allied to state estimators based on inverse models of these machines to avoid the use of costly sensors. During the research, however, it was found that the non-linearities and parametric variations inherent to these special machines could affect the performance of the control and estimation systems developed. In order to compensate for these limitations, from the 2000s onwards, new researches were developed using the main modern control and state estimation techniques successfully applied to conventional electrical machines. In this context, the development of microelectronics and new microcontrollers such as, for example, the DSP (Digital Signal Processor) was fundamental to enable the implementation of these new systems. To recover the history of these works and contribute to the development of future systems, this article presents an overview of the main control and estimation techniques applied to different types of machines without bearings and levitation systems developed so far, highlighting, critically, the advantages and limitations of each of the techniques discussed.

Bearingless, Induction motor, DSP, Estimation techniques, control techniques.

1 Introduction

Electric machines that operate with the radial magnetic suspension of the rotor shaft provide a significant reduction in mechanical wear and, consequently, in the number of predictive and corrective maintenance of this equipment [1]. This advantage over conventional electric machines has resulted in the study and development of new models with the most varied topologies over the past decades. In this context, several types of machines that use the magnetic suspension protocol in their operation have been developed over the years. Among them, we can list: Active Bearing Machines (AMB), Bearingless Induction Machine (BIM) with double winding and split winding bearingless motors, permanent magnet and bearingless motors of variable reluctance. Each of these engines has characteristics that fit specific applications. The projects and the performance improvement obtained with these special machines were only possible due to the advances of microelectronics together with the control and estimation techniques developed for conventional electrical machines. However, these machines have nonlinearities, which significantly limits the application only of classic controllers based on plants with constant and known parameters. In order to get around the nonlinearities of bearingless machines, many control systems use the combination of two or more control and/or estimation techniques with the classic methods (based on machine parameters) for tuning the controllers optimally. In the next sections, the basic characteristics and applications of the main estimation and control techniques applied to the various types of bearingless induction machines or Active Bearing Motors are presented, highlighting their respective advantages and disadvantages. Finally, a general analysis of these techniques is presented to contribute to the development and/or optimization of other control systems applied to these special machines.

2 General Model of Control System applied to bearingless machines

To start the discussions on the control and estimation strategies applied to machines without bearings and magnetic levitation systems, Figure (1) presents the universal block diagram, in which the main stages are present in most of the systems already implemented. The difference between them is evident from the techniques used to control variables and state estimation of the controlled machines. Among the most studied variables are forces and radial positioning, rotational speed, electrical torque, flux position, and machine parameters.

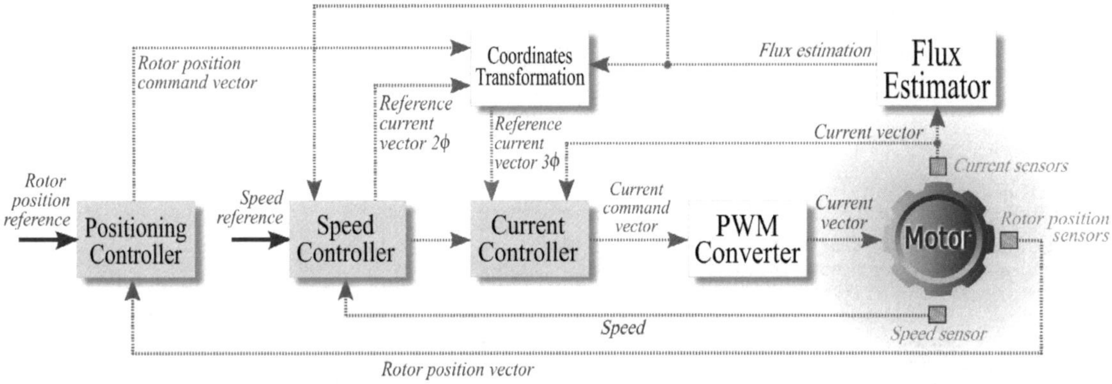

Figure 1: General diagram of a generic estimation and control system applied to bearingless machines and magnetic bearings.

This structure makes it possible to identify the following blocks: the radial positioning controller, the speed controller, the current controller, The PWM converter, the coordinate transformation block, and the flux estimator. The controllers most used in these stages are PI (Proportional-Integrative) or PD (Proportional-Derivative), or PID (Proportional-Integrative-Derivative). The estimator block is implemented by a flux-oriented vector model based on the inverse of the machine model. However, for plants with non-linear models with variable and unknown parameters, is necessary to replace or modify these control or estimation strategies with modern techniques based on artificial intelligence to improve the performance of these specialized machines. The PWM converter block is responsible for supplying the motor from the reference currents generated by the current controller; The coordinate transformation block converts two-phase control signals into three-phase reference signals, using the synchronous angle generated by the flux estimator. To this end, other strategies have been by using and implemented in recent years to assist the tuning of classic or modern controllers, making systems more robust to disturbances and parametric uncertainties. The estimation and control techniques covered in this article will be described in the following sections.

3 ESTIMATION TECHNIQUES APPLIED TO THE BEARINGLESS MACHINES

The states estimation and parameters of electrical machines aim to compensate for the difficulties of measuring these quantities in equipment with limited access or those which present significant variations under the influence of external phenomena, such as temperature and electrical and mechanical disturbances. The first estimation techniques developed were based on the internal model of the machine through which measurements of its inputs and outputs were made and, from the developed model, the values of the desired quantities were obtained. However, in most cases, the inverse models used for the estimation are linearized, operating at specific points. In addition, these models consider the machine parameters constant and immune to variations resulting from the influence of external agents. Several techniques have been studied and developed in the last decades to solve the limitations imposed by linearized models. One of those machines that have non-linearities is bearingless machines and machines with magnetic bearings. In order to understand the principle of operation and its respective applications, this article brought together the main estimation and control techniques applied to bearingless machines developed in the last decades. The description of the state of the art of these works as well as their applications will be presented in the following sections.

3.1 Estimation techniques based on the machine model

Each type of electric machine has a specific mathematical model, which depends essentially on its topology and construction technology. Thus, some machines have relatively simple and approximately linear models, such as DC machines. AC machines have more complex models with non-linear characteristics, as the machines are based on the principle of magnetic levitation. The following subsections describe the characteristics and applications of the main estimation techniques based on the bearingless machines inverse model.

3.1.1 Vector Model

The vector control techniques with flux guidance for induction machines, unlike scalar control of the V/F (voltage/frequency) type, present high performance both in the permanent regime and during the transients of these machines. The decoupling between the torque and flux controllers is achieved by transforming the model of the induction machine into a model similar to DC machines, whose control is much simpler and effective for both high and low speeds. The main requirement of vector controllers is to know the exact value of the magnitude and the position of the rotating flux. This factor generates the need for the use of flux sensors placed inside the machine, which is often not feasible for certain systems, either due to the difficulty of access or even the high cost of these sensors. For this reason, is very common to use flux estimators instead of sensors. In the vector control technique, a flux reference is chosen to obtain the model. The usual flux references are the stator flux vector the air gap flux vector, and the rotor flux vector. Each of these references has advantages and disadvantages in relation to the number of state equations and implementation difficulties. Initially, vector control was developed for the three-phase induction machine. However, features such as simplicity of equation and ease of implementation allowed the expansion of its applications to other types of machines. In [14], the vector model of an induction bearingless machine with split winding was developed. In this work, it was possible to prove that the model developed is similar to the vector model of a conventional three-phase induction machine when the rotor is centralized by the action of the radial position controller.

3.1.2 LSE based on zero-sequence voltage equations

The zero-sequence voltage equations to represent the models of three-phase electric induction machines are based on the representation in symmetric components widely used in multi-phase power systems. To make this representation possible, this model considers that all parameters of the induction machine are constant. From this model, it is possible to use the least-squares (LS) strategy to estimate the parameters through a regression equation composed by a prediction vector that depends on a regression vector and the parameter vector. To illustrate an application of this type of estimator, the work described in [30] presents an independent radial force control system on a double coiled induction bearing machine operating at high speed. In this work, the limitations of the control guided by the air gap flux are identified in relation to the delays inherent to the digital processing of the algorithm in a DSP for machines operating at high speeds. To compensate for these limitations, the developed estimator uses the least-squares technique (LSE) on the model of zero sequence equations of the machine. The proposed algorithm performs the estimation of stator parameters such as the mutual inductances between the windings and the resistance from the reading. With these estimated parameters, it is possible to independently control the electromagnetic torque and radial forces without direct dependence on the stator flux.

3.2 Artificial Neural Network (ANN)

Among the techniques for estimating nonlinear plants in evidence, estimators based on Artificial Neural Networks (ANNs) stand out [16]. This highlight is due to its wide range of applications and the ability to adapt and learn neural estimators concerning the behavior of the states to be estimated. One of the most promising research areas for implementing the artificial neural networks is in the control area and estimation of parameters and states of electrical machines. Induction machines have non-linearities in their models and suffer variations in the values of their parameters due to internal factors such as the rotor magnetic saturation and the temperature variation. In order to circumvent these problems, neural estimators can be trained under different load conditions and at various points of operation to contemplate and compensate for variations in machine parameters and maintain the desired performance for system controllers, whether they are classic or modern. Among the possible applications of RNAs in electrical machines is the induction bearingless machine with split winding. This machine type presents complex modeling in the function of the variation of the values of its parameters in function of the radial position and the non-linearities present in its equations.

Flux-oriented vector control depends on the exact position of the chosen reference flux, be it the stator, the rotor, or the air gap. However, the flux estimation through the conventional model of the machines does not consider the parametric variations due to the action of external agents, such as temperature variations. In order to compensate and allow a self-adjustment of the estimator in relation to these variations, a radial position and speed control system for a bearingless motor with split winding was implemented in [15] using the ANNs to estimate the rotor flux through Backpropagation networks with offline training. The results obtained showed that the performance of the system operating under the guidance of estimated rotor flux through the neural estimator was superior to the conventional estimator based on the machine model in terms of stability of the control of the radial position and the speed of the response of the speed control. In [18], by using neural network configurations were tested for speed control of the same type of machine, and the simulations proved that other RNA topologies could be used for flux estimation.

3.3 Adaptive Neuro-Fuzzy Inference System (ANFIS)

Neuro-Fuzzy networks combine the characteristics of Fuzzy Inference (FIS) systems with the ability to learn artificial neural networks (ANNs), resulting in a system known as Adaptive Neuro-Fuzzy Inference System (ANFIS). With the combination of the characteristics of the Fuzzy controllers and the Neural Networks, it is possible to develop estimators that adapt to the parametric variations of the plant under control. An application of the ANFIS networks is presented in [17] to estimate the rotor flux necessary for the vector control of a three-phase induction machine with split winding. In this work, the control system's performance using a conventional rotor flux estimator is compared with an ANFIS estimator. In this comparison, it is possible to observe that the flux estimated by the ANFIS network accurately follows the flux estimated by the conventional estimator, and its responses of speed, torque, and machine currents offer better performance. Thus, ANFIS-type networks emerge as an important tool for treating non-linearities in bearingless machines split-coil or in other machines with similar characteristics.

Each of the estimation techniques described above has its advantages and limitations (see Table (2)). For this reason, choosing the ideal technique for each type of machine is an important step towards achieving the desired dynamic performance. This choice depends on the machine's constructive characteristics and the performance specifications foreseen for each application. However, the good dynamic performance of the machine is only possible if there is a perfect connection between the chosen estimation technique and the most suitable control strategy for each type of control variable of machines without bearings or levitation systems. In this context, several control techniques have been proposed to compensate for the limitations imposed by the classical controllers widely used in the first works, as discussed in the next section.

4 CONTROL STRATEGIES APPLIED TO THE BEARINGLESS MACHINES AND LEVITATION SYSTEMS

As shown in Figure (1), it is common for control systems to have state estimators operating together with the controllers so that the machine without bearings or levitation system achieves the desired performance within the conditions defined in the project. For this, several types of controllers have been developed, tested, and compared to each other to overcome the limitations imposed by the controlled machine to guarantee operating stability, rejection to external disturbances, and robustness to parametric variations. The following sections present and discuss the basic design and operation principles of the main control techniques applied to bearingless machines and magnetic levitation systems.

4.1 Classic Controllers (PI, PD e PID)

Conventional PI, PD, and PID controllers were the first to be used in the control loops of systems applied to bearingless machines or even in magnetic levitation systems [1], [11] [13]. Because they are easy to implement and have consolidated theory, they are still widely used in the control systems of electrical machines that operate under constant torque and speed conditions. They are controllers with designs based on linear models with fixed parameters. However, due to non-linearities and parametric uncertainties, the tuning of these controllers becomes ineffective for some types of applications. In these cases, it is common to use a hybrid system composed of meshes implemented with PI, PD, and PID controllers, but with the aid of some modern estimation and control techniques for the online adjustment of its parameters [15], [26], [29], [30]. Thus, even with the emergence of modern control techniques, classical controllers should not be disregarded at first [5], [10], [12], [20], [21], [25], [27]. Initially, it is important to test them to identify possible limitations and define which estimation and/or control techniques should be added to compensate for them.

4.2 Convergent Control Algorithm

The convergent control algorithm (CC) is a control technique that operates in the frequency domain in which the Fourier coefficients of the controller input signal operating at excitation frequency are obtained by an adaptive control law [2]. If the excitation is composed of several frequencies, the convergent control algorithm must be applied to each excitation frequency separately through adaptive loops until the performance conditions are reached. Some works implement convergent control in machines without bearings to compensate for possible mechanical vibrations of the rotor at certain frequencies. The work presented in [3]. implements an active control of flexural vibration of the rotor using an adaptive harmonic controller LQR to attenuate the components of low-frequency vibrations, together with a convergent controller for radial positioning. The results obtained in the simulation show that the union of the techniques embedded in the system presents itself as a viable strategy for the control of vibrations at low frequencies in flexible rotors in an induction machine without cage bearings. In [2], the rotor's flexural vibrations with a two-pole cage rotor are studied. For this, the machine is equipped with a radial forces actuator and a supplementary winding located inside the stator slots. The model obtained from the machine uses the movement of the rotor coupled to the voltage and flux equations. The actuator built into the machine is a controller based on the Convergent Control Algorithm (CC) to attenuate periodic vibrations. The results showed that the controller used was able to attenuate the flexural vibrations of the rotor by up to 85%. In addition, the control technique implemented can also be used in the monitoring and diagnostics of machine failures.

4.3 Linear Quadratic Regulator (LQR)

The Optimal Control theory of a system considers that a good performance system depends on operating dynamics with minimal cost. Considering that a set of linear differential equations can represent the dynamics of the system and that the cost function is defined from a linear quadratic equation, the Quadratic Linear Regulator (LQR) is a control technique based on state feedback, stable and robust, whose main objective is to minimize the cost function. For this, the LQR control algorithm performs the adjustment of its coefficients through an iterative process. Some studies using LQR controllers have shown good results in the application of this control technique. In [28], the authors propose a decentralized LQR controller (LQRd) to control the radial position of a bearingless machine with double winding. In it, a direct comparison of the performance of the LQRd controller is made in relation to a conventional PD controller for radial positioning. The results obtained showed that the LQRd controller has a superior behavior to the PD controller and that the LQRd controller can be considered a PD controller with optimized parameters. Another work presented in [9] proposes an LQR controller combined with the theory of the Linear Inequality Matrix (LMI) for the optimization of the parameters of the Multiple Resonant controllers (MRC) to compensate the periodic oscillations generated by the disturbances on a multisector machine permanent magnet (MSPM). The results demonstrated that the control system combining the mentioned techniques presented a robust and efficient behavior in relation to oscillations in the radial position. Another application of the LQR controller is presented in [22]. In this work, a pole positioning controller is combined with the LQR controller to act in the radial position control of a permanent magnet dual machine with 4 degrees of freedom, double wound, and 10 kW. The article presents tests and simulations to check the stability of the radial position control at low speeds. When comparing the adopted controllers in relation to the PID controllers, it is shown that the LQR controller surpasses the performance in the radial positioning of the rotor. In addition, designing an LQR controller is simpler by using intuitive adjustments.

4.4 Decoupled torque and force control technique (FTD) and FTD with Harmonic Compensation (HFTD)

The articles that involve this technique use a machine structure without bearings with a permanent magnet rotor in which a single set of stator coils is responsible for generating the radial positioning forces and electromagnetic torque. In this machine model, the stator is divided into sectors, and a three-phase inverter controls each sector. The objective is to obtain an independent control of the x and y axes of radial position from the interaction of the magnetic fields resulting from each sector.

The FTD technique expresses the radial force F as a vector that can be decomposed into vectors on the x and y axes. These vectors, in turn, can be decomposed into three other axes (each axis corresponds to a spatial sector of the stator). The FTD technique uses the flux-oriented field control theory and shows that is possible to control radial and rotational forces in an uncoupled manner. In this technique, the "d" axis is responsible for controlling the radial positioning forces, and the "q" axis is responsible for the control of the electromagnetic torque. From this, as the intensity of the force F is controlled by the intensity of the component "d" of current of each sector, it has that the force F is maximum in the direction "d" of a determining sector when the resulting vector F is aligned with that "d" direction. The FTD technique controls the angle of force alignment F in relation to each sector's "d" components to carry out the radial position control. The control using the FTD technique does not include any compensation for synchronous disturbances. Therefore, the response of the position control system is highly interfered with by the action of an magnectic force F_m, resulting from the action of the centrifugal force caused by the unbalance of the rotor mass. To compensate for these synchronous disturbances, a modification in the FTD technique was proposed, resulting in the HFTD (Harmonic Force Torque Decoupled) technique. The HFTD technique alters the control law by including some parameters related to the synchronous frequency, which must be calculated using finite elements. Therefore, the HFTD control technique offers a harmonic compensation in a table form that must be updated according to the rotor position and the load torque to ensure the machine's stable operation. The work presented in [7] performs a direct comparison between the FTD, HFTD, and Pseudo Inverse Matrix (PIM) techniques, described in the following section. In this work, it was possible to prove that the FTD technique offers a good magnitude of radial forces and simplicity of computational implementation. However, the quality of these forces and the losses due to the Joule effect were significantly impaired by synchronous oscillations during the operation of the machine. When comparing the mentioned techniques, the HFTD technique provided an effective compensation for synchronous oscillations. However, its computational implementation is more complex than that required by the Pseudo Inverse Matrix technique (PIM).

4.5 Pseudo Inverse Matrix

The control technique through the Inverse Pseudo Matrix [7], [8], [10] was developed from the inverse model of a multi-sector permanent magnet bearingless motor (MSPM) with 18 slots and 6 poles with reconfigured winding. The mathematical modeling of the machine in question was based on the conventional permanent magnet machine in which the magnetic decoupling between sectors is considered, maintaining the geometric and symmetry characteristics. A matrix of K_E coefficients is found from the model obtained that characterizes the electro-magneto-mechanical outputs obtained by reading the currents responsible for the radial forces and torque. With these coefficients in place, the control action of the currents mentioned above is defined by obtaining the inverse K_E matrix. However, the matrix generates only three equations that differ from the number of state variables that depend on the number of sectors. In this way, the system of equations obtained presents infinite possibilities for solutions. Therefore, it is necessary to define a single solution that meets the minimum performance requirements. In order to minimize losses due to the joule effect, a solution using the Moore-Penrose pseudo matrix is proposed as a generalization of the inverse K_E matrix. In [7], the control technique is compared with the Decoupled Force and Torque (FTD) and Decoupled Force and Torque control techniques with Harmonic compensation (HFTD). The results obtained demonstrate that parameters such as: Magnitude and quality of radial forces minimize losses by Joule effect are equivalent between the HFTD and PIM techniques and better than FTD. However, the control technique based on the Pseudo Inverse Matrix (PIM) offers a simpler computational implementation than the HFTD technique. The only difficulty with the proposed technique is obtaining the Inverse Matrix K_E. For this, the computational technique SVD (Single Value Decomposition) is used with offline processing to minimize the computational effort during the Control System operation.

4.6 Support Vector Machine (SVM)

A support vector machine (SVM) is a concept in computer science for a set of supervised learning methods that analyze data and recognize patterns used for classification and regression analysis. The standard SVM takes a set of data as input and predicts each given input, which of two possible classes the input is part of, which makes SVM a non-probabilistic binary linear classifier. Given a set of training examples, each marked as belonging to one of two categories. An SVM training algorithm builds a model that assigns new examples to one category or another. An SVM model represents examples as points in space, mapped so that the examples in each category are divided into a clear space that is as wide as possible. The new examples are then mapped into the same space and predicted to belong to a category based on which side of the space they are placed on. The

SVM technique has been applied in some control systems of bearingless machines. In some cases, SVM is used in conjunction with other techniques. In [21], a decoupled control system based on the Least Squares Inverse Support Vector machine (LSSVMI) is proposed, applied on an induction machine of 5 degrees of freedom. This work presents the structure of the motor and its inverse mathematical model, which is obtained using the regression capacity of the nonlinear theory of the least squares support vector machine. In addition, the work proposes an optimization algorithm for the parameters, which improves the accuracy of the model's adjustment and prediction. The initial simulation results demonstrate the efficiency of the control strategy.

4.7 Active Disturbance Rejection Control (ADRC)

Active Disturbance Rejection Control (ADRC) is a robust controller based on non-linear functions that estimate and compensates for various disturbances and variations in plant parameters. For this, the equations of the ADRC controller consider the total disturbance (composed of unmodulated dynamics, nonlinearities, uncertainties, and load variations) as a new state of the system to be estimated in real-time through an extended state observer. These characteristics make the controller's performance independent to the plant model, which is ideal for controlling electrical machines. As a disadvantage for applying this technique, we can mention a large number of the controller parameters and the lack of a systematic method of tuning. As an application example of the ADRC controller, the work presented in [4] proposes an ADRC controller based on the inverse model of the rotor flux for the radial positioning and speed controls of a double winding induction machine, replacing the conventional PID controllers. The implemented ADRC controller considers the cross-couplings of torque and radial position, parametric variations, and loads as disturbances within the control law. An extended state observer (ESO) estimates the total disturbance and compensated by a non-linear state error controller. The simulation results showed that the proposed control system has the robustness and better performance in the responses of rotation speed and radial positioning in relation to the system implemented with conventional PI controllers. Another application of the ADRC controller is found in [19]. In this work, the performance of the proposed ADRC controller is compared to the performance of the PID controller. Among the control variables compared are: the response to the positioning step in the x and y directions, the radial displacement of the rotor, the speed response during the start and with the application of load variations. All the results obtained showed that the ADRC controller had a superior performance because they reduced the speed overshoot, quickness in the transient responses, and robustness to the parametric variations.

4.8 Sliding Mode Control (SMC)

Sliding mode control is a strategy with several applications in recent decades. It is indicated for plant control systems with strong nonlinear characteristics with great sensitivity to disturbances and parametric variations. The technique consists of direct control in two modes directing the system to a hyperplane, representing the plant's desired dynamic behavior. The trajectory of the outputs within this surface is known as operation in sliding modes. The technique has the advantages of quick response, negligible effects of unmodulated dynamics, and does not depend on plant parameters and external disturbances. The main problem with the SMC is that the control signal is applied at a high frequency to keep the output on a hyperplane to result in an unwanted phenomenon known as "Chattering". To decrease the Chattering effect, it is necessary to apply corrections to the control signal, replacing the discontinuous signal by a function with the saturation or the sigmoid function and limiting the signal level. Consequently, it is common for a finite error to appear permanently due to this correction of the control signal. In [23], a system for controlling the suspension of the rotor by sliding modes for an engine without variable reluctance bearings (BLSRM) is proposed. The results obtained in simulations with the proposed radial position control system demonstrated that the rotor remained stable both during acceleration and during the steady-state, with low vibrations. The rotor suspension and stabilization times were satisfactory.

4.9 Multi-Ressonant Controller (MRC)

Like the HFTD controllers applied to motors with magnetic bearings and machines without bearings, the Multiple Resonant Controllers (MRC) are used to mitigate the effects of periodic disturbances in the control of the radial position of the rotor. These disturbances arise mainly due to the misalignment of the geometric axis in relation to the center of mass of the rotor and the position sensing surface's constructive and magnetic imperfections. It is common to have cyclical disturbances of different frequencies in the same system, so the controller must be able to attenuate each one of them. In position control systems with MRC, there is usually a controller responsible for stabilizing the radial position operating in conjunction with the MRC controller dedicated to promoting the rejection of cyclic disturbances. This controller produces an infinite gain whenever the error has a frequency ωr. The reference tracking problem can be addressed by using n resonant controllers that lead to multiple resonances or multiple resonant controllers. The main idea is to generate an oscillating dynamic (sinusoidal mode) corresponding to the fundamental frequency of the signal of interest and its respective most significant harmonics. To exemplify an application of this type of controller, the system described in [9] implements an LQR type controller for a machine without a permanent magnet multi-sector bearing (MSPM) in which a mathematical model was developed to generate optimally current references. To minimize losses by Joule effect. In it, an MRC controller is incorporated into the system to compensate for periodic disturbances. The experimental results demonstrated the effective rejection of oscillations on the radial position, which significantly improved the performance of the magnetic levitation system.

4.10 Fuzzy Controller

The Fuzzy controller seeks to embody the human way of thinking in a control system. It emerged as an alternative to the classic control method as it does not require the knowledge of its mathematical modelling, thus facilitating control in non-linear systems that are difficult to modelling. The controller design is based on deductive reasoning, on the experience of the system operators, or on a desired behavior for the system. Fuzzy controllers are an important application of the fuzzy set theory (Fuzzy sets) and can be used in conjunction with other types of controllers such as PID and Neuro-Controllers. A recent proposal for applying Fuzzy controllers for a three-phase induction machine without bearings with split winding is presented in [5]. This article presents a Fuzzy controller with PD controller characteristics for radial positioning. The design of the controller took into account the knowledge of the desired performance of the system when using a PD controller to keep the rotor axis centralized. From this condition, the input and output pertinence functions were defined together with the rules of the Fuzzy controller. The work results showed that the Fuzzy controller with PD characteristics applied to the x and y axes offers good performance and robustness in the control of the radial position, as long as the operating frequency of the motor is close to the nominal frequency of the machine. However, for operation at frequencies below 60 Hz, the Fuzzy control performance was impaired, indicating the need for modifications in the adjustment of the controller parameters. Even with significant oscillations, the rotor has not exceeded the stability limits, which demonstrates the viability of the technique for some simpler applications.

4.11 H_∞ Controller

The H_∞ controller is a feedback controller of the same order as the plant. The H_∞ controller synthesis must be understood as a system optimization technique in the frequency domain. Its main objective is to minimize the norm of the system's frequency response when it is desired to guarantee its stability and good performance. The norm of a system is related to the maximum gain existing between the inputs and outputs in the entire spectrum of possible operating frequencies. A thesis paper was presented in [6]. In it, an H_∞ controller is proposed for the radial positioning of a bearingless motor in order to maintain stability and robustness even with parametric variations. The thesis provides a detailed description of the concepts necessary to understand the H_∞ control. Soon after, the work describes the theory, the design, and the synthesis of the controller H_∞ , performs several simulations to verify stability in radial positioning and its behavior under the action of harmonic disturbances, and presents the implementation and operation of the prototype. After analyzing the results of the tests, it was found that the H_∞ control system meets the specified performance requirements. Thus, the work concludes that the H_∞ controller is a very promising technique for application in bearingless motors because it is a complex and non-linear modeling plant.

4.12 μ-Synthesis

The μ-Synthesis control technique is an extension of the H_∞ method and presents itself as a robust controller for plants with a high degree of uncertainty, whether parametric or dynamic. However, for the application of this method, it is necessary to know the uncertainties of the plant in detail and in a structured way. In addition, the μ-Synthesis method generates a higher-order controller, which requires a recursive process of interactions to obtain the optimal solution, which must meet the performance conditions required for the plant. The robust control technique μ-Synthesis has some recent applications in levitation systems as described in [24]. In this work, the μ-Synthesis control technique acts in the recovery of levitation of an active magnetic bearing (AMB). The proposed strategy uses the model in the theory-Synthesis controller theory to find the controller parameters to provide the necessary forces that prevent the saturation of the control signals that occur during large deflections due to levitation problems. After the design and synthesis of the controller, experimental tests were carried out in which the levitation system was brought to instability simulating failures during operation. The work proposes that the μ-Synthesis controller is automatically activated in case of loss of support of the rotor. The results obtained demonstrated that the control system responded quickly to the recovery of levitation. However, the system was studied in a simplified model, which does not generate the need to expand research on machines with more complex models.

4.13 PWM modulation control techniques for inverters

To control the static switches of the inverters responsible for driving the electrical machines, is traditionally used the pulse width modulation (PWM) technique. Unfortunately, this high-frequency switching causes harmonic distortions that affect the quality of voltages and currents applied to electrical machines. In order to minimize these distortions and approximate these signals to perfect sine waves, several techniques were developed and implemented, significantly improving machine performance and control systems performance. The following subsections provide a brief description of these switching techniques.

4.13.1 Space Vector Modulation (SVM)

The technique known as Space Vector Modulation (SVM) is a strategy applied to the PWM (Pulse Width Modulation) signals that control the PWM inverters of an AC converter. Its operation can be defined by a specific sequence in the activation of the inverter switches so that the converter's output approaches the maximum of a sinusoidal signal. The main objective of this technique is to minimize the effect of harmonic components present in switched signals. Several works use the SVM technique combined with conventional control techniques to optimize the reference voltages and currents applied to the machines. An example of this combination of the SVM control technique applied to the PWM signals of the inverters is presented in [19]. Although this work is dedicated to the implementation of ADRC controllers for radial positioning, speed, and electromagnetic torque, the power stage uses SVM modulation in conjunction with a DTC (Direct Torque Controller) controller, resulting in a controller (SVM-DTC) for the supply of optimized voltage reference signals for the individual converters, for the supply of the two-pole windings (responsible for the radial positioning control) and four-pole windings (for the torque control) in order to guarantee the decoupling between the controllers and the reduction of high frequency harmonic components.

4.13.2 Direct Levitated Force Control

For the radial positioning control of bearingless machines, it is common to use flux-oriented vector control techniques synchronized with the torque and speed control. However, this type of control is strongly influenced by the machine parameters causing its radial levitation forces to be controlled indirectly. As a consequence of the indirect radial suspension force control, it is common the occurrence of lags on the flux angle that guides the radial positioning, resulting in the appearance of oscillations and sensitivity to disturbances. In order to ensure operating stability, minimize oscillations and make the positioning control more robust, the work presented in [12] proposes a new strategy for the direct control of levitation forces for an induction bearingless machine double-winding. This strategy executes an algorithm with predictive characteristics that calculates the future value of the angular position of the radial positioning flux from the difference between the current values of this angle and the angular position of the air gap flux, responsible for the torque and velocity control. For this, is necessary that in addition to the magnitude and phase values of the radial positioning flux predicted in the previous step, the magnitude and phase of the air gap flux are available, either through a sensor or calculated using a vector estimator flux. In this sense, to avoid using sensor flux, the cited work introduces an optimized model for the flux estimator responsible for the torque. For the optimized control of the inverter switching, the proposed system uses the SVPWM (Space Vectorial Pulse Width Modulation) technique.

The results obtained in the simulations and experimentally demonstrated that the technique of direct control of the levitation forces proposed improved the stability of the control of the radial position of the rotor, reduced the oscillations in the rotor levitation, and increased the disturbances rejection.

4.13.3 Direct Torque Control(DTC) and Direct Suspension Force Control (DSFC)

The Direct Torque Control (DTC) technique is a technique that allows direct control of stator flux and torque through optimal drive state selection using designer-defined switching tables. Due to the simplicity of its structure, it is common for the appearance of high levels of ripple in the torque signals, which significantly affects the performance of the DTC controller. Therefore, works such as the one presented in [19] propose adding other switching techniques added to the DTC controller. In this work, the SVM-DTC technique was proposed to smooth the torque ripple effect, resulting in a significant improvement in the performance of the torque and flux control system. In addition to torque and flux controls, control systems for bearingless machines must ensure efficient control of the radial positioning of the rotor. In general, radial positioning control uses the torque vector controller itself to generate the command of lifting forces necessary for radial positioning. However, most of the time, this over-radial positioning action only occurs after modulation of the torque controller, which results in indirect control. This characteristic makes it considerably difficult to calculate the exact suspension forces required for radial positioning, impairing the overall behavior of the control system. To solve this limitation, the work presented in [19] proposes a direct control algorithm based on the strategy presented in [12], called Direct Levitation Force Control, resulting in the control technique DSFC (Direct Suspension Force Control). With these techniques, suspension forces are calculated using the flux link read and updated in real-time. In this way, the algorithm uses the difference between the suspension forces to calculate the ideal flux for the machine windings responsible for radial positioning. Also, in [19], ADRC controllers are implemented to modulate the signals from SVM-DTC controllers for torque and flux and DSFC for radial positioning to replace PID controllers. The results obtained in the simulation confirm the efficiency of the proposed system.

5 General analysis of the techniques presented

By analyzing the various works listed in this article, essential aspects of implementation could be extracted as a way to collaborate in the development of future systems. Are they:

- Originally, most estimation systems were based on the inverse models of the machines under study. However, uncertainties related to parametric variations and nonlinearities led to the search for more robust estimation techniques to these limitations.
- Since the first works, the classic PI, PD, and PID controllers have been used due to their ease of implementation and many industrial applications in electrical machines. Therefore, even in advanced control systems, it is common to use them together with the proposed controllers.
- For the implementation of each estimation or control technique, it is essential to know the system's behavior operating with estimators based on the inverse model and the classic controllers mentioned above. This knowledge makes it possible to identify the limitations of classical strategies and the correct choice of the advanced estimation and/or control technique to be used, according to the individual characteristics of each type of bearingless machine or levitation system. of the estimation and/or control strategies demand a significant computational processing capacity. For this reason, before implementing systems on an industrial scale, it is necessary to analyze the resources and specifications of the hardware devices available at the time of the control system's development to make it possible to choose the most suitable model for each application.
- Some estimation techniques are dedicated to parameter estimation, others to state estimation. Therefore, knowledge and identification of the difficulties and limitations of each machine or system are one of the first steps in choosing the appropriate technique for that problem (see table 3).
- In general, advanced control techniques seek to compensate for some limitations of classic controllers (PI, PD, and PID) related to aspects such as: speed of response, stabilization time, robustness to parametric uncertainties and disturbances. For this reason, carrying out intensive tests of the machine under study, under the command of these controllers under the most diverse operating conditions, is essential for choosing the most appropriate control and/or estimation technique(s). (see table 4).

It is important to emphasize that this article does not intend to exhaust and address all the contents related to the estimation and control techniques developed so far and applied to machines without bearings and magnetic levitation systems. However, by relating the strategies presented in this work, it was possible to verify the importance and influence of advanced and technical techniques for other types of electrical machines in order to expand the possibilities of new tools in the coming years in view that, until the moment, there is no single type of controller or estimator, operating individually or together, that definitively compensates the performance limitations of all types of bearingless machines in their various applications.

Tables (1) and (2) with representation (Fig. (2) and Fig. (3)) show the chronological sequences of the works analyzed and their respective control and estimation techniques implemented. Next tables (3) and (4) with representation (Fig. (4) and Fig.(5)) present a general summary of the advantages and limitations of the control and estimation techniques analyzed in this article.

	2003	2004	2005	2006	2007	2008	2009	2010	2011	2012	2013	2014	2015	2016	2017	2018	2019	2020
Vector Model		[16]	[2]	[17]		[19]							[12]			[13]	[14]	
LSE based on zero sequence voltage equations													[3]					
Neuro-Fuzzy Networks (ANFIS)	[6]																[5]	[4]
Artificial Neural Network (ANN)										[7]								

Figure 2: Timeline: Main estimation techniques applied to machines bearingless.

	2003	2004	2005	2006	2007	2008	2009	2010	2011	2012	2013	2014	2015	2016	2017	2018	2019	2020
ClassicControllers (PI, PD and PID)		[3] [9]	[1] [2]	[10]		[5]		[11]	[12]		[7]	[13]		[14]	[16]		[15] [17] [27]	
Convergent Control Algoritm						[19]										[20]		
Linear Quadratic Regulator (LQR)						[21]							[6]				[22]	
Decoupled torque and force control technique (FTD)														[24]				
Decoupled torque and force control technique including Harmonic Compensation (HFTD)														[24]				
Pseudo Inverse Matrix														[24]	[25]			
Support Vector Machine (SVM)														[14]			[15]	
Active Disturbance Rejection Control (ADRC)															[26]		[27]	
Sliding Mode Control (SMC)																[28]		
Multi-Ressonant Controller (MRC)																	[22]	
Fuzzy Controllers																	[17]	
H∞ Controller											[29]							
μ-Synthesis															[30]			
Space Vector Modulation (SVM)																	[27]	
Direct Levitated Force Control																	[15]	
Direct Torque Control																	[27]	
Direct Suspension Force Control																	[27]	

Figure 3: Timeline : Main control techniques applied to the bearingless machines.

6 Conclusions

This article presented a general analysis of the main estimation and control techniques applied to machines that use the principle of magnetic levitation in their operation. Throughout the research, it was observed that most of the systems proposed in the analyzed works are based on control systems already developed and consolidated for conventional electrical machines. However, these special types of electrical machines have characteristics such has non-linearities and parametric uncertainties that

Estimation tecnique	Main advantages	Limitations
Vector Model	▸ The vector model approximates a direct-current machine, which facilitates its implementation. ▸ Provides good performance both in steady-state and during transient state.	▸ The model obtained depends heavily on knowledge of plant parameters. ▸ To ensure decoupling between flux and torque, it is necessary to have exact knowledge of the position and magnitude of the electromagnetic flux. ▸ The choice of the flow reference reflects the dimension and complexity of the model obtained.
LSE based on zero sequence voltage equations	▸ Its implementation only requires reading the voltages and currents applied to the machine windings. ▸ The accuracy of the estimation is not influenced by any other machine parameters.	▸ It demands high computational effort for real-time applications.
Neuro-Fuzzy Networks (ANFIS)	▸ Adapt to machine parametric variations and uncertainties. ▸ Allows learning and estimating non-linear plants.	▸ It demands high computational effort for real-time applications. ▸ The learning process can slow down plant dynamics.
Artificial Neural Network (ANN)	▸ It combines the learning characteristics of Neural Networks with the relationships and rules of the excellent performance of the Fuzzy controllers' system, allowing the representation of non-linear models with a high degree of uncertainty.	▸ It demands high computational effort for real-time applications. ▸ The process of learning and selfadjusting the rules can slow down the dynamics of the plant.

Figure 4: Main estimation techniques applied to machines bearingless.

make it difficult to implement estimators based on their models based on constant parameters and classical controllers (PI, PD, and PID). From this study, it was possible to gather and systematize the advantages and limitations of the strategies presented to collaborate with the development of new works.

7 Acknowledgments

This work was suported by Comissão de Aperfeiçoamento de Pessoal do Nível Superior (CAPES), Programa de Pós-graduação em Engenharia Elétrica e de Computação (PPgEEC/UFRN) and Conselho Nacional de Desenvolvimento Científico e Tecnológico (CNPq).

References

[1] O. Ichikawa M. Oshima M. Takemoto A. Chiba, T. Fukao and D. G. Dorrell. *Magnetic bearings and bearingless drives.* Elsevier, 2005.

[2] J. Orivuori Anssi Sinervo Kai Zenger A. Laiho, K. Tammi and Anter Arkkio. Electromechanical interaction in eccentric-rotor cage induction machine equipped with a self-bearing force actuator. In *11th International Symposium on Magnetic Bearings, ISMB 2008, Nara*, volume 1, 2008.

[3] K. Zenger A. Laiho, K. Tammi and A. Arkkio. A model-based flexural rotor vibration control in cage induction electrical machines by a built-in force actuator. *Electr Eng 90*, 34(4):407–421, 2008.

[4] W. S. Bu and Y. Q. Huang. Active disturbance rejection control of bearingless induction motor. *DEStech Transactions on Engineering and Technology Research*, 2017.

[5] E. R. L. Villarreal F. E. C. Souza L. P. D. S. Júnior J. S. B. Lopes E. A. D. F. Nunes, A. O. Salazar and J. C. C. Luque. Proposal of a fuzzy controller for radial position in a bearingless induction motor. *IEEE Access*, 7:114808–114816, 2019.

[6] A. R. Franco. *Posicionamento de eixo com motor mancal magnético, utilizando controle H_∞*. Programa de Pós-graduação em Engenharia Elétrica (COPPE), UFRJ, 2013.

[7] A. Formentini C. Gerada G. Valente, L. Papini and P. Zanchetta. Radial force control of multi-sector permanent magnet machines. In *2016 XXII International Conference on Electrical Machines (ICEM)*, volume 1, pages 2595–2601, 2016.

[8] A. Formentini C. Gerada G. Valente, L. Papini and P. Zanchetta. Radial force control of multisector permanent-magnet machines for vibration suppression. *IEEE Transactions on Industrial Electronics*, 65(7):5395–5405, 2018.

[9] L. Papini C. Gerada G. Valente, A. Formentini and P. Zanchetta. Performance improvement of bearingless multisector pmsm with optimal robust position control. *IEEE Transactions on Power Electronics*, 34(4):3575–3585, 2019.

[10] L. Papini P. Zanchetta G. Valente, A. Formentini and C. Gerada. Position control study of a bearingless multi-sector permanent magnet machine. In *IECON 2017-43rd Annual Conference of the IEEE Industrial Electronics Society*, volume 1, pages 8808–8813, 2017.

[11] F. E. F. Castro S. N. D. Lisboa A. O. Salazar J. A. De Paiva, J. M. D. S. Ferreira and A. L. Maitelli. New approach for a bearingless three-phase induction machine with divided winding. In *Ninth International Symposium on Magnetic Bearings, ISMB 2004, Lexongton-KY*, volume 1, 2004.

[12] X. Sun C. Bao J. Lu, Z. Yang and X. Chen. Direct levitation force control for a bearingless induction motor based on model prediction. *IEEE Access*, 7:65368–65378, 2019.

[13] A. 0. Salazar F. E. F. Castro J. M. S. Ferreira, J. A. De Paiva and S. N. D. Lisboas. Dsp utilization in radial positioning control of bearingless machine. In *2003 IEEE International Symposium on Industrial Electronics*, volume 1, 2003.

[14] A. O. Salazar J. M. S. Ferreira, M. Zucca and L. Donadio. Analysis of a bearingless machine with divided windings. *IEEE Transactions on Magnetics*, 41(10):3931–3933, 2005.

[15] A. L. Maitelli J. Á. De Paiva, A. O. Salazar and R. M. Stephan. Performance improvement of a split winding bearingless induction machine based on a neural network flux observer. In *11th International Symposium on Magnetic Bearings, ISMB 2008, Nara*, volume 1, 2008.

[16] A. O. Salazar J. Á. De Paiva and A. L. Maitelli. Review of control strategies and model estimation techniques applied to bearingless induction machine with divided winding. In *Proceedings 13th International Symposium on Magnetic Bearings, ISMB 2013, Rio de Janeiro*, volume 1, 2013.

[17] F.E. C. Souza L. P. D. Santos Junior J. A. De Paiva J.S. B. Lopes, J. D. Fernandes and A. O. Salazar. Adaptative neuro-fuzzy inference system for estimation of rotor flux of a bearingless induction motor applied to speed control. In *13th International Symposium on Magnetic Bearings, ISMB 2013, Rio de Janeiro*, volume 1, 2013.

[18] F.E. C. Souza L. P. D. Santos Junior J. A. De Paiva J.S. B. Lopes, J. D. Fernandes and A. O. Salazar. Using feed-forward back propagation for rotor flux estimation of a bearingless induction motor applied in the speed vector control. In *13th International Symposium on Magnetic Bearings, ISMB 2015, Kitayushu*, volume 1, pages 180–187, 2015.

[19] X. Sun D. Zhao K. Li, G. Cheng and Z. Yang. Direct torque and suspension force control for bearingless induction motors based on active disturbance rejection control scheme. *IEEE Access*, 7:86989–87001, 2019.

[20] J. T. D. Carvalho Neto V. F. Victor L. P. Junior, J. S. B. Lopes and A. O. Salazar. Robust speed control of a bearingless ac motor type divided winding, based on a conventional squirrel cage induction motor. In *22nd International Conference on Magnetically Levitated Systems and Linear Drives (MAGLEV)*, volume 1, pages 1–15, 2014.

[21] X. Sun Z. Chen L. Wan, Z. Yang and L. Chen. Decoupling control of a 5-degree-of- freedom bearingless induction motor based on least squares support vector machine inverse. *Advances in Mechanical Engineering*, 8(10):1–11, 2016.

[22] N. Nevaranta R. Jastrzebski P. Jaatinen, J. Vuojolainen and O. Pyrhonen. Control system commissioning of fully levitated bearingless machine. *Modeling, Identification and Control*, 40(1):27–39, 2019.

[23] C. V. N. Kumar P. V. S. Sobhan and P. V. R. Rao. Rotor suspension and stabilization of bearingless srm using sliding mode controller. *International Journal of Engineering & Technology, [S.l.]*, 7(1.8):214–218, 2018.

[24] A. H. Pesch and J. T. Sawicki. Active magnetic bearing online levitation recovery through μ-synthesis robust control. *Actuators*, 6(2):1–14, 2017.

[25] P. Jaatinen R. P. Jastrzebski and O. Pyrhonen. Modeling and control design simulations of a linear flux-switching permanent-magnet-levitated motor. In *Mechanical Engineering Journal*, volume 1, 2017.

[26] J. A. Santisteban R. R. Gomes and R. M. Stephan. Optimized performance of a motor -bearing. In *Symposium Series in Mechatronics*, volume 1, pages 417–424, 2006.

[27] E. F. Rodriguez and J. A. Santisteban. An improved control system for a split winding bearingless induction motor. *IEEE Transactions on Industrial Electronics*, 58(8):3401–3408, 2011.

[28] R. M. Stephan Domingos W. L. Kauss, A. C. D. N. Gomes and F. B. David. Lqr control of a bearingless macine implemented with a dsp. In *11th International Symposium on Magnetic Bearings, ISMB 2008, Nara*, volume 1, pages 2595–2601, 2008.

[29] Z. Wang and X. Liu. An improved rotor flux oriented control system of bearingless induction motors. In *2010 Chinese Control and Decision Conference*, volume 1, pages 2733–2737, 2010.

[30] X. Meng Z. Deng, X. Wang and Y. Yan. An independent controller of radial force subsystem for super-high-speed bearingless induction motors. In *Proceedings Ninth International Symposium on Magnetic Bearings, ISMB 2004, Lexongton-KY*, volume 1, 2004.

Control tecnique	Main advantages	Limitations
ClassicControllers (PI, PD and PID)	▸Consolidated theory in academia and industry. ▸Large number of applications in electrical machines. ▸It is easy to implement.	▸Difficulty of tuning in non-linear plants. ▸It has high sensitivity to parametric variations and disturbances.
Convergent Control Algoritm	▸Allows attenuation of vibrations generated by periodic power signals.	▸Needs accurate identification of frequencies responsible for vibrations.
Linear Quadratic Regulator (LQR)	▸High performance and simple implementation.	▸It is Sensitive to disturbances and parametric variations.
Decoupled torque and force control technique (FTD)	▸Provides decoupling of radial force and torque controllers ▸It is easy to implement computationally. ▸Provides high magnitude of radial forces.	▸Has low quality of radial force signals. ▸Generate high losses due to the joule effect. ▸It does not show good performance in the attenuation of total harmonic distortions (THD). ▸Generates considerable errors in the directions of radial forces compared to other techniques such as HFTD and PIM.
Decoupled torque and force control technique including Harmonic Compensation (HFTD)	▸Provides decoupling of radial force and torque controllers. ▸Provides high magnitude of radial forces. ▸Presents good quality of radial force signals. ▸Provides efficient attenuation of total harmonic distortions (THD). ▸Considerably reduces errors in radial force directions compared to FTD technique.	▸Needs accurate identification of frequencies responsible for vibrations. ▸The computational implementation is more complex.
Pseudo Inverse Matrix	▸Provides a reduction in computational effort for online operation compared to other techniques such as HFTD. ▸It is able to compensate for imbalances in radial forces. ▸Provides stable generation of radial forces without the need for additional control stages. ▸Allows the generalization of the technique to other plants of a higher order. ▸Allows the generation of optimized reference currents resulting in a reduction in copper joule losses.	▸Provides intermediate performance in reducing total harmonic distortion compared to other techniques such as HFTD. ▸It presents high computational effort for the online processing of the matrix.
Support VectorMachine (SVM)	▸It allows the supervised learning used to estimate a function that classifies the input data of the plants for the recognition of patterns with performance superior to the ANNs. ▸They are robust for large data sets.	▸It is sensitivity to the choice of parameter values and the difficulty of interpreting the model.
Active Disturbance Rejection Control (ADRC)	▸It does not require the exact model of the machine. ▸Estimates and cancels the disturbance effects. ▸High degree of adaptability and resistance to disturbance.	▸There is no a systematic tuning method. ▸The number of parameters makes difficult the controller tuning.
Sliding Mode Control (SMC)	▸Fast dynamic responses. ▸Disregards the effects of the plant nonmodeled dynamics. ▸Regardless of the parameters of the plant and the disturbances on it	▸It presents unwanted oscillations caused by the high-frequency switching (Chattering) of the inverters.
Multi-Ressonant Controller (MRC)	▸Provides the suppression of oscillations in the radial position caused by disturbances of a periodic nature.	▸Needs accurate identification of frequencies responsible for vibrations. ▸System dimension becomes large for a large amount of disturbance frequencies.
Fuzzy Controllers	▸It allows the representation of models with a high degree of uncertainty through rules of inputs and outputs.	▸They depend on the operator's knowledge for tuning the controllers.
H∞ Controller	▸It provides good dynamic response and robustness to parametric variations and rejection of disturbances.	▸Controller design is complex for multiple input and multiple output systems. ▸Tuning parameters (weights) and choosing controller functions tend to be difficult, especially in a high-order system. ▸The high computational effort for highorder controllers.
μ-Synthesis	▸Provides a reduced settling time. ▸It is suitable for plants with a high degree of uncertainty. ▸Allows the processing of multiple parametric uncertainties and their interactions. ▸It is robust to parametric variations and disturbances.	▸The large number of method iterations can result in long calculation times, which can degrade the overall performance of the system.
Space Vector Modulation (SVM)	▸Optimizes control of inverter switch states to make the output nearly sinusoidal. ▸Reduces harmonic distortions caused by switching. ▸Reduces heat losses and torque ripple. ▸Allows digital implementation of switch control, including for multilevel inverters.	▸Its implementation requires a little more processing than conventional PWM modulation. However, it can become easier when used in multilevel inverters.
Direct Levitated Force Control	▸It is a controller based on a predictive model that provides direct control of radial forces. ▸Minimizes the controller's dependence on machine parameters present in the traditional vector model.	▸There is still a strong dependence on the accuracy of flow observers (torque and levitation) used. ▸The computational effort for real-time operation of observers together with controllers tends to be high.
Direct Torque Control	▸The structure of the control algorithm is simple. ▸Allows direct control of stator torque and flux. ▸Allows the generation of approximately sinusoidal stator fluxes and currents. ▸Provides high dynamic performance. ▸No need for coordinate transformations. ▸No need for modulation blocks or the addition of other controllers for flux and torque. ▸Provides fast torque responses.	▸They can generate instability during machine startups. ▸They need flux and torque estimators, which creates the need for exact knowledge of the parameters. ▸Its torque and flux signals are ripple.
Direct Suspension Force Control	▸It is a controller based on a predictive model that provides direct control of radial forces. ▸It minimizes the controller's dependence on the machine parameters present in the vector model.	▸There is still a strong dependence on the accuracy of flow observers (torque and levitation) used. ▸The computational effort for real-time operation of observers together with controllers tends to be high.

Figure 5: Main control techniques applied to the bearingless machines.

AUTHOR INDEX